Edited by Niels Behrendt

Matrix Proteases in Health and Disease

WILEY-VCH

WILEY-VCH Verlag GmbH & Co. KGaA

The Editor

Dr. Niels Behrendt
The Finsen Laboratory
Dept. 37.35, Rigshospitalet/Biotech Research and Innovation Centre (BRIC), the University of Copenhagen
Ole Maaloes Vej 5
2200 Copenhagen N
Denmark

Library of Congress Card No.: applied for

British Library Cataloguing-in-Publication Data
A catalogue record for this book is available from the British Library.

Bibliographic information published by the Deutsche Nationalbibliothek
The Deutsche Nationalbibliothek lists this publication in the Deutsche Nationalbibliografie; detailed bibliographic data are available on the Internet at <http://dnb.d-nb.de>.

Cover Design Grafik-Design Schulz, Fußgönheim

Typesetting Laserwords Private Limited, Chennai, India

Printing and Binding Markono Print Media Pte Ltd, Singapore

Print ISBN: 978-3-527-32991-5
ePDF ISBN: 978-3-527-64935-8
ePub ISBN: 978-3-527-64934-1
mobi ISBN: 978-3-527-64933-4
oBook ISBN: 978-3-527-64932-7

Contents

Preface

This book about "Matrix Proteases in Health and Disease" is the result of an extraordinary effort by a large group of world-leading experts to provide a series of excellent, comprehensive, and very up-to-date reviews of central subjects in the field.

The proteolytic cascade systems operating in the extracellular matrix have been central research themes for several decades. These proteolytic systems are not only highly interesting and challenging subjects but also crucially important for normal physiology and critically associated with numerous severe pathological conditions. Following many years of intense investigation through the early period, more recent development along several lines of research has added new dimensions to the field but has also raised new important questions. Thus, the past 10–15 years have witnessed the unraveling of the physiological importance of several specific protease activities through studies with gene manipulation in mice, although not always enabling an understanding of the multicomponent properties of these systems. On the other hand, the determination of the complete genomes of man and several other higher organisms has provided complete catalogs of degradomes and pointed to very complicated proteolytic networks, yet without this providing functional information about individual components. The combination of these means of investigation, along with detailed biochemical elucidation of molecular properties, presents a major research challenge but should ultimately result in an understanding at a much higher level than previously possible.

However, for many matrix proteases, and for many associated physiological events, this type of understanding has already started to emerge. Simultaneously, the knowledge about disease-related aspects of matrix proteases has grown tremendously. Altogether, this has led to an overwhelming amount of valuable new information. These subjects are dealt with in considerable detail in the chapters of this book, a coverage that has been optimized due to the fact that all of the authors are centrally engaged in gathering the actual information through active, cutting-edge research in each of the areas covered.

I wish to thank all the authors for sharing their tremendous knowledge with the readers of this book.

Niels Behrendt

List of Contributors

Heike Allgayer
Ruprecht Karls University of Heidelberg
Department of Experimental Surgery and Molecular Oncology of Solid Tumors
Medical Faculty Mannheim
Collaboration Unit German Cancer Research Center-DKFZ-Heidelberg
Theodor-Kutzer-Ufer 1-3
68167 Mannheim
Germany

Peter A. Andreasen
Aarhus University
Department of Molecular Biology and Genetics
Danish-Chinese Centre for Proteases and Cancer
Gustav Wieds Vej 10
8000 Aarhus C
Denmark

Omer Bashir
Departments of Medicine and Research
Veterans Affairs Medical Center
79 Middleville Road
Northport
NY 11768
USA

and

Department of Medicine
Stony Brook University
School of Medicine
Stony Brook
NY 11794
USA

Niels Behrendt
The Finsen Laboratory
Rigshospitalet/Biotech Research and Innovation Centre (BRIC), the University of Copenhagen
Ole Maaloes Vej 5
2200 Copenhagen N
Denmark

Dieter Brömme
University of British Columbia
Department of Oral Biological
and Medical Sciences
Life Sciences institute
Faculty of Dentistry
2350 Health Research Mall
Vancouver
BC 6VT 1Z3
Canada

Emilie Buache
Université de Strasbourg
Département de Biologie du
Cancer
Institut de Génétique et de
Biologie Moléculaire et Cellulaire
(IGBMC)
CNRS UMR 7104
INSERM U964
1 rue L Fries
Illkirch 67404
France

Jian Cao
Departments of Medicine and
Research
Veterans Affairs Medical Center
79 Middleville Road
Northport
NY 11768
USA

and

Department of Medicine
Stony Brook University
School of Medicine
Stony Brook
NY 11794
USA

Ib J. Christensen
The Finsen Laboratory
Copenhagen University
Hospital/Biotech Research and
Innovation Centre (BRIC)
University of Copenhagen
Copenhagen Biocenter
Ole Maaløes Vej 5
2200 Copenhagen N
Denmark

Elena Deryugina
Department of Cell Biology, The
Scripps Research Institute
10550 North Torrey Pines Road
La Jolla
CA 92037
USA

Sarah Louise Dombernowsky
University of Copenhagen
Department of Biomedical
Sciences
BRIC
Ole Maaloes Vej 5
2200 Copenhagen N
Denmark

Vincent Ellis
University of East Anglia
School of Biological Sciences
Norwich Research Park
Norwich
NR4 7TJ
UK

Shan Gao
Aarhus University
Department of Molecular Biology
and Genetics
Danish-Chinese Centre for
Proteases and Cancer
Gustav Wieds Vej 10
8000 Aarhus C
Denmark

Ana Gutiérrez-Fernández
Universidad de Oviedo
Departamento de Bioquímica y Biología Molecular
Facultad de Medicina
Instituto Universitario de Oncología
C/Fernando Bongera sn
33006 Oviedo
Spain

Sven Hammerschmidt
Ernst Moritz Arndt University of Greifswald
Department of Genetics of Microorganisms
Interfaculty Institute for Genetics and Functional Genomics
Friedrich-Ludwig-Jahn-Strasse 15a
17487 Greifswald
Germany

Heiko Herwald
Lund University
Department of Clinical Sciences
Division of Infection Medicine
BMC
Tornavägen 10
22184 Lund
Sweden

Kenn Holmbeck
Matrix Metalloproteinase Section
Craniofacial and Skeletal Diseases Branch
National Institute of Dental and Craniofacial Research
NIH
30 Convent Drive
Bethesda
MD 20892
USA

Gunilla Høyer-Hansen
The Finsen Laboratory
Copenhagen University Hospital/Biotech Research and Innovation Centre (BRIC)
University of Copenhagen
Copenhagen Biocenter
Ole Maaløes Vej 5
2200 Copenhagen N
Denmark

Martin Illemann
The Finsen Laboratory
Copenhagen University Hospital/Biotech Research and Innovation Centre (BRIC)
University of Copenhagen
Copenhagen Biocenter
Ole Maaløes Vej 5
2200 Copenhagen N
Denmark

Jan K. Jensen
Aarhus University
Department of Molecular Biology and Genetics
Danish-Chinese Centre for Proteases and Cancer
Gustav Wieds Vej 10
8000 Aarhus C
Denmark

Tina M. Kousted
Aarhus University
Department of Molecular Biology and Genetics
Danish-Chinese Centre for Proteases and Cancer
Gustav Wieds Vej 10
8000 Aarhus C
Denmark

Marie Kveiborg
University of Copenhagen
Department of Biomedical Sciences
BRIC
Ole Maaloes Vej 5
2200 Copenhagen N
Denmark

Daniel A. Lawrence
University of Michigan Medical School
Department of Internal Medicine
1150 West Medical Center Drive
Ann Arbor
MI 48109
USA

Joerg Hendrik Leupold
Ruprecht Karls University of Heidelberg
Department of Experimental Surgery and Molecular Oncology of Solid Tumors
Medical Faculty Mannheim
Collaboration Unit German Cancer Research Center-DKFZ-Heidelberg
Theodor-Kutzer-Ufer 1-3
68167 Mannheim
Germany

Karin List
Wayne State University
Department of Pharmacology
Barbara Ann Karmanos Cancer Institute
School of Medicine
540 East Canfield
Detroit
Michigan 48201
USA

Carlos López-Otín
Universidad de Oviedo
Departamento de Bioquímica y Biología Molecular
Facultad de Medicina
Instituto Universitario de Oncología
C/Fernando Bongera sn
33006 Oviedo
Spain

Ida K. Lund
The Finsen Laboratory
Copenhagen University Hospital/Biotech Research and Innovation Centre (BRIC)
University of Copenhagen
Copenhagen Biocenter
Ole Maaløes Vej 5
2200 Copenhagen N
Denmark

Gregory S. Miller
Wayne State University
Department of Pharmacology
Barbara Ann Karmanos Cancer Institute
School of Medicine
540 East Canfield
Detroit
Michigan 48201
USA

Thomas Reinheckel
Albert-Ludwigs-University
Freiburg
Institute of Molecular Medicine
and Cell Research
Stefan Meier Str. 17
79104 Freiburg
Germany

and

Albert-Ludwigs-University
Ludwig-Heilmeyer
Comprehensive Cancer Center
and BIOSS Centre for Biological
Signalling Studies
Hebel Str. 25
79104 Freiburg
Germany

Marie-Christine Rio
Université de Strasbourg
Département de Biologie du
Cancer
Institut de Génétique et de
Biologie Moléculaire et Cellulaire
(IGBMC)
CNRS UMR 7104
INSERM U964
1 rue L Fries
Illkirch 67404
France

Kathrin Sachse
Albert-Ludwigs-University
Freiburg
Institute of Molecular Medicine
and Cell Research
Stefan Meier Str. 17
79104 Freiburg
Germany

and

Albert-Ludwigs-University
Freiburg
Faculty of Biology
Hauptstrasse 1
79104 Freiburg
Germany

Gerald Schielke
University of Michigan Medical
School
Department of Internal Medicine
1150 West Medical Center Drive
Ann Arbor
MI 48109
USA

Clara Soria-Valles
Universidad de Oviedo
Departamento de Bioquímica y
Biología Molecular
Facultad de Medicina
Instituto Universitario de
Oncología
C/Fernando Bongera sn
33006 Oviedo
Spain

Dorte Stautz
University of Copenhagen
Department of Biomedical
Sciences
BRIC
Ole Maaloes Vej 5
2200 Copenhagen N
Denmark

Tine Thurison
The Finsen Laboratory
Copenhagen University
Hospital/Biotech Research and
Innovation Centre (BRIC)
University of Copenhagen
Copenhagen Biocenter
Ole Maaløes Vej 5
2200 Copenhagen N
Denmark

Claudia Weber
Ernst Moritz Arndt University of
Greifswald
Department of Genetics of
Microorganisms
Interfaculty Institute for Genetics
and Functional Genomics
Friedrich-Ludwig-Jahn-
Strasse 15a
17487 Greifswald
Germany

Fee Werner
Albert-Ludwigs-University
Freiburg
Institute of Molecular Medicine
and Cell Research
Stefan Meier Str. 17
79104 Freiburg
Germany

and

Albert-Ludwigs-University
Freiburg
Spemann Graduate School of
Biology and Medicine (SGBM)
Albertstr. 19 A
79104 Freiburg
Germany

and

Albert-Ludwigs-University
Freiburg
Faculty of Biology
Hauptstrasse 1
79104 Freiburg
Germany

Gina L. Zoratti
Wayne State University
Department of Pharmacology
Barbara Ann Karmanos Cancer Institute
School of Medicine
540 East Canfield
Detroit
Michigan 48201
USA

Stanley Zucker
Departments of Medicine and Research
Veterans Affairs Medical Center
79 Middleville Road
Northport
NY 11768
USA

and

Department of Medicine
Stony Brook University
School of Medicine
Stony Brook
NY 11794
USA

Introduction

Niels Behrendt

The extracellular matrix of multicellular organisms provides the solid part of the cellular microenvironment and forms the framework of their macroscopic structure. Throughout development, and during several highly controlled events in the normal, adult body, this matrix is subject to refined processes of remodeling, resulting in an overwhelming variety of novel shapes, matrix composition, and residence opportunity for multiple cell types. Furthermore, these processes are not only a matter of dynamic structure. The remodeling of the matrix continuously allows for new cell–matrix contacts, enabling a whole array of communicative events, mediated through highly specialized cell–matrix interactions.

This book has its focus on the active players in these reactions: the world of "matrix proteases" in the pericellular and extracellular microenvironment that direct tissue remodeling. These proteases, belonging to several classes, are not only decisive for the physical rearrangement of the matrix but also for multiple substrate cleavages resulting in specific protein activation, release, or inactivation. The latter reactions are directed against enzymes, including other proteases, as well as cytokines and growth factors.

Not surprisingly, all of these proteolytic events are subject to stringent biological control. A multitude of regulatory mechanisms direct the specific expression of individual proteolytic components in time and space, with expression being distributed among a wide variety of cell types. Furthermore, regulatory events at the protein activity level, which include zymogen activation, protease inhibition, and specific localization and enhancement mechanisms, serve to tightly control proteolytic activity. These latter events are part of the conspicuous organization of several groups of matrix proteases in cascade systems in which proteases activate each other in strongly amplified processes, with associated mechanisms of negative control and localization of activity.

While proteolytic reactions in the extracellular matrix are often crucial for development and a healthy adult life, they are also in many cases associated with disease. Firstly, considering the tight regulatory mechanisms mentioned above, it is not surprising that a dysregulation of proteolytic activity can, in many cases, lead to pathological consequences. An extreme case of dysregulation is the complete lack

Matrix Proteases in Health and Disease, First Edition. Edited by Niels Behrendt.

of a certain, functional component, for example, due to mutation. In several cases, that situation is known to lead to disease in humans, with the same consequence often being mimicked in model systems in gene-manipulated mice. Secondly, proteases associated with particular tissues or cell types may become expressed and activated as a consequence of disease, with the resulting activity leading to a worsened condition. Examples of this situation include various inflammatory conditions in which exaggerative proteolytic activity, derived from immunoreactive cell types, leads to tissue and matrix destruction, and processes in which host-derived proteolytic activity turns out to facilitate infectious disease. The most thoroughly studied example, however, is the mechanism of cancer metastasis in which proteolytic activity, often resulting from the elusive tumor-derived stimulation of stromal cells, serves to enable the invasive migration of cancer cells.

The aim of this book is to address the role of several matrix proteases and proteolytic reactions in normal physiology and in various pathological conditions. These subjects, presented by leading experts, complement each other and the examples chosen comprise an important part of the current knowledge in the field. Starting out with a thorough and updated overview of the various groups of matrix proteases in the context of the entire degradome (Chapter 1), different classes of proteases, specific cascade systems, and in some cases single proteolytic components with defined roles in the extracellular matrix are treated with respect to their role in health and disease. Although it would not be possible to obtain a complete coverage of this very comprehensive group of active components, emphasis has been given both to several well-established active players in matrix biology and to a number of less well-known, or more recently discovered, proteases and regulators.

Among the serine proteases, the fibrinolytic system of plasminogen activation is thoroughly described with respect to enzymology and function in the normal organism (Chapter 2), as well as its role in cancer invasion (Chapter 9). An additional disease aspect of this proteolytic system with pronounced importance is its function in ischemic stroke, which has been treated in depth in a separate chapter (Chapter 6). The plasminogen activation system includes a number of well-established regulators that are also discussed in these chapters but, in addition, Chapter 10 treats the function of the serpin-type inhibitor, protease nexin-1, which plays a role both in this and in other proteolytic systems and also has likely roles in cancer.

An additional group of serine proteases that has received strong attention in recent years is the group of type-II transmembrane serine proteases (TTSPs). These enzymes, often with striking physiological roles and pathophysiological importance as shown in mouse models, are discussed in Chapter 5.

Within the class of metalloproteases, several aspects of the group of matrix metalloproteases (MMPs) have been addressed. More than any other group of proteases, the MMPs have been studied in relation to cancer invasion and metastasis and in Chapter 8, a comprehensive coverage of this theme is provided, with particular emphasis on experimental models used in this connection. A separate chapter has been dedicated to the membrane-associated matrix metalloproteases (the MT-MMPs), a group of MMPs that has attracted extraordinary attention since the discovery of these enzymes (Chapter 3). In addition to the MMPs, other metalloproteases such as the members of the ADAM group have crucial functions, both in the shedding of membrane proteins and other cellular processes. Chapter 12 provides a thorough coverage of the function of ADAMs and their role in cancer.

The third large class of proteases with well-defined matrix-related functions is the cysteine proteases, including the cysteine protease cathepsins. While the majority of these enzymes have predominantly intracellular functions in relation to protein degradation in the lysosomal compartment, the group also includes members with additional or predominant extracellular functions. The extracellular roles of these enzymes are discussed in Chapter 11, with particular emphasis on their role in cancer. A special, but physiologically very important, case in connection with extracellular cathepsin activity is the role of cathepsin K in bone remodeling, which has been thoroughly covered in a separate chapter (Chapter 4).

While many of these proteases have physiological functions that may be turned into pathological processes in the case of dysregulation, a completely different disease aspect lies in the fact that several pathogens have acquired the potential to "abuse" host matrix proteases as part of their invasion or other pathogenic processes. This very intriguing mechanism is discussed in Chapter 7.

The last part of the book is devoted to various very important, clinical aspects of matrix proteases. Thus, in addition to their functional importance in cancer, described in several chapters as mentioned above, some of these proteases and their regulators have proven very promising as cancer biomarkers, having potential for prognosis, prediction of treatment response, and even diagnostic utilization. This is the case for several components of the plasminogen activation system (Chapter 13) and for members of the MMP group and their inhibitors (Chapter 14). Finally, a major theme lying behind many studies of matrix proteases is their potential as therapeutic targets, notably in connection with cancer. The failure of early clinical trials focused on the targeting of MMP activity, often taken as an argument against this general strategy until recently, is less surprising today on the basis of our current understanding of the complexity and redundancy of these systems. This understanding, however, provides new prospects for novel, refined targeting strategies. In Chapter 15, this theme is discussed in detail, specifically addressing the targeting of MMPs.

Altogether, the 15 chapters in this book include a comprehensive and up-to-date coverage of a wide variety of subjects within this fascinating research area, addressing a whole array of components and processes with crucial importance in health and disease.

1
Matrix Proteases and the Degradome

Clara Soria-Valles, Carlos López-Otín, and Ana Gutiérrez-Fernández

1.1
Introduction

Proteases are defined as enzymes that have the ability to perform the hydrolysis of peptide bonds. Owing to this characteristic, proteases were initially described as nonspecific enzymes of protein catabolism, participating in processes such as tissue destruction or degradation of dietary proteins. More recently, a better understanding of their functions has allowed consideration of proteases as enzymes that perform highly specific reactions and take part in multiple biological processes such as DNA replication and transcription, cell proliferation, differentiation and migration, tissue morphogenesis and remodeling, heat shock and unfolded protein responses, neurogenesis, angiogenesis, ovulation, fertilization, wound repair, stem cell mobilization, coagulation, immunity, inflammation, senescence, autophagy, apoptosis, and necrosis [1]. According to the essential roles performed by proteases in all living organisms, alterations in their proteolytic activities may lead to important pathologies such as arthritis, cardiovascular alterations, neurodegenerative disorders, progeroid syndromes, and cancer [2, 3].

The biochemical reaction catalyzed by all proteases consists in the hydrolysis of a peptide bond through the nucleophilic attack at the carbonyl group. However, the way this reaction is performed differs between specific proteases. This characteristic feature has allowed the establishment of six different catalytic classes of proteases according to the group performing the nucleophilic attack: aspartic, metallo, cysteine, serine, and threonine proteases, as well as the most recently described group of glutamic proteases, which has only been found in some species of fungi and bacteria. In the case of aspartic, glutamic, and metalloproteases, a polarized water molecule located in the active center acts directly as a nucleophile, while in the other three classes, the reactive element is a hydroxyl (serine and threonine) or sulfhydryl (cysteine) group from the corresponding catalytic core [4]. Within each class, proteases can be further subdivided into different families and clans according to sequence conservation and three-dimensional structure similarities.

The diversity and complexity of proteases have made necessary the introduction of concepts and tools for their global analysis and characterization. Thus, the term

Matrix Proteases in Health and Disease, First Edition. Edited by Niels Behrendt.

degradome defines the complete set of protease genes expressed by a cell, tissue, or organism at a specific moment or circumstance [5]. Likewise, the degradome of a certain protease is the complete substrate repertoire of that specific enzyme. The ability to explore the complexity of proteases in an organism has been catalyzed by the impressive advances in the sequencing and annotation of complete genomes. In fact, since the completion in 1995 of the genome of *Haemophilus influenzae* (1 830 140 bp and 1740 genes) [6], the number of available genome sequences has continuously increased at a rapid pace. Nowadays, the genome sequence for most model organisms as well as several vertebrate species and thousands of microorganisms is publicly available [7–13]. This genomic progress has made the study of degradomes more accessible to the scientific community. The information obtained from the analysis of the degradome of a certain organism constitutes an important tool for the comprehension of biological and pathological processes and could be the key to find out new ways to diagnose, treat, or prevent human pathologies.

In this chapter, we discuss available bioinformatic tools for the construction and analysis of degradomes and present an overview of characteristic features and evolutionary aspects of several degradomes of biomedical interest. We also discuss human diseases of proteolysis and, finally, introduce the different classes of proteases with ability to degrade the extracellular matrix (ECM), which is the topic of this book.

1.2
Bioinformatic Tools for the Analysis of Complex Degradomes

The development of novel molecular technologies has significantly reduced the time and cost of generating a genome sequence. However, important parts of the subsequent analysis, including the annotation of functional elements in the genome and the integration of this information into biological processes, are still a difficult task [14]. The complexity of genomic information, in which the coding sequence is interrupted by the presence of large introns, or the existence of numerous pseudogenes with high sequence identity to bona fide genes, hampers the use of straightforward bioinformatic approaches for both the reliable identification of genes and the prediction of protein structure and function. Therefore, although many genes can be directly annotated by using bioinformatic approaches, current tools have limitations in distinguishing genes from pseudogenes. Accordingly, manually supervised annotation is still the most common method to annotate complex sequences such as the human genome and degradome [15]. Another consideration when dealing with homologous sequences, as in the case of protease-coding genes, is the importance of distinguishing between different types of gene relationships. Thus, homologous genes can be classified as either orthologous genes, which have originated as a result of a speciation event derived from a single ancestral gene in the last common ancestor of two given species, or paralogous genes, which originate as a result of a duplication event within the same genome [16]. Orthologous genes generally maintain the same ancestral function, while paralogous genes tend to evolve and acquire novel functions.

One of the main aims of genome sequencing studies is the comparison of the complement of genes between different species. Using this approach, it is possible to identify the presence of novel genes in one species, or the loss of other genes in another species, providing clues about the molecular mechanisms underlying some of the physiological differences between them. Despite the fact that genomes are sequences of nucleic acids, the comparison of the protease repertoire between two genomes is generally performed using protein sequences, as they are more conserved than DNA sequences, and, therefore, they are more informative and the searches are more sensitive [15]. Nucleotide sequences are only compared when analyzing noncoding regions or to determine evolutionary parameters using protein-coding regions.

Another issue to take into account for gene prediction when using available genome sequences is the possibility of having assembly artifacts or sequencing errors in the analyzed sequence [15]. Thus, assembly artifacts usually lead to the collapse of clusters of highly similar genes into an artificial single copy gene. Owing to the fact that about 20% of protease genes are located in clusters [17], careful examination of those regions and additional experimental approaches should be performed to correctly annotate protease clusters. On the other hand, the presence of sequencing errors could lead to the annotation of real genes as pseudogenes, although detailed examination of sequence traces or resequencing of specific regions can solve these problems. Pseudogenes are nonfunctional copies of a gene and depending on the structure and mechanism of generation we can distinguish two main types: conventional pseudogenes and processed pseudogenes. Conventional pseudogenes usually originate from a functional copy of a gene that has been inactivated by mutation and afterwards will degenerate through the accumulation of new mutations. By contrast, processed pseudogenes are derived from the mRNA copy of a gene that is retrotransposed into the genome. The creation of pseudogenes has been a major mechanism in the evolution of the mammalian degradomes [18]; therefore, their study is important to get a global picture when comparing different genomes.

Numerous databases, web pages, and programs are freely available and constitute valuable tools for the study and comparison of complete genomes. Owing to the extension of these resources, we briefly introduce some remarkable tools and databases that can be useful for the genomic study of proteases:

Degradome database (*http://degradome.uniovi.es/*): resource containing manually annotated information about all proteases and protease inhibitor genes from humans, the chimpanzee, mouse, and rat organized in catalytic classes and families. This database also provides a catalog of human hereditary diseases of proteolysis or degradomopathies and additional information about protease structures, ancillary domains present in proteases, and differences between mammalian degradomes [3, 17, 19, 20].

MEROPS (*http://merops.sanger.ac.uk*): comprehensive annotation of proteases and inhibitors in all sequenced organisms. Proteases are classified at the protein domain level and those that are statistically and significantly similar in amino acid sequences are grouped in a family. If the families share a common ancestor, they are grouped in a clan. Protease inhibitors are classified in the same way [21, 22].

Other resources for extracting protease information are Ensembl (*http://www.ensembl.org*), which developed a software for maintaining an automatic annotation and analysis on selected eukaryotic genomes; InterPro (*http://www.ebi.ac.uk/interpro*), which includes tools for predicting functional sites and protein domains [23] and alignment search tools such as BLAST (*http://www.ncbi.nlm.nih.gov/BLAST*) and BLAT (*http://genome.ucsc.edu*). However, in some cases, these methodologies are unable to identify distantly related protease homologs because of the high divergence of their sequences. In these situations, it is necessary to use more sensitive approaches, such as the hidden Markov models or position-specific score matrix, which are probabilistic models that collect information of specific positions from multiple sequence alignments and apply this information for the recognition of protein or DNA sequences in the genome [15]. In this regard, an important tool when working with different protease sequences is the usage of multiple sequence alignment methods developed under the principle of hierarchical clustering. Clustal (*http://www.ebi.ac.uk/clustal*) is one of the most commonly used methods for hierarchical multiple alignment [24].

1.3
Evolution of Mammalian Degradomes

The availability of complete sequences from several mammalian genomes opens the possibility of performing comparative studies of degradomes between species. This might lead to the identification of either highly conserved elements or genetic differences occurring during mammalian evolution and contribute to clarify the molecular basis of biological pathways and pathologies involving proteolytic systems [25, 26].

1.3.1
Human Degradome

Shortly after the completion of the human genome sequence draft, we performed an exhaustive bioinformatic analysis to try to find out new human protease-coding genes with sequence similarity to proteases already described in other organisms [3, 20]. By using this methodology, 570 proteases and protease-related genes, as well as 150 protease inhibitor genes, were identified, representing more than 2% of the total genes in the human genome. Interestingly, a total of 93 of these protease-related genes encode functional proteins that are catalytically inactive because of substitutions in one or more residues critical to their proteolytic activity. These nonprotease homologs may regulate the activation of other proteases or their access to substrates or inhibitors, although their precise contribution to human biology is still poorly known [1].

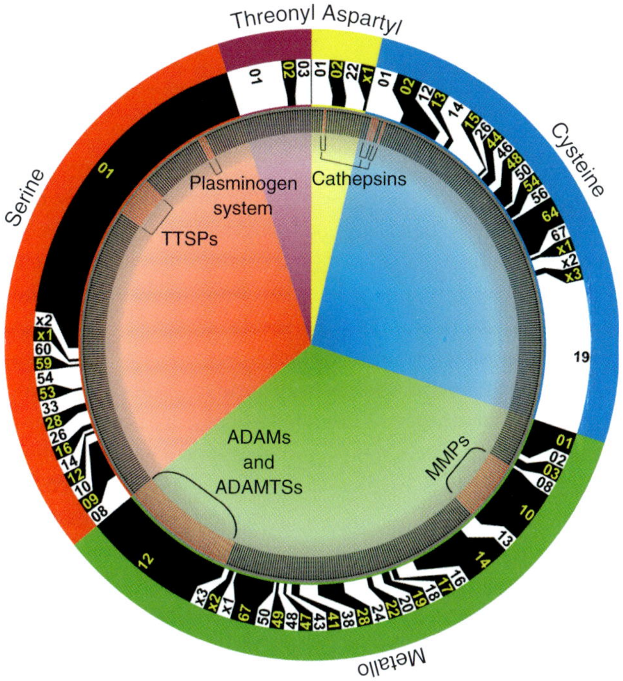

Figure 1.1 The human degradome wheel. Proteases are distributed in 5 catalytic classes and 68 different families. The code number for each protease family is indicated in the outer ring. Proteases discussed in this book are shown with red lines.

Human proteases are divided into 5 catalytic classes (aspartic, metallo, serine, cysteine, and threonine proteases) and 68 families (Figure 1.1). The most abundant classes are metalloproteases (191 members), serine proteases (178 members), and cysteine proteases (153 members). Threonine and aspartic proteases are more specialized classes and, therefore, they only have 27 and 21 members, respectively. From the 68 families of proteolytic enzymes, the most densely populated is the S01 family of serine proteases that includes enzymes of interest in the context of the present book such as plasmin, uPA (urokinase plasminogen activator), tPA (tissue-type plasminogen activator), and TTSPs (type II transmembrane serine proteases). Other representative families in the human degradome are the C01 family of cysteine proteases, mainly formed by cathepsins; the M10 metalloproteases containing more than 20 different matrix metalloproteinases (MMPs) and the M12 metalloproteases that include ADAMs (a disintegrin with metalloprotease domains) and ADAMTSs (ADAMs with thrombospondin repeats). Although the function of many of these proteases is still unknown, most of them are highly conserved between humans and other mammals, indicating that they appeared before the mammalian expansion and their conservation is probably due to their relevance in some biological function [27].

1.3.2
Rodent Degradomes

Mouse (*Mus musculus*) and rat (*Rattus norvegicus*) are the most frequently used models for the study of human physiological and pathological processes. However, the detailed analysis of their genomes has revealed important differences from the human genome, and this information should be taken into consideration when trying to interpret and extrapolate to humans the results obtained using these animal models. In the case of proteolytic enzymes, the availability of the mouse and rat genomes has allowed us to precisely define their similarities and differences with the human degradome. Surprisingly, the complement of proteases in rodents is more complex than that in humans (656 mouse proteases and 646 rat proteases) despite their genomes are roughly 15% smaller. These proteolytic differences are mainly due to the expansion in rodents of specific protease families involved in reproduction and immunological defense, or from specific pseudogenization events in the human degradome that have led to the inactivation of some protease-coding genes [3, 15]. The expansion of specific protease families in rodents can be illustrated by the comparison of the kallikrein cluster (S01 serine proteases). In humans, this cluster is formed by 15 protease genes, while in mouse and rat it has considerably expanded and it is now composed of 26 and 23 genes, respectively. Similarly, the SENP family of sentrin-specific proteases has 14 members in the mouse but only 7 in humans [3, 15]. There are also placental cathepsins and testases that are present in rodents but absent in the human genome [17].

Conversely, the reduced number of protease genes in the human genome is mainly due to the inactivation of specific protease genes during evolution. Some representative examples of protease genes functional in rodents but pseudogenized in humans include seven members of the ADAM family of metalloproteinases (ADAM-1a, -1b, -3b, -4, -4b, -5, and -6) proposed to be involved in ovum–sperm interaction, five testis-specific serine proteases (Tessp3, Tessp6 and Tesp1, 2, and 3), and five proteases with functional roles in digestion (chymosin, intestinal Disp, trypsins Try10 and Try15, and pancreatic elastase Ela1) [15].

1.3.3
Chimpanzee Degradome

The chimpanzee (*Pan troglodytes*) belongs to the hominid family together with humans, the orangutan, bonobo, and gorilla, and constitutes our closest relative. Humans and chimpanzees shared their last common ancestor 5–6 million years ago and have a 99.1% of identity in their genome sequences [28]. The chimpanzee degradome is virtually identical to that of humans, with 568 protease-coding genes, including protease-related genes. However, and despite the high conservation between chimpanzee and human orthologs, there are examples of differential proteases, or proteases showing a high degree of divergence between the two species. Most differences found in comparative genomic analysis of chimpanzees and humans are the consequence of a random genetic drift with a high number of neutral mutations. However, there are a few important functional changes that might be responsible for the phenotypic differences between these closely related species [29, 30]. Similar to the situation with rodents, most changes between human and chimpanzee proteases involve genes implicated in immune response or reproduction. Among them, there are some genes present in chimpanzees that are absent or have been pseudogenized in humans, such as *NAPSB*, *CASP12*, and *HPP*. Other protease genes including *GGTLA1*, *MMP23A*, *HPR*, and *PRSS33* are absent in the chimpanzee genome but are functional in humans [15]. An interesting case is *PRSS33*, which encodes a macrophage-specific serine protease conserved in most mammalian species [31], but is lost in the chimpanzee because of an Alu-mediated recombination [32]. Despite the identification of differential genes between the chimpanzee and humans, a number of proteases are identical between both species, including components of the proteasome or proteases implicated in neurological processes such as *PSEN1*, *BACE*, and *IMMP2L* [15].

1.3.4
Duck-Billed Platypus Degradome

The duck-billed platypus (*Ornithorhynchus anatinus*) is a monotrema that shared a common ancestor with humans more than 166 million years ago. This fascinating organism is an ideal source to better understand the evolution of mammalian genomes, as it shows some characteristics of reptiles and birds. The platypus genome has 536 protease genes, 90% of them being one-to-one orthologs with human genes [18, 33]. Although the number of proteases of humans and the platypus is similar, the phylogenetic position of the platypus has made it possible to identify some proteases of vertebrates conserved in the platypus but absent in eutherians and implicated in important biological processes such as apoptosis, immune response, tooth formation, reproduction, or digestion. As an example, the family alpha-aspartyl dipeptidases (AADs), represented in the platypus by a unique gene, appears in all the organisms except in eutherians, where it was lost by pseudogenization. The high expression of this gene in eggs, ovaries, and testes of Drosophila, zebrafish, frog, or chicken, suggests that it might have implications in reproduction. Something similar happens with nothepsin, an aspartyl protease conserved in oviparous animals such as fish, amphibians, reptiles, birds, and platypus, but whose gene has been pseudogenized in metatherians and eutherians [33]. It is believed that this protein might be related to the development of platypus eggs. The analysis of the platypus degradome has also revealed that similar to primates, its evolution has been mainly shaped by the loss of specific protease-coding genes. However, while in primates the inactivation of proteolytic genes is mainly due to pseudogenization caused by the accumulation of inactivating mutations, in the case of platypus most inactivated genes are completely lost due to gene deletion. In this regard, a remarkable situation is the absence in the platypus of important digestive proteases such as pepsinogens A, B, and C, conserved in other vertebrate species and responsible for the processing of dietary proteins [18].

1.3.5 Other Degradomes

The growing list of known degradomes also includes other important species such as orangutans (*Pongo pygmaeus* and *Pongo abelii*) and zebra finch (*Taeniopygia guttata*), which complement the information derived from the study of the above-described organisms. The genome sequencing of other evolutionary distant primates such as orangutan provides useful information for genomic comparison. In addition, the sequencing of the zebra finch genome might help understand its biological peculiarities and allow the comparison between birds and mammalian degradomes [34, 35].

As expected, the orangutan degradome is highly similar to that of the chimpanzee and humans, although there are several interesting differences that mainly affect the immune and reproductive systems [36]. In some cases, protease genes have been lost in the orangutan by pseudogenization. This is the case of *PRSS33*, which is a functional protease of the immune system in humans, but has been lost in the chimpanzee and pseudogenized in the orangutan because of the existence of a premature stop codon [36], or *HTRA4*, which acquired a stop codon in its first exon and seems to have consequences for the reproductive processes [36]. In other cases, some functional genes in orangutans have been pseudogenized in humans, including *CASP12*, a cysteine protease implicated in cytokine processing and involved in sepsis both in the chimpanzee and orangutan [19, 37]. In orangutans, CASP12 constitutes a fully functional protease, while its human ortholog is a pseudogene or encodes an inactive protease [38].

The degradome of the zebra finch contains about 460 proteases and differs from mammals in several protease-coding genes such as caspases, granzymes, acrosins, metalloproteases, and pepsinogens. These changes might be related to differences in apoptosis, immune system, bone development and reproduction between birds and mammals. For example, several members of the ADAM family, such as ADAM1–7 and ADAM30, which, in mammals, participate in fertilization are absent in the genomes of the zebra finch and chicken [34], while the gene encoding caspase-3, which has an important role in song-response habituation, has been specifically duplicated in the zebra finch genome [34, 39].

In summary, the completion of all these genome projects has provided a good overview of the proteolytic systems of different organisms. However, many proteases are still largely unknown and their identification and characterization through comparative genomics and bioinformatic approaches represent important future challenges in this field. In addition, the identification or prediction of substrates and inhibitors with this kind of bioinformatic tools, along with experimental investigation, will be necessary to facilitate a better understanding of normal and pathological functions of proteases and the discovery of new therapeutic targets [15].

1.4
Human Diseases of Proteolysis

The essential role of proteolytic enzymes in the control of cell behavior predicts that abnormal, either insufficient or excessive, protease activity may lead to important pathological processes. These diseases can be caused by direct alterations in protease genes or be due to alterations in other components of proteolytic systems. The first group can be further subdivided into two major groups: genetic disorders caused by mutation of protease genes or diseases caused by alterations in the spatiotemporal pattern of expression of proteases [3]. In relation to diseases of proteolysis caused by alterations in other components of proteolytic systems, we can mention that changes in substrates, inhibitors, regulatory factors, and transport systems of proteases generate important pathologies. For instance, mutations in the gene encoding APP enhance the processing of this protein by beta- or gamma secretases causing Alzheimer's disease [40]; similarly, removal of the protease processing site in a substrate can have dramatic consequences, generating life-threatening syndromes such as Hutchinson-Gilford progeria due to removal of the proteolytic processing site in prelamin A [41–43]. Likewise, different serpinopathies are caused by mutation in serine protease inhibitors [44], while hemophilia A, the most common form of hemophilia, is caused by mutations in the *F8* gene, a cofactor of the coagulation protease factor IX [45]. Finally, alterations in the subcellular localization of proteases due to changes in transport systems can also produce important pathologies. Thus, the transport of lysosomal proteases such as cathepsin Z is altered by mutations in *ERGIC53* that cause hematological diseases [46].

Among the best characterized diseases of proteolysis are those caused by mutations in protease-coding genes. The number of hereditary diseases of proteolysis has continuously grown during the past decade, especially since the completion of the human genome sequence and the introduction of novel tools for genomic analysis [15, 47]. To date, we have annotated a total of 89 human hereditary diseases caused by mutations in protease genes, what implies that more than 15% of the genes coding for human proteases are implicated in some form of hereditary pathology [20] (*http://www.uniovi.es/degradome*). Depending on the effect of the mutation on protease function, these pathologies can be classified as loss-of-function or gain-of-function mutations, being more frequent than those caused by loss-of-protease function [3].

Mutations in protease-coding genes may have multiple and diverse outcomes, generating hematological, neurological, immunological, digestive, reproductive, and bone pathologies. For further definition of the genes and gene products referred to in the following, the reader is referred to: *http://degradome.uniovi.es/diseases.html*. The hematological diseases caused by protease gene mutations mainly involve proteases associated with the coagulation system or implicated in blood homeostasis. Thus, hemophilia B, and factor VIIa, Xa, XIa, or XIIa deficiencies are caused by mutations in *F9, F7, F10, F11,* and *F12,* respectively [48–52], while thrombophilia is caused by defects in *PLG* or *PROC* [53]. Hyper- or hypoprothrombinemia can be originated by either gain or loss of *F2* expression [54], while mutations in *ADAMTS13* are responsible for thrombotic thrombocytopenic purpura [55], and cyclic hematopoiesis is caused by gain-of-function mutations in *ELA2* [56]. In addition, a wide range of neurological pathologies, from paraplegias and ataxias to degenerative disorders, are caused by mutations in proteases. Thus, gain-of-function mutations in *PSEN1* and *PSEN2* cause familial Alzheimer's disease [57, 58], while loss-of-function mutations in *SPG7* or *CLN2* originate spastic paraplegia or neuronal ceroid lipofuscinosis, respectively [59, 60]. Other neurological syndromes caused by mutations in protease genes include Gilles de la Tourette syndrome, Parkinson disease type V or nonsyndromic mental retardation, because of alterations in *IMMP2L, UCHL1,* or *PRSS12* [61–63]. Likewise, there are immunological diseases of proteolysis, including autoimmune lymphoproliferative syndromes (I and II), caused by mutation of either *CASP8* or *CASP10* [64, 65]. This group also includes pathologies associated with mutations in components of the complement system such as *C1R, C1S, C2, DF,* and *IF.*

In addition, some diseases of the digestive system are caused by loss-of-function mutations, such as *PRSS7* deficiency that results in failure to convert inactive trypsinogen into active trypsin [66], or by gain-of-function mutations such as hereditary pancreatitis caused by mutations in *PRSS1* [67]. Alterations in some proteases can also affect reproduction. Thus, mutations in *USP9Y* cause azoospermia [68], while gonadal dysgenesis is caused by mutations in *DHH* [69]. Finally, bone diseases can be developed by loss-of-function mutations in *MMP2,* which cause multicentric osteolysis with nodulosis and arthropathy [70]; or in *MMP9* and *MMP13,* causing recessive metaphyseal anadysplasia [71]; while activating mutations in *MMP13* cause spondyloepimetaphyseal dysplasia [72]. Moreover, alterations in *ADAMTS2, ADAMTS10,* or *ADAMTS17,* cause Ehlers-Danlos syndrome type VIIC [73] or Weill-Marchesani syndrome [74].

Besides this variety of hereditary pathologies, proteases are key players in cancer through their ability to participate in all steps of tumor progression and invasion [27]. Remarkably, other pathologies such as arthritis or cardiovascular diseases are also associated with important alterations in the spatiotemporal expression of protease genes. All these characteristics make proteolytic enzymes suitable as therapeutic targets for the development of drugs and as important prognosis biomarkers. Genomic, transcriptomic, proteomic, and degradomic studies, together with well-characterized animal models, may have the key to understand the origin and progression of pathological processes in which proteases are implicated.

1.5 Matrix Proteases and Their Inhibitors

As a prelude to the forthcoming chapters, we briefly describe in the context of the entire human degradome (Figure 1.1), the different ECM-degrading proteases and their inhibitors whose detailed analysis is addressed in this book. Among the numerous proteolytic enzymes with ECM-degrading capacity, attention was primarily focused on MMPs and the plasminogen system. Later, other metalloproteases, belonging to both ADAM and ADAMTS families, raised considerable interest because of their putative structural and functional links with MMPs. In addition, other proteases such as cathepsins acquired importance for their vital role in mammalian cellular turnover. More recently, new families of extra- or pericellular enzymes such as TTSPs have occupied a relevant place among the enzymes with ECM-degrading properties.

MMPs, ADAMs, and ADAMTSs belong to the catalytic class of metalloproteases, a large group of enzymes that can be classified according to their catalytic mechanism, substrates and structural homology [4]. The most characteristic feature of metalloproteases is that they use a metal ion to polarize a water molecule and perform the hydrolysis. Depending on the way each metalloprotease coordinates the metal ion and accommodates the substrate in a polypeptide scaffold, these enzymes can also be classified in different groups termed *clans* or *families*. The sequence HExxH is the most conserved zinc-binding motif in metalloproteases. Here, the two histidines coordinate the zinc ion and the glutamate acts as a general base in the catalytic reaction. This motif is characteristic of the MA clan, which contains two subclasses and 40 families of peptidases, including MMPs, ADAMs, and ADAMTSs [4].

MMPs are a group of more than 20 enzymes that are either secreted or membrane anchored. Their structural design is similar to reprolysins, mainly constituted by a peptidase motif in their catalytic domain, an N-terminal signal peptide, and a prodomain. In addition, in their archetypal domain organization there is a C-terminal hinge region followed by one hemopexin domain except for MMP-7, MMP-23A and -23B, and MMP-26 that lack this domain [75]. Some of these enzymes, such as stromelysin-3 or epilysin, have a furin-like target sequence inserted in their prodomains, and constitute the group of secreted MMPs activatable by furin convertases. Other MMPs are membrane anchored by a glycosylphosphatidylinositol or by type I or type II transmembrane segments, constituting a group called membrane-type matrix metalloproteinases (MT-MMPs). For many years, MMPs have been considered as proteases with the ability to degrade all major protein components of ECM and basement membranes, playing an important role in several physiological processes such as bone development, tissue resorption, or wound healing [76], and also in different pathological processes in which ECM remodeling was necessary, such as arthritis or tumor progression. Beyond this classic function of MMPs, further studies have shown that these enzymes target a large number of nonmatrix substrates, including growth factors, peptidase inhibitors, cytokines, receptors, adhesion molecules, clotting factors, and other proteases [77]. This variety of protease substrates allows these enzymes to regulate many other processes such as migration, apoptosis, proliferation, angiogenesis, and inflammation [78].

The expression of most MMPs is regulated by hormones, growth factors, cytokines, and tissue inhibitors of metalloproteinases (TIMPs). The mammalian TIMPs are a family of protease inhibitors constituted by four members: TIMP-1, TIMP-2, TIMP-3, and TIMP-4. They have two domains, an N-terminal domain of about 125 amino acids and a C-terminal domain of around 65 residues, and each one is stabilized by 3 disulfide bonds [79]. Expression of TIMPs and MMPs must be balanced for maintaining ECM metabolism and controlling the remodeling and physiological conditions of the tissues. In addition, TIMP-2 has been shown to be necessary for the binding and activation of proMMP2 by MT1-MMP, illustrating the complex relationships between these molecules.

The ADAM subfamily of reprolysins shares a structural domain organization based on a peptidase unit followed by a disintegrin domain, an epidermal growth factor (EGF)-like module, a transmembrane region, and a cytoplasmic tail. Most ADAMs are localized on the cell surface or are secreted, although some of them are situated at the Golgi. Their main function is the shedding of several transmembrane proteins such as cytokines and growth factor receptors from the cell surface [4]. In mammals, there are 38 different ADAMs; the mouse genome encodes 37 family members and the human genome 21. Interestingly, some ADAMs that are functional in rodents have become pseudogenes in humans, such as *ADAM1, 3, 4,* and *25* [3]. Conversely, *ADAM20* is a human-specific gene that is absent in rodents.

ADAMTSs are a group of secreted reprolysins that share some domains with ADAMs but lack the transmembrane and cytoplasmic tail and contain additional domains, being the most characteristic in a central thrombospondin (TS) type-1 and a series of C-terminal TS repeats [80, 81]. ADAMTS-7 and -12 also have a mucin domain, ADAMTS-20 and ADAMTS-9 a GON domain, ADAMTS-13 two CUB domains, and several other ADAMTSs a PLAC domain [82]. ADAMTSs are expressed in a wide range of adult and fetal tissues, and perform multiple physiological functions. As an example, ADAMTS-1, -4, -5, -8, -9, and -15 are aggrecanases with ability to degrade aggrecan, brevican, and versican, whereas ADAMTS-2, -3, and -14 are associated with the release of N-terminal propeptides of collagen.

Cathepsins were originally described as proteases functioning in acidic cellular compartments, just as lysosomes and endosomes, but they are present in the extracellular space as well as in the cytosol and nucleus [83]. Cathepsins can be classified based on their structure and catalytic type into aspartic (cathepsins D and E), serine (cathepsins A and G), and cysteine cathepsins (cathepsins B, C, F, H, K, L, S, O, L2, W, and Z) [84]. Among all these enzymes, cathepsins D, H, K, L, and S are capable of efficiently cleaving ECM components and other specific substrates and contribute to several physiological processes such as collagen turnover in bone and cartilage [85, 86], antigen presentation in the immune system [87], and neuropeptide and hormone processing [88, 89]. The main inhibitors of cathepsins are the proteins called *cystatins, stefins, thyropins,* and *serpins,* which bind to the target enzyme preventing substrate hydrolysis [90].

In addition to these matrix-degrading metalloproteases and cysteine proteases, there are several serine proteases implicated in the ECM metabolism including components of the plasminogen system and several members of the TTSPs. The plasminogen system is involved in numerous physiological and pathological processes because of its ability to degrade fibrin and several ECM proteins and to activate MMPs [91]. Plg is converted to the serine protease plasmin by one of the two plasminogen activators, tPA or uPA. The uPA receptor (uPAR) is a regulator of the plasminogen activator system. This glycosylphosphatidylinositol

(GPI)-bound protein acts as a membrane receptor for uPA and its zymogen form, pro-uPA [92]. When uPA is activated, it cleaves plasminogen, generating the protease plasmin, which, in a feedback process, cleaves and activates pro-uPA. Cell-associated plasminogen activation by uPA-uPAR can facilitate cell migration through ECM by pericellular proteolysis [92]. In turn, plasmin also cleaves ECM components, is essential for the development of fibrinolysis and activates MMPs, such as MMP3, MMP9, MMP12, and MMP13 [93]. The plasminogen activation system can be regulated by plasminogen activator inhibitor-1 (PAI-1; also known as *SERPINE1*) and -2 (PAI-2; also known as *SERPINB2*), that inhibit uPA and tPA preventing the activation of plasminogen into plasmin. Furthermore, the activity of plasmin can be also inhibited by α2-antiplasmin (SERPINF2).

The TTSPs are S01 serine proteases with a transmembrane domain close to the amino terminus of the protein that separates a short intracellular domain from a larger extracellular portion of the molecule containing a carboxy-terminal chymotrypsin domain (S1) and a variable internal stem region [94]. Members of this class include enteropeptidase, hepsin, spinesin, corin, matriptase, matriptase-2, TMPRSS2, TMPRSS3, and TMPRSS4. These enzymes have diverse roles in vertebrate physiology, being involved in the development and homeostasis of mammalian tissues such as the skin, heart, inner ear, placenta, and digestive tract [95]. The TTSPs are regulated by endogenous protease inhibitors, specifically Kunitz-domain-containing inhibitors and serpins. The transmembrane Kunitz-type inhibitor hepatocyte growth factor activator inhibitor-1 (HAI-1) is implicated in the inhibition of matriptase [96], hepsin [97], and prostasin [98]; and HAI-2 inhibits hepsin and matriptase [97, 99].

In summary, genomic and functional analyses have revealed the diversity and complexity of matrix proteases. Their roles in basic biological process as well as in several pathologies suggest the importance of these enzymes. This dual implication of proteases in life and disease should always be taken into consideration before dealing with clinical applications, but hopefully, the increasing knowledge of these enzymes will contribute to understand their diverse functional roles *in vivo* and to identify new therapeutic targets.

Acknowledgments

Our work is supported by grants from the Ministerio de Ciencia e Innovación-Spain, Fundación "M. Botín", and European Union (FP7 MicroEnviMet). The Instituto Universitario de Oncología is supported by Obra Social Cajastur and Acción Transversal del Cáncer-RTICC.

References

1. Lopez-Otin, C. and Bond, J.S. (2008) Proteases: multifunctional enzymes in life and disease. *J. Biol. Chem.*, **283**, 30433–30437.
2. Lopez-Otin, C. and Hunter, T. (2010) The regulatory crosstalk between kinases and proteases in cancer. *Nat. Rev. Cancer*, **10**, 278–292.

3. Puente, X.S., Sanchez, L.M., Overall, C.M., and Lopez-Otin, C. (2003) Human and mouse proteases: a comparative genomic approach. *Nat. Rev. Genet.*, **4**, 544–558.
4. Ugalde, A.P., Ordonez, G.R., Quiros, P.M., Puente, X.S., and Lopez-Otin, C. (2010) Metalloproteases and the degradome. *Methods Mol. Biol.*, **622**, 3–29.
5. Lopez-Otin, C. and Overall, C.M. (2002) Protease degradomics: a new challenge for proteomics. *Nat. Rev. Mol. Cell Biol.*, **3**, 509–519.
6. Fleischmann, R.D., Adams, M.D., White, O., Clayton, R.A., Kirkness, E.F., Kerlavage, A.R., Bult, C.J., Tomb, J.F., Dougherty, B.A., Merrick, J.M. *et al.* (1995) Whole-genome random sequencing and assembly of Haemophilus influenzae Rd. *Science*, **269**, 496–512.
7. Adams, M.D., Celniker, S.E., Holt, R.A., Evans, C.A., Gocayne, J.D., Amanatides, P.G., Scherer, S.E., Li, P.W., Hoskins, R.A., Galle, R.F. *et al.* (2000) The genome sequence of Drosophila melanogaster. *Science*, **287**, 2185–2195.
8. C.E.S.C. (1998) Genome sequence of the nematode C. elegans: a platform for investigating biology. *Science*, **282**, 2012–2018.
9. Gibbs, R.A., Weinstock, G.M., Metzker, M.L., Muzny, D.M., Sodergren, E.J., Scherer, S., Scott, G., Steffen, D., Worley, K.C., Burch, P.E. *et al.* (2004) Genome sequence of the Brown Norway rat yields insights into mammalian evolution. *Nature*, **428**, 493–521.
10. I.C.G.S.C. (2004) Sequence and comparative analysis of the chicken genome provide unique perspectives on vertebrate evolution. *Nature*, **432**, 695–716.
11. Lander, E.S., Linton, L.M., Birren, B., Nusbaum, C., Zody, M.C., Baldwin, J., Devon, K., Dewar, K., Doyle, M., FitzHugh, W. *et al.* (2001) Initial sequencing and analysis of the human genome. *Nature*, **409**, 860–921.
12. Venter, J.C., Adams, M.D., Myers, E.W., Li, P.W., Mural, R.J., Sutton, G.G., Smith, H.O., Yandell, M., Evans, C.A., Holt, R.A. *et al.* (2001) The sequence of the human genome. *Science*, **291**, 1304–1351.
13. Waterston, R.H., Lindblad-Toh, K., Birney, E., Rogers, J., Abril, J.F., Agarwal, P., Agarwala, R., Ainscough, R., Alexandersson, M., An, P. *et al.* (2002) Initial sequencing and comparative analysis of the mouse genome. *Nature*, **420**, 520–562.
14. Lewis, S., Ashburner, M., and Reese, M.G. (2000) Annotating eukaryote genomes. *Curr. Opin. Struct. Biol.*, **10**, 349–354.
15. Ordonez, G.R., Puente, X.S., Quesada, V., and Lopez-Otin, C. (2009) Proteolytic systems: constructing degradomes. *Methods Mol. Biol.*, **539**, 33–47.
16. Fitch, W.M. (2000) Homology a personal view on some of the problems. *Trends Genet.*, **16**, 227–231.
17. Puente, X.S. and Lopez-Otin, C. (2004) A genomic analysis of rat proteases and protease inhibitors. *Genome Res.*, **14**, 609–622.
18. Ordonez, G.R., Hillier, L.W., Warren, W.C., Grutzner, F., Lopez-Otin, C., and Puente, X.S. (2008) Loss of genes implicated in gastric function during platypus evolution. *Genome Biol.*, **9**, R81.
19. Puente, X.S., Gutierrez-Fernandez, A., Ordonez, G.R., Hillier, L.W., and Lopez-Otin, C. (2005a) Comparative genomic analysis of human and chimpanzee proteases. *Genomics*, **86**, 638–647.
20. Quesada, V., Ordonez, G.R., Sanchez, L.M., Puente, X.S., and Lopez-Otin, C. (2009) The Degradome database: mammalian proteases and diseases of proteolysis. *Nucleic Acids Res.*, **37**, D239–D243.
21. Rawlings, N.D., Morton, F.R., and Barrett, A.J. (2006) MEROPS: the peptidase database. *Nucleic Acids Res.*, **34**, D270–D272.
22. Rawlings, N.D., Tolle, D.P., and Barrett, A.J. (2004) MEROPS: the peptidase database. *Nucleic Acids Res.*, **32**, D160–D164.
23. Mulder, N.J., Apweiler, R., Attwood, T.K., Bairoch, A., Bateman, A., Binns, D., Bork, P., Buillard, V., Cerutti, L., Copley, R. *et al.* (2007) New developments in the InterPro database. *Nucleic Acids Res.*, **35**, D224–D228.

24. Thompson, J.D., Higgins, D.G., and Gibson, T.J. (1994) CLUSTAL W: improving the sensitivity of progressive multiple sequence alignment through sequence weighting, position-specific gap penalties and weight matrix choice. *Nucleic Acids Res.*, **22**, 4673–4680.
25. Page, M.J. and Di Cera, E. (2008) Evolution of peptidase diversity. *J. Biol. Chem.*, **283**, 30010–30014.
26. Puente, X.S., Sanchez, L.M., Gutierrez-Fernandez, A., Velasco, G., and Lopez-Otin, C. (2005b) A genomic view of the complexity of mammalian proteolytic systems. *Biochem. Soc. Trans.*, **33**, 331–334.
27. Edwards, D.R. *et al.* (eds) (2008) *The Cancer Degradome: Proteases and Cancer Biology*, Springer, New York.
28. C.S.A.C. (2005) Initial sequence of the chimpanzee genome and comparison with the human genome. *Nature*, **437**, 69–87.
29. Donaldson, I.J. and Gottgens, B. (2006) Evolution of candidate transcriptional regulatory motifs since the human-chimpanzee divergence. *Genome Biol.*, **7**, R52.
30. Varki, A. and Altheide, T.K. (2005) Comparing the human and chimpanzee genomes: searching for needles in a haystack. *Genome Res.*, **15**, 1746–1758.
31. Chen, C., Darrow, A.L., Qi, J.S., D'Andrea, M.R., and Andrade-Gordon, P. (2003) A novel serine protease predominately expressed in macrophages. *Biochem. J.*, **374**, 97–107.
32. Johnson, M.E., Cheng, Z., Morrison, V.A., Scherer, S., Ventura, M., Gibbs, R.A., Green, E.D., and Eichler, E.E. (2006) Recurrent duplication-driven transposition of DNA during hominoid evolution. *Proc. Natl. Acad. Sci. U.S.A.*, **103**, 17626–17631.
33. Warren, W.C., Hillier, L.W., Marshall Graves, J.A., Birney, E., Ponting, C.P., Grutzner, F., Belov, K., Miller, W., Clarke, L., Chinwalla, A.T. *et al.* (2008) Genome analysis of the platypus reveals unique signatures of evolution. *Nature*, **453**, 175–183.
34. Quesada, V., Velasco, G., Puente, X.S., Warren, W.C., and Lopez-Otin, C. (2010) Comparative genomic analysis of the zebra finch degradome provides new insights into evolution of proteases in birds and mammals. *BMC Genomics*, **11**, 220.
35. Warren, W.C., Clayton, D.F., Ellegren, H., Arnold, A.P., Hillier, L.W., Kunstner, A., Searle, S., White, S., Vilella, A.J., Fairley, S. *et al.* (2010) The genome of a songbird. *Nature*, **464**, 757–762.
36. Locke, D.P., Hillier, L.W., Warren, W.C., Worley, K.C., Nazareth, L.V., Muzny, D.M., Yang, S.P., Wang, Z., Chinwalla, A.T., Minx, P. *et al.* (2011) Comparative and demographic analysis of orang-utan genomes. *Nature*, **469**, 529–533.
37. Roy, S., Sharom, J.R., Houde, C., Loisel, T.P., Vaillancourt, J.P., Shao, W., Saleh, M., and Nicholson, D.W. (2008) Confinement of caspase-12 proteolytic activity to autoprocessing. *Proc. Natl. Acad. Sci. U.S.A.*, **105**, 4133–4138.
38. Xue, Y., Daly, A., Yngvadottir, B., Liu, M., Coop, G., Kim, Y., Sabeti, P., Chen, Y., Stalker, J., Huckle, E. *et al.* (2006) Spread of an inactive form of caspase-12 in humans is due to recent positive selection. *Am. J. Hum. Genet.*, **78**, 659–670.
39. Huesmann, G.R. and Clayton, D.F. (2006) Dynamic role of postsynaptic caspase-3 and BIRC4 in zebra finch song-response habituation. *Neuron*, **52**, 1061–1072.
40. Bertram, L., Lill, C.M., and Tanzi, R.E. (2010) The genetics of Alzheimer disease: back to the future. *Neuron*, **68**, 270–281.
41. De Sandre-Giovannoli, A., Bernard, R., Cau, P., Navarro, C., Amiel, J., Boccaccio, I., Lyonnet, S., Stewart, C.L., Munnich, A., Le Merrer, M. *et al.* (2003) Lamin a truncation in Hutchinson-Gilford progeria. *Science*, **300**, 2055.
42. Eriksson, M., Brown, W.T., Gordon, L.B., Glynn, M.W., Singer, J., Scott, L., Erdos, M.R., Robbins, C.M., Moses, T.Y., Berglund, P. *et al.* (2003) Recurrent de novo point mutations in lamin A cause Hutchinson-Gilford progeria syndrome. *Nature*, **423**, 293–298.
43. Varela, I., Pereira, S., Ugalde, A.P., Navarro, C.L., Suarez, M.F., Cau, P.,

Cadinanos, J., Osorio, F.G., Foray, N., Cobo, J. *et al.* (2008) Combined treatment with statins and aminobisphosphonates extends longevity in a mouse model of human premature aging. *Nat. Med.*, **14**, 767–772.

44. Ekeowa, U.I., Freeke, J., Miranda, E., Gooptu, B., Bush, M.F., Perez, J., Teckman, J., Robinson, C.V., and Lomas, D.A. (2010) Defining the mechanism of polymerization in the serpinopathies. *Proc. Natl. Acad. Sci. U.S.A.*, **107**, 17146–17151.
45. Summers, R.J., Meeks, S.L., Healey, J.F., Brown, H.C., Parker, E.T., Kempton, C.L., Doering, C.B., and Lollar, P. (2011) Factor VIII A3 domain substitution N1922S results in hemophilia A due to domain-specific misfolding and hyposecretion of functional protein. *Blood*, **117** (11), 3190–3198.
46. Hauri, H.P., Kappeler, F., Andersson, H., and Appenzeller, C. (2000) ERGIC-53 and traffic in the secretory pathway. *J. Cell Sci.*, **113** (Pt 4), 587–596.
47. I.H.G.S.C. (2004) Finishing the euchromatic sequence of the human genome. *Nature*, **431**, 931–945.
48. Bowen, D.J. (2002) Haemophilia A and haemophilia B: molecular insights. *Mol. Pathol.*, **55**, 127–144.
49. Kravtsov, D.V., Wu, W., Meijers, J.C., Sun, M.F., Blinder, M.A., Dang, T.P., Wang, H., and Gailani, D. (2004) Dominant factor XI deficiency caused by mutations in the factor XI catalytic domain. *Blood*, **104**, 128–134.
50. Millar, D.S., Elliston, L., Deex, P., Krawczak, M., Wacey, A.I., Reynaud, J., Nieuwenhuis, H.K., Bolton-Maggs, P., Mannucci, P.M., Reverter, J.C. *et al.* (2000) Molecular analysis of the genotype-phenotype relationship in factor X deficiency. *Hum. Genet.*, **106**, 249–257.
51. Perry, D.J. (2002) Factor VII deficiency. *Br. J. Haematol.*, **118**, 689–700.
52. Soria, J.M., Almasy, L., Souto, J.C., Bacq, D., Buil, A., Faure, A., Martinez-Marchan, E., Mateo, J., Borrell, M., Stone, W. *et al.* (2002) A quantitative-trait locus in the human factor XII gene influences both plasma factor XII levels and susceptibility to thrombotic disease. *Am. J. Hum. Genet.*, **70**, 567–574.
53. Gehring, N.H., Frede, U., Neu-Yilik, G., Hundsdoerfer, P., Vetter, B., Hentze, M.W., and Kulozik, A.E. (2001) Increased efficiency of mRNA 3' end formation: a new genetic mechanism contributing to hereditary thrombophilia. *Nat. Genet.*, **28**, 389–392.
54. Meeks, S.L. and Abshire, T.C. (2008) Abnormalities of prothrombin: a review of the pathophysiology, diagnosis, and treatment. *Haemophilia*, **14**, 1159–1163.
55. Levy, G.G., Nichols, W.C., Lian, E.C., Foroud, T., McClintick, J.N., McGee, B.M., Yang, A.Y., Siemieniak, D.R., Stark, K.R., Gruppo, R. *et al.* (2001) Mutations in a member of the ADAMTS gene family cause thrombotic thrombocytopenic purpura. *Nature*, **413**, 488–494.
56. Horwitz, M., Benson, K.F., Person, R.E., Aprikyan, A.G., and Dale, D.C. (1999) Mutations in ELA2, encoding neutrophil elastase, define a 21-day biological clock in cyclic haematopoiesis. *Nat. Genet.*, **23**, 433–436.
57. Citron, M., Westaway, D., Xia, W., Carlson, G., Diehl, T., Levesque, G., Johnson-Wood, K., Lee, M., Seubert, P., Davis, A. *et al.* (1997) Mutant presenilins of Alzheimer's disease increase production of 42-residue amyloid beta-protein in both transfected cells and transgenic mice. *Nat. Med.*, **3**, 67–72.
58. Esler, W.P. and Wolfe, M.S. (2001) A portrait of Alzheimer secretases--new features and familiar faces. *Science*, **293**, 1449–1454.
59. Casari, G., De Fusco, M., Ciarmatori, S., Zeviani, M., Mora, M., Fernandez, P., De Michele, G., Filla, A., Cocozza, S., Marconi, R. *et al.* (1998) Spastic paraplegia and OXPHOS impairment caused by mutations in paraplegin, a nuclear-encoded mitochondrial metalloprotease. *Cell*, **93**, 973–983.
60. Sleat, D.E., Donnelly, R.J., Lackland, H., Liu, C.G., Sohar, I., Pullarkat, R.K., and Lobel, P. (1997) Association of mutations in a lysosomal protein with classical late-infantile neuronal ceroid lipofuscinosis. *Science*, **277**, 1802–1805.

61. Leroy, E., Boyer, R., Auburger, G., Leube, B., Ulm, G., Mezey, E., Harta, G., Brownstein, M.J., Jonnalagada, S., Chernova, T. *et al.* (1998) The ubiquitin pathway in Parkinson's disease. *Nature*, **395**, 451–452.
62. Mitsui, S., Yamaguchi, N., Osako, Y., and Yuri, K. (2007) Enzymatic properties and localization of motopsin (PRSS12), a protease whose absence causes mental retardation. *Brain Res.*, **1136**, 1–12.
63. Petek, E., Windpassinger, C., Vincent, J.B., Cheung, J., Boright, A.P., Scherer, S.W., Kroisel, P.M., and Wagner, K. (2001) Disruption of a novel gene (IMMP2L) by a breakpoint in 7q31 associated with Tourette syndrome. *Am. J. Hum. Genet.*, **68**, 848–858.
64. Chun, H.J., Zheng, L., Ahmad, M., Wang, J., Speirs, C.K., Siegel, R.M., Dale, J.K., Puck, J., Davis, J., Hall, C.G. *et al.* (2002) Pleiotropic defects in lymphocyte activation caused by caspase-8 mutations lead to human immunodeficiency. *Nature*, **419**, 395–399.
65. Wang, J., Zheng, L., Lobito, A., Chan, F.K., Dale, J., Sneller, M., Yao, X., Puck, J.M., Straus, S.E., and Lenardo, M.J. (1999) Inherited human Caspase 10 mutations underlie defective lymphocyte and dendritic cell apoptosis in autoimmune lymphoproliferative syndrome type II. *Cell*, **98**, 47–58.
66. Holzinger, A., Maier, E.M., Buck, C., Mayerhofer, P.U., Kappler, M., Haworth, J.C., Moroz, S.P., Hadorn, H.B., Sadler, J.E., and Roscher, A.A. (2002) Mutations in the proenteropeptidase gene are the molecular cause of congenital enteropeptidase deficiency. *Am. J. Hum. Genet.*, **70**, 20–25.
67. Kereszturi, E., Szmola, R., Kukor, Z., Simon, P., Weiss, F.U., Lerch, M.M., and Sahin-Toth, M. (2009) Hereditary pancreatitis caused by mutation-induced misfolding of human cationic trypsinogen: a novel disease mechanism. *Hum. Mutat.*, **30**, 575–582.
68. Sun, C., Skaletsky, H., Birren, B., Devon, K., Tang, Z., Silber, S., Oates, R., and Page, D.C. (1999) An azoospermic man with a de novo point mutation in the Y-chromosomal gene USP9Y. *Nat. Genet.*, **23**, 429–432.
69. Canto, P., Soderlund, D., Reyes, E., and Mendez, J.P. (2004) Mutations in the desert hedgehog (DHH) gene in patients with 46,XY complete pure gonadal dysgenesis. *J. Clin. Endocrinol. Metab.*, **89**, 4480–4483.
70. Tuysuz, B., Mosig, R., Altun, G., Sancak, S., Glucksman, M.J., and Martignetti, J.A. (2009) A novel matrix metalloproteinase 2 (MMP2) terminal hemopexin domain mutation in a family with multicentric osteolysis with nodulosis and arthritis with cardiac defects. *Eur. J. Hum. Genet.*, **17**, 565–572.
71. Lausch, E., Keppler, R., Hilbert, K., Cormier-Daire, V., Nikkel, S., Nishimura, G., Unger, S., Spranger, J., Superti-Furga, A., and Zabel, B. (2009) Mutations in MMP9 and MMP13 determine the mode of inheritance and the clinical spectrum of metaphyseal anadysplasia. *Am. J. Hum. Genet.*, **85**, 168–178.
72. Kennedy, A.M., Inada, M., Krane, S.M., Christie, P.T., Harding, B., Lopez-Otin, C., Sanchez, L.M., Pannett, A.A., Dearlove, A., Hartley, C. *et al.* (2005) MMP13 mutation causes spondyloepimetaphyseal dysplasia, Missouri type (SEMD(MO). *J. Clin. Invest.*, **115**, 2832–2842.
73. Colige, A., Nuytinck, L., Hausser, I., van Essen, A.J., Thiry, M., Herens, C., Ades, L.C., Malfait, F., Paepe, A.D., Franck, P. *et al.* (2004) Novel types of mutation responsible for the dermatosparactic type of Ehlers-Danlos syndrome (Type VIIC) and common polymorphisms in the ADAMTS2 gene. *J. Invest. Dermatol.*, **123**, 656–663.
74. Morales, J., Al-Sharif, L., Khalil, D.S., Shinwari, J.M., Bavi, P., Al-Mahrouqi, R.A., Al-Rajhi, A., Alkuraya, F.S., Meyer, B.F., and Al Tassan, N. (2009) Homozygous mutations in ADAMTS10 and ADAMTS17 cause lenticular myopia, ectopia lentis, glaucoma, spherophakia, and short stature. *Am. J. Hum. Genet.*, **85**, 558–568.
75. Folgueras, A.R., Pendas, A.M., Sanchez, L.M., and Lopez-Otin, C. (2004) Matrix metalloproteinases in cancer: from new functions to improved inhibition strategies. *Int. J. Dev. Biol.*, **48**, 411–424.

76. Fanjul-Fernandez, M., Folgueras, A.R., Cabrera, S., and Lopez-Otin, C. (2010) Matrix metalloproteinases: evolution, gene regulation and functional analysis in mouse models. *Biochim. Biophys. Acta*, **1803**, 3–19.

77. Rodriguez, D., Morrison, C.J., and Overall, C.M. (2010) Matrix metalloproteinases: what do they not do? New substrates and biological roles identified by murine models and proteomics. *Biochim. Biophys. Acta*, **1803**, 39–54.

78. Kessenbrock, K., Plaks, V., and Werb, Z. (2010) Matrix metalloproteinases: regulators of the tumor microenvironment. *Cell*, **141**, 52–67.

79. Brew, K., Dinakarpandian, D., and Nagase, H. (2000) Tissue inhibitors of metalloproteinases: evolution, structure and function. *Biochim. Biophys. Acta*, **1477**, 267–283.

80. Cal, S., Obaya, A.J., Llamazares, M., Garabaya, C., Quesada, V., and Lopez-Otin, C. (2002) Cloning, expression analysis, and structural characterization of seven novel human ADAMTSs, a family of metalloproteinases with disintegrin and thrombospondin-1 domains. *Gene*, **283**, 49–62.

81. Llamazares, M., Cal, S., Quesada, V., and Lopez-Otin, C. (2003) Identification and characterization of ADAMTS-20 defines a novel subfamily of metalloproteinases-disintegrins with multiple thrombospondin-1 repeats and a unique GON domain. *J. Biol. Chem.*, **278**, 13382–13389.

82. Porter, S., Clark, I.M., Kevorkian, L., and Edwards, D.R. (2005) The ADAMTS metalloproteinases. *Biochem. J.*, **386**, 15–27.

83. Reiser, J., Adair, B., and Reinheckel, T. (2010) Specialized roles for cysteine cathepsins in health and disease. *J. Clin. Invest.*, **120**, 3421–3431.

84. Santamaria, I., Velasco, G., Pendas, A.M., Fueyo, A., and Lopez-Otin, C. (1998) Cathepsin Z, a novel human cysteine proteinase with a short propeptide domain and a unique chromosomal location. *J. Biol. Chem.*, **273**, 16816–16823.

85. Deal, C. (2009) Potential new drug targets for osteoporosis. *Nat. Clin. Pract. Rheumatol.*, **5**, 20–27.

86. Stoch, S.A. and Wagner, J.A. (2008) Cathepsin K inhibitors: a novel target for osteoporosis therapy. *Clin. Pharmacol. Ther.*, **83**, 172–176.

87. Honey, K. and Rudensky, A.Y. (2003) Lysosomal cysteine proteases regulate antigen presentation. *Nat. Rev. Immunol.*, **3**, 472–482.

88. Friedrichs, B., Tepel, C., Reinheckel, T., Deussing, J., von Figura, K., Herzog, V., Peters, C., Saftig, P., and Brix, K. (2003) Thyroid functions of mouse cathepsins B, K, and L. *J. Clin. Invest.*, **111**, 1733–1745.

89. Funkelstein, L., Toneff, T., Mosier, C., Hwang, S.R., Beuschlein, F., Lichtenauer, U.D., Reinheckel, T., Peters, C., and Hook, V. (2008) Major role of cathepsin L for producing the peptide hormones ACTH, beta-endorphin, and alpha-MSH, illustrated by protease gene knockout and expression. *J. Biol. Chem.*, **283**, 35652–35659.

90. Turk, B., Turk, D., and Salvesen, G.S. (2002) Regulating cysteine protease activity: essential role of protease inhibitors as guardians and regulators. *Curr. Pharm. Des.*, **8**, 1623–1637.

91. Behrendt, N. (2004) The urokinase receptor (uPAR) and the uPAR-associated protein (uPARAP/Endo180): membrane proteins engaged in matrix turnover during tissue remodeling. *Biol. Chem.*, **385**, 103–136.

92. Smith, H.W. and Marshall, C.J. (2010) Regulation of cell signalling by uPAR. *Nat. Rev. Mol. Cell Biol.*, **11**, 23–36.

93. Carmeliet, P., Moons, L., Lijnen, R., Baes, M., Lemaitre, V., Tipping, P., Drew, A., Eeckhout, Y., Shapiro, S., Lupu, F. *et al.* (1997) Urokinase-generated plasmin activates matrix metalloproteinases during aneurysm formation. *Nat. Genet.*, **17**, 439–444.

94. Bugge, T.H., Antalis, T.M., and Wu, Q. (2009) Type II transmembrane serine proteases. *J. Biol. Chem.*, **284**, 23177–23181.

95. Szabo, R. and Bugge, T.H. (2008) Type II transmembrane serine proteases in development and disease. *Int. J. Biochem. Cell Biol.*, **40**, 1297–1316.
96. Lin, C.Y., Anders, J., Johnson, M., and Dickson, R.B. (1999) Purification and characterization of a complex containing matriptase and a Kunitz-type serine protease inhibitor from human milk. *J. Biol. Chem.*, **274**, 18237–18242.
97. Kirchhofer, D., Peek, M., Lipari, M.T., Billeci, K., Fan, B., and Moran, P. (2005) Hepsin activates pro-hepatocyte growth factor and is inhibited by hepatocyte growth factor activator inhibitor-1B (HAI-1B) and HAI-2. *FEBS Lett.*, **579**, 1945–1950.
98. Fan, B., Wu, T.D., Li, W., and Kirchhofer, D. (2005) Identification of hepatocyte growth factor activator inhibitor-1B as a potential physiological inhibitor of prostasin. *J. Biol. Chem.*, **280**, 34513–34520.
99. Szabo, R., Hobson, J.P., List, K., Molinolo, A., Lin, C.Y., and Bugge, T.H. (2008) Potent inhibition and global co-localization implicate the transmembrane Kunitz-type serine protease inhibitor hepatocyte growth factor activator inhibitor-2 in the regulation of epithelial matriptase activity. *J. Biol. Chem.*, **283**, 29495–29504.

2
The Plasminogen Activation System in Normal Tissue Remodeling

Vincent Ellis

2.1
Introduction

The plasminogen activation system was one of the first proteolytic systems to be identified, its activity first being observed by the French physiologist Albert Dastre in 1893, some 10 years before Pavlov's discovery of enterokinase activity in the duodenum. Dastre observed the spontaneous lysis of fibrin clots, a process that he termed *fibrinolysis*. This term is still used today, and the process of the proteolytic breakdown of fibrin is still thought to be the central role of the plasminogen activation system, although this role is not necessarily restricted to intravascular hemostasis.

One of the key features of this proteolytic system, and one that marks it out from all the other proteases and proteolytic systems described in this volume, is that its principal component, plasminogen, is extremely abundant and constitutively expressed. Plasminogen is synthesized primarily in the liver and circulates in the blood at a concentration of approximately 2.4 μM, representing a total of greater than 1 g of plasminogen in the average body. This represents the largest source of proteolytic potential in the body, not just because of the abundance of plasminogen. For example, prothrombin, the terminal protease of the blood clotting cascade, is also present in the circulation at a similar concentration, yet has relatively little potential for proteolysis. The reason for this difference lies in the respective substrate specificities of these two very abundant enzymes.

Both plasminogen and prothrombin, as almost all other serine proteases, are synthesized as zymogens or precursor forms, essentially devoid of catalytic activity. Their activity is dependent on proteolytic conversion to their active forms, plasmin and thrombin, respectively. Both proteases have trypsin-like substrate specificity, meaning that they cleave peptide bonds C-terminally of the basic residues, lysine or arginine, the so-called primary specificity of the protease. However, thrombin has an extremely restricted substrate specificity, meaning that it will cleave very few protein substrates in addition to fibrinogen, and this is dictated by features of both its active-site region and parts of the enzyme distant from the active site. Therefore, its activity is limited by its specificity. By contrast, plasmin can be considered to be

Matrix Proteases in Health and Disease, First Edition. Edited by Niels Behrendt.

a much less sophisticated protease, having a much broader substrate specificity, and in the right circumstances is able to cleave proteins at virtually any available lysine/arginine residue in much the same way as trypsin itself.

These characteristics have led to plasmin, and the plasminogen activation system, having a very wide variety of roles attributed to it. *In vitro* and in purified protein systems, plasmin can be encouraged to engage in reactions with many substrates that possibly are not biologically relevant substrates *in vivo*, either in physiological or pathological situations. However, over the past two decades, the combination of traditional enzymological approaches with observations and experiments made possible by the generation of mice deficient in each of the main components of the plasminogen activation system [1–6] has allowed insight into the true functions of this proteolytic system.

Historically, research into the plasminogen activation system has its roots in two distinct areas. This is in part a result of the existence of two distinct proteolytic activators of plasminogen, urokinase plasminogen activator (uPA) and tissue-type plasminogen activator (tPA), and their discovery. uPA, after its initial isolation from urine, was found to be expressed by many types of cells and became established as the protease involved in the extravascular activation of plasminogen, with intense interest in its role in the development and progression of cancer – a concept consolidated in the landmark review by Danø and colleagues [7] (Chapter 9). tPA, although first isolated from tissues, as suggested by its name, became established as the intravascular plasminogen activator. This clear distinction was reinforced by subsequent molecular analysis of their function, with the activity of uPA enhanced by its binding to the surface of cells and tPA activity by its binding to fibrin. Once again, studies in mice deficient in the two plasminogen activators has had a significant impact, and blurred these boundaries, and there is now thought to be considerable overlap in the biological roles of the two plasminogen activators.

This chapter focuses on how the activity of the plasminogen activation system contributes to tissue remodeling and other related processes. In light of the issues discussed above, the key aspects considered are the target protein substrates in these situations and how the activity of this proteolytic system is regulated.

2.2
Biochemical and Enzymological Fundamentals

The biochemical characteristics of the enzymatic and regulatory components of the plasminogen activation proteolytic system have been extensively and comprehensively reviewed elsewhere [8–10]. The description here is limited to the fundamentals that underpin current understanding of the biological roles of this proteolytic system *in vivo*, both in normal physiology and in pathological situations. The components that constitute the plasminogen activation system as defined here are plasminogen itself, the two plasminogen activators uPA and tPA, the serpin-family protease inhibitors α_2-antiplasmin, and PAI-1 (plasminogen activator inhibitor-1) and uPA receptor (uPAR), the cellular receptor for uPA.

As will become apparent, other protein components can also interact with this basic system, potentially regulating its activity.

The broad substrate specificity of plasmin has already been mentioned, and it is this that gives the plasminogen activation system the potential to have target substrates other than its undoubted primary target fibrin. However, this broad substrate specificity, together with the considerable catalytic efficiency of plasmin, has proved to be a problem for the biochemical study of the function of this system. *In vivo*, in the absence of opposing mechanisms, the activity of plasmin has an extremely short half-life because of its inhibition by α_2-antiplasmin. This inhibitor is possibly the fastest acting of the serpin-family inhibitors, operating at close to the theoretical diffusionally controlled rate limit, and the calculated half-life for the inhibition of plasmin in blood plasma is less than 0.01 s. Many *in vitro* studies have essentially ignored this, and have studied the effects of unopposed plasmin generation and activity, with the result that several activities attributed to plasmin are possibly artifacts. This conclusion has tended to be supported by *in vivo* studies. Nevertheless, biochemical and other *in vitro* experiments have made a very important contribution to our understanding of the plasminogen activation system, particularly when these have been carried out under carefully controlled conditions, paying due attention to the fundamental principles of enzymology and enzyme kinetics. The combination of these two very different approaches has proved to be highly complementary, with each approach throwing up some surprises that might not have been predicted from the other.

2.2.1
Plasminogen

In common with many serine proteases, plasminogen is a complex modular protein. In the majority of these proteases, this modular structure is involved in some way in regulating the function or activity of the enzyme, and the three protease components of the plasminogen activation system are among the best studied examples of this form of regulation. Plasminogen contains six independent protein modules, an N-terminal module and five homologous kringle modules, in addition to the C-terminal catalytic domain. The similarity in the proteolytic activities of plasmin and trypsin have been highlighted here, but the presence of these accessory domains in plasmin(ogen) regulates its function in two fundamental ways that are not available to the much simpler trypsin(ogen). Firstly, interdomain interactions regulate the overall topology of the protein, allowing it to adopt a very compact conformation that is essentially resistant to proteolytic activation [11]. Secondly, several of the kringle modules contain binding sites for lysine and related aminocarboxylic acids. These binding sites allow plasmin(ogen) to bind to a variety of proteins, in particular fibrin, a variety of cell-surface proteins on mammalian cells, the so-called plasminogen receptors [12], and on pathogenic bacteria [13]. This binding has several consequences, all acting in a concerted manner to promote the activation and activity of plasmin. Binding releases the conformational constraint on plasminogen, so that it is available for activation; the binding of plasminogen is often in juxtaposition to a bound plasminogen activator, thereby enhancing activation by a catalytic template mechanism; and bound plasmin is protected to a variable extent from inhibition by α_2-antiplasmin. Therefore, this binding facilitates the activation of plasminogen, allows plasmin activity to persist, and potentially localizes this activity.

Activation from the inactive zymogen form is accomplished by a single proteolytic cleavage at the canonical activation site Arg^{561}-Val^{562} (equivalent to the Arg^{15}-Ile^{16} bond of chymotrypsin). It is worth noting here that plasminogen can also be activated by an unusual, although not unique, nonproteolytic mechanism. This is one of several mechanisms used by the previously mentioned infectious microorganisms, particularly strains of *Streptococcus*. In essence, streptokinase, the protein responsible for this effect, binds to plasminogen and causes a conformational change in its catalytic domain such that it closely resembles that of the proteolytically activated protease. In addition to becoming proteolytically active, substrate specificity is altered and the complex acquires the ability to activate additional plasminogen molecules in a conventional proteolytic manner [14, 15]. This intriguing mechanism has been exploited in the use of streptokinase as a therapeutic thrombolytic agent, for example, in acute myocardial infarction. However, despite the elegance of this activation mechanism, it seems that it is only used by microorganisms in conjunction with host plasminogen. There is no evidence for such conformational activation of plasminogen, or indeed any other serine protease zymogen, as an endogenous mechanism in vertebrates, and no proteins with any homology to streptokinase, or any of the several other bacterial activators, are found in eukaryotes. The reason for mentioning this is that it has been speculated that such mechanisms do exist, but any supporting evidence has largely been based on poorly designed experiments.

2.2.2 Regulation of the Plasminogen Activation System

The proteolytic potential of the plasminogen activation system makes it evident that its activity needs to be tightly regulated. As a clear illustration of this, exogenous "overactivation" of the plasminogen activation system in the circulation by acute overadministration of the bacterial plasminogen activator streptokinase leads to complete consumption of α_2-antiplasmin, unregulated plasmin activity, and systemic defibrinogenation. However, possibly to avoid similar consequences, both of the mammalian plasminogen activators are rather inefficient enzymes. tPA is an extreme example of this, to the extent that, uniquely among serine proteases, its activity is not regulated by zymogen activation. The single-chain zymogen form of tPA has an activity close to that of the proteolytically activated, two-chain form, but both are very poor plasminogen activators. Fibrin acts as a cofactor for both forms of tPA, greatly increasing their activity both by the previously mentioned catalytic template mechanism and also by effects on the active site of the protease [16, 17]. Therefore, the fundamental level of regulation of tPA activity is by cofactor binding, with tPA essentially only active when bound to fibrin. Fibrin, of course, is the primary target of the generated plasmin, and therefore the generation of fibrin stimulates its subsequent degradation. It is this aspect of the regulation of tPA activity that pointed to it being the primary fibrinolytic protease, with plasmin generated by uPA being ascribed distinct roles, for example, degradation of extracellular matrix proteins and growth factor activation.

Although the tPA/fibrin axis is extremely elegant in its self-sufficiency, the situation is complicated by the observation that a range of other proteins can stimulate the activity of tPA, albeit to a variable extent and with varying degrees of specificity. These proteins include a variety of denatured and aggregated proteins, but more interestingly include the neuropathological proteins amyloid-β [18] and PrP, the prion protein [19]. The common feature shared between these proteins, and fibrin itself, has been proposed to be their ability to form fibrillar structures that adopt the so-called cross-β conformation [20].

By contrast to tPA, the activity of uPA is strictly regulated by zymogen activation. The zymogen pro-uPA can be activated by a variety of trypsin-like serine proteases, including plasmin itself [21–23]. This means that plasmin generated by the activity of uPA can activate pro-uPA in a feedback activation process. Because uPA can activate plasminogen, and plasmin can activate pro-uPA, this is known as *reciprocal zymogen activation*. This mechanism suggests that plasmin is the physiological activator of pro-uPA, but raises the question of how the first molecules of plasmin are generated to initiate the activity of the system. Pro-uPA has a very low, but detectable, level of intrinsic proteolytic activity and it has been demonstrated in mechanistic model systems that the assembly of ternary complexes containing pro-uPA and plasminogen is sufficient to initiate plasmin generation, by greatly increasing the efficiency of activation of both pro-plasminogen activator and plasminogen [24]. This therefore provides a mechanism both for the initiation of plasminogen activation and to increase the relatively poor catalytic efficiency of uPA. In biological systems, uPAR, the specific cellular receptor for uPA, is thought to provide the same advantages.

uPAR is a protein expressed on a wide variety of cell types. It binds uPA, via its N-terminal epidermal growth factor (EGF)-like domain, with high affinity ($K_d < 1$ nM) in an interaction that is now well understood at the structural level [25, 26]. The binding of pro-uPA to uPAR has been shown to greatly stimulate the activation of pro-uPA by plasmin [27] and to increase the catalytic efficiency of plasminogen activation by bound uPA [28, 29]. The assembly of this catalytic complex on the cell surface also protects the generated plasmin from inactivation by α_2-antiplasmin [28], but allows efficient inhibition of uPA by PAI-1 [30], enabling PAI-1 to modulate the activity of the system [31]. However, none of these effects are a direct consequence of the uPA–uPAR interaction *per se*, as they cannot be mimicked by soluble uPAR [32], but they can by mutants of uPA directly associated with the cell membrane by a glycolipid anchor [33].

These studies led to the conclusion that uPAR facilitates the assembly of a proteolytic complex to localize and promote the generation and activity of plasmin on the surface of uPAR expressing cells, and that this should be the preferential site of uPA-catalyzed plasminogen activation *in vivo*. However, mice deficient in uPAR were found not to display the same phenotype as uPA-deficient mice [3, 34], casting doubt on the notion that uPAR plays a significant role in promoting plasminogen activation *in vivo*. Nevertheless, there is substantial evidence that increased levels of uPAR play an important role in pathological situations, particularly in cancer [35], and it is now apparent that uPAR plays important, although perhaps more subtle, roles in normal physiological situations. The role of uPAR is discussed in detail at the end of this chapter.

The discovery and characterization of uPAR stimulated interest in other mechanisms for promoting cell-surface plasminogen activation, particularly by tPA. Annexin-II was identified as a protein that could bind both tPA and plasminogen on endothelial cells [36]. Mice deficient in annexin-II have been shown to deposit fibrin in their microvasculature and to have impaired angiogenesis in several experimental models [37]. These observations suggest that annexin-II is a regulator of plasmin generation *in vivo*, although, as with uPAR, it is clearly not essential. Other tPA-binding proteins have been identified on other cell types, such as p63/cytoskeleton-associated protein 4 (CKAP4) on vascular smooth muscle cells [38–40]. Although this protein clearly regulated both tPA activity and its inhibition *in vitro*, its role *in vivo* has not been established.

2.3
Biological Roles of the Plasminogen Activation System

A large body of experimental evidence accumulated over many years in biochemical, cell biological, and *ex vivo* systems was widely interpreted as demonstrating that this proteolytic system played fundamental biological roles and indeed that it would be indispensible. Therefore, during the 1990s, it came as a huge surprise when mice with genetic ablation of the components of this system, including plasminogen itself, were found not only to be born and viable but also to be reasonably healthy. Much of the research in the field has since been focused on determining what really are the biological roles of plasminogen activation, and, in particular, the more subtle effects that it may have.

At the biochemical level, the activity of this system is involved in the direct breakdown of protein substrates, such as fibrin and other extracellular matrix proteins, the activation of other proteases and proteolytic pathways, and in regulating the activity and bioavailability of other bioactive proteins, including cytokines and growth factors. At the level of the organism, the activity of the plasminogen activation system has been associated with a very wide range of biological processes, the dysregulation of this system is thought to contribute to a wide range of disease processes, and its activity is exploited in a variety of ways by infectious microorganisms (Chapter 7).

Before the advent of gene-targeting technology, several fundamental physiological roles had been ascribed to the plasminogen activation system, most notably in ovulation, fertilization, embryogenesis, and development. This was largely as a result of exquisitely regulated expression patterns of the plasminogen activators assessed by immunohistochemical and *in situ* hybridization studies in spermatozoa, oocytes, trophoblasts, and the developing mouse embryo [41–43]. However, these processes appear to be completely unaffected in mice lacking plasminogen, or for that matter any other component of the plasminogen activation system, with the animals being born at the expected frequency and Mendelian inheritance, and having no overt phenotypic abnormalities at birth [3, 44]. A minor caveat to this is that breeding from homozygous plasminogen-deficient mice is not usually possible, but this is an indirect consequence of the general ill health of these animals (as discussed subsequently), and in particular the presence of fibrotic lesions in the female genital tract. Nevertheless, these observations demonstrate that the plasminogen activation system does not have a role in these processes, or at least suggest a high level of redundancy in these processes, with other proteases or mechanisms able to overcome or compensate for the lack of plasmin activity.

However, before the advent of these technologies, some insight into the functions of the plasminogen activation system *in vivo* had been gleaned from the study of individuals with congenital deficiencies of components of the plasminogen activation system. By contrast with the blood coagulation system, where mutations have been identified in all of the components, and in many cases led to their original identification, mutations in the plasminogen activation system are less common, have not been identified in all of the components, and have in general been less informative.

2.3.1 Congenital Plasminogen Deficiencies

The lack of an overtly dramatic effect of ablation of the plasminogen gene in mice was perhaps not so surprising when the consequences of known plasminogen mutations in humans are considered. In 1978, a Japanese patient with recurrent thrombosis was identified as being apparently heterozygous for a functional mutation in plasminogen [45]. Subsequently, a point mutation leading to the substitution $Ala^{601} > Thr(cAla^{55}$ in chymotrypsin, a conserved residue) was found to be responsible [46, 47]. This plasminogen can be activated normally by plasminogen activators, but the activity of the resultant plasmin is severely compromised. Estimates of its activity range from 14% in plasma-based fibrinolytic assays [48] to 8% against amidolytic substrates using purified protein [49]. Ala^{601} lies proximal to one of the catalytic triad residues, His^{603} ($cHis^{57}$), and the low activity of this mutant protein is consistent with a disrupted orientation of this residue, which is essential for deprotonation of the catalytic serine residue ($cSer^{195}$) during catalysis, or alternatively with an effect on the catalytic triad residue Asp^{646} ($cAsp^{102}$) which also lies close to the side chain of Ala^{601}.

Although this mutation, and a variety of others mutations [47, 50, 51], were initially identified in patients suffering from recurrent thrombosis, the link between dysfunctional plasminogen and thrombosis is not fully established. A defective fibrinolytic system might be expected to have this consequence, but many individuals carrying this mutation, including homozygotes, do not suffer from thrombotic events. The $Ala^{601} > Thr$ mutation occurs at a high frequency (>2%) in the Japanese population, and as such is considered polymorphic. Nevertheless, in interpreting these observations it has to be remembered that plasminogen in these individuals still has a significant level of activity, and that relatively normal function may be possible under these conditions.

2.3.2 Intravascular Fibrinolysis

Despite the lack of a clear thrombotic tendency in individuals with congenital plasminogen deficiency, there seems little doubt that the plasminogen activation system plays a role in maintaining vascular patency in normal individuals. Observations made in plasminogen-deficient mice are similar in this respect, as these animals do not suffer from overt spontaneous venous thrombosis. However, thrombotic occlusion of the microvasculature is common and underlies the chronic pathologies observed in these animals, which include ulceration of the gastric and colonic mucosa, and rectal prolapse as a result of ulceration, inflammation, and necrosis [3, 44]. Conversely, mice deficient in PAI-1 display a mildly hyperfibrinolytic state, determined *ex vivo*, and are relatively resistant to experimentally induced venous thrombosis, but do not have impaired general hemostasis [52].

An important consideration is that the relative lack of effect of plasminogen deficiency is not due to any form of developmental compensation, as treatment of wild-type or tPA-deficient mice with an inhibitory monoclonal antibody to uPA has similar effects [53, 54]. There has long been some suggestion that plasminogen-independent fibrinolytic mechanisms exist, for example, mediated by neutrophil elastase [55], and this may possibly contribute to the relatively mild phenotype of the plasminogen-deficient mice. However, there is little support for this from observations in plasminogen-deficient mice, although experimental thrombi in these animals have been shown to accumulate increased numbers of neutrophils and *in vitro* these cells lysed clots after extended incubation times [56].

A remarkable observation that has been made in plasminogen-deficient mice is that the chronic pathologies, which include wasting, can be compensated for by an absence of fibrinogen [57]. Mice with fibrinogen-deficiency alone had previously been shown to survive well with only relatively mild spontaneous bleeding [58]. One clear message from these experiments is that fibrin is undoubtedly not the only major substrate for plasmin, but in a relatively unchallenged situation it may be the only relevant substrate.

2.3.3 Extravascular Fibrinolysis – Ligneous Conjunctivitis

Perhaps the most interesting development from studies in mice deficient in components of the plasminogen activation system is the concept that a major role of this system is in extravascular fibrinolysis, and what has been termed *fibrin surveillance* [34]. As has been mentioned, mice with complete deficiency of plasminogen develop significant deposition of fibrin in various tissues throughout the body starting from an early age, rather than developing overt venous thrombosis. Therefore, individuals suffering from recurrent venous thrombosis may not be the best place to identify plasminogen deficiency states in humans.

This turns out to be true, as the identification of further functional mutations in plasminogen has come from an unexpected source: the condition ligneous conjunctivitis. This is a rare and unusual form of chronic conjunctivitis in which pseudomembranes form on the surfaces of the eyelids and other mucosal surfaces. These can become woodlike (ligneous) and progress to disfiguring nodular masses that replace the normal mucosa. These lesions have a histologically disrupted epithelium, that is, replaced by a fibrin-rich, hyaline-like matrix with an infiltration of inflammatory cells [59]. Some individuals with this condition are also found to be affected by occlusive hydrocephalus. In 1997, two unrelated patients of Turkish decent suffering from ligneous conjunctivitis were found to be severely plasminogen deficient [60]. One mutation was identified as $Arg^{216} > His$ in the kringle-2 domain of plasminogen, which is not a functional mutation (i.e., this residue it is not a part of the lysine-binding site) but results in greatly reduced levels of plasminogen. The other was $Trp^{597} > STOP$ in the catalytic domain. Remarkably, this patient was homozygous for the mutation, and had undetectable levels of plasminogen, determined both antigenically and in activity assays.

Subsequently, many more cases of plasminogen deficiency associated with ligneous conjunctivitis, and other related chronic mucosal inflammatory conditions, have been identified in diverse populations and with varying degrees of severity [61]. Several of these patients have been treated successfully with intravenous replacement or topical application of plasminogen [62, 63], and in some cases these patients have been successfully treated using anticoagulant therapy, for example, with the direct thrombin inhibitor argatroban [64].

Interestingly, reinvestigation of the plasminogen-deficient mice revealed that these animals also develop conjunctival lesions indistinguishable from those in the ligneous conjunctivitis patients, and these animals also occasionally suffer occlusive hydrocephalus [65]. Unsurprisingly, these symptoms were not apparent in the combined plasminogen/fibrinogen-deficient mice, but nevertheless emphasizing the pathological role of fibrin in this debilitating condition.

2.3.4
Congenital Inhibitor Deficiencies

Undetectable levels of PAI-1 (both activity and antigen) were first observed in a patient with a recurrent lifelong bleeding disorder [66], but the molecular defect was not determined, as is also the case for several other individuals subsequently identified. However, the defect has been determined in one large kindred, with a frameshift mutation confirming the lack of a functional inhibitor, and at least seven individuals being homozygous for the null allele [67, 68]. As with the original patient, all suffered from significant bleeding episodes, which were never observed in heterozygous individuals. However, no other physiological abnormalities have been detected, clearly demonstrating the role of PAI-1 in regulating plasminogen activator activity, and once again emphasizing that the preeminent role of the plasminogen activation system is in fibrinolysis.

Congenital deficiencies of α_2-antiplasmin have also been observed, and again these are associated with rare bleeding disorders [69]. Similar to mice with PAI-1 deficiency [52], α_2-antiplasmin-deficient mice do not display a bleeding tendency; however, they do have enhanced fibrinolysis as they have been demonstrated to be protected from endotoxin-induced thrombosis [6]. Given the strong emphasis placed on the role of α_2-antiplasmin in controlling the activity of the powerful protease plasmin, these observations are particularly surprising. However, in addition to the lessons to be learnt from all of these deficiency states in both humans and mice, this particular deficiency emphasizes the importance of the regulation of the mechanisms involved in the activation of plasminogen.

2.4 Tissue Remodeling Processes

2.4.1 Wound Healing

The healing of skin wounds is an essential part of normal homeostasis. It is the process that best exemplifies tissue remodeling in the adult organism and, as an experimental model, its use is well established in the study of pathological tissue remodeling. The first response to wounding of the skin is activation of the primary and secondary phases of the hemostatic system. Blood loss from the damaged vasculature is rapidly plugged by platelets aided by vasoconstriction, and the plug is subsequently stabilized by the generation and polymerization of fibrin in response to the blood coagulation cascade of serine proteases. The formation of this fibrin clot, in addition to plugging the wound, is involved in its subsequent healing; a complex process leading to re-epithelialization. In this context, the stabilized fibrin clot acts as a form of extracellular matrix, allowing the movement of cells through and across it, and for this reason it is also commonly known as the *provisional matrix*.

Inflammatory cells are the first to arrive in the provisional matrix, initially neutrophils followed by macrophages. In addition to immune functions at the wound site, these cells also send signals for the migration and proliferation of fibroblasts, leading to the formation of granulation tissue to replace the provisional fibrin matrix, a process that is aided by neovascularization of the newly formed granulation tissue. The crucial event for re-epithelialization of the wound is the migration of basal keratinocytes from the edges of the wound and their subsequent proliferation, a process that can precede the formation of granulation tissue depending on the severity of the wound [70].

The activity of the plasminogen activation system has been implicated at several stages of the wound-healing process, and detailed immunohistochemical and *in situ* hybridization studies in mouse models have shown that expression of components of the plasminogen activation system is highly regulated in migrating keratinocytes and also in fibroblasts and macrophages in granulation tissue [71]. The functional involvement of the plasminogen activation system in the overall process of skin wound healing was demonstrated by the observation that the complete healing of full-thickness skin wounds is severely delayed in plasminogen-deficient mice, increasing from 16–19 to 43–55 days [72]. This was largely due to impaired keratinocyte migration with the tip of the migrating wedge of keratinocytes being observed to have a "blunt" appearance, with fibrin accumulating in front of and below the migrating sheet of cells [72]. Consistent with this fibrinolytic role, mice deficient in PAI-1 have been observed to heal skin wounds at an accelerated rate, particularly during the early stages, presumably due to increased plasmin generation [73]. The role of unresolved fibrin in impeding wound healing has been directly demonstrated, as the healing of wounds in plasminogen-deficient mice can be restored to essentially normal levels by the combined deficiency of fibrinogen [57]. Facilitating keratinocyte migration during re-epithelialization of the wound appears to be the main mechanism by which the plasminogen activation system affects wound healing, as the formation of granulation tissue and wound angiogenesis are largely unaffected [57]. This is not limited to skin wound healing, as a very similar situation has also been observed in corneal wound healing [74]. Here, despite a lack of blood vessels and bleeding, the fibrin deposits are still the main obstacle to the migrating keratinocytes.

One intriguing observation made in these skin wound-healing experiments is that in mice with a double deficiency in uPA and tPA the wounds healed significantly better than in plasminogen-deficient mice [75]. This suggests the plasmin can be generated *in vivo* in the absence of the two established plasminogen activators. Plasma kallikrein was proposed to be the protease that fulfilled this role based on indirect experimental evidence and the previously reported ability of plasma kallikrein to activate plasminogen in a purified system, albeit very inefficiently [76]. This observation remains to be fully tested, but it does suggest that the generation of very low levels of plasmin activity *in vivo* can have biologically significant effects.

2.4.2 Vascular Remodeling

Vascular remodeling is the process by which arteries adapt to both physiological and pathological stresses and involves changes in both shape and composition. As a tissue remodeling process, most of the evidence that the plasminogen activation system plays a significant role has come from the study of various models of vessel damage leading to stenosis, models of atherosclerosis, or aortic aneurysm in gene-ablated mice.

Initial observations in models of vascular injury showed that arterial repair and intimal hyperplasia were reduced by uPA-deficiency, but not by tPA-deficiency [77], and that PAI-1 deficiency had the opposite effect [78]. Together with similar studies in plasminogen-deficient mice [79, 80], the overall conclusion is that plasmin generation contributes to favorable arterial remodeling primarily involving fibrin clearance.

One particularly interesting aspect to emerge from these and other studies of vascular remodeling is that they provide strong *in vivo* evidence that plasmin activity is involved in the activation of certain matrix metalloproteases (MMPs). For example, in uPA-deficient mice in the ApoE$^{-/-}$ model of atherosclerosis, the vessel wall was protected against aneurysm formation [81]. The vascular erosion leading to aneurysm formation involves degradation of elastin and collagen, which are not substrates for plasmin. Macrophages from these mice, in contrast to wild-type macrophages, were unable to activate MMP-3, -9, -12, and -13 *ex vivo*. Subsequent experiments in MMP-3 (stromelysin-1)-deficient mice, which are similarly protected from aneurysms, are consistent with the interpretation that plasmin is required for MMP activation in this situation [82].

2.4.3
Fibrosis

In addition to its role in normal wound healing, the plasminogen activation system is also involved in the fibrotic process accompanying dysregulated wound healing. *Fibrosis* is defined as a fibroproliferative disorder, in which fibroblasts become hyperactivated, leading to the pathological manifestation of excessive deposition and accumulation of extracellular matrix – the very processes that are normally counteracted by the generation of plasmin activity. However, *PAI-1* is one of the genes most significantly upregulated in response to transforming growth factor-β (TGF-β), the major profibrotic cytokine, via the canonical Smad signaling pathway [83]. Consequently, PAI-1 levels are elevated in fibrotic tissue. This leads to a reduced activity of the plasminogen activation system, and further promotion of fibrosis. This is not only observed in skin fibrosis but also in other tissue fibrosis including that in the lung, liver, and kidney, as recently reviewed [84].

Plasmin also has an additional activity that might be seen as contributing to the fibrotic process. TGF-β is secreted in complex with the latency-associated peptide (LAP), which helps target TGF-β to the extracellular matrix. *In vitro*, several mechanisms lead to the activation of latent TGF-β, including proteolytic processing by plasmin [85]. The observation that PAI-1 contributes to fibrosis, by decreasing plasmin generation, might suggest that plasmin does not contribute significantly to the activation of TGF-β *in vivo*. However, the absence of PAI-1 in mice, rather than suppressing fibrosis, has been observed to promote cardiac fibrosis [86, 87]. Fibrosis is also increased in these mice in a model of glomerulonephritis, and this has been directly linked to an increased activation of TGF-β [88], thus supporting the notion that plasmin is involved in the processing of latent TGF-β.

Therefore, it appears from these observations that the balance between plasmin generation and its suppression by PAI-1 can modulate the fibrotic process through at least two opposing mechanisms. Plasmin can contribute to the proteolytic degradation of the extracellular matrix responsible for the manifestations of fibrosis, and proteolytic activation of latent TGF-β by plasmin contributes to profibrotic signaling.

Another mechanism by which the activity of the plasminogen activation system might counteract the fibrotic process is by increasing the activity of HGF/SF (hepatocyte growth factor/scatter factor). This is one of two plasminogen-related growth factors, and similar to plasminogen and other serine protease zymogens, it requires canonical proteolytic activation for biological activity, which in this case is signaling through the receptor tyrosine kinase c-Met [89]. Owing to its close homology with plasminogen, it was presumed that uPA was a key proteolytic activator of HGF/SF, and several studies appeared to demonstrate this [90, 91]. However, detailed enzyme kinetic studies, which directly compared a variety of proposed activators under a wide range of experimental conditions, found no evidence for the direct involvement of uPA activity [92]. In contrast, two other serine proteases, hepatocyte growth factor activator (HGFA) and matriptase, have been shown to be extremely efficient activators, both *in vitro* [92, 93] and *in vivo* [94, 95]. Nevertheless, evidence for a relationship between the plasminogen activation system and HGF/SF activity persists. HGF/SF is known to bind to heparan sulfate proteoglycan in the extracellular matrix [96], and although this binding promotes the interaction between HGF/SF and c-Met [97], plasmin-mediated degradation of the matrix might also be expected to increase the availability of HGF/SF. This has indeed been shown to be the case in models of lung fibrosis.

In the bleomycin-induced model of lung fibrosis, plasminogen activation acts to decrease fibrosis. This is apparent from the observed effects of PAI-1, which acts to increase fibrosis in transgenic mice overexpressing PAI-1, while fibrosis is less severe in PAI-1-deficient mice [98]. Similarly, deficiency of uPA or plasminogen increases fibrosis in this model [99]. These observations concur with the elevated level of PAI-1 consistently observed in fibrotic tissues, as described above, and suggest the same plasmin-mediated, fibrinolytic mechanism of protection. Furthermore, in these models, protection from fibrosis is paralleled by decreased levels of fibrin and increased levels of fibrin degradation products. Surprisingly, however, similar experiments in fibrinogen-deficient mice have revealed no difference in the levels of fibrosis to wild-type animals, and therefore that in this model the presence or absence of fibrin in the lung does not affect disease progression [100]. Although these observations cannot be interpreted as indicating that fibrin does not contribute to the fibrotic process, it clearly demonstrates that plasmin has alternative substrates in this process. Using PAI-1-deficient mice, it was subsequently shown that increased levels of HGF/SF were present in the bronchoalveolar lavage fluid after bleomycin treatment and that this increase was plasmin dependent [101]. Consistent with the interpretation that this increase was due to the release of HGF/SF from the extracellular matrix, this was directly observed in cell culture models by the same investigators [102]. Finally, it was observed that pharmacological inhibition of c-Met tyrosine kinase signaling prevented the protective effects of PAI-1 deficiency and subsequent plasmin generation [103].

These observations, in a range of fibrotic conditions, not only highlight the potentially opposing effects mediated by the generation of plasmin activity but also that these effects may involve mechanisms distinct from fibrinolysis. This is particularly apparent in the compelling evidence that plasmin activity is responsible for increasing the bioavailability of both profibrotic (TGF-β) and antifibrotic (HGF/SF) growth factors/cytokines by their release from the extracellular matrix.

2.4.4 Nerve Injury

The plasminogen activation system is known to have important roles in both the normal function and pathology of the central nervous system, some of which are independent of fibrinogen. These topics are covered elsewhere in this book (Chapter 6). However, the plasminogen activation system also has important functions in the peripheral nervous system and in particular in its response to injury. Sciatic nerve injury in mice causes the release of tPA from Schwann cells, the glial cells of the peripheral nervous system responsible for the formation and maintenance of the myelin sheath. This tPA release was found to be beneficial, as mice deficient in either tPA or plasminogen were more severely affected after sciatic nerve injury, with increased demyelination of axons [104]. Fibrin was observed to accumulate in the wound area, and was again found to be the target substrate of plasmin in this situation, as the effects of injury to both fibrinogen-deficient and fibrinogen/plasminogen-deficient mice were similar to wild-type mice [104]. Therefore, in wild-type mice, tPA-mediated plasmin generation was sufficient to overcome the pathological effects of fibrin accumulation.

These observations also suggest that the plasminogen activation system may have a significant protective role in inflammatory diseases of the peripheral nervous system, including multiple sclerosis, in which elevated levels of tPA have been observed [105]. Supporting this, tPA deficiency has been shown to exacerbate neuronal dysfunction in experimental autoimmune encephalomyelitis, a mouse model of multiple sclerosis [106, 107].

2.4.5 Rheumatoid Arthritis

In addition to involving tissue remodeling, most of the situations mentioned also have a significant inflammatory component in which the accumulation of fibrin appears to be part of the process leading to pathology, and in some cases the presence of fibrin is actually the cause of the inflammatory response. However, a model in which the generation of plasmin activity leads to the removal of fibrin deposits, and is therefore broadly beneficial is clearly oversimplistic. In this respect, observations made in animal models of rheumatoid arthritis, an inflammatory, pathological tissue condition, are particularly informative regarding the different, and often opposing, contributions that the activity of the plasminogen activation system may have at different stages of inflammatory conditions.

Initial studies in both uPA- and plasminogen-deficient mice demonstrated an increase in disease severity accompanying a significant accumulation of fibrin in monoarticular antigen-induced arthritis [108], suggesting that plasmin activity is necessary for the clearance of fibrin in this model. Similarly, tPA deficiency has also been observed to exacerbate disease in this model [109]. However, using models such as the collagen-induced arthritis model, in which systemic effects are responsible for the development of polyarticular arthritis, different effects are observed. Using this model, it was observed that uPA-deficient mice had a greatly reduced incidence and severity of disease, and that disease was completely absent in plasminogen-deficient mice [110]. Similar observations have been made in both uPA- and plasminogen-deficient mice in other systemic arthritis models [111]. The apparently contradictory roles of the plasminogen activation system in different animal models of arthritis have been proposed to be related to differences in pathogenic mechanisms operating in these models [111, 112], each of which reflects aspects of the complex human disease.

Antigen-induced arthritis involves a degree of local trauma and, in responding to this, has a significant "wound healing" component. Fibrin accumulation in the synovium appears to be a pathogenic factor under these conditions. The systemic nature of collagen-induced arthritis involves an adaptive immune response and, in contrast to antigen-induced arthritis, is a persistent condition. In this model, the deposition of fibrin is comparatively minor and destruction of articular matrix predominates. Plasmin could clearly contribute to this tissue degradation directly, but it has also been proposed that it might act through activation of the complement system [110], which is known to be a contributory factor both in this model and in human disease.

These experimental models, and others that combine their various aspects, emphasize the contrasting roles of the plasminogen activation system. Initially, plasmin activity is needed to promote inflammatory cell recruitment. Subsequently, or where local trauma to the joint has a significant role, plasmin generation mediates fibrinolysis and this has an opposing, protective role.

However, it is also apparent that the type and cellular source of the plasminogen activator involved is important, rather than the simple presence of plasmin activity. In the collagen-induced arthritis model, in contrast to the protection observed in uPA-deficient mice, tPA deficiency has been shown to have a detrimental effect [113]. This may reflect the major cellular sources of the two activators; uPA expression in inflammatory cells supporting their recruitment to the synovium, and tPA expression within the synovium having a protective role in the clearance of fibrin. These experimental observations correlate well with measurements of plasminogen activator activity in the synovium of affected joints in patients suffering from rheumatoid arthritis, with increased levels of uPA and decreased levels of tPA [114], suggesting that the observed alteration in the plasminogen activator profile contributes to disease progression.

2.4.6
Complex Tissue Remodeling

All of the above-mentioned processes involve some degree of tissue remodeling, but the plasminogen activation system also has a prominent role in complex, large-scale tissue remodeling processes. This is best exemplified by the processes that occur in the mammary gland during pregnancy and lactation, and the even more extensive remodeling process of mammary gland involution that subsequently occurs during weaning [115].

Early studies in mice indicated that plasminogen activator activity was transiently, but strongly, increased during the onset and early stages of mammary gland involution [116]. Later studies, in which mRNA levels were also measured, showed that uPA levels were greatly reduced during lactation and returned to normal on weaning [117], and components of the plasminogen activation system are known to be hormonally regulated at the level of transcription [118, 119].

The functional involvement of the plasminogen activation system in these processes has been demonstrated in plasminogen-deficient mice. Firstly, it was observed that these animals have a greatly reduced lactational competence, primarily due to poor development of the mammary glands, a subsequent lack of correctly organized involution, and consequent failure to support further litters [120]. Interestingly, heterozygous plasminogen-deficient mice also demonstrate a less severe form of this phenotype, demonstrating a plasminogen dose dependency not generally observed in these animals. Fibrin has been observed to accumulate in the mammary glands of plasminogen-deficient mice during lactation, and the effects on the overall process of lactation can be overcome by simultaneous heterozygous deficiency of fibrinogen [121]. Therefore, the most likely explanation for the effects of plasminogen deficiency on lactation competence is that they are due to the accumulation of fibrin in the mammary ducts. Blockage of the ducts in this way will cause milk stasis during lactation, promoting premature apoptosis and involution.

These observations highlight once again that, at least in plasminogen-deficient mice, the principal defect is an impaired or absent degradation of fibrin, and that the accumulation of fibrin can then set in train a series of pathological processes. The corollary of this is that under normal physiological conditions, the primary role of the plasminogen activation system is to prevent the accumulation of fibrin, and that this process may be just as important in tissues as it would be expected to be in the vasculature.

2.4.7
Angiogenesis

Angiogenesis is a process that accompanies many tissue remodeling processes, as well as involving the remodeling of tissues in its own right, that is, to allow the growth of blood vessels into existing tissue. Although the plasminogen activation system is clearly not necessary for vasculogenesis (from the many observations of normal development in mice deficient in components of the system), there is evidence that it plays a significant role in both physiological and pathological angiogenesis.

Plasminogen activator activity was first directly linked to the angiogenic process by the observation that topical application of uPA to rabbit corneas led to neovascularization [122], although the presence of fibrinolytic activity associated with capillary sprouting had been observed much earlier. Since then, evidence has accumulated that the plasminogen activation system is involved in various aspects of angiogenesis including basement membrane degradation, cell migration/extracellular matrix degradation, and capillary lumen formation. The only component of the plasminogen activation system expressed in quiescent endothelial cells *in vivo* is tPA [123], but during angiogenesis uPA, PAI-1, and uPAR are all expressed, largely due to the activities of fibroblast growth factor (FGF)-2 and vascular endothelial growth factor (VEGF) [124, 125]. Much of the evidence for the involvement of these components comes from *ex vivo* or *in vitro* models of angiogenesis. However, despite the evidence from these studies, mice with deficiency of plasminogen do not display any phenotype attributable to defective angiogenesis and, as previously mentioned, wound angiogenesis is also unaffected.

The situation is thought to be different in pathological angiogenesis, and in particular tumor angiogenesis [126]. Although this is outside the scope of this chapter, it is relevant to consider what mechanisms might be involved. Release of HGF/SF from the extracellular matrix by plasmin has been mentioned in the context of fibrosis, but HGF/SF is also a powerful driver of angiogenesis [127, 128], so this proteolytic mechanism may also come into play here. Similarly, FGF-2 can be released from the extracellular matrix in a plasminogen-dependent manner by endothelial-cell-associated uPA activity [129]. PAI-1 has been shown to play a particularly important role in tumor angiogenesis as, paradoxically, PAI-1-deficient mice are unable to vascularize transplanted tumors [130]. Several explanations are possible here, as PAI-1 has interactions with several other proteins, including vitronectin [131]. However, the evidence points to a direct protease inhibitory effect of PAI-1 [132], suggesting that a very closely controlled level of plasmin-mediated proteolysis may be necessary for effective angiogenesis.

An example of angiogenesis where the plasminogen activation system has been directly linked to the turnover of fibrin is in a mouse model of oxygen-induced retinopathy, which mimics the neonatal clinical condition known as *retinopathy of prematurity*. In this model, hyperoxia causes endothelial damage, vaso-obliteration, and retinal ischemia, leading to an angiogenic response. This has been shown to lead to the hypoxia-inducible factor (HIF)-1α-dependent upregulation of annexin-II, which, as mentioned previously, is thought to act as a template for the assembly of tPA and plasminogen on endothelial cells, promoting plasmin generation. In annexin-II-deficient mice, retinopathy is accompanied by a large increase in fibrin deposition and a reduction in angiogenesis [133]. Therefore, it appears that annexin-II is necessary for the clearance of fibrin, and that in the absence of this clearance, angiogenesis is inhibited. Whether these effects are due to fibrin presenting a physical barrier to the migration of endothelial cells during angiogenesis is unclear. The reason for this is that fibrin, and various degradation products of both fibrinogen and fibrin, have been reported to have a variety of activities that might influence the angiogenic process, including vascular contraction [134], stimulation of endothelial cells [135], and vascular smooth muscle cell proliferation [136]. In addition to this, the fibrin matrix might sequester growth factors that are released on proteolysis, in much the same way as is established for components of the extracellular matrix [137].

2.4.8
uPAR – Cinderella Finds Her Shoe

One component of the plasminogen activation system that has not been considered here in detail thus far in the context of the biological roles of the plasminogen activation system is uPAR, the glycolipid-anchored cellular receptor for uPA. Over the past 25 years of research in this area, this is perhaps the most interesting component to emerge. There is overwhelming biochemical, cell biological, and other *in vitro* evidence pointing to uPAR having a fundamental role in promoting uPA-dependent plasminogen activation. However, it cannot be ignored that the generation of uPAR-deficient mice provided little or no evidence to support these *in vitro* observations [4, 34], making this molecule truly the Cinderella of the field.

Since its initial identification as the specific cellular binding site for uPA, uPAR has received much attention as a potential multiligand receptor. It has been reported to interact with a wide variety of other proteins, including extracellular matrix proteins, cell adhesion receptors, internalization receptors of the low-density lipoprotein receptor (LDLR)-family, receptor tyrosine kinases, and G-protein-coupled receptors, leading to great interest in a potential role for uPAR in cell signaling (reviewed in [138, 139]). In this context, uPAR has been proposed to be a "signaling orchestrator," involved in a wide range of cellular processes, including cell adhesion, chemotaxis, migration, proliferation, differentiation, and survival [138]. There is good evidence from *in vitro* experiments that an interplay between uPAR and cell adhesion receptors of the integrin family can influence both cell adhesion and integrin "outside-in" signaling [140]. This evidence is particularly compelling for $\alpha_5\beta_1$, $\alpha_M\beta_2$, and $\alpha_v\beta_3$ integrins, where uPAR-dependent conformational changes in the integrin ectodomains have been shown to lead to rearrangements in the integrin transmembrane and cytoplasmic domains [141–143].

Despite this intense interest, there is only very limited direct evidence that uPAR has a role in any of these cellular processes *in vivo*. There have been several reports of phenotypic abnormalities in uPAR-deficient mice, particularly in the behavior of leukocytes in various models of inflammation and innate immune responses [144–147]. However, some of these effects, for example, have not been reproduced in studies carefully controlled with respect to the genetic background of the animals [148], while others may be secondary to other potential effects of uPAR deficiency.

This has been highlighted in an extremely elegant study of the role of the uPA/uPAR interaction *in vivo*. Making use of the interspecies specificity of this interaction, Bugge and colleagues [148] engineered mice to express endogenous uPA containing four amino acid substitutions that completely abolished its binding to uPAR, without affecting its proteolytic activity or interfering with any other potential function of uPAR: *Plau*$^{GFDhu/GFDhu}$ mice. As alluded to above, two previous studies of uPAR-deficient mice did not observe any defect in fibrinolysis, in sharp contrast to the extensive fibrin deposits found in plasminogen-deficient mice [4, 149]. However, *Plau*$^{GFDhu/GFDhu}$ mice were observed to accumulate fibrin in several tissues, although not until the mice were one year old, significantly older than the animals analyzed in the previous studies. These experiments demonstrate unequivocally that the uPA/uPAR interaction promotes plasminogen activation *in vivo*. Therefore, the primary, and possibly sole, physiological role of uPAR is in the regulation of proteolysis, presumably by promoting cell–surface plasminogen activation on leukocytes and thereby facilitating fibrinolysis and the process of fibrin surveillance.

Interestingly, despite the previous paucity of experimental evidence for a physiological role of uPAR in regulating uPA activity, several lines of evidence have clearly demonstrated pathological and other roles *in vivo*. Much of this has been in the context of cancer cell dissemination, metastasis, and invasion and is outside the scope of this chapter. However, the evidence includes the effects of various antagonists of the uPA/uPAR interaction and siRNA inhibition of uPAR expression, and has recently been reviewed [35, 150, 151]. An additional example is in transgenic mice overexpressing uPA and uPAR in the basal epidermis, which displayed significant epidermal thickening and subepidermal blistering [152]. This required the presence of both of the proteins and uPA proteolytic activity. In a nontransgenic system, the proteolytic activation of an engineered uPA-activatable anthrax toxin was found to be dependent on the presence of uPAR both *in vitro* and *in vivo* [153, 154]. Both these experiments unequivocally support the role of uPAR in promoting the activity of uPA, as now demonstrated physiologically using the *Plau*$^{GFDhu/GFDhu}$ mice, and particularly in situations where proteolytic activity needs to be generated directly at the cell surface.

Now that it has been established beyond doubt that uPAR-mediated plasminogen activation is a physiologically relevant process, interactions between uPAR and other cell-surface molecules mentioned above might be of biological relevance if they can regulate plasminogen activation. This could potentially involve effects directly on the binding of uPA to uPAR, or on the ability of the uPA/uPAR complex to activate plasminogen. In relation to the former possibility, it has been speculated that uPAR may exist in a latent form, unable to bind uPA [155]. This is based on strong evidence from X-ray crystallography, which has shown surprisingly large differences in the conformation of the uPA-binding-site in the presence of various ligands, due to interdomain flexibility in uPAR [25, 26, 156, 157].

Circumstantial evidence for such an effect with cell-surface uPAR comes from experiments in which expression of the tetraspanin protein CD82 was observed to lead to a reduction in cell-surface plasminogen activation concomitant with a reduction in uPA binding [158]. This effect correlated with a redistribution of uPAR to regions of the cell surface where it colocalized with the $\alpha_5\beta_1$ integrin. Although it was not possible to demonstrate direct interactions between any of these proteins, these observations might suggest that interactions between $\alpha_5\beta_1$ and uPAR lead to a reduced binding of uPA – or, in other words, to a latent form of uPAR. Subsequent experiments in which the expression of $\alpha_5\beta_1$ was manipulated in a different cell type did not support this model, as plasminogen activation was found to be increased in the presence of $\alpha_5\beta_1$ and also to be affected by the conformational "activation" status of the integrin [159]. However, both studies clearly point to a potential role for integrins in modulating uPA binding and plasminogen activation.

In the context of uPAR-mediated plasminogen activation and other proteins that may influence this reaction, the adhesive protein vitronectin stands out. The binding of vitronectin to uPAR was first observed as a cell adhesion phenomenon [160]. It has been studied extensively since then and is well understood at the structural level. The interaction is now known to have profound effects on cell morphology, for example, the formation of lamellipodia and this is regulated by the binding of uPA [161–163]. However, how these interactions may affect the primary functions of uPA and uPAR in mediating plasminogen activation is yet to be determined.

2.5 Conclusions

This chapter has focused on the physiological role of the plasminogen activation system in various tissue remodeling processes. Although the plasminogen activation system is clearly not necessary for the tissue remodeling processes involved in reproduction and development, it is crucial for many postnatal tissue remodeling processes and this largely involves fibrinolysis. A reduced activity of this system results in a range of fibrotic and inflammatory conditions that result from an accumulation of fibrin in tissues, and these effects are more pronounced than intravascular effects, for example, increased thrombotic tendency. Therefore, the primary role of the plasminogen activation system appears to be the removal of potentially pathogenic fibrin deposits from tissues, that is, fibrin surveillance.

Surprisingly, considering the apparent advantages of tPA as a fibrin-specific plasminogen activator, uPA has a major role in this fibrin surveillance. In addition to the involvement of the two plasminogen activators, several different mechanisms or pathways are involved: uPA can function in both a uPAR-dependent and an independent manner, tPA can function bound to either fibrin, annexin-II, or possibly other proteins, and there is some evidence for an alternative pathway of plasminogen activation. It is also clear that there is a great deal of redundancy between these pathways. This perhaps reflects an evolutionary advantage in maintaining a robust system for the generation of plasmin activity. The price paid for this robustness is that dysregulation of the plasminogen activation system has pathological consequences. The very rare situation where the system lacks activity, that is, congenital plasminogen deficiency, leads to the condition ligneous conjunctivitis, but much more importantly dysregulation due to altered expression of the components of this system is involved in a variety of pathologies, most notably cancer (Chapter 9), but also cardiovascular and neurological/neurodegenerative disease, among others.

Fibrin has been the main focus here as the target substrate for the plasminogen activation system, but there is now very good *in vivo* evidence to support the predictions from *in vitro* experiments that plasmin has other targets. These are probably much more limited than originally supposed, but clearly include activation of several MMPs, and several mechanisms leading to either the activation of growth factors or an increase in their bioavailability. The latter most likely involves degradation of components of the extracellular matrix, although it is not clearly defined which components are affected. The extent of the involvement of the plasminogen activation system in directly degrading components of the extracellular matrix is also not clear, and the large family of MMPs, other metalloproteases, and other families of proteolytic enzymes probably have a more significant role. However, although there is thought to be a considerable degree of overlap and some redundancy in the function of these proteases, it is also clear that the plasminogen activation system alone is responsible for the essential process of fibrin surveillance, without which a wide variety of inflammatory and other pathological conditions can arise.

References

1. Carmeliet, P., Kieckens, L., Schoonjans, L., Ream, B., van Nuffelen, A., Prendergast, G., Cole, M., Bronson, R., Collen, D., and Mulligan, R.C. (1993a) Plasminogen activator inhibitor-1 gene-deficient mice. I. Generation by homologous recombination and characterization. *J. Clin. Invest.*, **92**, 2746–2755.
2. Carmeliet, P., Schoonjans, L., Kieckens, L., Ream, B., Degen, J., Bronson, R., De Vos, R., van den Oord, J.J., Collen, D., and Mulligan, R.C. (1994) Physiological consequences of loss of plasminogen activator gene function in mice. *Nature*, **368**, 419–424.
3. Bugge, T.H., Flick, M.J., Daugherty, C.C., and Degen, J.L. (1995a) Plasminogen deficiency causes severe thrombosis but is compatible with development and reproduction. *Genes Dev.*, **9**, 794–807.
4. Bugge, T.H., Suh, T.T., Flick, M.J., Daugherty, C.C., Rømer, J., Solberg, H., Ellis, V., Danø, K., and Degen, J.L. (1995b) The receptor for urokinase-type plasminogen activator is not essential for mouse development or fertility. *J. Biol. Chem.*, **270**, 16886–16894.
5. Ploplis, V.A., French, E.L., Carmeliet, P., Collen, D., and Plow, E.F. (1998) Plasminogen deficiency differentially affects recruitment of inflammatory cell populations in mice. *Blood*, **91**, 2005–2009.
6. Lijnen, H.R., Okada, K., Matsuo, O., Collen, D., and Dewerchin, M. (1999) α2-Antiplasmin gene deficiency in mice is associated with enhanced fibrinolytic potential without overt bleeding. *Blood*, **93**, 2274–2281.
7. Danø, K., Andreasen, P.A., Grøndahl-Hansen, J., Kristensen, P., Nielsen, L.S., and Skriver, L. (1985) Plasminogen activators, tissue degradation and cancer. *Adv. Cancer Res.*, **44**, 139–266.
8. Ellis, V. (2003a) Plasminogen activation at the cell surface. *Curr. Top. Dev. Biol.*, **54**, 263–312.
9. Ellis, V. (2003b) in *Plasminogen: Structure, Activation and Regulation* (ed. D.M. Waisman), Kluwer Academic/ PlenumPublishers, pp. 19–45.
10. Castellino, F.J. and Ploplis, V.A. (2005) Structure and function of the

plasminogen/plasmin system. *Thromb. Haemost.*, **93**, 647–654.

11. Mangel, W.F., Lin, B.H., and Ramakrishnan, V. (1990) Characterization of an extremely large, ligand-induced conformational change in plasminogen. *Science*, **248**, 69–73.
12. Miles, L.A., Hawley, S.B., Baik, N., Andronicos, N.M., Castellino, F.J., and Parmer, R.J. (2005) Plasminogen receptors: the sine qua non of cell surface plasminogen activation. *Front Biosci.*, **10**, 1754–1762.
13. Walker, M.J., McArthur, J.D., McKay, F., and Ranson, M. (2005) Is plasminogen deployed as a Streptococcus pyogenes virulence factor? *Trends Microbiol.*, **13**, 308–313.
14. Wang, X., Lin, X., Loy, J.A., Tang, J., and Zhang, X.C. (1998) Crystal structure of the catalytic domain of human plasmin complexed with streptokinase. *Science*, **281**, 1662–1665.
15. Boxrud, P.D., Verhamme, I.M.A., Fay, W.P., and Bock, P.E. (2001) Streptokinase triggers conformational activation of plasminogen through specific interactions of the amino-terminal sequence and stabilizes the active zymogen conformation. *J. Biol. Chem.*, **276**, 26084–26089.
16. Hoylaerts, M., Rijken, D.C., Lijnen, H.R., and Collen, D. (1982) Kinetics of the activation of plasminogen by human tissue plasminogen activator. Role of fibrin. *J. Biol. Chem.*, **257**, 2912–2919.
17. Petersen, L.C., Johannessen, M., Foster, D., Kumar, A., and Mulvihill, E. (1988) The effect of polymerised fibrin on the catalytic activities of one-chain tissue-type plasminogen activator as revealed by an analogue resistant to plasmin cleavage. *Biochim. Biophys. Acta.*, **952**, 245–254.
18. Kingston, I.B., Castro, M.J., and Anderson, S. (1995) In vitro stimulation of tissue-type plasminogen activator by Alzheimer amyloid beta-peptide analogues. *Nat. Med.*, **1**, 138–142.
19. Ellis, V., Daniels, M., Misra, R., and Brown, D.R. (2002) Plasminogen activation is stimulated by prion protein and regulated in a copper-dependent manner. *Biochemistry*, **41**, 6891–6896.
20. Kranenburg, O., Bouma, B., Kroon-Batenburg, L.M., Reijerkerk, A., Wu, Y.P., Voest, E.E., and Gebbink, M.F. (2002) Tissue-type plasminogen activator is a multiligand cross-beta structure receptor. *Curr. Biol.*, **12**, 1833–1839.
21. Skriver, L., Nielsen, L.S., Stephens, R., and Danø, K. (1982) Plasminogen activator released as inactive proenzyme from murine cells transformed by sarcoma virus. *Eur. J. Biochem.*, **124**, 409–414.
22. Ellis, V., Scully, M.F., and Kakkar, V.V. (1987) Plasminogen activation by single-chain urokinase in functional isolation. A kinetic study. *J. Biol. Chem.*, **262**, 14998–15003.
23. Kilpatrick, L.M., Harris, R.L., Owen, K.A., Bass, R., Ghorayeb, C., Bar-Or, A., and Ellis, V. (2006) Initiation of plasminogen activation on the surface of monocytes expressing the type II transmembrane serine protease matriptase. *Blood*, **108**, 2616–2623.
24. Ellis, V. and Danø, K. (1993) Potentiation of plasminogen activation by an anti-urokinase monoclonal antibody due to ternary complex formation. A mechanistic model for receptor-mediated plasminogen activation. *J. Biol. Chem.*, **268**, 4806–4813.
25. Llinas, P., Le Du, M.H., Gårdsvoll, H., Danø, K., Ploug, M., Gilquin, B., Stura, E.A., and Menez, A. (2005) Crystal structure of the human urokinase plasminogen activator receptor bound to an antagonist peptide. *EMBO J.*, **24**, 1655–1663.
26. Huai, Q., Mazar, A.P., Kuo, A., Parry, G.C., Shaw, D.E., Callahan, J., Li, Y., Yuan, C., Bian, C., Chen, L., Furie, B., Furie, B.C., Cines, D.B., and Huang, M. (2006) Structure of human urokinase plasminogen activator in complex with its receptor. *Science*, **311**, 656–659.
27. Ellis, V., Scully, M.F., and Kakkar, V.V. (1989) Plasminogen activation initiated by single-chain urokinase-type plasminogen activator. Potentiation by U937 monocytes. *J. Biol. Chem.*, **264**, 2185–2188.

28. Ellis, V., Behrendt, N., and Danø, K. (1991) Plasminogen activation by receptor-bound urokinase. A kinetic study with both cell-associated and isolated receptor. *J. Biol. Chem.*, **266**, 12752–12758.
29. Ellis, V., Whawell, S.A., Werner, F., and Deadman, J.J. (1999) Assembly of urokinase receptor-mediated plasminogen activation complexes involves direct, non-active-site interactions between urokinase and plasminogen. *Biochemistry*, **38**, 651–659.
30. Ellis, V., Wun, T.-C., Behrendt, N., Rønne, E., and Danø, K. (1990) Inhibition of receptor-bound urokinase by plasminogen-activator inhibitors. *J. Biol. Chem.*, **265**, 9904–9908.
31. Behrendt, N., List, K., Andreasen, P.A., and Danø, K. (2003) The pro-urokinase plasminogen-activation system in the presence of serpin-type inhibitors and the urokinase receptor: rescue of activity through reciprocal pro-enzyme activation. *Biochem. J.*, **371**, 277–287.
32. Ellis, V. (1996) Functional analysis of the cellular receptor for urokinase in plasminogen activation. *J. Biol. Chem.*, **271**, 14779–14784.
33. Lee, S.W., Ellis, V., and Dichek, D.A. (1994) Characterization of plasminogen activation by glycosylphosphatidylinositol-anchored urokinase. *J. Biol. Chem.*, **269**, 2411–2418.
34. Bugge, T.H., Flick, M.J., Danton, M.J., Daugherty, C.C., Rømer, J., Danø, K., Carmeliet, P., Collen, D., and Degen, J.L. (1996a) Urokinase-type plasminogen activator is effective in fibrin clearance in the absence of its receptor or tissue-type plasminogen activator. *Proc. Natl. Acad. Sci. U.S.A.*, **93**, 5899–5904.
35. Rømer, J., Nielsen, B.S., and Ploug, M. (2004) The urokinase receptor as a potential target in cancer therapy. *Curr. Pharm. Des.*, **10**, 2359–2376.
36. Cesarman, G.M., Guevara, C.A., and Hajjar, K.A. (1994) An endothelial cell receptor for plasminogen/tissue plasminogen activator (t-PA). II. Annexin II-mediated enhancement of t-PA-dependent plasminogen activation. *J. Biol. Chem.*, **269**, 21198–21203.
37. Ling, Q., Jacovina, A.T., Deora, A., Febbraio, M., Simantov, R., Silverstein, R.L., Hempstead, B., Mark, W.H., and Hajjar, K.A. (2004) Annexin II regulates fibrin homeostasis and neoangiogenesis in vivo. *J. Clin. Invest.*, **113**, 38–48.
38. Ellis, V. and Whawell, S.A. (1997) Vascular smooth muscle cells potentiate plasmin generation by both urokinase and tissue plasminogen activator dependent mechanisms: evidence for a specific tPA receptor on these cells. *Blood*, **90**, 2312–2322.
39. Werner, F., Razzaq, T.M., and Ellis, V. (1999) Tissue plasminogen activator binds to human vascular smooth muscle cells by a novel mechanism – evidence for a reciprocal linkage between inhibition of catalytic activity and cellular binding. *J. Biol. Chem.*, **274**, 21555–21561.
40. Razzaq, T.M., Bass, R., Vines, D.J., Werner, F., Whawell, S.A., and Ellis, V. (2003) Functional regulation of tissue plasminogen activator on the surface of vascular smooth muscle cells by the type-II transmembrane protein p63 (CKAP4). *J. Biol. Chem.*, **278**, 42679–42685.
41. Huarte, J., Belin, D., and Vassalli, J.D. (1985) Plasminogen activator in mouse and rat oocytes: induction during meiotic maturation. *Cell*, **43**, 551–558.
42. Huarte, J., Belin, D., Bosco, D., Sappino, A.P., and Vassalli, J.D. (1987) Plasminogen activator and mouse spermatozoa: urokinase synthesis in the male genital tract and binding of the enzyme to the sperm cell surface. *J. Cell Biol.*, **104**, 1281–1289.
43. Sappino, A.P., Huarte, J., Belin, D., and Vassalli, J.D. (1989) Plasminogen activators in tissue remodeling and invasion: mRNA localization in mouse ovaries and implanting embryos. *J. Cell Biol.*, **109**, 2471–2479.
44. Ploplis, V.A., Carmeliet, P., Vazirzadeh, S., Van Vlaenderen, I., Moons, L., Plow, E.F., and Collen, D. (1995) Effects of disruption of the plasminogen gene on thrombosis, growth, and health in mice. *Circulation*, **92**, 2585–2593.
45. Aoki, N., Moroi, M., Sakata, Y., Yoshida, N., and Matsuda, M. (1978)

Abnormal plasminogen. A hereditary molecular abnormality found in a patient with recurrent thrombosis. *J. Clin. Invest.*, **61**, 1186–1195.

46. Miyata, T., Iwanaga, S., Sakata, Y., and Aoki, N. (1982) Plasminogen Tochigi: inactive plasmin resulting from replacement of alanine-600 by threonine in the active site. *Proc. Natl. Acad. Sci. U.S.A.*, **79**, 6132–6136.

47. Ichinose, A., Espling, E.S., Takamatsu, J., Saito, H., Shinmyozu, K., Maruyama, I., Petersen, T.E., and Davie, E.W. (1991) Two types of abnormal genes for plasminogen in families with a predisposition for thrombosis. *Proc. Natl. Acad. Sci. U.S.A.*, **88**, 115–119.

48. Yamaguchi, M., Doi, S., and Yoshimura, M. (1989) Plasminogen phenotypes in a Japanese population. Four new variants including one with a functional defect. *Hum. Hered.*, **39**, 356–360.

49. Sakata, Y. and Aoki, N. (1980) Molecular abnormality of plasminogen. *J. Biol. Chem.*, **255**, 5442–5447.

50. Wohl, R.C., Summaria, L., and Robbins, K.C. (1979) Physiological activation of the human fibrinolytic system. Isolation and characterization of human plasminogen variants, Chicago I and Chicago II. *J. Biol. Chem.*, **254**, 9063–9069.

51. Soria, J., Soria, C., Bertrand, O., Dunn, F., Drouet, L., and Caen, J.P. (1983) Plasminogen Paris I: congenital abnormal plasminogen and its incidence in thrombosis. *Thromb. Res.*, **32**, 229–238.

52. Carmeliet, P., Stassen, J.M., Schoonjans, L., Ream, B., van den Oord, J.J., De Mol, M., Mulligan, R.C., and Collen, D. (1993b) Plasminogen activator inhibitor-1 gene-deficient mice. II. Effects on hemostasis, thrombosis, and thrombolysis. *J. Clin. Invest.*, **92**, 2756–2760.

53. Lund, I.K., Jogi, A., Rono, B., Rasch, M.G., Lund, L.R., Almholt, K., Gardsvoll, H., Behrendt, N., Romer, J., and Hoyer-Hansen, G. (2008) Antibody-mediated targeting of the urokinase-type plasminogen activator proteolytic function neutralizes fibrinolysis in vivo. *J. Biol. Chem.*, **283**, 32506–32515.

54. Jögi, A., Rønø, B., Lund, I.K., Nielsen, B.S., Ploug, M., Høyer-Hansen, G., Rømer, J., and Lund, L.R. (2010) Neutralisation of uPA with a monoclonal antibody reduces plasmin formation and delays skin wound healing in tPA-deficient mice. *PLoS ONE*, **5**, e12746.

55. Plow, E.F. (1982) Leukocyte elastase release during blood coagulation. A potential mechanism for activation of the alternative fibrinolytic pathway. *J. Clin. Invest.*, **69**, 564–572.

56. Zeng, B., Bruce, D., Kril, J., Ploplis, V., Freedman, B., and Brieger, D. (2002) Influence of plasminogen deficiency on the contribution of polymorphonuclear leukocytes to fibrin/ogenolysis: studies in plasminogen knock-out mice. *Thromb. Haemost.*, **88**, 805–810.

57. Bugge, T.H., Kombrinck, K.W., Flick, M.J., Daugherty, C.C., Danton, M.J.S., and Degen, J.L. (1996b) Loss of fibrinogen rescues mice from the pleiotropic effects of plasminogen deficiency. *Cell*, **87**, 709–719.

58. Suh, T.T., Holmback, K., Jensen, N.J., Daugherty, C.C., Small, K., Simon, D.I., Potter, S., and Degen, J.L. (1995) Resolution of spontaneous bleeding events but failure of pregnancy in fibrinogen-deficient mice. *Genes Dev.*, **9**, 2020–2033.

59. Eagle, R.C. Jr., Brooks, J.S., Katowitz, J.A., Weinberg, J.C., and Perry, H.D. (1986) Fibrin as a major constituent of ligneous conjunctivitis. *Am. J. Ophthalmol.*, **101**, 493–494.

60. Schuster, V., Mingers, A.M., Seidenspinner, S., Nussgens, Z., Pukrop, T., and Kreth, H.W. (1997) Homozygous mutations in the plasminogen gene of two unrelated girls with ligneous conjunctivitis. *Blood*, **90**, 958–966.

61. Klammt, J., Kobelt, L., Aktas, D., Durak, I., Gokbuget, A., Hughes, Q., Irkec, M., Kurtulus, I., Lapi, E., Mechoulam, H., Mendoza-Londono, R., Palumbo, J.S., Steitzer, H., Tabbara, K.F., Ozbek, Z., Pucci, N., Sotomayor, T., Sturm, M.,

Drogies, T., Ziegler, M., and Schuster, V. (2011) Identification of three novel plasminogen (PLG) gene mutations in a series of 23 patients with low PLG activity. *Thromb. Haemost.*, **105**, 454–460.

62. Schott, D., Dempfle, C.E., Beck, P., Liermann, A., Mohr-Pennert, A., Goldner, M., Mehlem, P., Azuma, H., Schuster, V., Mingers, A.M., Schwarz, H.P., Kramer, M.D., Liesenhoff, H., and Niessen, K.H. (1998) Therapy with a purified plasminogen concentrate in an infant with ligneous conjunctivitis and homozygous plasminogen deficiency. *New Engl. J. Med.*, **339**, 1679–1686.
63. Watts, P., Suresh, P., Mezer, E., Ells, A., Albisetti, M., Bajzar, L., Marzinotto, V., Andrew, M., Massicotle, P., and Rootman, D. (2002) Effective treatment of ligneous conjunctivitis with topical plasminogen. *Am. J. Ophthalmol.*, **133**, 451–455.
64. Suzuki, T., Ikewaki, J., Iwata, H., Ohashi, Y., and Ichinose, A. (2009) The first two Japanese cases of severe type I congenital plasminogen deficiency with ligneous conjunctivitis: successful treatment with direct thrombin inhibitor and fresh plasma. *Am. J. Hematol.*, **84**, 363–365.
65. Drew, A.F., Kaufman, A.H., Kombrinck, K.W., Danton, M.J.S., Daugherty, C.C., Degen, J.L., and Bugge, T.H. (1998) Ligneous conjunctivitis in plasminogen-deficient mice. *Blood*, **91**, 1616–1624.
66. Dieval, J., Nguyen, G., Gross, S., Delobel, J., and Kruithof, E.K. (1991) A lifelong bleeding disorder associated with a deficiency of plasminogen activator inhibitor type 1. *Blood*, **77**, 528–532.
67. Fay, W.P., Shapiro, A.D., Shih, J.L., Schleef, R.R., and Ginsburg, D. (1992) Brief report: complete deficiency of plasminogen-activator inhibitor type 1 due to a frame-shift mutation. *N. Engl. J. Med.*, **327**, 1729–1733.
68. Fay, W.P., Parker, A.C., Condrey, L.R., and Shapiro, A.D. (1997) Human plasminogen activator inhibitor-1 (PAI-1) deficiency: characterization of a large kindred with a null mutation in the PAI-1 gene. *Blood*, **90**, 204–208.
69. Carpenter, S.L. and Mathew, P. (2008) Alpha2-antiplasmin and its deficiency: fibrinolysis out of balance. *Haemophilia*, **14**, 1250–1254.
70. Singer, A.J. and Clark, R.A.F. (1999) Cutaneous wound healing. *New. Engl. J. Med.*, **341**, 738–746.
71. Rømer, J., Lund, L.R., Ralfkiær, E., Zeheb, R., Gelehrter, T.D., Danø, K., and Kristensen, P. (1991) Differential expression of urokinase-type plasminogen activator and its type-1 inhibitor during healing of mouse skin wounds. *J. Invest. Dermatol.*, **97**, 803–811.
72. Rømer, J., Bugge, T.H., Pyke, C., Lund, L.R., Flick, M.J., Degen, J.L., and Danø, K. (1996) Impaired wound healing in mice with a disrupted plasminogen gene. *Nat. Med.*, **2**, 287–292.
73. Chan, J.C., Duszczyszyn, D.A., Castellino, F.J., and Ploplis, V.A. (2001) Accelerated skin wound healing in plasminogen activator inhibitor-1-deficient mice. *Am. J. Pathol.*, **159**, 1681–1688.
74. Kao, W.W., Kao, C.W., Kaufman, A.H., Kombrinck, K.W., Converse, R.L., Good, W.V., Bugge, T.H., and Degen, J.L. (1998) Healing of corneal epithelial defects in plasminogen- and fibrinogen-deficient mice. *Invest. Ophthalmol. Vis. Sci.*, **39**, 502–508.
75. Lund, L.R., Green, K.A., Stoop, A.A., Ploug, M., Almholt, K., Lilla, J., Nielsen, B.S., Christensen, I.J., Craik, C.S., Werb, Z., Danø, K., and Rømer, J. (2006) Plasminogen activation independent of uPA and tPA maintains wound healing in gene-deficient mice. *EMBO J.*, **25**, 2686–2697.
76. Colman, R.W. (1969) Activation of plasminogen by human plasma kallikrein. *Biochem. Biophys. Res. Commun.*, **35**, 273–279.
77. Carmeliet, P., Moons, L., Herbert, J.-M., Crawley, J., Lupu, F., Lijnen, H.R., and Collen, D. (1997a) Urokinase but not tissue plasminogen activator mediates arterial neointima formation in mice. *Circ. Res.*, **81**, 829–839.
78. Carmeliet, P., Moons, L., Lijnen, R., Janssens, S., Lupu, F., Collen, D., and Gerard, R.D. (1997c) Inhibitory role

of plasminogen activator inhibitor-1 in arterial wound healing and neointima formation – A gene targeting and gene transfer study in mice. *Circulation*, **96**, 3180–3191.

79. Carmeliet, P., Moons, L., Ploplis, V.A., Plow, E.F., and Collen, D. (1997d) Impaired arterial neointima formation in mice with disruption of the plasminogen gene. *J. Clin. Invest.*, **99**, 200–208.

80. Drew, A.F., Tucker, H.L., Kombrinck, K.W., Simon, D.I., Bugge, T.H., and Degen, J.L. (2000) Plasminogen is a critical determinant of vascular remodeling in mice. *Circ. Res.*, **87**, 133–139.

81. Carmeliet, P., Moons, L., Lijnen, R., Baes, M., Lemaitre, V., Tipping, P., Drew, A., Eeckhout, Y., Shapiro, S., Lupu, F., and Collen, D. (1997b) Urokinase-generated plasmin activates matrix metalloproteinases during aneurysm formation. *Nat. Genet.*, **17**, 439–444.

82. Heymans, S., Luttun, A., Nuyens, D., Theilmeier, G., Creemers, E., Moons, L., Dyspersin, G.D., Cleutjens, J.P., Shipley, M., Angellilo, A., Levi, M., Nube, O., Baker, A., Keshet, E., Lupu, F., Herbert, J.M., Smits, J.F., Shapiro, S.D., Baes, M., Borgers, M., Collen, D., Daemen, M.J., and Carmeliet, P. (1999) Inhibition of plasminogen activators or matrix metalloproteinases prevents cardiac rupture but impairs therapeutic angiogenesis and causes cardiac failure. *Nat. Med.*, **5**, 1135–1142.

83. Dennler, S., Itoh, S., Vivien, D., ten Dijke, P., Huet, S., and Gauthier, J.M. (1998) Direct binding of Smad3 and Smad4 to critical TGF beta-inducible elements in the promoter of human plasminogen activator inhibitor-type 1 gene. *EMBO J.*, **17**, 3091–3100.

84. Ghosh, A.K. and Vaughan, D.E. (2011) PAI-1 in tissue fibrosis. *J. Cell. Physiol.*, doi: 10.1002/jcp.22783

85. Rifkin, D.B., Mazzieri, R., Munger, J.S., Noguera, I., and Sung, J. (1999) Proteolytic control of growth factor availability. *APMIS*, **107**, 80–85.

86. Moriwaki, H., Stempien-Otero, A., Kremen, M., Cozen, A.E., and Dichek, D.A. (2004) Overexpression of urokinase by macrophages or deficiency of plasminogen activator inhibitor type 1 causes cardiac fibrosis in mice. *Circ. Res.*, **95**, 637–644.

87. Ghosh, A.K., Bradham, W.S., Gleaves, L.A., De Taeye, B., Murphy, S.B., Covington, J.W., and Vaughan, D.E. (2010) Genetic deficiency of plasminogen activator inhibitor-1 promotes cardiac fibrosis in aged mice: involvement of constitutive transforming growth factor-beta signaling and endothelial-to-mesenchymal transition. *Circulation*, **122**, 1200–1209.

88. Hertig, A., Berrou, J., Allory, Y., Breton, L., Commo, F., Costa De Beauregard, M.A., Carmeliet, P., and Rondeau, E. (2003) Type 1 plasminogen activator inhibitor deficiency aggravates the course of experimental glomerulonephritis through overactivation of transforming growth factor beta. *FASEB J.*, **17**, 1904–1906.

89. Birchmeier, C., Birchmeier, W., Gherardi, E., and Vande Woude, G.F. (2003) Met, metastasis, motility and more. *Nat. Rev. Mol. Cell Biol.*, **4**, 915–925.

90. Naldini, L., Tamagnone, L., Vigna, E., Sachs, M., Hartmann, G., Birchmeier, W., Daikuhara, Y., Tsubouchi, H., Blasi, F., and Comoglio, P.M. (1992) Extracellular proteolytic cleavage by urokinase is required for activation of hepatocyte growth factor/scatter factor. *EMBO J.*, **11**, 4825–4833.

91. Mars, W.M., Zarnegar, R., and Michalopoulos, G.K. (1993) Activation of hepatocyte growth factor by the plasminogen activators uPA and tPA. *Am. J. Pathol.*, **143**, 949–958.

92. Owen, K.A., Qiu, D., Alves, J., Schumacher, A.M., Kilpatrick, L.M., Li, J., Harris, J.L., and Ellis, V. (2010) Pericellular activation of hepatocyte growth factor by the transmembrane serine proteases matriptase and hepsin, but not by the membrane-associated protease uPA. *Biochem. J.*, **426**, 219–228.

93. Shimomura, T., Miyazawa, K., Komiyama, Y., Hiraoka, H., Naka, D., Morimoto, Y., and Kitamura, N. (1995) Activation of hepatocyte growth factor by two homologous proteases, blood-coagulation factor XIIa and hepatocyte growth factor activator. *Eur. J. Biochem.*, **229**, 257–261.

94. Miyazawa, K., Shimomura, T., and Kitamura, N. (1996) Activation of hepatocyte growth factor in the injured tissues is mediated by hepatocyte growth factor activator. *J. Biol. Chem.*, **271**, 3615–3618.

95. Szabo, R., Rasmussen, A.L., Moyer, A.B., Kosa, P., Schafer, J.M., Molinolo, A.A., Gutkind, J.S., and Bugge, T.H. (2011) c-Met-induced epithelial carcinogenesis is initiated by the serine protease matriptase. *Oncogene*, **30**, 2003–2016.

96. Lyon, M., Deakin, J.A., Mizuno, K., Nakamura, T., and Gallagher, J.T. (1994) Interaction of hepatocyte growth factor with heparan sulfate. Elucidation of the major heparan sulfate structural determinants. *J. Biol. Chem.*, **269**, 11216–11223.

97. Deakin, J.A. and Lyon, M. (1999) Differential regulation of hepatocyte growth factor/scatter factor by cell surface proteoglycans and free glycosaminoglycan chains. *J. Cell Sci.*, **112** (Pt 12), 1999–2009.

98. Eitzman, D.T., McCoy, R.D., Zheng, X.X., Fay, W.P., Shen, T.L., Ginsburg, D., and Simon, R.H. (1996) Bleomycin-induced pulmonary fibrosis in transgenic mice that either lack or overexpress the murine plasminogen activator inhibitor-1 gene. *J. Clin. Invest.*, **97**, 232–237.

99. Swaisgood, C.M., French, E.L., Noga, C., Simon, R.H., and Ploplis, V.A. (2000) The development of bleomycin-induced pulmonary fibrosis in mice deficient for components of the fibrinolytic system. *Am. J. Pathol.*, **157**, 177–187.

100. Hattori, N., Degen, J.L., Sisson, T.H., Liu, H., Moore, B.B., Pandrangi, R.G., Simon, R.H., and Drew, A.F. (2000) Bleomycin-induced pulmonary fibrosis in fibrinogen-null mice. *J. Clin. Invest.*, **106**, 1341–1350.

101. Hattori, N., Mizuno, S., Yoshida, Y., Chin, K., Mishima, M., Sisson, T.H., Simon, R.H., Nakamura, T., and Miyake, M. (2004) The plasminogen activation system reduces fibrosis in the lung by a hepatocyte growth factor-dependent mechanism. *Am. J. Pathol.*, **164**, 1091–1098.

102. Matsuoka, H., Sisson, T.H., Nishiuma, T., and Simon, R.H. (2006) Plasminogen-mediated activation and release of hepatocyte growth factor from extracellular matrix. *Am. J. Respir. Cell Mol. Biol.*, **35**, 705–713.

103. Bauman, K.A., Wettlaufer, S.H., Okunishi, K., Vannella, K.M., Stoolman, J.S., Huang, S.K., Courey, A.J., White, E.S., Hogaboam, C.M., Simon, R.H., Toews, G.B., Sisson, T.H., Moore, B.B., and Peters-Golden, M. (2010) The antifibrotic effects of plasminogen activation occur via prostaglandin E2 synthesis in humans and mice. *J. Clin. Invest.*, **120**, 1950–1960.

104. Akassoglou, K., Kombrinck, K.W., Degen, J.L., and Strickland, S. (2000) Tissue plasminogen activator-mediated fibrinolysis protects against axonal degeneration and demyelination after sciatic nerve injury. *J. Cell Biol.*, **149**, 1157–1166.

105. Akenami, F.O., Siren, V., Koskiniemi, M., Siimes, M.A., Teravainen, H., and Vaheri, A. (1996) Cerebrospinal fluid activity of tissue plasminogen activator in patients with neurological diseases. *J. Clin. Pathol.*, **49**, 577 580.

106. Lu, W., Bhasin, M., and Tsirka, S.E. (2002) Involvement of tissue plasminogen activator in onset and effector phases of experimental allergic encephalomyelitis. *J. Neurosci.*, **22**, 10781–10789.

107. East, E., Baker, D., Pryce, G., Lijnen, H.R., Cuzner, M.L., and Gveric, D. (2005) A role for the plasminogen activator system in inflammation and neurodegeneration in the central nervous system during experimental allergic encephalomyelitis. *Am. J. Pathol.*, **167**, 545–554.

108. Busso, N., Peclat, V., Van Ness, K., Kolodziesczyk, E., Degen, J., Bugge, T., and So, A. (1998) Exacerbation of antigen-induced arthritis in urokinase-deficient mice. *J. Clin. Invest.*, **102**, 41–50.

109. Yang, Y.H., Carmeliet, P., and Hamilton, J.A. (2001) Tissue-type plasminogen activator deficiency exacerbates arthritis. *J. Immunol.*, **167**, 1047–1052.
110. Li, J., Ny, A., Leonardsson, G., Nandakumar, K.S., Holmdahl, R., and Ny, T. (2005a) The plasminogen activator/plasmin system is essential for development of the joint inflammatory phase of collagen type II-induced arthritis. *Am. J. Pathol.*, **166**, 783–792.
111. De Nardo, C.M., Lenzo, J.C., Pobjoy, J., Hamilton, J.A., and Cook, A.D. (2010) Urokinase-type plasminogen activator and arthritis progression: contrasting roles in systemic and monoarticular arthritis models. *Arthritis. Res. Ther.*, **12**, R199.
112. Li, J., Guo, Y., Holmdahl, R., and Ny, T. (2005b) Contrasting roles of plasminogen deficiency in different rheumatoid arthritis models. *Arthritis Rheum.*, **52**, 2541–2548.
113. Cook, A.D., Braine, E.L., Campbell, I.K., and Hamilton, J.A. (2002) Differing roles for urokinase and tissue-type plasminogen activator in collagen-induced arthritis. *Am. J. Pathol.*, **160**, 917–926.
114. Busso, N., Peclat, V., So, A., and Sappino, A.P. (1997) Plasminogen activation in synovial tissues: differences between normal, osteoarthritis, and rheumatoid arthritis joints. *Ann. Rheum. Dis.*, **56**, 550–557.
115. Green, K.A. and Lund, L.R. (2005) ECM degrading proteases and tissue remodelling in the mammary gland. *Bioessays*, **27**, 894–903.
116. Ossowski, L., Biegel, D., and Reich, E. (1979) Mammary plasminogen activator: correlation with involution, hormonal modulation and comparison between normal and neoplastic tissue. *Cell*, **16**, 929–940.
117. Busso, N., Huarte, J., Vassalli, J.D., Sappino, A.P., and Belin, D. (1989) Plasminogen activators in the mouse mammary gland. Decreased expression during lactation. *J. Biol. Chem.*, **264**, 7455–7457.
118. Irigoyen, J.P., Munoz-Canoves, P., Montero, L., Koziczak, M., and Nagamine, Y. (1999) The plasminogen activator system: biology and regulation. *Cell. Mol. Life Sci.*, **56**, 104–132.
119. Nagamine, Y., Medcalf, R.L., and Munoz-Canoves, P. (2005) Transcriptional and posttranscriptional regulation of the plasminogen activator system. *Thromb. Haemost.*, **93**, 661–675.
120. Lund, L.R., Bjorn, S.F., Sternlicht, M.D., Nielsen, B.S., Solberg, H., Usher, P.A., Osterby, R., Christensen, I.J., Stephens, R.W., Bugge, T.H., Dano, K., and Werb, Z. (2000) Lactational competence and involution of the mouse mammary gland require plasminogen. *Development*, **127**, 4481–4492.
121. Green, K.A., Nielsen, B.S., Castellino, F.J., Rømer, J., and Lund, L.R. (2006) Lack of plasminogen leads to milk stasis and premature mammary gland involution during lactation. *Dev. Biol.*, **299**, 164–175.
122. Berman, M., Winthrop, S., Ausprunk, D., Rose, J., Langer, R., and Gage, J. (1982) Plasminogen activator (urokinase) causes vascularization of the cornea. *Invest. Ophthalmol. Vis. Sci.*, **22**, 191–199.
123. Levin, E.G. and Del Zoppo, G.J. (1994) Localization of tissue plasminogen activator in the endothelium of a limited number of vessels. *Am. J. Pathol.*, **144**, 855–861.
124. Mandriota, S.J., Seghezzi, G., Vassalli, J.D., Ferrara, N., Wasi, S., Mazzieri, R., Mignatti, P., and Pepper, M.S. (1995) Vascular endothelial growth factor increases urokinase receptor expression in vascular endothelial cells. *J. Biol. Chem.*, **270**, 9709–9716.
125. Pepper, M.S. (2001) Role of the matrix metalloproteinase and plasminogen activator-plasmin systems in angiogenesis. *Arterioscler. Thromb. Vasc. Biol.*, **21**, 1104–1117.
126. Rakic, J.M., Maillard, C., Jost, M., Bajou, K., Masson, V., Devy, L., Lambert, V., Foidart, J.M., and Noel, A. (2003) Role of plasminogen activator-plasmin system in tumor angiogenesis. *Cell. Mol. Life. Sci.*, **60**, 463–473.
127. Bussolino, F., Di Renzo, M.F., Ziche, M., Bocchietto, E., Olivero, M., Naldini, L., Gaudino, G., Tamagnone, L., Coffer,

A., and Comoglio, P.M. (1992) Hepatocyte growth factor is a potent angiogenic factor which stimulates endothelial cell motility and growth. *J. Cell Biol.*, **119**, 629–641.

128. Ding, S., Merkulova-Rainon, T., Han, Z.C., and Tobelem, G. (2003) HGF receptor up-regulation contributes to the angiogenic phenotype of human endothelial cells and promotes angiogenesis in vitro. *Blood*, **101**, 4816–4822.

129. Ribatti, D., Leali, D., Vacca, A., Giuliani, R., Gualandris, A., Roncali, L., Nolli, M.L., and Presta, M. (1999) In vivo angiogenic activity of urokinase: role of endogenous fibroblast growth factor-2. *J. Cell Sci.*, **112** (Pt. 23), 4213–4221.

130. Bajou, K., Noel, A., Gerard, R.D., Masson, V., Brunner, N., Holst-Hansen, C., Skobe, M., Fusenig, N.E., Carmeliet, P., Collen, D., and Foidart, J.M. (1998) Absence of host plasminogen activator inhibitor 1 prevents cancer invasion and vascularization. *Nat. Med.*, **4**, 923–928.

131. Stefansson, S. and Lawrence, D.A. (1996) The serpin PAI-1 inhibits cell migration by blocking integrin $\alpha_v\beta_3$ binding to vitronectin. *Nature*, **383**, 441–443.

132. Bajou, K., Masson, V., Gerard, R.D., Schmitt, P.M., Albert, V., Praus, M., Lund, L.R., Frandsen, T.L., Brunner, N., Dano, K., Fusenig, N.E., Weidle, U., Carmeliet, G., Loskutoff, D., Collen, D., Carmeliet, P., Foidart, J.M., and Noel, A. (2001) The plasminogen activator inhibitor PAI-1 controls in vivo tumor vascularization by interaction with proteases, not vitronectin. Implications for antiangiogenic strategies. *J. Cell Biol.*, **152**, 777–784.

133. Huang, B., Deora, A.B., He, K.L., Chen, K., Sui, G., Jacovina, A.T., Almeida, D., Hong, P., Burgman, P., and Hajjar, K.A. (2011) Hypoxia-inducible factor-1 drives annexin A2 system-mediated perivascular fibrin clearance in oxygen-induced retinopathy in mice. *Blood*, **118** (10), 2918–2929.

134. Buluk, K. and Malofiejew, M. (1969) The pharmacological properties of fibrinogen degradation products. *Br. J. Pharmacol.*, **35**, 79–89.

135. Bootle-Wilbraham, C.A., Tazzyman, S., Thompson, W.D., Stirk, C.M., and Lewis, C.E. (2001) Fibrin fragment E stimulates the proliferation, migration and differentiation of human microvascular endothelial cells in vitro. *Angiogenesis*, **4**, 269–275.

136. Naito, M., Stirk, C.M., Smith, E.B., and Thompson, W.D. (2000) Smooth muscle cell outgrowth stimulated by fibrin degradation products. The potential role of fibrin fragment E in restenosis and atherogenesis. *Thromb. Res.*, **98**, 165–174.

137. Schonherr, E. and Hausser, H.J. (2000) Extracellular matrix and cytokines: a functional unit. *Dev. Immunol.*, **7**, 89–101.

138. Blasi, F. and Carmeliet, P. (2002) uPAR: a versatile signalling orchestrator. *Nat. Rev. Mol. Cell Biol.*, **3**, 932–943.

139. Smith, H.W., and Marshall, C.J. (2010) Regulation of cell signalling by uPAR. *Nat. Rev. Mol. Cell Biol.*, **11**, 23–36.

140. Wei, Y., Lukashev, M., Simon, D.I., Bodary, S.C., Rosenberg, S., Doyle, M.V., and Chapman, H.A. (1996) Regulation of integrin function by the urokinase receptor. *Science*, **273**, 1551–1555.

141. Wei, Y., Czekay, R.P., Robillard, L., Kugler, M.C., Zhang, F., Kim, K.K., Xiong, J.P., Humphries, M.J., and Chapman, H.A. (2005) Regulation of $\alpha5\beta1$ integrin conformation and function by urokinase receptor binding. *J. Cell Biol.*, **168**, 501–511.

142. Tang, M.L., Vararattanavech, A., and Tan, S.M. (2008) Urokinase-type plasminogen activator receptor induces conformational changes in the integrin alphaMbeta2 headpiece and reorientation of its transmembrane domains. *J. Biol. Chem.*, **283**, 25392–25403.

143. Wei, C., Moller, C.C., Altintas, M.M., Li, J., Schwarz, K., Zacchigna, S., Xie, L., Henger, A., Schmid, H., Rastaldi, M.P., Cowan, P., Kretzler, M., Parrilla, R., Bendayan, M., Gupta, V., Nikolic, B., Kalluri, R., Carmeliet, P., Mundel, P., and Reiser, J. (2008) Modification of kidney barrier function by the urokinase receptor. *Nat. Med.*, **14**, 55–63.

144. May, A.E., Kanse, S.M., Lund, L.R., Gisler, R.H., Imhof, B.A., and Preissner, K.T. (1998) Urokinase receptor (CD87) regulates leukocyte recruitment via beta 2 integrins in vivo. *J. Exp. Med.*, **188**, 1029–1037.
145. Gyetko, M.R., Sud, S., Kendall, T., Fuller, J.A., Newstead, M.W., and Standiford, T.J. (2000) Urokinase receptor-deficient mice have impaired neutrophil recruitment in response to pulmonary Pseudomonas aeruginosa infection. *J. Immunol.*, **165**, 1513–1519.
146. Renckens, R., Roelofs, J.J., Florquin, S., and van der Poll, T. (2006) Urokinase-type plasminogen activator receptor plays a role in neutrophil migration during lipopolysaccharide-induced peritoneal inflammation but not during Escherichia coli-induced peritonitis. *J. Infect. Dis.*, **193**, 522–530.
147. Wiersinga, W.J., Kager, L.M., Hovius, J.W., van der Windt, G.J., De Vos, A.F., Meijers, J.C., Roelofs, J.J., Dondorp, A., Levi, M., Day, N.P., Peacock, S.J., and van der Poll, T. (2010) Urokinase receptor is necessary for bacterial defense against pneumonia-derived septic melioidosis by facilitating phagocytosis. *J. Immunol.*, **184**, 3079–3086.
148. Connolly, B.M., Choi, E.Y., Gardsvoll, H., Bey, A.L., Currie, B.M., Chavakis, T., Liu, S., Molinolo, A., Ploug, M., Leppla, S.H., and Bugge, T.H. (2010) Selective abrogation of the uPA-uPAR interaction in vivo reveals a novel role in suppression of fibrin-associated inflammation. *Blood*, **116**, 1593–1603.
149. Dewerchin, M., Nuffelen, A.V., Wallays, G., Bouche, A., Moons, L., Carmeliet, P., Mulligan, R.C., and Collen, D. (1996) Generation and characterization of urokinase receptor-deficient mice. *J. Clin. Invest.*, **97**, 870–878.
150. Mazar, A.P., Ahn, R.W., and O'Halloran, T.V. (2011) Development of novel therapeutics targeting the urokinase plasminogen activator receptor (uPAR) and their translation toward the clinic. *Curr. Pharm. Des.*, **17** (19), 1970–1978.
151. Kriegbaum, M.C., Persson, M., Haldager, L., Alpizar-Alpizar, W., Jacobsen, B., Gardsvoll, H., Kjaer, A., and Ploug, M. (2011) Rational targeting of the urokinase receptor (uPAR): development of antagonists and non-invasive imaging probes. *Curr. Drug. Targets*, **12** (12), 1711–1728.
152. Zhou, H.M., Nichols, A., Meda, P., and Vassalli, J.D. (2000) Urokinase-type plasminogen activator and its receptor synergize to promote pathogenic proteolysis. *EMBO J.*, **19**, 4817–4826.
153. Liu, S., Bugge, T.H., and Leppla, S.H. (2001) Targeting of tumor cells by cell surface urokinase plasminogen activator-dependent anthrax toxin. *J. Biol. Chem.*, **276**, 17976–17984.
154. Liu, S., Aaronson, H., Mitola, D.J., Leppla, S.H., and Bugge, T.H. (2003) Potent antitumor activity of a urokinase-activated engineered anthrax toxin. *Proc. Natl. Acad. Sci. U.S.A.*, **100**, 657–662.
155. Yuan, C. and Huang, M. (2007) Does the urokinase receptor exist in a latent form? *Cell. Mol. Life Sci.*, **64**, 1033–1037.
156. Barinka, C., Parry, G., Callahan, J., Shaw, D.E., Kuo, A., Bdeir, K., Cines, D.B., Mazar, A., and Lubkowski, J. (2006) Structural basis of interaction between urokinase-type plasminogen activator and its receptor. *J. Mol. Biol.*, **363**, 482–495.
157. Huai, Q., Zhou, A., Lin, L., Mazar, A.P., Parry, G.C., Callahan, J., Shaw, D.E., Furie, B., Furie, B.C., and Huang, M. (2008) Crystal structures of two human vitronectin, urokinase and urokinase receptor complexes. *Nat. Struct. Mol. Biol.*, **15**, 422–423.
158. Bass, R., Werner, F., Odintsova, E., Sugiura, T., Berditchevski, F., and Ellis, V. (2005) Regulation of urokinase receptor proteolytic function by the tetraspanin CD82. *J. Biol. Chem.*, **280**, 14811–14818.
159. Bass, R. and Ellis, V. (2009) Regulation of urokinase receptor function and pericellular proteolysis by the integrin $\alpha 5\beta 1$. *Thromb. Haemost.*, **101**, 954–962.
160. Wei, Y., Waltz, D.A., Rao, N., Drummond, R.J., Rosenberg, S., and Chapman, H.A. (1994) Identification of the urokinase receptor as an adhesion

receptor for vitronectin. *J. Biol. Chem.*, **269**, 32380–32388.

161. Madsen, C.D., Ferraris, G.M., Andolfo, A., Cunningham, O., and Sidenius, N. (2007) uPAR-induced cell adhesion and migration: vitronectin provides the key. *J. Cell Biol.*, **177**, 927–939.

162. Hillig, T., Engelholm, L.H., Ingvarsen, S., Madsen, D.H., Gardsvoll, H., Larsen, J.K., Ploug, M., Dano, K., Kjoller, L., and Behrendt, N. (2008) A composite role of vitronectin and urokinase in the modulation of cell morphology upon expression of the urokinase receptor. *J. Biol. Chem.*, **283**, 15217–15223.

163. Gårdsvoll, H., Jacobsen, B., Kriegbaum, M.C., Behrendt, N., Engelholm, L.H., ⊘stergaard, S., and Ploug, M. (2011) Conformational regulation of urokinase receptor (uPAR) function. Impact of receptor occupancy and epitope-mapped monoclonal antibodies on lamellipodia induction. *J. Biol. Chem.*, **M111**, 220087.

3
Physiological Functions of Membrane-Type Metalloproteases

Kenn Holmbeck

3.1
Introduction

The membrane-type matrix metalloproteinases (MT-MMPs) are the most recent addition to the large family of zinc endopeptidases initially identified as secreted collagenases and now commonly referred to as the *MMPs* [1]. The earliest description and characterization of an MT-type MMP, MT1-MMP, was the culmination of the search for the endogenous activator of proMMP2 [2, 3]. The perceived physiological significance of MT1-MMP and its molecular relatives (MT2, -3, -4, -5, -6) has been shaped by this initial finding and only recently has a more complex role for MT-MMPs emerged [3–9]. This chapter is aimed at reviewing parts of our current understanding of the physiological roles of MT-type MMPs, where one such exists. As the preceding chapters describe specific roles of some MMPs, including MT-type MMPs, in development and homeostasis of specific tissues and additionally the role these molecules may play in pathophysiological conditions, this chapter touches only peripherally on those aspects of MT-MMP biology. The content here therefore is narrowed to a description of select functions maintained by MT-MMP activity and only features a few accounts of the specific implications for tissue development and homeostasis.

3.2
Historical Perspective

In 1962, Jerome Gross and Charles Lapiere [10] for the first time defined a proteolytic activity in the tadpole tail capable of degrading collagen. A similar activity was isolated in tissue from humans and through combined efforts the collagenolytic activity was identified and assigned to an expanding group of related zinc-dependent endopeptidases, which would later be known as *matrix-degrading metalloproteinases* [1, 11–13]. Considerable energy subsequently went toward understanding the mechanisms responsible for activation of the latent forms of MMPs. In the early 1990s several independent studies reported that cell membranes harbored the

Matrix Proteases in Health and Disease, First Edition. Edited by Niels Behrendt.

potential to activate the purified zymogen of the type IV collagen-cleaving enzyme MMP2 and importantly this activator was sensitive to MMP-specific inhibitors such as tissue inhibitors of metalloproteinase 2 (TIMP-2) or orthophenanthroline [2, 14, 15]. These findings strongly suggested that the activator was a metalloproteinase itself and this assumption was additionally supported by the observation that a range of other inhibitors or purified preparations of their targets had no effect on the ability of membrane preparations or whole cells to convert the MMP2 zymogen to active enzyme [16, 17]. In 1994, the identity of the membrane-associated MMP activator was revealed when the laboratory of Dr. Motoharu Seiki [3] reported the isolation and cloning of a membrane-type MMP aptly named MT-MMP. This type I transmembrane (TM) enzyme shared a high degree of homology with secreted MMPs and importantly retained the ability to convert the zymogen of MMP2 into an active enzyme. It was also the first example of an MMP bound to the cell membrane by a formal TM domain, and as such not only had the previously elusive membrane-associated MMP activator been identified but a whole new type of MMP had also been discovered. The name of this MMP was quickly amended as five additional members of this expanding new subfamily were discovered in rapid succession [4–6, 8, 9]. A more detailed characterization of these related molecules revealed that MT1, -2, -3, and MT5-MMP were true type I TM proteinases, whereas MT4-MMP and MT6-MMP lacked the TM domain and maintained membrane tethering through glycosylphosphatidylinositol (GPI) anchorage (Figure 3.1) [18, 19]. The observation that all MT-MMPs except MT4-MMP were capable of converting

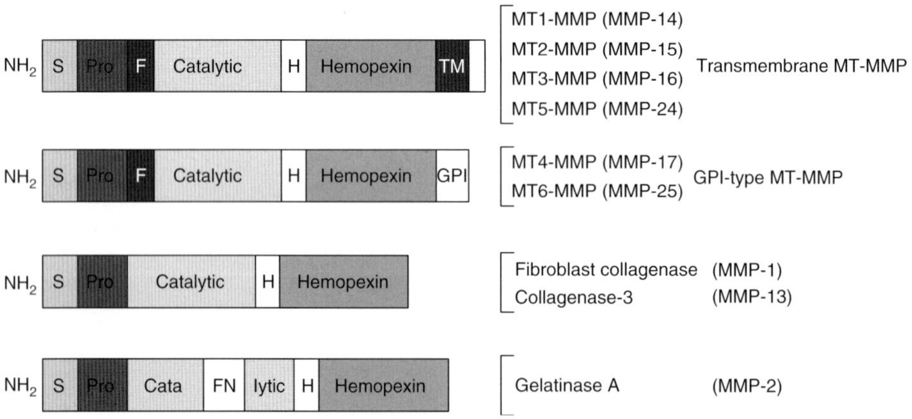

Figure 3.1 Diagrammatic representation of domain structure in MT-MMPs. Outline of the structural similarities and differences in secreted and MT-type MMPs. S, signal peptide; Pro, prodomain responsible for latency of the zymogen; F, furin/PACE cleavage site present in all MT-type MMPs and MMP-11; Catalytic: catalytic domain of the protease; FN, fibronectin type-II-like repeat present within the catalytic domain of MMP-2 and MMP-9; H: hinge domain; Hemopexin, hemopexin-like domain responsible for substrate interaction; TM, transmembrane domain – this domain is followed by a short cytoplasmic tail; GPI, glycosylphosphatidylinositol anchor present exclusively in MT4-MMP and MT6-MMP.

proMMP2 into MMP2 early led to the general assumption that MT-MMPs chiefly were serving the role of cell-associated activators of other secreted MMPs [3–5, 7, 8, 20–23].

3.3 Activation of the Activator

Some of the first questions arising from the discovery of MT1-MMP were what led to activation of this "activating" MMP. From early studies employing domain swapping experiments it was convincingly demonstrated that membrane association was absolutely required for proper MT1-MMP function [24]. It was further shown that the pro form of MT1-MMP contained a sequence in the prodomain, RXKR, sensitive to the pro-protein convertases similar to the sequence found in MMP11 (Figure 3.1) [25–28]. MT-MMPs are like MMP11 therefore activated in the trans-Golgi network and reach the plasma membrane in an active state ready to interact with susceptible substrates [4, 29–31]. From the plasma membrane MT1-MMP is internalized and recycled to the plasma membrane, again by a dynamin-dependent endocytic process [32]. This process has also been described for MT3-MMP and is likely to be a general mode of exposure and recycling of MT-MMP-type MMPs. This notion is supported further by the observation that all type I membrane MT-MMPs contain residues in the cytoplasmic domain responsible for interaction with the PDZ domain protein Mint-3 [33]. Recycling of the protease is at least in part regulated by the availability of susceptible substrates; collagen will thus abrogate the recycling of MT1-MMP back into the cells [34]. On the basis of this observation, it is conceivable that the protease is recycled on a constant basis and retained whenever a susceptible substrate engages with the enzyme. Aside from the retention of enzyme by the substrate, mobilization of additional enzyme also takes place by exposure of cells to substrate. Accordingly, MT1-MMP levels are upregulated by fibrillar collagen in the pericellular environment through integrin-mediated signaling [35, 36]. Attenuation of MT1-MMP is achieved in part by inhibitor interaction and partly by MMP-dependent processing to a 44 kDa inactive form and internalization [37–39].

3.4 Potential Roles of MT-MMPs and Discovery of a Human MMP Mutation

Beyond the mode of activation and recycling of the MT-type MMPs the bigger question driving research into their biology is the active role they play in physiology. In the case of the MT-type MMPs, their role as obligate activators of secreted MMPs certainly could justify their existence alone; however, is that the sole function? As a preface to the remaining chapters and an overview of our understanding of MT-MMP activity, a brief summary of our current knowledge of their interaction with other MMPs of the secreted type is probably in its place.

The role of MT-MMPs as MMP2 zymogen activator and pro-MMP13 activator was earlier considered to be the primary role of MT1-MMP [3, 7, 17, 40, 41]. This function has now been affirmed by several independent studies outlining the role of TIMP-2 in a ternary activation complex and more recently the requirement for oligomerization of MT1-MMP to attain pro-MMP2-activating properties [42, 43].

Before the generation of mouse models of MT-MMP deficiency, Dr. Itohara's laboratory [44] generated mice deficient for MMP2. Despite the potent gelatinolytic activity of MMP2 *in vitro*, mice devoid of this enzyme activity fared relatively well compared to the anticipated outcome. This finding is notably at odds with the clinical manifestations in human subjects deficient in MMP2 who endure a severe skeletal and soft connective tissue disease referred to as *Torg-Winchester syndrome* and multicentric osteolysis [45, 46]. Closer analysis and contrasting of clinical findings with the mouse model would later reveal that the mouse shares aspects of the disease; however, they are less severe than in the human condition associated with MMP2 deficiency [47].

This example serves to illustrate an important point about our evaluation of gene function in humans based on ablation of mouse ortholog genes. A direct extrapolation is not always possible and in the absence of human null mutations for MT-type MMPs the interpretation of animal data should be tempered by the example offered by the divergent consequences of MMP2 deficiency in man and mouse. In this context it is further worth considering that some of the *MMP* genes assigned prominent theoretical importance in humans, such as MMP1 (interstitial collagenase), are not found in mouse and the closest orthologs, mColA and mColB, described by the laboratory of Dr. Carlos Lopez-Otin [48] display a very restricted expression pattern, which precludes a universal role as collagen remodeling enzymes in the way that MMP1 is considered to work in humans.

That being said, the deletion-mutant studies in animals offer a powerful tool for predicting the role of selected genes, and in the absence of spontaneous human mutations they constitute the only viable tool for assessing the larger role of a given candidate gene product on the tissue and organism level.

3.5
MT-MMP Function?

Following analysis of the MMP2-deficient mouse model, the expectation was that ablation of MT1-MMP might be much less severe – especially because several other MT-type MMPs shared the capacity to activate proMMP2 and thus could compensate for the lost MT1-MMP-mediated MMP2 activation [44].

These predictions were, however, confounded somewhat by *in vitro* studies demonstrating the capacity of soluble mutant forms of MT1-MMP to convert non-MMP substrates [49–51]. One caveat in these analyses was that true MT-MMP activity is intimately tied to the membrane association of the enzyme [24, 52]. The relevance of substrate degradation in cell-free assays by soluble deletion mutants of a membrane enzyme was therefore hard to assess in a biological context. The physiological role of MT1-MMP was furthermore clouded by the earliest studies that mostly assigned expression of MT1-MMP to malignant cells. An expression pattern of this nature made a likely role in cancer progression tangible, but it did not reveal the normal physiological function served by MT1-MMP [3, 53, 54]. As of today, little data pertaining to nonpathogenic MT-MMP function in humans have been derived from studies of genetic aberrations although in the MT3-MMP encoding sequence small nucleotide polymorphisms coincide with bronchopulmonary dysplasia. At present, this finding does not identify MT3-MMP as the causative agent in the disease and the precise significance of this study requires further analysis [55]. Naturally occurring mutations in all MT-type MMPs have been reported in the NCBI Entrez SNP database, yet none have been homozygous. Our current understanding of the true physiological role of these proteases is therefore largely based on extrapolation of data derived from *in vitro* experiments and from observations made in animal models carrying selective gene ablations in *MT-MMP* loci [56–63].

3.6 Physiological Roles of MT1-MMP in the Mouse

When mouse models of MT1-MMP deficiency were generated, they demonstrated severe effects of the null mutation (dwarfism, joint disease, reproductive failure, muscle abnormalities, and premature death) that were highly surprising and unanticipated. It quickly became evident that most of the fibrillar collagen remodeling had been disabled with the loss of MT1-MMP and that this defect formed the basis for the phenotypic changes observed [57, 63, 64]. Mesenchymal cells isolated from MT1-MMP-deficient mice failed to degrade native fibrillar type I collagen when plated on this matrix and stimulated with cytokines known to invoke collagen degradation in wild-type cells [57, 65]. A detailed test of various MMPs under these conditions convincingly demonstrated that none of the secreted downstream targets of MT1-MMP, even in an activated form, conferred a collagenolytic phenotype on cells [52]. Only cells retaining MT1-MMP were competent in the cleavage of fibrillar collagen type I, II, and III in the mouse, MT1-MMP can therefore be assigned as a major collagenase [52, 57, 63]. Despite the loss of this important proteolytic capacity, mice can survive without MT1-MMP expression for some time and in doing so offer a veritable map of collagen remodeling in the organism and also illustrate the consequences of losing the ability to metabolize this important constituent of the pericellular environment [57, 63, 66, 67]. Significant parts of the body with high collagen content thus suffer from dysfunctions that are discussed in greater detail in the subsequent chapters; however, in brief, these defects can all be explained by the loss of a cell autonomous mediator of pericellular matrix degradation and cell motility required for dynamic matrix remodeling. Accordingly, several tissues in the mouse, such as organs, with high collagen content endure few, if any, consequences from the lack of MT1-MMP because they are subject to essentially no remodeling once they are laid down in an appropriate manner. On the other hand, tissues with high collagen content and rapid matrix turnover from growth-related remodeling or remodeling dictated by high mechanical stress rates are heavily affected. This is to a degree so they display progressive scarring over time, which ultimately leads to failure of the tissue. These effects are most apparent in the skeleton and more specifically in the part of the skeleton that contains collagen that persists as unmineralized, such as certain cartilages, periosteum, tendons, and ligaments. In unmineralized cartilages of the skull vault, Meckel's cartilage, and articular cartilages, collagen accumulates because of a lack of MT1-MMP-mediated degradation induced during a unique developmental remodeling of cartilage directly into the bone. In articular cartilages, MT1-MMP is required for persistent remodeling of the tissue to accommodate growth and tissue maintenance. A similar property is required by resident cells of the periosteum, tendons, and ligaments to maintain extracellular matrix homeostasis. Once the catabolic response to collagen synthesis by cells in these tissues is disabled, the net gain in extracellular matrix leads to tissue failure. For preosteoblasts of the periosteum, failure to degrade their resident matrix leads to entrapment and inability to migrate to the mineralized surface of the bone and sustain osteoid/bone deposition. In contrast to the unmineralized matrix, collagen associated with mineral such as the bone proper, dentin, and mineralized cartilage is largely unaffected by MT1-MMP deficiency; however, the continuity of these tissue compartments ultimately lead to defects in both locations [57, 63, 66].

Because the role of MT-MMPs as MMP2 or MMP13 activators has shaped a lot of the perceptions of MT-MMP function, our current knowledge of this interaction with potential downstream substrates of MMP origin should be summed up before a further review of MT-MMP function. Comparatively, single mouse null-mutations of MMP2 and MMP13 share very few similarities with the single MT1-MMP null mutation, which results in severe pleiotropic effects in postnatal development and growth in mice [44, 57, 63, 68, 69]. In other words, loss of the presumed activating enzyme is different and in some physiological aspects more severe than loss of any of the potential downstream targets. Cell-based studies of the fibrillar collagen processing capacity of various MMPs – secreted and membrane-associated – have revealed that MT-type MMPs are the principal collagenases *in vitro* [52, 70]. However, a significant study by Egeblad *et al.* [57, 63, 71–74] also demonstrated that MMP2 deficiency in an animal background mutated in the classic 3/4–1/4 collagenase cleavage site leads to a severe disease that closely resembles the consequences of MT1-MMP deficiency. Our current understanding of collagen remodeling based on this study is that MMP2 can process the bulk of denatured collagen initially cleaved by MT-type MMPs in the pericellular environment, but is dispensable to some extent. When collagen is processed alternatively – presumably at a slower rate and at alternate cleavage sites – MMP2 becomes more essential [73, 75]. MMP2 thus serves an important role in a tiered collagen-processing pathway, but is partially dispensable in the presence of collagenase-sensitive collagen [44].

These hypotheses are further corroborated by the observation that single mutations of MT1-MMP or MMP2 are compatible with postnatal survival in mice, whereas double deficiency leads to uniform perinatal lethality [76]. These animals additionally display severe developmental defects in a variety of organs and muscles, indicative of inability to process both collagenous and noncollagenous substrates during the later part of development. In summary, the roles of MT1-MMP and MMP2 are partially overlapping such that loss of MMP2 is tolerated, and for the most part is remedied by the activity of MT1-MMP. In the reverse scenario, MT1-MMP deficiency is tolerated very poorly, even when mitigated by MMP2 activity and ultimately leads to severe disease and premature death. Under these circumstances, MMP2 activity is perceived to be important, as survival is dependent on this secreted enzyme and loss of both MMP2 alleles results in uniform perinatal demise. This experiment also highlights another important point of MT1-MMP and MMP2 biology. MT1-MMP is not the only biologically relevant and obligate activator of MMP2 – rather sufficient proteolytic processing from other MT-type MMPs and possible unidentified non-MMP activators exists to facilitate MMP2 activation and survival of MT1-MMP deficient mice [76–78].

3.7 MT1-MMP Function in Lung Development

In MT1-MMP-deficient mice, one of the more conspicuous examples of a physiological process governed by MT1-MMP is the prominent postnatal developmental defect observed in the lungs. Consistent with the conspicuous expression of MT1-MMP in lung parenchyma, MT1-MMP-deficient mice develop severe postnatal emphysema. At birth, MT1-MMP-deficient mice display essentially normal proximal airways with unaffected branching. Soon after, however, the distal airways display signs of defective alveolization and aberrant Clara cell morphology [79]. These defects are not associated with overt accumulation of prospective substrates of MT1-MMP or substrates susceptible to any of the prospective downstream proteases subject to MT1-MMP-dependent activation. Importantly, analyses of lung tissue from MMP2-deficient mice fail to demonstrate phenotypically identical findings. This leads to the suggestion that the physiological shortcomings in lung development derive from a failure in the affected tissue directly caused by MT1-MMP and not from a failure to activate MMP2 [79]. One important clue to what this failure may be comes from the observation that alveoli in mutant mice are devoid of normal pores of Kohn [80]. These pore structures are thought to equalize the pressure between individual alveoli and ensure the proper inflation throughout the entire lung, which is important during the alveolization process. Despite intensive research to identify any molecular defects, no overt signs of matrix accumulation can be documented nor does the number and distribution of alveolar macrophages in MT1-MMP-deficient lungs lend credence to the possibility that macrophage function is directly connected to the aberrant morphology of pores and Clara cells. One confounding property of the pore defect in the context of possible insufficient substrate processing is that the amount of matrix required for obstruction of the pore structures is infinitely small. Neither the semiquantitative property of cytochemistry nor mass spectrometry analysis suffices to document accumulation of specific MT1-MMP substrates in these quantities.

Despite the inconsistency between a normal number of macrophages, identical matrix distribution, but an aberrant pore phenotype, the observation elsewhere that MT1-MMP−/− macrophages display diminished elastolytic activity in aneurysms provides a possible explanation for the inability of alveolar macrophages to keep the pore structures open [81]. Further dissection of the elastolytic defect could serve to possibly account for the pore phenotype and emphysema.

MT1-MMP-deficient mice additionally display a dramatic defect in the repair of distal airways following experimental lung injury with naphthalene [82]. Resolution of the epithelial injury is substantially reduced because of the failure of epithelial cells to proliferate at rates comparable to those found in wild-type mice. Moreover, MT1-MMP-deficient lung epithelial cells display a marked reduction in the expression fibroblast growth factor receptor 2b (FGFR2b) (keratinocyte growth factor receptor (KGFR)), leading to the suggestion that the proliferative deficit is indirectly or directly tied to disrupted FGF7 signaling in the lung repair process [82].

3.8
MT1-MMP Is Required for Root Formation and Molar Eruption

In dissecting the various tissue defects in the MT1-MMP-deficient mice, one of the more prominent morphogenetic events disrupted by the gene ablation is the root formation and eruption in molar teeth [83]. Detailed analysis of the tooth eruption reveals that molars in the absence of MT1-MMP do develop both enamel and cusps. However, the formation of roots and their elongation are disrupted because of a lack of collagen remodeling and a sequestration of cells in the fibrotic tissue of the periodontium from where they normally are recruited into the genesis of the root proper and the periodontal ligament. In the absence of proper collagen turnover in the periodontium, the persistence of unerupted mandibular and maxillary molars is observed and the teeth reside permanently under the gingival connective tissue and overlying oral mucosa. How then are these findings reconciled with the cellular and molecular deficit resulting from ablation of MT1-MMP? Expression analysis reveals that the periodontal ligament under normal physiological conditions is the site of most intense MT1-MMP expression [84]. The absence of MT1-MMP is histologically evident by a prominent lack of Sharpey's fibers, and analysis elsewhere in MT1-MMP mutant mice reveal that MT1-MMP activity is a prerequisite for Sharpey's fibers formation; as such, they are the anatomical footprint representing a high collagen turnover phenotype [66, 85].

3.9
Identification of Cooperative Pathways for Collagen Metabolism

The periodontium in MT1-MMP-deficient mice also displays prominent intracellular abnormalities, which offers a unique opportunity to gain appreciation for the modes of collagen processing utilized by cells under conditions of high matrix turnover rates. In the periodontal ligament, the extensive tissue remodeling is in part due to initial needs for growth and development and later to accommodate the renewal of the connective tissue as it is subjected to the physical forces associated with mastication that leads to mechanical weakening. In the absence of the pericellular proteolytic activity associated with MT1-MMP, fibroblasts in the periodontal ligament display a 50-fold increase in collagen acquired by phagocytosis [83, 86]. This phagocytosis of collagen is a normal parallel physiological remodeling mechanism for many connective tissue cell types and is mediated by the mannose receptor family member urokinase plasminogen activator receptor associated protein (uPARAP/Endo180), which is coexpressed with MT1-MMP [66, 67, 87, 88]. In the absence of MT1-MMP, phagocytosis mediates collagen uptake and eventual processing by cathepsin-mediated proteolysis [89]. Normally, the load of intracellular collagen is modest, but ablation of MT1-MMP-mediated pericellular proteolysis necessitates rerouting and degradation of collagen by the intracellular machinery to enable matrix remodeling.

The synergistic interaction of these two collagen degradation pathways is demonstrated very vividly by a combined loss of the loci encoding the MT1-MMP and uPARAP [90]. The reduced viability of MT1-MMP-deficient mice is further compromised, more than 50% of double-deficient mice perish within 10 days of birth, and by weaning the remaining mice are dead. The phenotypic changes recorded with loss of MT1-MMP are greatly exacerbated by the further loss of uPARAP, and the dwarfism and wasting are prominent. Because these mice do not live long, some of the age-dependent histological findings associated with loss of MT1-MMP alone are not evident in the double mutants; however, evaluation of dynamic cell function in skeletal cells demonstrates significantly reduced cell viability and matrix deposition due to entrapment of cells in the pericellular matrix of the connective tissues and disruption of collagen phagocytosis [90].

In the case of MT1-MMP/uPARAP-deficient mice, the concept that dysfunction of cells in a subset of connective tissues can result in reduced viability is not straightforward to explain mechanistically, nor are the specific molecular deficits it imparts on various processes clear. These may be related to intrinsic cellular functions and/or possible processing of messengers affecting neighboring and more distant cells. However, a simple experiment in mice does reveal that restoration of MT1-MMP expression in only a subset of cells dramatically increases viability of otherwise unviable MT1-MMP-deficient young mice [91]. We can therefore assign the diminished viability to a certain tissue compartment and designate that as essential for early viability in mice (in this case, the skeleton).

3.10 MT-MMP Activity in the Hematopoietic Environment

As mentioned previously, the role of MMP activity in bone remodeling and homeostasis has proved to be important. In the case of MT1-MMP, one of the more enigmatic findings is a defect in the ability of skeletal progenitor cells to support a hematopoietic environment in postnatal ectopic ossicles [57, 91]. Interestingly, reconstitution of MT1-MMP expression in merely a subset of skeletal progenitor cells restores the ability of the marrow stroma environment to support hematopoiesis again. MT1-MMP thus appears central to engraftment and homing of hematopoietic stem cells and their progeny in a postnatal repopulation of the hematopoietic environment. The specific molecular and cellular properties that favor engraftment in the presence of MT1-MMP and prevent hematopoiesis in the absence of proteolytic activity still remain unidentified. Interestingly, these observations coincide with the ability of MT1-MMP to process CD44, which is proposed as an important step in induced mobilization of hematopoietic stem cells from the marrow to the circulation [92, 93]. At present, it is unresolved whether this protease-dependent mobilization phenotype is also invoked in migrating clonal progeny of hematopoietic stem cells during colonization of prospective new hematopoietic territories in postnatal marrow formation.

3.11 Physiological Role of MT2-MMP

MT1-MMP is, with a few exceptions such as injury repair in the lung, largely expressed in the dense connective tissues of the mouse [66, 94, 95]. In contrast, MT2-MMP is expressed abundantly in some epithelia such as mammary epithelial cells [96]. In addition, MT2-MMP is expressed in the trophoblasts of the human and mouse placenta and in mouse salivary gland epithelium [61, 97, 98]. This prompts the question if MT2-MMP is an epithelial counterpart to MT1-MMP and MT3-MMP in terms of matrix remodeling. If MT2-MMP indeed serves a role of this nature, the physiological significance of its function is not immediately evident as mice devoid of MT2-MMP expression are grossly unaffected [61]. Mice with haploinsufficiency for MT1-MMP and MT2-MMP are likewise unaffected as are haploinsufficient MT1-MMP mice in an MT2-MMP-deficient background. However, interbreeding of these mice fail to produce double-deficient offspring despite detection of viable embryos at E9.5 with an expected Mendelian distribution. Immediately after, most double-deficient embryos display apparent failure to thrive and at E11.5 all embryos are dead. This finding constitutes the first example of MMP activity required in early development and identifies MT1-MMP and MT2-MMP as interchangeable mediators of an essential physiological process in gestation [61].

Closer analysis of the defect associated with the demise of double-deficient embryos pinpoints a failure to establish a functional labyrinth in the placenta. This observation is consistent with the expression of both MT1-MMP and MT2-MMP in the placenta at the time of embryo demise. A more careful investigation reveals that the embryonic vasculature emanating from the allantoic mesenchyme fails to penetrate into the prospective labyrinth and establish a vascular exchange with the maternal blood sinuses. As in the underdeveloped lung of MT1-MMP-deficient mice, there is no overt accumulation of extracellular matrix to suggest that the cause of the developmental defect is abrogation of gross matrix remodeling. By ultrastructural analysis it is evident that the perivascular environment of the embryonic vessel is compromised by disruption of the cell fusion of chorionic trophoblasts that leads to syncytiotrophoblast formation [61].

While this study reveals a critical role for MT1-MMP or MT2-MMP in placental development, it does not address the wider implication of combined deficiency for these two proteinases beyond the point of placental maturation. Fortunately, conditional ablation strategies can resolve this issue and reveal that combined loss of MT1-MMP and MT2-MMP before placental labyrinth formation is incompatible with gestation as previously observed. In contrast, loss of both protease activities following labyrinth formation is compatible with gestation to term, and pups are indistinguishable from mice with MT1-MMP deficiency. The loss of both proteases in adulthood is moreover compatible with life and importantly does not affect the tissue remodeling processes associated with involution of the mammary gland, as might be expected from the abundant expression of both proteases there [61, 96]. In summary, these studies reveal that MT1-MMP or MT2-MMP govern an essential step in the placental development and syncytiotrophoblast formation. When compared with other double-deficient models of MMP deficiency, a unique role of MT1-MMP and MT2-MMP can be assigned in development. The physiological processes affected by the combined loss of these two MT-MMPs are notably different from those affected by the activities of MT1-MMP and MMP2 in concert. MT2-MMP must therefore be assigned a role in physiology that is at least partially independent of its ability to activate MMP2 and be recognized as a matrix-degrading enzyme in its own right.

The advent of an inducible system for ablation of MT1-MMP in a cell-specific and temporal manner has also enabled a more detailed analysis of the compound phenotype, which characterizes the unconditional loss of MT1-MMP in the mouse. In relation to the role of the protease in specific cell types, much is yet to be learned about gene function in development, homeostasis, and disease. This research will hopefully serve to resolve some of the issues confounded by the severe disease associated with MT1-MMP deficiency from conception. Ongoing experiments with this system thus reveal that MT1-MMP serves two major functions in the mouse by enabling developmentally regulated tissue remodeling and secondarily is essential in connective tissue homeostasis of constantly remodeling tissues in adulthood (Shi and Holmbeck, unpublished results).

3.12
MT-Type MMPs Work in Concert to Execute Matrix Remodeling

The observation that collagen remodeling is governed by both extracellular and intracellular pathways suggests that these processes are so important in cell and organism viability that positive selection for such properties has led to several independent and redundant pathways. The presence of several related MT-MMP types, of which MT1-MMP and MT3-MMP share a prominent overlap in temporal and spatial expression in several tissues, prompts the question whether a functional redundancy has been acquired between these two molecules at some point during evolution [60, 94–96, 99]. Loss of MT3-MMP in mice leads to a modest but measurable retardation of growth and is not associated with any further overt impediments to development such as, for instance, pulmonary hyperplasia [55, 60]. This prompted the question whether MT3-MMP serves a role in collagen remodeling at all and if the function of this proteinase serves a particular function over and above that of MT1-MMP. Mice doubly deficient for MT1-MMP and MT3-MMP are not viable and thus reveal that MT1-MMP deficiency can be mitigated by the activity of MT3-MMP. Specifically, double-deficient mice present with a general exacerbation of the physiological defects associated with the loss of MT1-MMP, notably in an allele-dosage-dependent manner. Accordingly, haploinsufficiency for both loci is compatible with normal development; however, additional loss of the remaining MT3-MMP allele leads to growth retardation and changes in the gross appearance of the mice. MT3-MMP haploinsufficiency in an MT1-MMP null background further exacerbates the physiological defects associated with MT1-MMP ablation and mice die without exception between birth and weaning, thereby demonstrating that MT3-MMP homozygosity is required for survival of mice devoid of MT1-MMP activity. The further loss of the remaining MT3-MMP allele leads to the uniform demise of all mice at birth. The mortality is associated with markedly diminished size indicative of severe growth retardation in the latter part of gestation and anatomical defects in connective tissues with particularly conspicuous consequences for remodeling and morphogenetic processes leading to secondary palate formation. This defect is tied to diminished cell proliferation, but deviates from other observed developmental impairments by affecting a tissue that is mostly devoid of type I and II collagens, but rich in type III collagen. The loss of gene products from the two loci further displays their combined function in matrix remodeling associated with angiogenesis. Despite a prominent expression in perivascular tissue, the absence of MT1-MMP activity is otherwise only linked with defects in vascular canal formation in the prospective secondary ossification centers of long bones and with vascular invasion following artificial induction of angiogenesis in a highly collagenous environment [57, 62, 63]. However, MT1-MMP/MT3-MMP-deficient mice additionally display a role for MT-MMP activity in a specific matrix remodeling process associated with formation of vasculature in the highly collagenous tissues of the longitudinal cartilage septae of the metaphysis [60]. This and other findings associated with combined MT1-MMP/MT3-MMP deficiency are qualitatively different from those encountered in other double-deficient strains of mice discussed here (MT1-MMP/MMP2, MT1-MMP/uPARAP, and MT1-MMP/MT2-MMP). Despite some common traits, each mouse line displays separate and nonoverlapping defects, suggesting that specific developmental and postnatal remodeling processes governed by MMP2, uPARAP, MT2-MMP, or MT3-MMP can be functionally compensated for by MT1-MMP, but this is only revealed in the absence of the pleiotropic *MT1-MMP* gene. As with uPARAP function, the role of MT3-MMP was questioned in the context of its possible role in matrix remodeling. Previous reports documented the ability of MT3-MMP expressing cells to form tubules in collagen gels, as observed previously for MT1-MMP [22, 100]. Dissection of the molecular defect associated with the loss of MT3-MMP reveals that this proteinase in the absence of MT1-MMP will degrade native (fibrillar) high-density collagen films. Importantly, this is not restricted only to type I and II collagens but also to type III collagen. These observations are consistent with the defects in tissue remodeling, proliferation, and viability of cells populating connective tissues abundant in these matrix components [60]. On the basis of these properties of MT3-MMP, it is assigned a role as a major collagenase in the mouse and is the likely product of a gene duplication facilitating the processing of bulk fibrillar collagen matrices in the organism in conjunction with other pericellular protease and intracellular matrix processing mechanisms.

3.13
MT4-MMP – an MT-MMP with Elusive Function

As most other MT-MMPs, MT4-MMP is found in the central nervous system and particularly in the granular cells of the dentate gyrus [6, 59] and (Holmbeck *et al.*, unpublished data). MT4-MMP is further detected in the testis, lungs, and uterus; however, here, the particular cell types that serve as the source of MT4-MMP have not been identified, but are likely cells of the monocyte macrophage lineage. MT4-MMP-deficient mice are overtly normal and display no defects in tumor necrosis factor-αTNF-α activation despite findings to support this *in vitro* [21]. It is, however, possible that a subset of cells in the central nervous system utilize MT4-MMP preferentially for TNF-α conversion in place of a disintegrin and metalloproteinase 17/TNF-α converting enzyme (ADAM17/TACE) [101]. As a further unique feature separating it from the remaining MT-type MMPs, MT4-MMP is incapable of activating MMP2 zymogen to active MMP2 [21]. More recently, MT4-MMP has been implicated in osteoarthritis based on the findings of MT4-MMP immunoreactivity in postmortem articular cartilage from human subjects. Consistent with this observation, MT4-MMP-deficient mice release reduced amounts of glycosamino glycans into the synovial fluid following intra-articular injection of interleukin-1β(IL-1β) [102]. Interestingly, MT4-MMP expression has not been reported in articular chondrocytes in the mouse and this prompts the question if the cellular source of MT4-MMP in the joint may be of monocyte/macrophage origin.

3.14
MT5-MMP Modulates Neuronal Growth and Nociception

MT5-MMP is abundant in the central and peripheral nervous system and associated with growing neurons; however, the precise role in this process is not fully appreciated. *In vitro*, MT5-MMP expression can relieve the inhibitory effects of proteoglycans on neurite growth and MT5-MMP is found to bind to α-amino-3-hydroxy-5-methyl-4-isoxazolepropionic acid (AMPA) receptor-binding proteins and glutamate receptor-binding proteins and degrade cadherins *in vitro* [103, 104]. MT5-MMP contains a secondary furin/PACE sensitive site in the C-terminus and can be shed to the pericelluar space in an active form [105].

MT5-MMP-deficient mice appear normal, but display enhanced motor skills on rotorod tests and diminished stress response in water maze tests. Furthermore, enhanced resistance to mechanical allodynia following sciatic nerve injury is observed in MT5-MMP-deficient mice and this diminished susceptibility to neuropathic pain coincides with absence of Aβ nerve fiber ingrowth into the dorsal horn of the spinal cord [58].

In contrast to the growth restriction observed in dorsal root ganglia, MT5-MMP deficiency is observed to affect peripheral nerve growth positively. Accordingly, MT5-MMP-deficient mice display hyperinnervation of the footpad skin and neonatal dorsal root ganglia neurons display a grater degree of elaboration when cultured on collagen [56]. In contrast to the decreased sensitivity to mechanical allodynia, MT5-MMP-deficient mice are more sensitive to thermal nociception; however, inflammation-induced hyperalgesia is reduced and mutant mice show resistance to hyperalgesia following challenge, with a number of classic inflammatory growth factors and cytokines [56]. Consistent with these findings, MT5-MMP-deficient mice display resistance to induced mast cell degranulation and abnormal neuronal N-cadherin distribution, which suggests that MT5-MMP is required for proper N-cadherin processing in inflammatory nociceptive response.

3.15 Summary and Concluding Remarks

On the basis of a large body of work since the first characterization of MT-MMPs, the current understanding of the function of this subfamily of MMPs has developed from the initial appreciation of these molecules as activators of secreted MMPs to the current realization that they, in addition, work as essential effectors of proteolysis. The information extracted from genetic models has highlighted a central role in connective tissue homeostasis and pericellular proteolysis for MT1-MMP. It is now also evident that there is a great deal of functional redundancy between MT1-MMP, MT2-MMP, and MT3-MMP, but only through ablation of MT1-MMP is it possible to appraise the functional role of the remaining two proteases and their reciprocal ability to singly rescue the molecular functions normally undertaken by the concerted activity. In the case of MT2-MMP, a clear understanding of the physiological function above and beyond the role exerted in conjunction with MT1-MMP still has to be achieved. The function undertaken by these proteases in placental morphogenesis is clearly indispensable. However, because the MT1-MMP null mutation leads to premature death and lack of reproduction, the natural selection against MT1-MMP homozygosity is complete. MT2-MMP is therefore never required solely for its function in placental morphogenesis, and the selective pressure to maintain the allele is derived from other requirements that have not been assessed so far. Further analysis of the MT2-MMP-deficient mouse strain will therefore be needed to tease out additional roles of MT2-MMP in physiology. The function of the molecular relative MT3-MMP is easier to appreciate as the single mutation leads to measurable physiological consequences in mice. Double mutations for MT1-MMP and MT3-MMP further demonstrate that MT3-MMP is a collagenase in its own right and is required in development and postnatal growth.

MT5-MMP is restricted largely to the nervous system and here it serves to affect sensory input and matrix remodeling in response to the injury and inflammation. Despite its more restricted nature, MT5-MMP displays true matrix remodeling properties in the pericellular environment of the nervous system and affects specific cellular functions even in a tissue where expression of the other MT-MMP members is abundant.

The physiological roles of MT4-MMP and MT6-MMP are yet to be determined as is the role of the type II membrane MMP, MMP23, or cysteine array matric mettaloproteinase (CA-MMP) [106, 107]. The latter is an odd fellow in the company of MT-MMPs as the processing to an active form results in the release of the mature protein from the cell. Future work and the availability of animal models to assess the role of MT6-MMP and CA-MMP should enhance our knowledge of these proteases.

The remaining challenge beyond continued analysis of available animal models will be to focus on finding the role of these molecules in human physiology. Here, multiple unanswered questions remain – the most pertinent of which may be whether we can equate the role of MT1-MMP as a principal collagenase in the mouse with a similar function in humans. An important part of this work is to gain a thorough appreciation of the physiological roles mediated by the MT-MMP subfamily given their prominent expression in many pathophysiological conditions where they constitute, at least for the time being, a potential target for intervention. As the anti-MMP therapeutic strategies are pursued, it is imperative that the consequences can be anticipated and weighed against the welfare of the patient. If this is ignored, then any renewed attempts to intervene in diseases with targeted MMP inhibitor compounds will face the same failure as clinical trials based on MMP intervention already have suffered [108, 109].

Acknowledgment

This research was supported in part by the Intramural research Program of the NIH, NIDCR.

References

1. Okada, Y., Nagase, H., and Harris, E.D. Jr. (1987) Matrix metalloproteinases 1, 2, and 3 from rheumatoid synovial cells are sufficient to destroy joints. *J. Rheumatol.*, **14** (Spec. No.), 41–42.
2. Liotta, L.A., Tryggvason, K., Garbisa, S., Robey, P.G., and Abe, S. (1981) Partial purification and characterization of a neutral protease which cleaves type IV collagen. *Biochemistry*, **20**, 100–104.
3. Sato, H., Takino, T., Okada, Y., Cao, J., Shinagawa, A., Yamamoto, E., and Seiki, M. (1994) A matrix metalloproteinase expressed on the surface of invasive tumour cells. *Nature*, **370**, 61–65.
4. Pei, D. (1999b) Identification and characterization of the fifth membrane-type matrix metalloproteinase MT5-MMP. *J. Biol. Chem.*, **274**, 8925–8932.

5. Pei, D. (1999c) Leukolysin/MMP25/MT6-MMP: a novel matrix metalloproteinase specifically expressed in the leukocyte lineage. *Cell Res.*, **9**, 291–303.
6. Puente, X.S., Pendas, A.M., Llano, E., Velasco, G., and Lopez-Otin, C. (1996) Molecular cloning of a novel membrane-type matrix metalloproteinase from a human breast carcinoma. *Cancer Res.*, **56**, 944–949.
7. Strongin, A.Y., Collier, I., Bannikov, G., Marmer, B.L., Grant, G.A., and Goldberg, G.I. (1995) Mechanism of cell surface activation of 72-kDa type IV collagenase. Isolation of the activated form of the membrane metalloprotease. *J. Biol. Chem.*, **270**, 5331–5338.
8. Takino, T., Sato, H., Shinagawa, A., and Seiki, M. (1995) Identification of the second membrane-type matrix metalloproteinase (MT-MMP-2) gene from a human placenta cDNA library. MT-MMPs form a unique membrane-type subclass in the MMP family. *J. Biol. Chem.*, **270**, 23013–23020.
9. Will, H. and Hinzmann, B. (1995) cDNA sequence and mRNA tissue distribution of a novel human matrix metalloproteinase with a potential transmembrane segment. *Eur. J. Biochem.*, **231**, 602–608.
10. Gross, J. and Lapiere, C.M. (1962) Collagenolytic activity in amphibian tissues: a tissue culture assay. *Proc. Natl. Acad. Sci. U.S.A.*, **48**, 1014–1022.
11. Evanson, J.M., Jeffrey, J.J., and Krane, S.M. (1967) Human collagenase: identification and characterization of an enzyme from rheumatoid synovium in culture. *Science*, **158**, 499–502.
12. Fullmer, H.M. and Gibson, W. (1966) Collagenolytic activity in gingivae of man. *Nature*, **209**, 728–729.
13. Fullmer, H.M., Gibson, W.A., Lazarus, G., and Stam, A.C. Jr. (1966) Collagenolytic activity of the skin associated with neuromuscular diseases including amyotrophic lateral sclerosis. *Lancet*, **1**, 1007–1009.
14. Goldberg, G.I., Marmer, B.L., Grant, G.A., Eisen, A.Z., Wilhelm, S., and He, C.S. (1989) Human 72-kilodalton type IV collagenase forms a complex with a tissue inhibitor of metalloproteases designated TIMP-2. *Proc. Natl. Acad. Sci. U.S.A.*, **86**, 8207–8211.
15. Stetler-Stevenson, W.G., Krutzsch, H.C., and Liotta, L.A. (1989) Tissue inhibitor of metalloproteinase (TIMP-2). A new member of the metalloproteinase inhibitor family. *J. Biol. Chem.*, **264**, 17374–17378.
16. Okada, Y., Morodomi, T., Enghild, J.J., Suzuki, K., Yasui, A., Nakanishi, I., Salvesen, G., and Nagase, H. (1990) Matrix metalloproteinase 2 from human rheumatoid synovial fibroblasts. Purification and activation of the precursor and enzymic properties. *Eur. J. Biochem.*, **194**, 721–730.
17. Ward, R.V., Atkinson, S.J., Slocombe, P.M., Docherty, A.J., Reynolds, J.J., and Murphy, G. (1991) Tissue inhibitor of metalloproteinases-2 inhibits the activation of 72 kDa progelatinase by fibroblast membranes. *Biochim. Biophys. Acta*, **1079**, 242–246.
18. Itoh, Y., Kajita, M., Kinoh, H., Mori, H., Okada, A., and Seiki, M. (1999) Membrane type 4 matrix metalloproteinase (MT4-MMP, MMP-17) is a glycosylphosphatidylinositol-anchored proteinase. *J. Biol. Chem.*, **274**, 34260–34266.
19. Kojima, S., Itoh, Y., Matsumoto, S., Masuho, Y., and Seiki, M. (2000) Membrane-type 6 matrix metalloproteinase (MT6-MMP, MMP-25) is the second glycosyl-phosphatidyl inositol (GPI)-anchored MMP. *FEBS Lett.*, **480**, 142–146.
20. Butler, G.S., Will, H., Atkinson, S.J., and Murphy, G. (1997) Membrane-type-2 matrix metalloproteinase can initiate the processing of progelatinase A and is regulated by the tissue inhibitors of metalloproteinases. *Eur. J. Biochem.*, **244**, 653–657.
21. English, W.R., Puente, X.S., Freije, J.M., Knauper, V., Amour, A., Merryweather, A., Lopez-Otin, C., and Murphy, G. (2000) Membrane type 4 matrix metalloproteinase (MMP17) has tumor necrosis factor-alpha

convertase activity but does not activate pro-MMP2. *J. Biol. Chem.*, **275**, 14046–14055.
22. Kang, T., Yi, J., Yang, W., Wang, X., Jiang, A., and Pei, D. (2000) Functional characterization of MT3-MMP in transfected MDCK cells: progelatinase A activation and tubulogenesis in 3-D collagen lattice. *FASEB J.*, **14**, 2559–2568.
23. Kolkenbrock, H., Hecker-Kia, A., Orgel, D., Ulbrich, N., and Will, H. (1997) Activation of progelatinase A and progelatinase A/TIMP-2 complex by membrane type 2-matrix metalloproteinase. *Biol. Chem.*, **378**, 71–76.
24. Cao, J., Sato, H., Takino, T., and Seiki, M. (1995) The C-terminal region of membrane type matrix metalloproteinase is a functional transmembrane domain required for pro-gelatinase A activation. *J. Biol. Chem.*, **270**, 801–805.
25. Imai, K., Ohuchi, E., Aoki, T., Nomura, H., Fujii, Y., Sato, H., Seiki, M., and Okada, Y. (1996) Membrane-type matrix metalloproteinase 1 is a gelatinolytic enzyme and is secreted in a complex with tissue inhibitor of metalloproteinases 2. *Cancer Res.*, **56**, 2707–2710.
26. Pei, D. and Weiss, S.J. (1995) Furin-dependent intracellular activation of the human stromelysin-3 zymogen. *Nature*, **375**, 244–247.
27. Pei, D. and Weiss, S.J. (1996) Transmembrane-deletion mutants of the membrane-type matrix metalloproteinase-1 process progelatinase A and express intrinsic matrix-degrading activity. *J. Biol. Chem.*, **271**, 9135–9140.
28. Sato, H., Kinoshita, T., Takino, T., Nakayama, K., and Seiki, M. (1996) Activation of a recombinant membrane type 1-matrix metalloproteinase (MT1-MMP) by furin and its interaction with tissue inhibitor of metalloproteinases (TIMP)-2. *FEBS Lett.*, **393**, 101–104.
29. Kang, T., Nagase, H., and Pei, D. (2002) Activation of membrane-type matrix metalloproteinase 3 zymogen by the proprotein convertase furin in the trans-Golgi network. *Cancer Res.*, **62**, 675–681.
30. Tanaka, S.S., Mariko, Y., Mori, H., Ishijima, J., Tachi, S., Sato, H., Seiki, M., Yamanouchi, K., Tojo, H., and Tachi, C. (1997) Cell-cell contact down-regulates expression of membrane type metalloproteinase-1 (MT1-MMP) in a mouse mammary gland epithelial cell line. *Zool. Sci.*, **14**, 95–99.
31. Yana, I. and Weiss, S.J. (2000) Regulation of membrane type-1 matrix metalloproteinase activation by proprotein convertases. *Mol. Biol. Cell*, **11**, 2387–2401.
32. Jiang, A., Lehti, K., Wang, X., Weiss, S.J., Keski-Oja, J., and Pei, D. (2001) Regulation of membrane-type matrix metalloproteinase 1 activity by dynamin-mediated endocytosis. *Proc. Natl. Acad. Sci. U.S.A.*, **98**, 13693–13698.
33. Wang, X., Ma, D., Keski-Oja, J., and Pei, D. (2004) Co-recycling of MT1-MMP and MT3-MMP through the trans-Golgi network. Identification of DKV582 as a recycling signal. *J. Biol. Chem.*, **279**, 9331–9336.
34. Lafleur, M.A., Mercuri, F.A., Ruangpanit, N., Seiki, M., Sato, H., and Thompson, E.W. (2006) Type I collagen abrogates the clathrin-mediated internalization of membrane type 1 matrix metalloproteinase (MT1-MMP) via the MT1-MMP hemopexin domain. *J. Biol. Chem.*, **281**, 6826–6840.
35. Haas, T.L., Davis, S.J., and Madri, J.A. (1998) Three-dimensional type I collagen lattices induce coordinate expression of matrix metalloproteinases MT1-MMP and MMP-2 in microvascular endothelial cells. *J. Biol. Chem.*, **273**, 3604–3610.
36. Ruangpanit, N., Chan, D., Holmbeck, K., Birkedal-Hansen, H., Polarek, J., Yang, C., Bateman, J.F., and Thompson, E.W. (2001) Gelatinase A (MMP-2) activation by skin fibroblasts: dependence on MT1-MMP expression and fibrillar collagen form. *Matrix Biol.*, **20**, 193–203.

37. Lehti, K., Lohi, J., Valtanen, H., and Keski-Oja, J. (1998) Proteolytic processing of membrane-type-1 matrix metalloproteinase is associated with gelatinase A activation at the cell surface. *Biochem. J.*, **334** (Pt 2), 345–353.
38. Stanton, H., Gavrilovic, J., Atkinson, S.J., D'Ortho, M.P., Yamada, K.M., Zardi, L., and Murphy, G. (1998) The activation of ProMMP-2 (gelatinase A) by HT1080 fibrosarcoma cells is promoted by culture on a fibronectin substrate and is concomitant with an increase in processing of MT1-MMP (MMP-14) to a 45 kDa form. *J. Cell Sci.*, **111** (Pt 18), 2789–2798.
39. Zucker, S., Hymowitz, M., Conner, C., DeClerck, Y., and Cao, J. (2004) TIMP-2 is released as an intact molecule following binding to MT1-MMP on the cell surface. *Exp. Cell Res.*, **293**, 164–174.
40. Cowell, S., Knauper, V., Stewart, M.L., D'Ortho, M.P., Stanton, H., Hembry, R.M., Lopez-Otin, C., Reynolds, J.J., and Murphy, G. (1998) Induction of matrix metalloproteinase activation cascades based on membrane-type 1 matrix metalloproteinase: associated activation of gelatinase A, gelatinase B and collagenase 3. *Biochem. J.*, **331** (Pt 2), 453–458.
41. Knauper, V., Will, H., Lopez-Otin, C., Smith, B., Atkinson, S.J., Stanton, H., Hembry, R.M., and Murphy, G. (1996) Cellular mechanisms for human procollagenase-3 (MMP-13) activation. Evidence that MT1-MMP (MMP-14) and gelatinase a (MMP-2) are able to generate active enzyme. *J. Biol. Chem.*, **271**, 17124–17131.
42. Ingvarsen, S., Madsen, D.H., Hillig, T., Lund, L.R., Holmbeck, K., Behrendt, N., and Engelholm, L.H. (2008) Dimerization of endogenous MT1-MMP is a regulatory step in the activation of the 72-kDa gelatinase MMP-2 on fibroblasts and fibrosarcoma cells. *Biol. Chem.*, **389**, 943–953.
43. Itoh, Y., Ito, N., Nagase, H., and Seiki, M. (2008) The second dimer interface of MT1-MMP, the transmembrane domain, is essential for ProMMP-2 activation on the cell surface. *J. Biol. Chem.*, **283**, 13053–13062.
44. Itoh, T., Ikeda, T., Gomi, H., Nakao, S., Suzuki, T., and Itohara, S. (1997) Unaltered secretion of beta-amyloid precursor protein in gelatinase A (matrix metalloproteinase 2)-deficient mice. *J. Biol. Chem.*, **272**, 22389–22392.
45. Martignetti, J.A., Aqeel, A.A., Sewairi, W.A., Boumah, C.E., Kambouris, M., Mayouf, S.A., Sheth, K.V., Eid, W.A., Dowling, O., Harris, J. *et al.* (2001) Mutation of the matrix metalloproteinase 2 gene (MMP2) causes a multicentric osteolysis and arthritis syndrome. *Nat. Genet.*, **28**, 261–265.
46. Zankl, A., Pachman, L., Poznanski, A., Bonafe, L., Wang, F., Shusterman, Y., Fishman, D.A., and Superti-Furga, A. (2007) Torg syndrome is caused by inactivating mutations in MMP2 and is allelic to NAO and Winchester syndrome. *J. Bone Miner. Res.*, **22**, 329–333.
47. Mosig, R.A., Dowling, O., DiFeo, A., Ramirez, M.C., Parker, I.C., Abe, E., Diouri, J., Aqeel, A.A., Wylie, J.D., Oblander, S.A. *et al.* (2007) Loss of MMP-2 disrupts skeletal and craniofacial development and results in decreased bone mineralization, joint erosion and defects in osteoblast and osteoclast growth. *Hum. Mol. Genet.*, **16**, 1113–1123.
48. Balbin, M., Fueyo, A., Knauper, V., Lopez, J.M., Alvarez, J., Sanchez, L.M., Quesada, V., Bordallo, J., Murphy, G., and Lopez-Otin, C. (2001) Identification and enzymatic characterization of two diverging murine counterparts of human interstitial collagenase (MMP-1) expressed at sites of embryo implantation. *J. Biol. Chem.*, **276**, 10253–10262.
49. D'Ortho, M.P., Will, H., Atkinson, S., Butler, G., Messent, A., Gavrilovic, J., Smith, B., Timpl, R., Zardi, L., and Murphy, G. (1997) Membrane-type matrix metalloproteinases 1 and 2 exhibit broad-spectrum proteolytic capacities comparable to many matrix metalloproteinases. *Eur. J. Biochem.*, **250**, 751–757.

50. Fosang, A.J., Last, K., Fujii, Y., Seiki, M., and Okada, Y. (1998) Membrane-type 1 MMP (MMP-14) cleaves at three sites in the aggrecan interglobular domain. *FEBS Lett.*, **430**, 186–190.
51. Ohuchi, E., Imai, K., Fujii, Y., Sato, H., Seiki, M., and Okada, Y. (1997) Membrane type 1 matrix metalloproteinase digests interstitial collagens and other extracellular matrix macromolecules. *J. Biol. Chem.*, **272**, 2446–2451.
52. Hotary, K.B., Allen, E.D., Brooks, P.C., Datta, N.S., Long, M.W., and Weiss, S.J. (2003) Membrane type I matrix metalloproteinase usurps tumor growth control imposed by the three-dimensional extracellular matrix. *Cell*, **114**, 33–45.
53. Tokuraku, M., Sato, H., Murakami, S., Okada, Y., Watanabe, Y., and Seiki, M. (1995) Activation of the precursor of gelatinase A/72 kDa type IV collagenase/MMP-2 in lung carcinomas correlates with the expression of membrane-type matrix metalloproteinase (MT-MMP) and with lymph node metastasis. *Int. J. Cancer*, **64**, 355–359.
54. Yamamoto, M., Mohanam, S., Sawaya, R., Fuller, G.N., Seiki, M., Sato, H., Gokaslan, Z.L., Liotta, L.A., Nicolson, G.L., and Rao, J.S. (1996) Differential expression of membrane-type matrix metalloproteinase and its correlation with gelatinase A activation in human malignant brain tumors in vivo and in vitro. *Cancer Res.*, **56**, 384–392.
55. Hadchouel, A., Decobert, F., Franco-Montoya, M.L., Halphen, I., Jarreau, P.H., Boucherat, O., Martin, E., Benachi, A., Amselem, S., Bourbon, J. *et al.* (2008) Matrix metalloproteinase gene polymorphisms and bronchopulmonary dysplasia: identification of MMP16 as a new player in lung development. *PLoS ONE*, **3**, e3188.
56. Folgueras, A.R., Valdes-Sanchez, T., Llano, E., Menendez, L., Baamonde, A., Denlinger, B.L., Belmonte, C., Juarez, L., Lastra, A., Garcia-Suarez, O. *et al.* (2009) Metalloproteinase MT5-MMP is an essential modulator of neuro-immune interactions in thermal pain stimulation. *Proc. Natl. Acad. Sci. U.S.A.*, **106**, 16451–16456.
57. Holmbeck, K., Bianco, P., Caterina, J., Yamada, S., Kromer, M., Kuznetsov, S.A., Mankani, M., Robey, P.G., Poole, A.R., Pidoux, I. *et al.* (1999) MT1-MMP-deficient mice develop dwarfism, osteopenia, arthritis, and connective tissue disease due to inadequate collagen turnover. *Cell*, **99**, 81–92.
58. Komori, K., Nonaka, T., Okada, A., Kinoh, H., Hayashita-Kinoh, H., Yoshida, N., Yana, I., and Seiki, M. (2004) Absence of mechanical allodynia and Abeta-fiber sprouting after sciatic nerve injury in mice lacking membrane-type 5 matrix metalloproteinase. *FEBS Lett.*, **557**, 125–128.
59. Rikimaru, A., Komori, K., Sakamoto, T., Ichise, H., Yoshida, N., Yana, I., and Seiki, M. (2007) Establishment of an MT4-MMP-deficient mouse strain representing an efficient tracking system for MT4-MMP/MMP-17 expression in vivo using beta-galactosidase. *Genes Cells*, **12**, 1091–1100.
60. Shi, J., Son, M.Y., Yamada, S., Szabova, L., Kahan, S., Chrysovergis, K., Wolf, L., Surmak, A., and Holmbeck, K. (2008) Membrane-type MMPs enable extracellular matrix permissiveness and mesenchymal cell proliferation during embryogenesis. *Dev. Biol.*, **313**, 196–209.
61. Szabova, L., Son, M.Y., Shi, J., Sramko, M., Yamada, S.S., Swaim, W.D., Zerfas, P., Kahan, S., and Holmbeck, K. (2010) Membrane-type MMPs are indispensable for placental labyrinth formation and development. *Blood*, **116**, 5752–5761.
62. Yana, I., Sagara, H., Takaki, S., Takatsu, K., Nakamura, K., Nakao, K., Katsuki, M., Taniguchi, S., Aoki, T., Sato, H. *et al.* (2007) Crosstalk between neovessels and mural cells directs the site-specific expression of MT1-MMP to endothelial tip cells. *J. Cell Sci.*, **120**, 1607–1614.
63. Zhou, Z., Apte, S.S., Soininen, R., Cao, R., Baaklini, G.Y., Rauser, R.W., Wang, J., Cao, Y., and Tryggvason, K. (2000) Impaired endochondral ossification

and angiogenesis in mice deficient in membrane-type matrix metalloproteinase I. *Proc. Natl. Acad. Sci. U.S.A.*, **97**, 4052–4057.

64. Ohtake, Y., Tojo, H., and Seiki, M. (2006) Multifunctional roles of MT1-MMP in myofiber formation and morphostatic maintenance of skeletal muscle. *J. Cell Sci.*, **119**, 3822–3832.
65. Havemose-Poulsen, A., Holmstrup, P., Stoltze, K., and Birkedal-Hansen, H. (1998) Dissolution of type I collagen fibrils by gingival fibroblasts isolated from patients of various periodontitis categories. *J. Periodontal Res.*, **33**, 280–291.
66. Holmbeck, K., Bianco, P., Chrysovergis, K., Yamada, S., and Birkedal-Hansen, H. (2003) MT1-MMP-dependent, apoptotic remodeling of unmineralized cartilage: a critical process in skeletal growth. *J. Cell Biol.*, **163**, 661–671.
67. Holmbeck, K., Bianco, P., Pidoux, I., Inoue, S., Billinghurst, R.C., Wu, W., Chrysovergis, K., Yamada, S., Birkedal-Hansen, H., and Poole, A.R. (2005) The metalloproteinase MT1-MMP is required for normal development and maintenance of osteocyte processes in bone. *J. Cell Sci.*, **118**, 147–156.
68. Inada, M., Wang, Y., Byrne, M.H., Rahman, M.U., Miyaura, C., Lopez-Otin, C., and Krane, S.M. (2004) Critical roles for collagenase-3 (Mmp13) in development of growth plate cartilage and in endochondral ossification. *Proc. Natl. Acad. Sci. U.S.A.*, **101**, 17192–17197.
69. Stickens, D., Behonick, D.J., Ortega, N., Heyer, B., Hartenstein, B., Yu, Y., Fosang, A.J., Schorpp-Kistner, M., Angel, P., and Werb, Z. (2004) Altered endochondral bone development in matrix metalloproteinase 13-deficient mice. *Development*, **131**, 5883–5895.
70. Sabeh, F., Ota, I., Holmbeck, K., Birkedal-Hansen, H., Soloway, P., Balbin, M., Lopez-Otin, C., Shapiro, S., Inada, M., Krane, S. *et al.* (2004) Tumor cell traffic through the extracellular matrix is controlled by the membrane-anchored collagenase MT1-MMP. *J. Cell Biol.*, **167**, 769–781.
71. Egeblad, M., Shen, H.C., Behonick, D.J., Wilmes, L., Eichten, A., Korets, L.V., Kheradmand, F., Werb, Z., and Coussens, L.M. (2007) Type I collagen is a genetic modifier of matrix metalloproteinase 2 in murine skeletal development. *Dev. Dyn.*, **236**, spc1. 1683–1693.
72. Gross, J. and Nagai, Y. (1965) Specific degradation of the collagen molecule by tadpole collagenolytic enzyme. *Proc. Natl. Acad. Sci. U.S.A.*, **54**, 1197–1204.
73. Liu, X., Wu, H., Byrne, M., Jeffrey, J., Krane, S., and Jaenisch, R. (1995) A targeted mutation at the known collagenase cleavage site in mouse type I collagen impairs tissue remodeling. *J. Cell Biol.*, **130**, 227–237.
74. Sakai, T. and Gross, J. (1967) Some properties of the products of reaction of tadpole collagenase with collagen. *Biochemistry*, **6**, 518–528.
75. Krane, S.M., Byrne, M.H., Lemaitre, V., Henriet, P., Jeffrey, J.J., Witter, J.P., Liu, X., Wu, H., Jaenisch, R., and Eeckhout, Y. (1996) Different collagenase gene products have different roles in degradation of type I collagen. *J. Biol. Chem.*, **271**, 28509–28515.
76. Oh, J., Takahashi, R., Adachi, E., Kondo, S., Kuratomi, S., Noma, A., Alexander, D.B., Motoda, H., Okada, A., Seiki, M. *et al.* (2004) Mutations in two matrix metalloproteinase genes, MMP-2 and MT1-MMP, are synthetic lethal in mice. *Oncogene*, **23**, 5041–5048.
77. Koo, B.H., Kim, H.H., Park, M.Y., Jeon, O.H., and Kim, D.S. (2009a) Membrane type-1 matrix metalloprotease-independent activation of pro-matrix metalloprotease-2 by proprotein convertases. *FEBS J.*, **276**, 6271–6284.
78. Koo, B.H., Park, M.Y., Jeon, O.H., and Kim, D.S. (2009b) Regulatory mechanism of matrix metalloprotease-2 enzymatic activity by factor Xa and thrombin. *J. Biol. Chem.*, **284**, 23375–23385.
79. Atkinson, J.J., Holmbeck, K., Yamada, S., Birkedal-Hansen, H., Parks,

W.C., and Senior, R.M. (2005) Membrane-type 1 matrix metalloproteinase is required for normal alveolar development. *Dev. Dyn.*, **232**, 1079–1090.

80. Mazzone, R.W. and Kornblau, S. (1981) Size of pores of Kohn: influence of transpulmonary and vascular pressures. *J. Appl. Physiol.*, **51**, 739–745.
81. Xiong, W., Knispel, R., MacTaggart, J., Greiner, T.C., Weiss, S.J., and Baxter, B.T. (2009) Membrane-type 1 matrix metalloproteinase regulates macrophage-dependent elastolytic activity and aneurysm formation in vivo. *J. Biol. Chem.*, **284**, 1765–1771.
82. Atkinson, J.J., Toennies, H.M., Holmbeck, K., and Senior, R.M. (2007) Membrane type 1 matrix metalloproteinase is necessary for distal airway epithelial repair and keratinocyte growth factor receptor expression after acute injury. *Am. J. Physiol. Lung Cell Mol. Physiol.*, **293**, L600–L610.
83. Beertsen, W., Holmbeck, K., Niehof, A., Bianco, P., Chrysovergis, K., Birkedal-Hansen, H., and Everts, V. (2002) On the role of MT1-MMP, a matrix metalloproteinase essential to collagen remodeling, in murine molar eruption and root growth. *Eur. J. Oral. Sci.*, **110**, 445–451.
84. Holmbeck, K., Bianco, P., Yamada, S., and Birkedal-Hansen, H. (2004) MT1-MMP: a tethered collagenase. *J. Cell. Physiol.*, **200**, 11–19.
85. Johnson, R.B. (1987) A classification of Sharpey's fibers within the alveolar bone of the mouse: a high-voltage electron microscope study. *Anat. Rec.*, **217**, 339–347.
86. Melcher, A.H. and Chan, J. (1981) Phagocytosis and digestion of collagen by gingival fibroblasts in vivo: a study of serial sections. *J. Ultrastruct. Res.*, **77**, 1–36.
87. Engelholm, L.H., List, K., Netzel-Arnett, S., Cukierman, E., Mitola, D.J., Aaronson, H., Kjoller, L., Larsen, J.K., Yamada, K.M., Strickland, D.K. *et al.* (2003) uPARAP/Endo180 is essential for cellular uptake of collagen and promotes fibroblast collagen adhesion. *J. Cell Biol.*, **160**, 1009–1015.
88. Engelholm, L.H., Nielsen, B.S., Dano, K., and Behrendt, N. (2001) The urokinase receptor associated protein (uPARAP/endo180): a novel internalization receptor connected to the plasminogen activation system. *Trends Cardiovasc. Med.*, **11**, 7–13.
89. Everts, V., Hou, W.S., Rialland, X., Tigchelaar, W., Saftig, P., Bromme, D., Gelb, B.D., and Beertsen, W. (2003) Cathepsin K deficiency in pycnodysostosis results in accumulation of non-digested phagocytosed collagen in fibroblasts. *Calcif. Tissue Int.*, **73**, 380–386.
90. Wagenaar-Miller, R.A., Engelholm, L.H., Gavard, J., Yamada, S.S., Gutkind, J.S., Behrendt, N., Bugge, T.H., and Holmbeck, K. (2007) Complementary roles of intracellular and pericellular collagen degradation pathways in vivo. *Mol. Cell. Biol.*, **27**, 6309–6322.
91. Szabova, L., Yamada, S.S., Wimer, H., Chrysovergis, K., Ingvarsen, S., Behrendt, N., Engelholm, L.H., and Holmbeck, K. (2009) MT1-MMP and type II collagen specify skeletal stem cells and their bone and cartilage progeny. *J. Bone Miner. Res.*, **24**, 1905–1916.
92. Kajita, M., Itoh, Y., Chiba, T., Mori, H., Okada, A., Kinoh, H., and Seiki, M. (2001) Membrane-type 1 matrix metalloproteinase cleaves CD44 and promotes cell migration. *J. Cell Biol.*, **153**, 893–904.
93. Vagima, Y., Avigdor, A., Goichberg, P., Shivtiel, S., Tesio, M., Kalinkovich, A., Golan, K., Dar, A., Kollet, O., Petit, I. *et al.* (2009) MT1-MMP and RECK are involved in human CD34+ progenitor cell retention, egress, and mobilization. *J. Clin. Invest.*, **119**, 492–503.
94. Apte, S.S., Fukai, N., Beier, D.R., and Olsen, B.R. (1997) The matrix metalloproteinase-14 (MMP-14) gene is structurally distinct from other MMP genes and is co-expressed with the TIMP-2 gene during mouse embryogenesis. *J. Biol. Chem.*, **272**, 25511–25517.
95. Kinoh, H., Sato, H., Tsunezuka, Y., Takino, T., Kawashima, A., Okada, Y.,

and Seiki, M. (1996) MT-MMP, the cell surface activator of proMMP-2 (pro-gelatinase A), is expressed with its substrate in mouse tissue during embryogenesis. *J. Cell Sci.*, **109** (Pt 5), 953–959.

96. Szabova, L., Yamada, S.S., Birkedal-Hansen, H., and Holmbeck, K. (2005) Expression pattern of four membrane-type matrix metalloproteinases in the normal and diseased mouse mammary gland. *J. Cell. Physiol.*, **205**, 123–132.

97. Bjørn, S.F., Hastrup, N., Larsen, J.F., Lund, L.R., and Pyke, C. (2000) Messenger RNA for membrane-type 2 matrix metalloproteinase, MT2-MMP, is expressed in human placenta of first trimester. *Placenta*, **21**, 170–176.

98. Rebustini, I.T., Myers, C., Lassiter, K.S., Surmak, A., Szabova, L., Holmbeck, K., Pedchenko, V., Hudson, B.G., and Hoffman, M.P. (2009) MT2-MMP-dependent release of collagen IV NC1 domains regulates submandibular gland branching morphogenesis. *Dev. Cell*, **17**, 482–493.

99. Nuttall, R.K., Sampieri, C.L., Pennington, C.J., Gill, S.E., Schultz, G.A., and Edwards, D.R. (2004) Expression analysis of the entire MMP and TIMP gene families during mouse tissue development. *FEBS Lett.*, **563**, 129–134.

100. Kadono, Y., Shibahara, K., Namiki, M., Watanabe, Y., Seiki, M., and Sato, H. (1998) Membrane type 1-matrix metalloproteinase is involved in the formation of hepatocyte growth factor/scatter factor-induced branching tubules in madin-darby canine kidney epithelial cells. *Biochem. Biophys. Res. Commun.*, **251**, 681–687.

101. Black, R.A., Rauch, C.T., Kozlosky, C.J., Peschon, J.J., Slack, J.L., Wolfson, M.F., Castner, B.J., Stocking, K.L., Reddy, P., Srinivasan, S. *et al.* (1997) A metalloproteinase disintegrin that releases tumour-necrosis factor-alpha from cells. *Nature*, **385**, 729–733.

102. Clements, K.M., Flannelly, J.K., Tart, J., Brockbank, S.M., Wardale, J., Freeth, J., Parker, A.E., and Newham, P. (2011) Matrix metalloproteinase 17 is necessary for cartilage aggrecan degradation in an inflammatory environment. *Ann. Rheum. Dis.*, **70**, 683–689.

103. Hayashita-Kinoh, H., Kinoh, H., Okada, A., Komori, K., Itoh, Y., Chiba, T., Kajita, M., Yana, I., and Seiki, M. (2001) Membrane-type 5 matrix metalloproteinase is expressed in differentiated neurons and regulates axonal growth. *Cell Growth Differ.*, **12**, 573–580.

104. Monea, S., Jordan, B.A., Srivastava, S., DeSouza, S., and Ziff, E.B. (2006) Membrane localization of membrane type 5 matrix metalloproteinase by AMPA receptor binding protein and cleavage of cadherins. *J. Neurosci.*, **26**, 2300–2312.

105. Wang, X. and Pei, D. (2001) Shedding of membrane type matrix metalloproteinase 5 by a furin-type convertase: a potential mechanism for down-regulation. *J. Biol. Chem.*, **276**, 35953–35960.

106. Pei, D. (1999a) CA-MMP: a matrix metalloproteinase with a novel cysteine array, but without the classic cysteine switch. *FEBS Lett.*, **457**, 262–270.

107. Pei, D., Kang, T., and Qi, H. (2000) Cysteine array matrix metalloproteinase (CA-MMP)/MMP-23 is a type II transmembrane matrix metalloproteinase regulated by a single cleavage for both secretion and activation. *J. Biol. Chem.*, **275**, 33988–33997.

108. Coussens, L.M., Fingleton, B., and Matrisian, L.M. (2002) Matrix metalloproteinase inhibitors and cancer: trials and tribulations. *Science*, **295**, 2387–2392.

109. Zucker, S. and Cao, J. (2009) Selective matrix metalloproteinase (MMP) inhibitors in cancer therapy: ready for prime time? *Cancer Biol. Ther.*, **8**, 2371–2373.

4
Bone Remodeling: Cathepsin K in Collagen Turnover

Dieter Brömme

4.1
Introduction

The human skeleton undergoes a lifelong process of gradual resorption and formation called *bone remodeling*. The remodeling process is necessary to repair microfractures and to provide a physiological level of Ca^{2+} ions within the blood stream [1]. It is estimated that the entire skeleton is renewed every 7–10 years. The bone remodeling cycle is initiated by the activation of bone-resorbing osteoclasts at prospective repair sites. The exact mechanism of site recognition still remains unclear. The remodeling process is likely initiated by mechanical stress, the apoptosis of bone-embedded osteocytes, and biochemical signaling from the extracellular matrix. The osteoclast initially forms a sealing zone that allows the formation of the ruffled border underneath the osteoclast. The first step in bone resorption is characterized by the secretion of protons via the V-ATPase in the ruffled border, which acidifies the area below the osteoclast, called *resorption lacunae*. This leads to the demineralization of the matrix and exposure of collagen fibrils. Type I collagen constitutes 90% of the total organic matrix of bones. The second step includes the secretion of proteolytic enzymes that primarily act as potent collagenases. The removal of the collagen fibrils yields the formation of a resorption pit, which is subsequently refilled with fresh bone material produced by invading osteoblasts. It is thought that the whole process of remodeling takes about three weeks [2]. The site of osteoclast/osteoblast interplay responsible for bone resorption and bone apposition is called a basic multicellular unit (BMU) [3] (Figure 4.1).

The bone remodeling process is well regulated. Bone formation during the growth phase is dominant over the resorption, and a balance between resorption and formation is reached in adulthood. In later age, the resorption process gains importance, leading to a net loss of bone. Peak bone mass is reached at the age of 30–35. Pathological loss of bone mass is called *osteoporosis*, and it affects women in menopause and postmenopause (type I osteoporosis) and both genders at advanced ages (type II or senile osteoporosis). Secondary osteoporosis can occur as an accompanying factor of diseases such as arthritis, bone tumors, or prolonged use of medications such as glycocorticoides.

Matrix Proteases in Health and Disease, First Edition. Edited by Niels Behrendt.

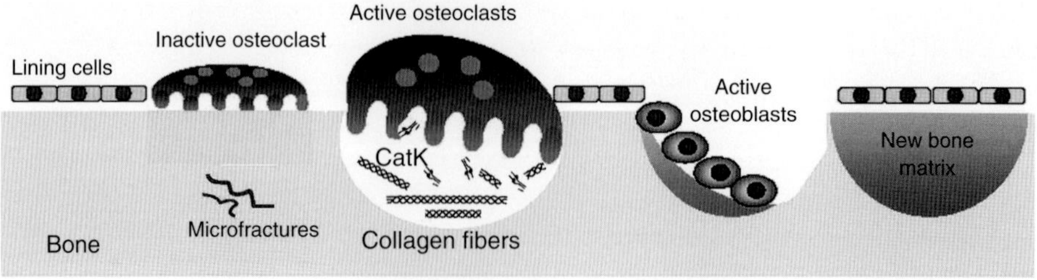

Figure 4.1 Basic multicellular units contain individual osteoclasts and osteoblasts that mediate the bone remodeling cycle. Osteoclasts sense sites of prospective repair and remove the damaged bone matrix by demineralization and collagenolysis. The freshly formed resorption pit is subsequently occupied by osteoblasts, which refill the pit with freshly synthesized bone matrix. It is estimated that the interplay between osteoclasts and osteoblasts replaces the entire skeletal matrix every 7–10 years. catK, cathepsin K.

Major features of osteoporosis are the loss of bone mineral density (BMD), the deterioration of the bone fine structure with both leading to increased fracture risks. Therefore, the inhibition of the bone resorption process is a primary aim of pharmaceutical intervention. Various antiresorptive strategies have been recently discussed in detail [4, 5]. This review focuses on the contribution of cysteine cathepsins and, in particular, cathepsin K to the degradation of the bone matrix.

4.2
Proteolytic Machinery of Bone Resorption and Cathepsin K

As the main constituent of bone is type I collagen, attention has been focused on collagenase activities of the osteoclast. Historically, matrix metalloproteases (MMPs) have been the center of interest as they were the first protease group with a defined collagenase activity. However, MMPs are proteases active at neutral pH and thus unlikely to be active in an acidified resorption lacuna. Therefore, lysosomal cysteine protease were considered and cysteine protease inhibitors such as leupeptin and E64 proved effective to inhibit bone resorption in tissue cultures and isolated osteoclasts [6, 7]. These early findings indicated that lysosomal cysteine cathepsins might be the main culprits of collagen degradation, but a specific cathepsin responsible for bone degradation was not yet identified. Only three mammalian cysteine proteases, cathepsin B, H, and L, were known with none of them exhibiting a triple-helical collagenase activity. In 1994, a rabbit-osteoclast-specific cDNA of a novel cathepsin, OC-2, was discovered [8]. The human ortholog cDNA was cloned shortly thereafter, published under several names (cathepsin O, cathepsin O2, cathepsin X, and cathepsin K) [9–13] and finally named cathepsin K in 1996. It was shown that cathepsin K is the major protein expressed in osteoclasts

(4% of random EST clones sequenced represented the cathepsin K cDNA and these accounted for about 98% of total cathepsin ESTs in this cell type [10]). Northern blot analyses revealed very low expression levels for cathepsins L and S in osteoclasts [9]. The characterization of the recombinant protease revealed that cathepsin K is a potent endopeptidase and exhibits a unique collagenase activity, which, in contrast to MMPs, cleaves at multiple sites within triple-helical collagens of types I and II. Zn-dependent collagenases such as MMP1, 8, and 13 and MT1-MMP cleave only at one site and release typical $^3/_4$ and $^1/_4$ fragments [14, 15]. The critical role of cathepsin K in bone resorption became clear when cathepsin K deficiency was identified as the cause of the autosomal recessive bone disorder, pycnodysostosis [16]. Pycnodysostosis is characterized by short stature, generalized osteosclerosis, dysmorphic appearance, and pathologic fractures on the clinical level [17] and by disordered growth plates and the accumulation of undigested collagen fibrils within osteoclasts and fibroblasts on the tissue and cellular level [18, 19]. Cathepsin-K-deficient mice revealed defects in bone development: bones were osteopetrotic, the cartilage in the growth plate was hypercalcified, and osteoclast activity was severely impaired [20–23]. However, murine cathepsin K deficiency had an overall milder phenotype when compared to the human phenotype as neither dwarfism nor intracellular collagen fibril accumulation was observed. This may indicate at least partial redundancy in collagen-degrading activities in mice [24]. Moreover, long and vertebral bones showed increased bone strength and high bone formation rates [25]. This might be in contradiction to the observed increased brittleness and thus higher fragility in bones of these mice [26]. The increased bone formation rates could be caused by increased levels of bone formation factors such as insulin-like growth factor-1 (IGF-1) and transforming growth factor-β1(TGF-β1), which are the substrates of cathepsin K [27, 28].

Highly relevant to the bone resorption function of cathepsin K is its regulation by estrogens. Estrogen has been identified as a negative regulator of cathepsin K expression. Northern blot analysis revealed a 50% downregulation of cathepsin K mRNA expression in mature osteoclasts [29] and a reduction in the depth of osteoclast-mediated resorption pits [30]. Ovariectomized mice showed an increase in cathepsin K protein expression by 200% when compared to sham-operated littermates. Treatment with 17β-estradiol reduced cathepsin K expression to baseline levels of the sham-operated mice [31]. RANKL, the main osteoclastogenesis factor, increased the expression of cathepsin K in osteoclasts [32] but is downregulated by estrogen [33]. These findings may explain the elevated bone degradation rates in postmenopausal women.

The degradation of the collagen fibrils likely takes place in the resorption lacuna as well as intracellularly in the lysosomal compartment. The extracellular degradation pathway is supported by the finding that cathepsin K is secreted into the resorption lacuna [34, 35] and can be found in culture supernatant in its proenzyme as well as mature form in fibroblast cultures [36]. The proenzyme is efficiently processed into the mature form within lysosomes [34, 37], and the intracellular degradation of collagen was confirmed by the accumulation of undigested collagen fibrils in lysosomes of cathepsin-K-deficient or cysteine-protease-inhibitor-treated osteoclasts and fibroblasts [18, 19, 38].

Cathepsin K is not only responsible for the bulk degradation of type I collagen within the resorption lacunae of osteoclasts and the intracelluar hydrolysis of phagocytosed collagen fibrils but also appears to be crucial for the initiation of the resorption process. Before resorption, the osteoclast needs to interact with the collageneous bone surface, which leads to the formation of a sealing zone and the ruffled border. The formation of the so-called actin ring is equated with the activation status of an osteoclast [39]. The actin ring formation is mediated by $\alpha v\beta 3$ intergrin interactions with RGD peptide motifs in collagen fibrils of the extracellular matrix. We have shown that cathepsin K activity is required to expose cryptic RGD motifs and to initiate actin ring formation. Cathepsin-K-deficient osteoclasts and cathepsin-K-inhibitor-treated cells displayed significantly less actin rings than wild-type or untreated cells. However, pretreatment of the collagen matrix with recombinant cathepsin K restored the formation of actin rings in osteoclasts when grown on collagen. This was not achieved by pretreatment of the collagen with cathepsin L or MMP 1 [40]. Moreover, cathepsin K activity may also directly regulate the duration of the resorption process by an individual osteoclast via the generation of RGD peptides in the resorption lacuna. Type I collagen contains seven RGD peptide motifs in its α-chains, which might be released during its hydrolysis by cathepsin K. Adding excess amounts of RGD peptides into the cell culture terminated osteoclast activity, as demonstrated by the loss of actin rings. Inhibitory RGD peptide concentrations were achieved by predigesting either type I collagen or bone powder with cathepsin K but not with cathepsin L or MMP1 [40].

4.3
Specificity and Mechanism of Collagenase Activity of Cathepsin K

Cathepsin K belongs to the cathepsin L subgroup of papainlike cysteine proteases, and its three-dimensional structure is highly similar to those of other members of this group [41]. As all other cathepsins, it is synthesized as a prepropeptide (Figure 4.2a). The signal peptide is lost during the passage into the endoplasmic reticulum, and the inhibitory propeptide is cleaved off in the acidic environment of the lysosomal compartment. Cathepsin K is active in the acidic pH range between 5 and 6 [15] and exhibits a potent endopeptidase activity with a preference for nonaromatic hydrophobic residues such as Leu, Ile, Met, and Val in its S2 binding pocket. In contrast to cathepsin L, bulky aromatic residues are excluded from the S2 subsite. Uniquely, among cathepsins, cathepsin K preferentially accepts proline at this site. The acceptance of proline as a P2 residue is a requirement for the collagenolytic activity of cathepsin K. Mutations at the S2 subsite, which change the specificity of this site into a cathepsin-L-like specificity, abolished the ability of cathepsin K to cleave triple-helical collagen [42, 43]. The S3 and S4 subsites are characterized by a limited specificity: positively charged as well as neutral small or aromatic hydrophobic residues are accepted [14, 15, 44, 45] (Figure 4.2b,c). The

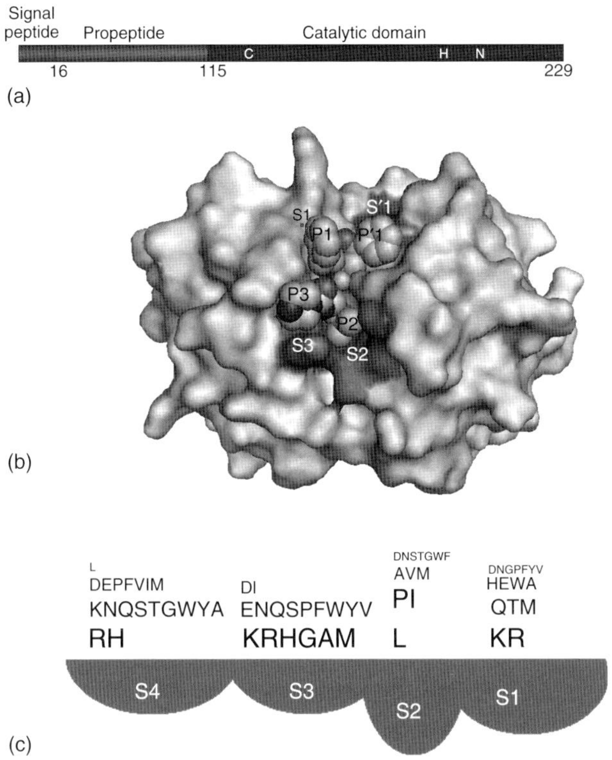

Figure 4.2 Structure and subsites specificity of cathepsin K: (a) Cathepsin K is synthesized as a 331 amino acid long precursor protein that contains a signal peptide, a propeptide, and the catalytically active domain. The signal peptide is removed in the endoplasmic reticulum, and the propeptide is cleaved off in the acidified lysosomal compartment. (b) Surface structure of cathepsin K with the subsite and active site areas highlighted. (c) S1–S4 subsite specificity of cathepsin K. The letter size of the single-letter amino acid abbreviations corresponds to the acceptance level of these residues to the appropriate subsites.

occupation of the S′ sites strongly affects the binding of residues in the S1 subsite. While P4–P1 peptide libraries indicate an exclusion of Gly from the S1 subsite [45], prime-site-extended peptide substrates revealed a preferred acceptance of a P1Gly residue [46].

Cleavage sites within the triple helix of type I and II collagens have been described [47–49]. These sites represent initial cleavage sites and are located at the N- and C-terminal regions of the α-chains (Figure 4.3a). The overall efficacy of cathepsin-K-mediated solubilization of collagen from cortical bone is about three to five times more potent than that of MMP1, -9, and -13 [47]. Other demonstrated substrates of cathepsin K in the bone matrix are osteopontin, osteocalcin, aggrecan, and SPARC [14, 27, 50, 51]. Cathepsin K also exerts a potent elastolytic activity [15], which has been implicated in cardiovascular diseases [52–54].

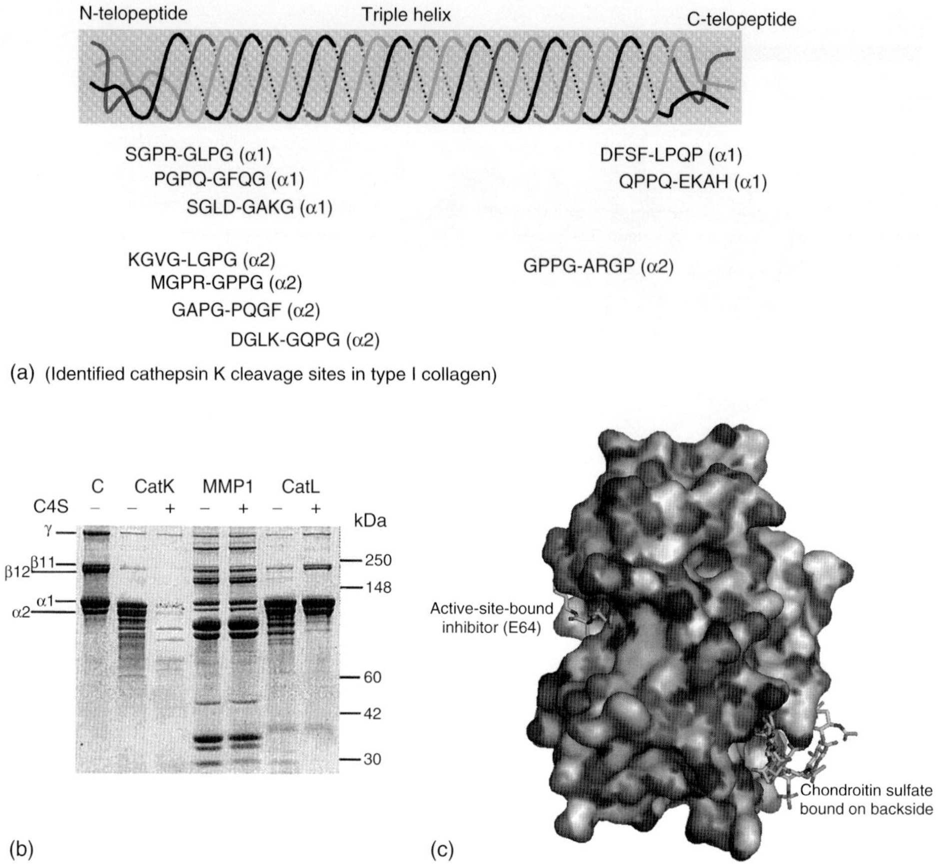

Figure 4.3 Collagenolytic activity of cathepsin K and complex formation with chondroitin sulfate. (a) Known cleavage sites of cathepsin K in type I collagen. (b) sodium dodecyl sulfate polyacrylamide gel electrophoresis (SDS-PAGE) of type I collagen cleavage by cathepsin K in comparison with cathepsin L and MMP1 in the absence and presence of chondroitin sulfate (C4S). (c) Binding of chondroitin sulfate to cathepsin K on the back of the protease (Protein Data Bank (PDB) file: 3C9E) (*http://www.pdb.org/pdb/home/home.do*). ((a) Source: From Garnero *et al.*, 1998; Sassi *et al.*, 2000.)

Although the elucidated subsite specificity of cathepsin K supports the capability of the protease to cleave at typical collagen amino acid sequences, it does not allow the understanding of the mechanism of triple-helical collagen degradation by this protease. The active site entrance of cathepsin K and any other cysteine cathepsin is about 5 Å wide and insufficient to accommodate a collagen triple helix of 15 Å in diameter. Before the binding of collagen α-chains in the substrate-binding region of the protease, triple-helical collagen requires a partial unfolding. This appears to be mediated by the formation of an oligomeric cathepsin K molecule complex with bone- and cartilage-resident glycosaminoglycans such as chondroitin sulfate [55, 56]. We have previously shown that cathepsin K as a monomer and in the absence of glycosaminoglycans is unable to cleave triple-helical collagens (Figure 4.3b) [56]. Crystallographic studies revealed that chondroitin sulfate specifically binds on the back of the L-domain of cathepsin K in a positively charged surface area [57] (Figure 4.3c). The structure also revealed a unique bead on a strand-like organization of individual cathepsin K molecules along an extended chondroitin sulfate chain. However, enzymatic data indicate that the "beads on a strand" structure (multiple cathepsin K molecules associate on a single chondroitin sulfate chain) is unlikely to represent the active conformation of the collagenase activity. Maximal activity of cathepsin K/glycosaminoglycan complexes is achieved at stoichiometries of 2 : 1 for cathepsin K–chondroitin sulfate mixtures. Excess of cathepsin K or chondroitin sulfate leads to the loss of collagenolytic activity without affecting the general proteolytic activities of cathepsin K (unpublished data, DB). The observed interaction sites between the protease and chondroitin sulfate are crucial for the collagenase activity, as mutations at these sites in cathepsin K lead to a significant reduction of the collagenase activity. In line with the assumption that the complex is specifically required for the collagenase activity, no inhibitory effect on the peptidolytic and gelatinase activities was observed in the mutant proteins [58]. It should also be noted that the active cathepsin K–chondroitin sulfate complex requires specific protein–protein interaction sites. The crystal structure of the complex revealed specific dimer interactions between two cathepsin K molecules [58]. Interestingly, the dimer interface contains residue Tyr98. The replacement of this amino acid residue by a cysteine has been previously identified as a pycnodysostosis-causing mutation [59]. The Tyr98Cys mutant of cathepsin K displayed no collagenase activity but retained its peptidolytic activity [59]. Furthermore, we showed that this mutant was incapable of forming higher molecular weight complexes with chondroitin sulfate as typical for the wild-type enzyme [56]. All these data indicate that a specific cathepsin K–glycosaminoglycan complex is required for the effective hydrolysis of collagens. It is likely that this complex acts as a helicase to unfold triple-helical collagen. The formation of a collagenolytically active complex between a cathepsin and glycosaminoglycans appears to be specific to cathepsin K. Neither cathepsins L, S, or F formed similar complexes [60]. Besides cathepsin K, only cathepsin V is capable of forming a complex but does not exhibit any collagenase activity [61].

The formation of a "helicase" complex opens up the possibility of specifically inhibiting the hydrolysis of collagens without blocking the active site. We have shown that negatively charged oligonucleotides and polypeptides specifically inhibit the collagenase activity of cathepsin K [62]. In contrast, neutrally or positively charged polypeptides had no effect. We are presently searching for low-molecular-weight inhibitors that specifically prevent the complex formation. This kind of inhibitors may have an advantage over classical active-site-directed compounds as they would not inhibit collagenase-independent proteolytic activities of cathepsin K.

4.4
Role of Glycosaminoglycans in Bone Diseases

The inhibition of the collagenase activity of cathepsin K by excess glycosaminoglycans appears to be a contributing factor to the severe skeletal phenotypes seen in mucopolysaccharidoses (MPSs). MPSs are autosomal recessive lysosomal storage disorders caused by defects in the expression of various glycosaminoglycan-degrading hydrolases [63]. These diseases are classified by their enzyme deficiencies, which include defects in sulfatase and glycosidase activities. These enzyme deficiencies result in a massive accumulation of undigested glycosaminoglycans (mucopolysaccharides) in various tissues including bone. Patients are generally characterized by short stature, dysostosis multiplex, kyphosis, loose joints, swelling of the liver and spleen, cardiac and respiratory problems, mental retardation, and a significantly reduced life span [64]. Bones show developmental problems in their growth plates, which appear to be disorganized and lack normal ossification [65, 66]. Considering the dependency of the collagenase activity of cathepsin K on specific glycosaminoglycan concentrations and its central role in the bone remodeling cycle, it is likely that cathepsin K is inhibited in MPS diseases. We have demonstrated that MPS I mice lack the degradation of subepiphysial cartilage by osteoclasts [67]. The cartilage-stained area in the subepiphysial region of MPS I mice doubled when compared with wild-type littermates. A similar increase in nonresorbed cartilage was observed in the long bones of cathepsin-K-deficient mice [67]. Double staining for heparan sulfate and cathepsin K showed the colocalization of both components throughout the area underneath the epiphysial growth plate. Using a specific antibody that recognizes a cathepsin-K-mediated cleavage site in type II collagen, no degradation of this collagen was observed in the MPS I mutant mice [67, 68]. The removal of cartilaginous matrix by osteoclasts is a requirement of the ossification process during bone growth. The lack of collagenase activity was specifically related to the inhibition of the collagenase activity of cathepsin K and not to problems in osteoclastogenesis. Contrarily, the number of osteoclasts was increased in MPS I bones, which can be interpreted as a compensation for the lack of collagenolytically active cathepsin K [68]. Furthermore, we could demonstrate that MPS I osteoclasts were unable to form actively resorbing cells as indicated by their lack of actin ring formation. As discussed above, actin rings are formed as a consequence of the recognition of RGD motifs in the collagen matrix by $\alpha v\beta 3$ integrins expressed on the surface of osteoclasts. We have shown that glycosaminoglycans, such as dermatan sulfate, significantly reduce cathepsin-K-mediated actin ring formation in wild-type osteoclasts in a concentration range as seen in MPS tissues. On the other hand, the actin ring could be restored when MPS I osteoclasts were incubated on cathepsin K predigested type I collagen indicating that the dysfunction of MPS I osteoclasts was indeed due to a deficiency in cathepsin-K-dependent collagenase activity [67].

4.5
Development of Specific Cathepsin K Inhibitors and Clinical Trials

With the identification of cathepsin K as a target for the treatment of osteoporosis, major efforts were initiated to synthesize specific and potent inhibitors against this protease. At least four cathepsin K inhibitors entered clinical trials: (i) balicatib (AAE581) is a nitrile-based inhibitor developed by Novartis with a high selectivity and potency for cathepsin K [69]; (ii) relacatib, a peptidomimetic-based inhibitor from GlaxoSmithKline with a surprisingly poor specificity [70, 71]; (iii) odanacatib, a Merck-designed nitrile compound with a high potency and selectivity for cathepsin K [72]; and (iv) ONO-5334, a yet undisclosed compound from Ono Pharmaceuticals (Figure 4.4). Balicatib was withdrawn after phase II clinical trials because of potential adverse effects with scleroderma-like symptoms [74, 75]. Relacatib showed problems in phase I trials because of drug–drug interactions (GSK Protocol summary PSB-462795/008). ONO-5334 completed the phase II trial but no public information about next steps is available [76]. ONO-5334 was well tolerated, revealed no severe adverse events, and showed increased BMD similar to the bisphosphonate, alendronate, in the lumbar spine and better improvements in the femur. Balicatib and ONO-5334 had little effect on bone formation parameters, which represents a significant advantage of cathepsin K inhibitors over other antiresorptive therapies such as hormone replacement therapy, selective estrogen receptor modulators, and bisphosphonates. Interestingly, in studies using ovariectomized rhesus monkeys, balicatib showed a stimulatory effect on periosteal bone formation [77].

The most advanced compound, to date, is odanacatib, which is presently in phase III clinical trial evaluation. Odanacatib is a nonbasic and thus nonlysosomotropic compound exhibiting a dose-dependent inhibition of bone resorption without suppressing bone formation. The compound is orally applicable and has a half-life of up to 93 h. In placebo-controlled [72] phase I studies, two groups of healthy postmenopausal women received either daily ($n = 49$) or weekly ($n = 30$) doses of odanacatib for three weeks. At 50 mg per week, about 60% reductions in serum CTX and urinary NTX were observed (NTX and CTX are N- and C-terminal telopeptide cross-links of type I collagen used as biomarkers for bone degradation). Slightly larger reductions in both resorption markers were recorded in the 2.5 mg day^{-1} regimen [78]. Phase II trials included about 400 postmenopausal women with low BMD and participants were treated for 12 months and a possible 12 months' extension at four different doses administered weekly (3, 10, 25, and 50 mg). All but the lowest dose increased lumbar spine and femoral BMD in a dose-dependent manner. Furthermore, urinary NTX and serum CTX were decreased by 60% from baseline after 12 months and somewhat less after 24 months. At 10 and 25 mg week^{-1}, bone formation factors such as serum P1NP remained at the levels as of the control placebo group after 6 months of treatment, whereas the 50 mg treatment group showed about a 30% decrease after 12 months and a 20% decrease after 24 months [79]. Two hundred and eighty women enrolled into a third-year extension of the trial that resulted in further improvement of the BMD. After three

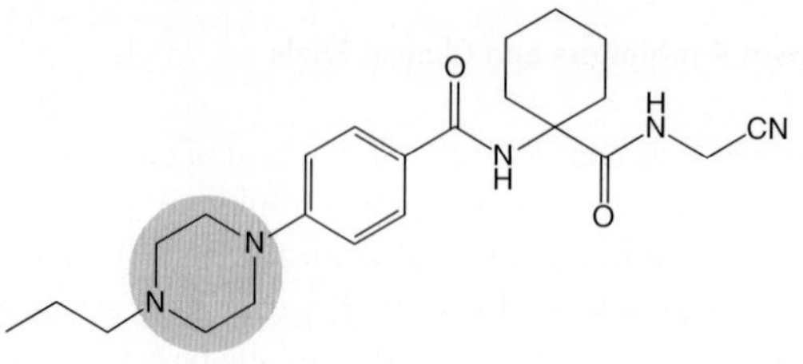

(a)

Balicatib
Lysosomotropic
IC_{50} (nM)
catK = 1.4
catB = 2900
catL = 18 000
catS = 28 000

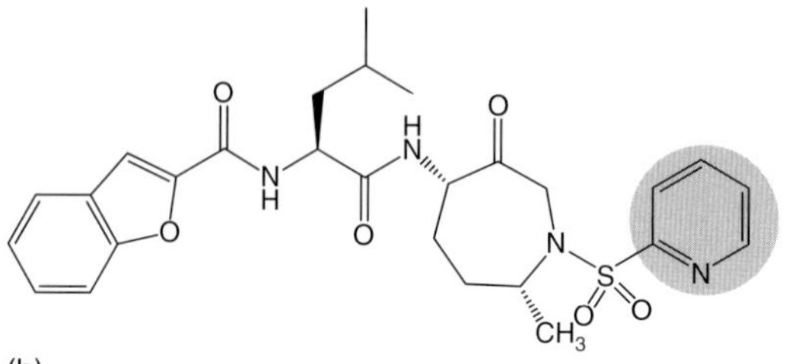

(b)

Relicatib
Lysosomotropic
$K_{i,app}$ (nM)
catK = 0.041
catB = 13
catL = 0.068
catS = 1.6
catV = 0.063

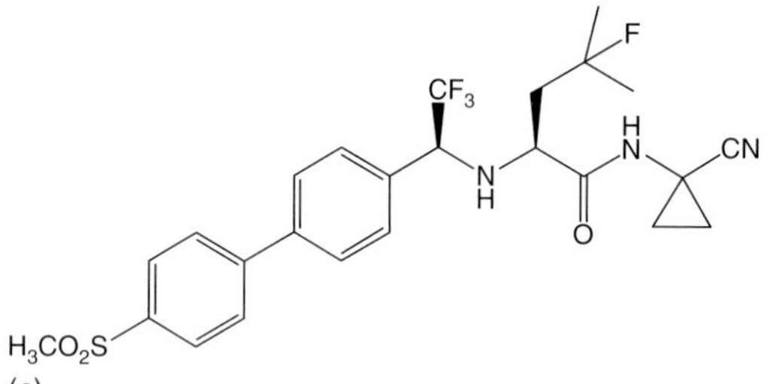

(c)

Odanacatib
Nonlysosomotropic
IC_{50} (nm)
catK = 0.2
catB = 1034
catL = 2995
catS = 60

Figure 4.4 Structure and inhibition parameters of cathepsin K inhibitors in a clinical trial for osteoporosis. (a) Balicatib, a lysosomotropic and highly potent selective inhibitor for cathepsin K developed by Novartis (withdrawn after phase II trial). (b) Relicatib, a lysosomotropic and highly potent but poorly selective inhibitor developed by GlaxoSmithKline (withdrawn after phase I trials). (c) Odanacatib, a nonlysosomotropic and potent selective cathepsin K inhibitor developed by Merck & Co. (presently evaluated in phase III trials). (Source: (a) from Black and Percival, 2006; (b) from Yamashita *et al.*, 2006; (c) from Gauthier *et al.*, 2008.) catK, cathepsin K; catB, cathepsin B; catL, cathepsin L; catS, cathepsin S; catV, cathepsin V.

years, the average increase in BMD was between +7.9 and 5.8% for the lumbar spine and the femoral neck sites. Urinary NTX remained −50% after two years of treatment. Bone formation parameters such as serum BSAP and P1NP either stayed slightly increased or fell to baseline. Bone biopsies did not reveal any abnormalities [80]. One arm of the third-year extension served as drug discontinuation group. Withdrawal from drug treatment led to a rapid increase in the resorption markers, urinary NTX and serum CTX, to 50 and 120%, respectively, above baseline levels.

They returned to values between 30 and 10% above baseline levels after one year of drug withdrawal. BMD values also decreased and returned to baseline by the end of the third year [80, 81].

For phase III clinical trial, Merck & Co has recruited more than 16 000 subjects for the clinical evaluation of odanacatib. The primary end point of the study is the reduction in incidence rates of radiographic spine and other bone fractures.

In addition to the treatment of osteoporosis, odanacatib is also considered for the intervention in bone tumors associated with breast and prostate cancers. Clinical trials in breast cancer patients with established bone metastases showed promising results in reducing urinary NTX by 77% after four weeks of 5 mg oral per day. For comparison, zoledronic acid, a bisphosphonate, achieved a reduction in NTX by 72%. No specific side effects due to the drug treatment were recorded [82].

4.6
Off-Target and Off-Site Inhibition

In addition to being effective against the pharmaceutical targets, a perfect drug will (i) inhibit the target specifically without cross-inhibition of unrelated and related targets, (ii) it will be specific to the target at the pharmaceutically relevant site of action in the body (here in bone), and (iii) it should not interfere with other pathways of the target. As it appears, relacatib failed most likely in all three parameters (i–iii), balicatib in two (ii and iii), and odanacatib may potentially have a problem in one (iii).

An advantage of odanacatib appears to be that in contrast to clinically failed balicatib and relacatib, the compound is not lysosomotropic [73, 83, 84]. This might initially appear counterintuitive as cathepsin K is a lysosomal protease, and the accumulation of an inhibitor at the site of its target would be expected to be beneficial. This strategy was discussed by Fuller and coworkers [85]. However, this approach would only be effective if the osteoclast is the only cellular target of lysosomotropic inhibitors and if the inhibitors are highly selective for cathepsin K. One problem with lysosomotropic compounds is the off-target inhibition of related cathepsins due to the buildup of the compound in lysosomes. The second and potentially more severe problem is that lysosomotropic inhibitors also accumulate in cell types such as fibroblasts, macrophages, and epithelial cells, which are known to express cathepsin K. This could explain the fibrotic phenotype seen in the failure of balicatib, and it is unlikely that this is an off-target effect but rather an off-site effect. The expression of cathepsin K in cells other than osteoclasts has been unfortunately neglected for some time.

It is tempting to speculate that a purely nonlysosomotropic inhibitor may primarily prevent the extracellular activity of cathepsin K, which is required for the activation of the osteoclast. The inhibition of cathepsin-K-mediated RGD peptide motif exposure and thus the prevention of integrin-mediated osteoclast activation are sufficient to terminate or attenuate bone resorption by the osteoclasts [40]. This may explain the lack of reported side effects of odanacatib in skin as it may act less

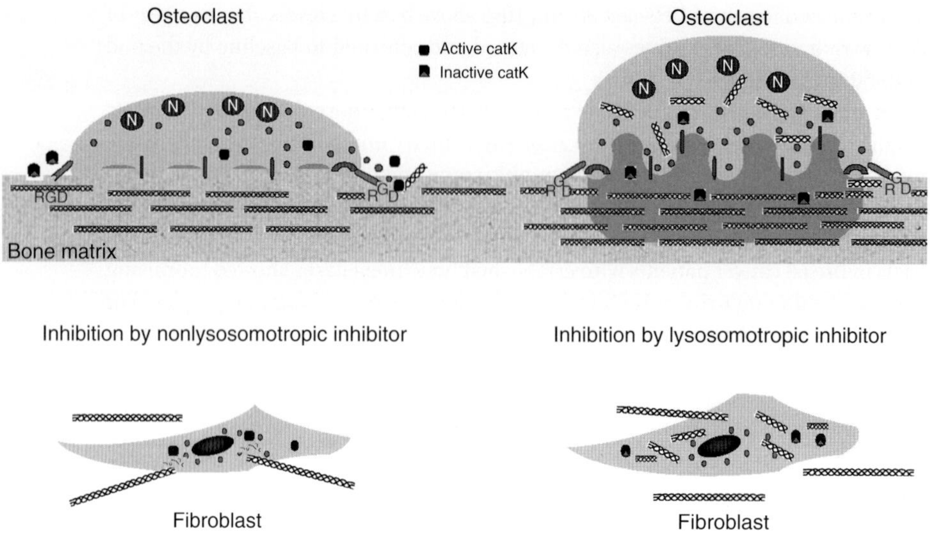

Figure 4.5 Scheme of extra- and intracellular osteoclast (upper panel) and fibroblast (lower panel) inhibition by nonlysosomotropic and lysosomotropic cathepsin K inhibitors. Nonlysosomotropic drugs primarily inhibit extracellular cathepsin K activity, which will prevent osteoclast activation but not block collagen degradation by fibroblasts after phagocytosis. Lysosomotropic drugs primarily block lysosomal collagen degradation in osteoclasts and fibroblasts and lead to the intracellular accumulation of undigested collagen fibers. catK, cathepsin K.

efficiently intracellularly in fibroblasts and thus may not inhibit cathepsin K in these cells. In contrast, lysosomotropic balicatib caused fibrotic skin changes consistent with a failure of collagen metabolism in fibroblasts [75]. It has been previously shown that either the lack of cathepsin K expression or its inhibition leads to an accumulation of undigested collagen fibrils in fibroblasts [19, 38]. However, a recent report about the mechanism of action of odanacatib indicates that this drug not only acts in the resorption lacuna but also inhibits intracellular bone protein degradation. Despite being nonlysosomotropic, odanacatib is likely phago/pinocytosed together with partially degraded collagen fibers from the resorption lacuna and thus will also act at least in part intracellularly [86]. The difference between balicatib and odanacatib regarding skin fibrotic side effects might be in the degree of intracellular inhibitor accumulation. Figure 4.5 summarizes the potential outcomes of lysosomotropic and nonlysosomotropic inhibitors.

It should be also noted that cathepsin K deficiency increases the severity of fibrosis in bleomycin-induced pulmonary fibrosis [87] and that overexpression of cathepsin K is protective against silica-induced lung fibrosis [88]. Moreover, we have recently shown that cathepsin K deficiency in mice causes abnormalities in airway development. The airway epithelium was disorganized; the surrounding tissue of airways showed increased collagen and glycosaminoglycan deposition and increased proliferation rates of epithelial cells and fibroblasts [28]. The latter was associated with increased TGF-β1 concentrations in the lung tissue, which may also act in a profibrotic manner [89]. Cathepsin K was able to specifically degrade TGF-β1 in lung tissues [28].

On the other hand, off-site inhibition of cathepsin K could, in some cases, be beneficial [90]. Cathepsin-K-deficient mice showed a stabilization of atherosclerotic plaques by increased collagen deposition, thicker fibrous caps, and less elastin breaks in the tunica media [54]. However, lack of cathepsin K activity has been also described as being potentially proatherosclerotic by promoting foam cell formation [52]. Inhibition of cathepsin K in synovial fibroblasts, macrophages, and chondrocytes would be also beneficial in arthritic diseases as cathepsin K expression in these cells has been implicated in articular cartilage erosion [91, 92]. Cathepsin K is capable of breaking down the two major organic constituents of cartilage, type II collagen and aggrecan [48, 50]. It is thought that the activity of cathepsin K in these cells is primarily intracellular, and thus lysosomotropic inhibitors would be more effective but would potentially cause off-site adverse effects, as discussed above. On the other hand, the inhibition of osteoclast activity is likely beneficial for arthritis patients as this disease is also characterized by a significant osteoporotic component.

4.7 Conclusion

Cathepsin K plays a key role in osteoclast-driven bone resorption. It has a unique collagenase activity and constitutes the major proteolytic activity in osteoclasts. To exert its collagenolytic activity, cathepsin K forms oligomeric complexes with bone- and cartilage-resident glycosaminoglycans. These complexes are likely required for the unfolding of triple-helical collagens before proteolytic cleavage. Highly potent cathepsin K inhibitors were developed, which entered clinical trials. Inhibitors with lysosomotropic features failed because of off-target and off-site effects. Odanacatib, a nonlysosomotropic cathepsin K inhibitor, is presently in phase III clinical trials, and so far has evaded causing severe adverse effects.

Acknowledgment

The work was supported by a Canada Research Chair Award and grants from the Canadian Institutes of Health Research (MOP 89974 and 86586).

References

1. Taylor, D. and Lee, T.C. (2003) Microdamage and mechanical behaviour: predicting failure and remodelling in compact bone. *J. Anat.*, **203**, 203–211.
2. Delaisse, J.M. and Vaes, G. (1992) *Mechanism of Mineral Solubilization and Matrix Degradation in Osteoclastic Bone Resorption*, CRC Press, Boca Raton, FL, Ann Arbor, MI, London, Tokyo.
3. Matsuo, K. and Irie, N. (2008) Osteoclast-osteoblast communication. *Arch. Biochem. Biophys.*, **473**, 201–209.

4. Rachner, T.D., Khosla, S., and Hofbauer, L.C. (2011) Osteoporosis: now and the future. *Lancet*, **377**, 1276–1287.
5. Rejnmark, L. and Mosekilde, L. (2011) New and emerging antiresorptive treatments in osteoporosis. *Curr. Drug Saf.*, **6**, 75–88.
6. Everts, V., Beertsen, W., and Schroder, R. (1988) Effects of proteinase inhibitors leupeptin and E-64 on osteoclastic bone resorption. *Calcif. Tissue. Int.*, **43**, 172–178.
7. Everts, V., Beertsen, W., and Tigchelaar-Gutter, W. (1985b) The digestion of phagocytosed collagen is inhibited by the proteinase inhibitors leupeptin and E-64. *Coll. Relat. Res.*, **5**, 315–336.
8. Tezuka, K., Tezuka, Y., Maejima, A., Sato, T., Nemoto, K., Kamioka, H., Hakeda, Y., and Kumegawa, M. (1994) Molecular cloning of a possible cysteine proteinase predominantly expressed in osteoclasts. *J. Biol. Chem.*, **269**, 1106–1109.
9. Bromme, D. and Okamoto, K. (1995) Human cathepsin O2, a novel cysteine protease highly expressed in osteoclastomas and ovary molecular cloning, sequencing and tissue distribution. *Biol. Chem. Hoppe–Seyler*, **376**, 379–384.
10. Drake, F.H., Dodds, R.A., James, I.E., Connor, J.R., Debouck, C., Richardson, S., Lee-Rykaczewski, E., Coleman, L., Rieman, D., Barthlow, R. *et al.* (1996) Cathepsin K, but not cathepsins B, L, or S, is abundantly expressed in human osteoclasts. *J. Biol. Chem.*, **271**, 12511–12516.
11. Inaoka, T., Bilbe, G., Ishibashi, O., Tezuka, K.I., Kumegawa, M., and Kokubo, T. (1995) Molecular cloning of human cDNA for cathepsin K: novel cysteine proteinase predominantly expressed in bone. *Biochem. Biophys. Res. Commun.*, **206**, 89–96.
12. Li, Y.-P., Alexander, M., Wucherpfennig, A.L., Yelick, P., Chen, W., and Shashenko, P. (1995) Cloning and complete coding sequence of a novel human cathepsin expressed in giant cells of osteoclastomas. *J. Bone Miner. Res.*, **10**, 1197–1202.
13. Shi, G.P., Chapman, H.A., Bhairi, S.M., DeLeeuw, C., Reddy, V.Y., and Weiss, S.J. (1995) Molecular cloning of human cathepsin O, a novel endoproteinase and homologue of rabbit OC2. *FEBS Lett.*, **357**, 129–134.
14. Bossard, M.J., Tomaszek, T.T., Thompson, S.K., Amegadzies, B.Y., Hannings, C.R., Jones, C., Kurdyla, J.T., McNulty, D.E., Drake, F.H., Gowen, M., and Levy, M.A. (1996) Proteolytic activity of human osteoclast cathepsin K. Expression, purification, activation, and substrate identification. *J. Biol. Chem.*, **271**, 12517–12524.
15. Brömme, D., Okamoto, K., Wang, B.B., and Biroc, S. (1996b) Human cathepsin O2, a matrix protein-degrading cysteine protease expressed in osteoclasts. Functional expression of human cathepsin O2 in Spodoptera frugiperda and characterization of the enzyme. *J. Biol. Chem.*, **271**, 2126–2132.
16. Gelb, B.D., Shi, G.P., Chapman, H.A., and Desnick, R.J. (1996) Pycnodysostosis, a lysosomal disease caused by cathepsin K deficiency. *Science*, **273**, 1236–1238.
17. Gelb, B.D., Brömme, D., and Desnick, R.J. (2001) in *The Metabolic and Molecular Bases of Inherited Diseases* (eds C.R. Sriver, A.L. Beaudet, D. Valle, and W.C.S. Sly), McGraw-Hill. Inc., New York, St. Louis, MO, San Francisco, CA, pp. 3453–3468.
18. Everts, V., Aronson, D.C., and Beertsen, W. (1985a) Phagocytosis of bone collagen by osteoclasts in two cases of pycnodysostosis. *Calcif. Tissue. Int.*, **37**, 25–31.
19. Everts, V., Hou, W.S., Rialland, X., Tigchelaar, W., Saftig, P., Bromme, D., Gelb, B.D., and Beertsen, W. (2003) Cathepsin K deficiency in pycnodysostosis results in accumulation of non-digested phagocytosed collagen in fibroblasts. *Calcif. Tissue. Int.*, **73**, 380–386.
20. Boskey, A.L., Gelb, B.D., Pourmand, E., Kudrashov, V., Doty, S.B., Spevak, L., and Schaffler, M.B. (2009) Ablation of cathepsin k activity in the young mouse causes hypermineralization of long bone

and growth plates. *Calcif. Tissue. Int.*, **84**, 229–239.

21. Dodds, R.A., Cristiano, F., Feild, J., Kapadia, R., Liang, X., Debouck, C., Kola, I., and Gowen, M. (1998) Cathepsin K knockout mice develop osteopetrosis due to lack of full function in their osteoclasts. *Bone*, **32** (Suppl.), S164.

22. Hofbauer, L.C. and Heufelder, A.E. (1999) Osteopetrosis in cathepsin K-deficient mice. *Eur. J. Endocrinol.*, **140**, 376–377.

23. Saftig, P., Wehmeyer, O., Hunziker, E., Jones, S., Boyde, A., Rommerskirch, W., and von Figura, K. (1998) Impaired osteoclastic bone resorption leads to osteopetrosis in cathepsin K-deficient mice. *Proc. Natl. Acad. Sci. U.S.A.*, **95**, 13453–13458.

24. Kiviranta, R., Morko, J., Alatalo, S.L., NicAmhlaoibh, R., Risteli, J., Laitala-Leinonen, T., and Vuorio, E. (2005) Impaired bone resorption in cathepsin K-deficient mice is partially compensated for by enhanced osteoclastogenesis and increased expression of other proteases via an increased RANKL/OPG ratio. *Bone*, **36**, 159–172.

25. Pennypacker, B., Shea, M., Liu, Q., Masarachia, P., Saftig, P., Rodan, S., Rodan, G., and Kimmel, D. (2009) Bone density, strength, and formation in adult cathepsin K (-/-) mice. *Bone*, **44**, 199–207.

26. Li, C.Y., Jepsen, K.J., Majeska, R.J., Zhang, J., Ni, R., Gelb, B.D., and Schaffler, M.B. (2006) Mice lacking cathepsin K maintain bone remodeling but develop bone fragility despite high bone mass. *J. Bone Miner. Res.*, **21**, 865–875.

27. Fuller, K., Lawrence, K.M., Ross, J.L., Grabowska, U.B., Shiroo, M., Samuelsson, B., and Chambers, T.J. (2008) Cathepsin K inhibitors prevent matrix-derived growth factor degradation by human osteoclasts. *Bone*, **42**, 200–211.

28. Zhang, D., Leung, N., Weber, E., Saftig, P., and Bromme, D. (2011) The effect of cathepsin K deficiency on airway development and TGF-beta1 degradation. *Respir. Res.*, **12**, 72.

29. Mano, H., Yuasa, T., Kameda, T., Miyazawa, K., Nakamura, Y., Shiokawa, M., Mori, Y., Yamada, T., Miyata, K., Shindo, H. *et al.* (1996) Mammalian mature osteoclasts as estrogen target cells. *Biochem. Biophys. Res. Commun.*, **223**, 637–642.

30. Parikka, V., Lehenkari, P., Sassi, M.L., Halleen, J., Risteli, J., Harkonen, P., and Vaananen, H.K. (2001) Estrogen reduces the depth of resorption pits by disturbing the organic bone matrix degradation activity of mature osteoclasts. *Endocrinology*, **142**, 5371–5378.

31. Furuyama, N. and Fujisawa, Y. (2000) Regulation of collagenolytic cysteine protease synthesis by estrogen in osteoclasts. *Steroids*, **65**, 371–378.

32. Fujisaki, K., Tanabe, N., Suzuki, N., Kawato, T., Takeichi, O., Tsuzukibashi, O., Makimura, M., Ito, K., and Maeno, M. (2007) Receptor activator of NF-kappaB ligand induces the expression of carbonic anhydrase II, cathepsin K, and matrix metalloproteinase-9 in osteoclast precursor RAW264.7 cells. *Life Sci.*, **80**, 1311–1318.

33. Robinson, L.J., Yaroslavskiy, B.B., Griswold, R.D., Zadorozny, E.V., Guo, L., Tourkova, I.L., and Blair, H.C. (2009) Estrogen inhibits RANKL-stimulated osteoclastic differentiation of human monocytes through estrogen and RANKL-regulated interaction of estrogen receptor-alpha with BCAR1 and Traf6. *Exp. Cell Res.*, **315**, 1287–1301.

34. Dodds, R.A., James, I.E., Rieman, D., Ahern, R., Hwang, S.M., Connor, J.R., Thompson, S.D., Veber, D.F., Drake, F.H., Holmes, S. *et al.* (2001) Human osteoclast cathepsin K is processed intracellularly prior to attachment and bone resorption. *J. Bone Miner. Res.*, **16**, 478–486.

35. Xia, L., Kilb, J., Wex, H., Lipyansky, A., Breuil, V., Stein, L., Palmer, J.T., Dempster, D.W., and Brömme, D. (1999) Localization of rat cathepsin K in osteoclasts and resorption pits: inhibition of bone resorption cathepsin K-activity by peptidyl vinyl sulfones. *Biol. Chem.*, **380**, 679–687.

36. Hou, W.-S., Li, W., Keyszer, G., Weber, E., Levy, R., Klein, M.J., Gravallese,

E.M., Goldring, S.R., and Bromme, D. (2002) Comparison of cathepsins K and S expression within the rheumatoid and osteoarthritic synovium. *Arthritis Rheum.*, **46**, 663–674.

37. Rieman, D.J., McClung, H.A., Dodds, R.A., Hwang, S.M., Holmes, M.W., James, I.E., Drake, F.H., and Gowen, M. (2001) Biosynthesis and processing of cathepsin K in cultured human osteoclasts. *Bone*, **28**, 282–289.
38. Hou, W.S., Li, Z., Gordon, R.E., Chan, K., Klein, M.J., Levy, R., Keysser, M., Keyszer, G., and Bromme, D. (2001) Cathepsin K is a critical protease in synovial fibroblast-mediated collagen degradation. *Am. J. Pathol.*, **159**, 2167–2177.
39. Gay, C.V. and Weber, J.A. (2000) Regulation of differentiated osteoclasts. *Crit. Rev. Eukaryot. Gene Expr.*, **10**, 213–230.
40. Wilson, S.R., Peters, C., Saftig, P., and Bromme, D. (2009b) Cathepsin K activity-dependent regulation of osteoclast actin ring formation and bone resorption. *J. Biol. Chem.*, **284**, 2584–2592.
41. McGrath, M.E. (1999) The lysosomal cysteine proteases. *Annu. Rev. Biophys. Biomol. Struct.*, **28**, 181–204.
42. Lecaille, F., Choe, Y., Brandt, W., Li, Z., Craik, C.S., and Bromme, D. (2002) Selective inhibition of the collagenolytic activity of human cathepsin K by altering its S2 subsite specificity. *Biochemistry*, **41**, 8447–8454.
43. Lecaille, F., Chowdhury, S., Purisima, E., Bromme, D., and Lalmanach, G. (2007) The S2 subsites of cathepsins K and L and their contribution to collagen degradation. *Protein Sci.*, **16**, 662–670.
44. Brömme, D., Klaus, J.L., Okamoto, K., Rasnick, D., and Palmer, J.T. (1996a) Peptidyl vinyl sulphones: a new class of potent and selective cysteine protease inhibitors: S2P2 specificity of human cathepsin O2 in comparison with cathepsins S and L. *Biochem. J.*, **315**, 85–89.
45. Choe, Y., Leonetti, F., Greenbaum, D.C., Lecaille, F., Bogyo, M., Bromme, D., Ellman, J.A., and Craik, C.S. (2006) Substrate profiling of cysteine proteases using a combinatorial peptide library identifies functionally unique specificities. *J. Biol. Chem.*, **281**, 12824–12832.
46. Alves, M.F., Puzer, L., Cotrin, S.S., Juliano, M.A., Juliano, L., Bromme, D., and Carmona, A.K. (2003) S3 to S3' subsite specificity of recombinant human cathepsin K and development of selective internally quenched fluorescent substrates. *Biochem. J.*, **373**, 981–986.
47. Garnero, P., Borel, O., Byrjalsen, I., Ferreras, M., Drake, F.H., McQueney, M.S., Foged, N.T., Delmas, P.D., and Delaisse, J.M. (1998) The collagenolytic activity of cathepsin K is unique among mammalian proteinases. *J. Biol. Chem.*, **273**, 32347–32352.
48. Kafienah, W., Bromme, D., Buttle, D.J., Croucher, L.J., and Hollander, A.P. (1998) Human cathepsin K cleaves native type I and II collagens at the N-terminal end of the triple helix. *Biochem. J.*, **331**, 727–732.
49. Sassi, M.L., Eriksen, H., Risteli, L., Niemi, S., Mansell, J., Gowen, M., and Risteli, J. (2000) Immunochemical characterization of assay for carboxyterminal telopeptide of human type I collagen: loss of antigenicity by treatment with cathepsin K. *Bone*, **26**, 367–373.
50. Hou, W.S., Li, Z., Buttner, F.H., Bartnik, E., and Bromme, D. (2003) Cleavage site specificity of cathepsin K toward cartilage proteoglycans and protease complex formation. *Biol. Chem.*, **384**, 891–897.
51. Podgorski, I., Linebaugh, B.E., Koblinski, J.E., Rudy, D.L., Herroon, M.K., Olive, M.B., and Sloane, B.F. (2009) Bone marrow-derived cathepsin K cleaves SPARC in bone metastasis. *Am. J. Pathol.*, **175**, 1255–1269.
52. Lutgens, E., Lutgens, S.P., Faber, B.C., Heeneman, S., Gijbels, M.M., de Winther, M.P., Frederik, P., van der Made, I., Daugherty, A., Sijbers, A.M. *et al.* (2006) Disruption of the cathepsin K gene reduces atherosclerosis progression and induces plaque fibrosis but accelerates macrophage foam cell formation. *Circulation*, **113**, 98–107.
53. Lutgens, S.P., Cleutjens, K.B., Daemen, M.J., and Heeneman, S. (2007) Cathepsin cysteine proteases in cardiovascular disease. *FASEB J.*, **21**, 3029–3041.

54. Samokhin, A.O., Wong, A., Saftig, P., and Bromme, D. (2008) Role of cathepsin K in structural changes in brachiocephalic artery during progression of atherosclerosis in apoE-deficient mice. *Atherosclerosis*, **200**, 58–68.
55. Li, Z., Hou, W.S., and Bromme, D. (2000) Collagenolytic activity of cathepsin K is specifically modulated by cartilage-resident chondroitin sulfates. *Biochemistry*, **39**, 529–536.
56. Li, Z., Hou, W.S., Escalante-Torres, C.R., Gelb, B.D., and Bromme, D. (2002) Collagenase activity of cathepsin K depends on complex formation with chondroitin sulfate. *J. Biol. Chem.*, **277**, 28669–28676.
57. Li, Z., Kienetz, M., Cherney, M.M., James, M.N., and Bromme, D. (2008) The crystal and molecular structures of a cathepsin K:chondroitin sulfate complex. *J. Mol. Biol.*, **383**, 78–91.
58. Cherney, M.M., Lecaille, F., Kienitz, M., Nalleseth, F., James, M.N.G., and Brömme, D. (2011) Activity and structure of cathepsin K variant M5 in complex with chondroitin sulfate. *J. Biol. Chem.*, **18**, 8988–8998.
59. Hou, W.-S., Brömme, D., Zhao, Y., Mehler, E., Dushey, C., Weinstein, H., Miranda, C.S., Fraga, C., Greig, F., Carey, J. *et al.* (1999) Cathepsin K: characterization of novel Mutations in the pro and mature polypeptide regions causing pycnodysostosis. *J. Clin. Invest.*, **103**, 731–738.
60. Li, Z., Yasuda, Y., Li, W., Bogyo, M., Katz, N., Gordon, R.E., Fields, G.B., and Bromme, D. (2004) Regulation of collagenase activities of human cathepsins by glycosaminoglycans. *J. Biol. Chem.*, **279**, 5470–5479.
61. Yasuda, Y., Li, Z., Greenbaum, D., Bogyo, M., Weber, E., and Bromme, D. (2004) Cathepsin V, a novel and potent elastolytic activity expressed in activated macrophages. *J. Biol. Chem.*, **279**, 36761–36770.
62. Selent, J., Kaleta, J., Li, Z., Lalmanach, G., and Bromme, D. (2007) Selective inhibition of the collagenase activity of cathepsin K. *J. Biol. Chem.*, **282**, 16492–16501.
63. Clarke, L.A. (2008) The mucopolysaccharidoses: a success of molecular medicine. *Expert Rev. Mol. Med.*, **10**, e1.
64. Neufeld, E.F. and Muenzer, J. (2001) *The Mucopolysaccharidoses*, McGraw-Hill, New York, St. Louis, MO, San Francisco, CA, Auckland, etc.
65. Abreu, S., Hayden, J., Berthold, P., Shapiro, I.M., Decker, S., Patterson, D., and Haskins, M. (1995) Growth plate pathology in feline mucopolysaccharidosis VI. *Calcif. Tissue. Int.*, **57**, 185–190.
66. McClure, J., Smith, P.S., Sorby-Adams, G., and Hopwood, J. (1986) The histological and ultrastructural features of the epiphyseal plate in Morquio type A syndrome (mucopolysaccharidosis type IVA). *Pathology*, **18**, 217–221.
67. Wilson, S., Hashamiyan, S., Clarke, L., Saftig, P., Mort, J., Dejica, V.M., and Bromme, D. (2009a) Glycosaminoglycan-mediated loss of cathepsin K collagenolytic activity in MPS I contributes to osteoclast and growth plate abnormalities. *Am. J. Pathol.*, **175**, 2053–2062.
68. Wilson, S. and Bromme, D. (2010) Potential role of cathepsin K in the pathophysiology of mucopolysaccharidoses. *J. Pediatr. Rehabil. Med.*, **3**, 139–146.
69. Missbach, M., Altmann, E., Betschart, C., Buh, I.T., Gamse, R., Gasser, J.A., Green, J.R., Ishihara, H., Jerome, C., Kometani, M. *et al.* (2005) AAE581, a potent and highly specific cathepsin K inhibitor, prevents bone resorption after oral treatment in rat and monkey. *J. Bone Miner. Res.*, **20**, S251.
70. Yamashita, D.S., Marquis, R.W., Xie, R., Nidamarthy, S.D., Oh, H.J., Jeong, J.U., Erhard, K.F., Ward, K.W., Roethke, T.J., Smith, B.R. *et al.* (2006) Structure activity relationships of 5-, 6-, and 7-methyl-substituted azepan-3-one cathepsin K inhibitors. *J. Med. Chem.*, **49**, 1597–1612.
71. Kumar, S., Dare, L., Vasko-Moser, J.A., James, I.E., Blake, S.M., Rickard, D.J., Hwang, S.M., Tomaszek, T., Yamashita, D.S., Marquis, R.W. *et al.* (2007) A highly potent inhibitor of cathepsin K (relacatib) reduces biomarkers of bone

resorption both in vitro and in an acute model of elevated bone turnover in vivo in monkeys. *Bone*, **40**, 122–131.

72. Gauthier, J.Y., Chauret, N., Cromlish, W., Desmarais, S., Duong le, T., Falgueyret, J.P., Kimmel, D.B., Lamontagne, S., Leger, S., LeRiche, T. *et al.* (2008) The discovery of odanacatib (MK-0822), a selective inhibitor of cathepsin K. *Bioorg. Med. Chem. Lett.*, **18**, 923–928.
73. Black, W.C. and Percival, M.D. (2006) The consequences of lysosomotropism on the design of selective cathepsin K inhibitors. *Chembiochem*, **7**, 1525–1535.
74. Adami, S., Supronik, J., Hala, T., Brown, J.P., Garnero, P., Haemmerle, S. *et al.* (2006) Effect of one year treatment with the cathepsin-K inhibitor, balicatib,on bone mineral density (BMD) in postmenopausal women with osteopenia/osteoporosis (abstract). *J. Bone Miner. Res.*, **21**, S24.
75. Runger, T.M., Adami, S., Benhamou, C.L., Czerwinski, E., Farrerons, J., Kendler, D.L., Mindeholm, L., Realdi, G., Roux, C., and Smith, V. (2012) Morphea-like skin reactions in patients treated with the cathepsin K inhibitor balicatib. *J.Invest. Dermatol.* **66**, 89–96.
76. Eastell, R., Nagase, S., Ohyama, M., Small, M., Sawyer, J., Boonen, S., Spector, T., Kuwayama, T., and Deacon, S. (2011) Safety and efficacy of the Cathepsin K inhibitor, ONO-5334, in postmenopausal osteoporosis – the OCEAN study. *J. Bone Miner. Res.* **26**, 1303–1312.
77. Jerome, C., Missbach, M., and Gamse, R. (2011) Balicatib, a cathepsin K inhibitor, stimulates periosteal bone formation in monkeys. *Osteoporos Int.* **22**, 3001–3011.
78. Stoch, S.A., Zajic, S., Stone, J., Miller, D.L., Van Dyck, K., Gutierrez, M.J., De Decker, M., Liu, L., Liu, Q., Scott, B.B. *et al.* (2009) Effect of the cathepsin K inhibitor odanacatib on bone resorption biomarkers in healthy postmenopausal women: two double-blind, randomized, placebo-controlled phase I studies. *Clin. Pharmacol. Ther.*, **86**, 175–182.
79. Bone, H.G., McClung, M.R., Roux, C., Recker, R.R., Eisman, J.A., Verbruggen, N., Hustad, C.M., DaSilva, C., Santora, A.C., and Ince, B.A. (2010) Odanacatib, a cathepsin-K inhibitor for osteoporosis: a two-year study in postmenopausal women with low bone density. *J. Bone Miner. Res.*, **25**, 937–947.
80. Eisman, J.A., Bone, H.G., Hosking, D.J., McClung, M.R., Reid, I.R., Rizzoli, R., Resch, H., Verbruggen, N., Hustad, C.M., DaSilva, C. *et al.* (2011) Odanacatib in the treatment of postmenopausal women with low bone mineral density: three-year continued therapy and resolution of effect. *J. Bone Miner. Res.*, **26**, 242–251.
81. Bauer, D.C. (2011) Discontinuation of odanacatib and other osteoporosis treatments: here today and gone tomorrow? *J. Bone Miner. Res.*, **26**, 239–241.
82. Jensen, A.B., Wynne, C., Ramirez, G., He, W., Song, Y., Berd, Y., Wang, H., Mehta, A., and Lombardi, A. (2010) The cathepsin K inhibitor odanacatib suppresses bone resorption in women with breast cancer and established bone metastases: results of a 4-week, double-blind, randomized, controlled trial. *Clin. Breast Cancer*, **10**, 452–458.
83. Desmarais, S., Black, W.C., Oballa, R., Lamontagne, S., Riendeau, D., Tawa, P., Duong le, T., Pickarski, M., and Percival, M.D. (2008) Effect of cathepsin k inhibitor basicity on in vivo off-target activities. *Mol. Pharmacol.*, **73**, 147–156.
84. Falgueyret, J.P., Desmarais, S., Oballa, R., Black, W.C., Cromlish, W., Khougaz, K., Lamontagne, S., Masse, F., Riendeau, D., Toulmond, S., and Percival, M.D. (2005) Lysosomotropism of basic cathepsin K inhibitors contributes to increased cellular potencies against off-target cathepsins and reduced functional selectivity. *J. Med. Chem.*, **48**, 7535–7543.
85. Fuller, K., Lindstrom, E., Edlund, M., Henderson, I., Grabowska, U., Szewczyk, K.A., Moss, R., Samuelsson, B., and Chambers, T.J. (2010) The resorptive apparatus of osteoclasts supports lysosomotropism and increases potency of basic versus non-basic

inhibitors of cathepsin K. *Bone*, **46**, 1400–1407.

86. Leung, P., Pickarski, M., Zhuo, Y., Masarachia, P.J., and Duong, L.T. (2011) The effects of the cathepsin K inhibitor odanacatib on osteoclastic bone resorption and vesicular trafficking. *Bone*, **49**, 623–635.
87. Buhling, F., Rocken, C., Brasch, F., Hartig, R., Yasuda, Y., Saftig, P., Bromme, D., and Welte, T. (2004) Pivotal role of cathepsin K in lung fibrosis. *Am. J. Pathol.*, **164**, 2203–2216.
88. Srivastava, M., Steinwede, K., Kiviranta, R., Morko, J., Hoymann, H.G., Langer, F., Buhling, F., Welte, T., and Maus, U.A. (2008) Overexpression of cathepsin K in mice decreases collagen deposition and lung resistance in response to bleomycin-induced pulmonary fibrosis. *Respir. Res.*, **9**, 54.
89. Halwani, R., Al-Muhsen, S., Al-Jahdali, H., and Hamid, Q. (2011) Role of transforming growth factor-beta in airway remodeling in asthma. *Am. J. Respir. Cell. Mol. Biol.*, **44**, 127–133.
90. Podgorski, I. (2009) Future of anti-cathepsin K drugs: dual therapy for skeletal disease and atherosclerosis? *Future Med. Chem.*, **1**, 21–34.
91. Dejica, V.M., Mort, J.S., Laverty, S., Percival, M.D., Antoniou, J., Zukor, D.J., and Poole, A.R. (2008) Cleavage of type II collagen by cathepsin K in human osteoarthritic cartilage. *Am. J. Pathol.*, **173**, 161–169.
92. McDougall, J.J., Schuelert, N., and Bowyer, J. (2010) Cathepsin K inhibition reduces CTXII levels and joint pain in the guinea pig model of spontaneous osteoarthritis. *Osteoarthr. Cartil.*, **18**, 1355–1357.

5
Type-II Transmembrane Serine Proteases: Physiological Functions and Pathological Aspects

Gregory S. Miller, Gina L. Zoratti, and Karin List

5.1
Introduction

Over the past decade, type II transmembrane serine proteases (TTSPs) have emerged as an important component of the human degradome, garnering significant attention for their roles in epithelial biology, regulation of metabolic homeostasis, and deregulation in cancer [1–4]. TTSPs were identified as a unique group of serine proteases (SPs) that localize to the plasma membrane, thus having potential roles in cell surface proteolysis [2]. Genome analysis has revealed 17 mammalian TTSPs conserved between mice and humans, with only two identifiable *Drosophila* orthologs, indicating considerable expansion of this family of proteins during vertebrate evolution [1, 5]. Annotation of protein homology has resulted in the distinction of four TTSP subfamilies; however, this grouping does not always correlate with significant overlap in biological function. TTSPs encompass diverse and divergent areas of physiology, with canonical roles in digestion and blood pressure regulation, as well as emerging roles in epidermal development and homeostasis, iron metabolism, hearing, and oncogenesis, and many have characterized mutant phenotypes in humans and/or model organisms (Table 5.1) [1, 6, 7].

5.2
Functional/Structural Properties of TTSPs

TTSPs are all synthesized as single-chain zymogens and share a conserved N-terminal transmembrane domain, a "stem region" that is composed of a variable number of domains that belong to one of six conserved motifs, and a C-terminal SP domain (Figure 5.1) [29]. Depending on the organization of the stem region, TTSPs are classified into the HAT/DESC (*h*uman *a*irway *t*rypsin-like protease/*d*ifferentially *e*xpressed in *s*quamous cell *c*arcinoma) subfamily, the hepsin/TMPRSS (*trans*membrane *p*rotease/*s*erine) subfamily, the matriptase subfamily, or the corin subfamily (Figure 5.1). The HAT/DESC subfamily has the simplest stem configuration, containing a single sea urchin sperm protein, enteropeptidase,

Matrix Proteases in Health and Disease, First Edition. Edited by Niels Behrendt.

Table 5.1 Type II membrane serine proteases in development and disease.

TTSP	Physiological function	Human mutations/phenotypes	Mutant mouse models/phenotypes	Zebrafish model/ phenotypes	Physiological substrates
Hepsin	Cochlear development	NR	Transgenic mice/overexpression of hepsin in prostate epithelium/disorganized basement membrane and increased metastasis [8]. Null mice/abnormal cochlear development, thickening of the tectorial membrane, thyroxine deficiency [9]	NR	NR
TMPRSS2	NR	NR	Null mice/no apparent phenotype [10]	NR	NR
TMPRSS3	Cochlear development	Missense and deletion mutations/nonsyndromic and autosomal recessive deafness [11]	NR	NR	NR
Enteropeptidase	Intestinal digestion	Missense and point mutations/failure to thrive, diarrhea, congenital enteropeptidase deficiency [12]	NR	NR	Intestinal hydrolases
Matriptase	Epidermal and thymic development epithelial homeostasis	Missense, point and deletion mutations/reduced prostasin activation and profilaggrin processing. Reduced epidermal barrier function, ichtyosis, hypotrichosis, abnormal permanent teeth, follicular atrophoderma [13]	Null, hypomorphic, and conditional knockout mice/reduction or loss of prostasin activation and profilaggrin processing, impaired epidermal barrier function, impaired intestinal barrier with enlarged colon and decreased mucinogen production, abnormal tooth enamel, salivary gland, thymic hypoplasia, ichthyosis, and hypotrichosis [7, 14–17]. Transgenic mice/spontaneous squamous cell carcinomas which is rescued by epidermal C-Met deficiency [18, 19].	NR	Pro-Prostasin, Pro-HGF

Matriptase-2	Inhibitor of hepcidin to regulate body iron	Insertion, deletion, and point and missense mutations/increased hepcidin levels lead to iron deficiency anemia [20]	Null and hypomorphic mice/increased hepcidin levels lead to iron deficiency anemia [21]	Mutation leads to iron deficiency anemia [22]	Hemojuvelin
Corin	Regulation of blood pressure and volume	Point mutations/impaired zymogen activation, increased levels of pro-ANP and pro-BNP, hypertension, cardiac hypertrophy, [23, 24]	Null mice/no pro-ANP or pro-BNP processing, increased body weight, abnormal hair pigmentation, cardiac hypertrophy, hypertension, and sodium retention that correlates with proteinuric kidney disease [25–27]	NR	Pro-ANP, pro-BNP

NR: none reported.
Note: the table includes references [10, 28], among others cited within the text.

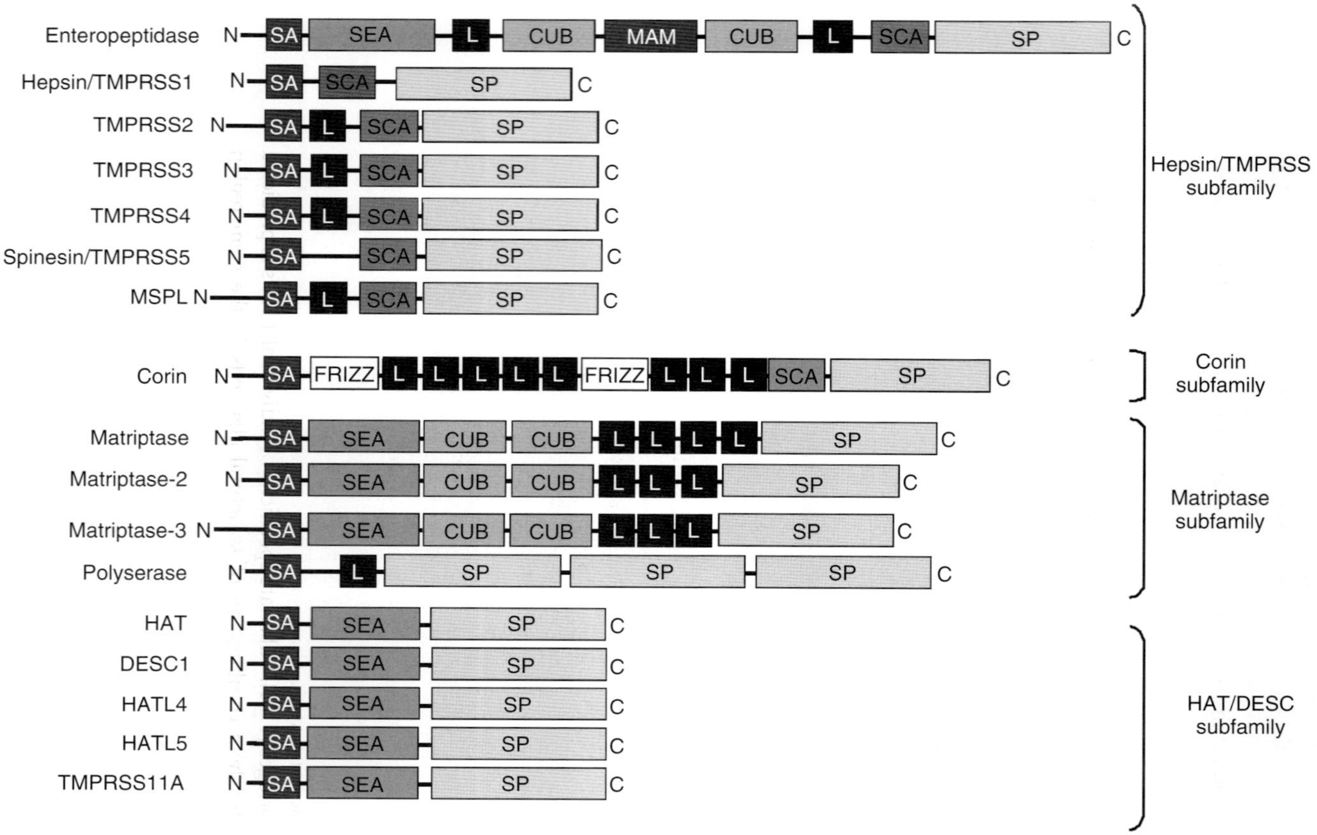

Figure 5.1 Structural organization of human type II membrane serine proteases. TTSPs are grouped into four subfamilies according to the predicted domain architecture of the stem region and protease domain(s), including the hepsin/TMPRSS subfamily, the corin subfamily, the matriptase subfamily, and the HAT/DESC subfamily. Domain/structural abbreviations: SA, signal anchor; SEA, SEA domain; L, LDLA domain; CUB, CUB domain; MAM, MAM domain; SCA, group A scavenger domain; FRIZZ, frizzled domain; and SP, serine protease domain.

agrin (SEA) domain. The members of this subfamily include HAT, DESC, TMPRSS11A, and HAT-like (HATL) 4 and 5. All the genes encoding HAT/DESC transcripts occur within a single gene cluster in both mice and humans, suggesting that gene duplication events gave rise to the various individual loci [30]. The hepsin/TMPRSS subfamily is composed of hepsin, TMPRSS2-4, spinesin, MSPL (*m*osaic *s*erine *p*rotease *l*arge-form), and enteropeptidase. Hepsin and spinesin contain a single group A scavenger receptor domain preceding the SP domain, while TMPRSS2-4 and MSPL contain an LDLA (low density lipoprotein receptor class A) domain toward the N-terminal of the scavenger domain. Enteropeptidase contains seven motifs before SP, which from the N to C terminus include an SEA domain, an LDLA domain, a CUB (Cls/Clr, urchin embryonic growth factor, bone morphogenic protein-1) domain, an MAM (meprin/A5 antigen/receptor protein phosphatase mu) domain, a second CUB domain, a second LDLA domain, and the scavenger domain. The matriptase subfamily includes matriptase, matriptases-2 and 3, and polyserase-1. Matriptase-2 and 3 have identical stem region organization, beginning with one SEA domain, two CUB domains, and three LDLA domains before the protease domain. The founding member matriptase has one additional LDLA domain before the C-terminal SP. Polyserase-1 is unique among TTSPs in that it contains a single LDLA domain followed by three SP domains, with the C-terminal-most domain being catalytically inactive, and is included in the matriptase subfamily because of the high sequence similarity of the protease domains [31]. The corin subfamily solely consists of the corin protease, which contains a frizzled domain, five LDLA domains, a second frizzled domain, three additional LDLA domains, and a scavenger domain followed by the SP domain.

In order to be catalytically active, TTSPs require proteolytic cleavage at an Arg or Lys residue within a highly conserved activation domain at the end of the stem region and preceding the SP domain. Following activation cleavage, the protease domain remains tethered to the plasma membrane via a disulfide bridge that covalently links it with the stem region; however, shed forms of members including enteropeptidase, corin, and matriptase have been isolated *in vivo* [2, 32, 33]. Unlike the majority of secreted SPs, several TTSPs are capable of efficient autoactivation *in vitro*, raising the possibility that they may act as initiators of proteolytic signaling cascades at the cellular surface [34–37]. Biochemical analysis has revealed a preference of TTSP proteolysis for Arg or Lys residues, suggesting that TTSPs that do not autoactivate may be activated by other family members *in vivo* [1, 38]. All TTSPs have a highly conserved chymotrypsin S1 SP domain that contains a catalytic amino acid triad (histidine, aspartic acid, and serine), which performs peptide bond hydrolysis, and six cysteine residues, which are predicted to form disulfide bridges [1, 39]. Substrate specificity of individual TTSPs is predicted to be coordinated by a combination of structural elements in the stem region and protease domain and may be affected by interactions with local activators and inhibitors. Unique expression patterns that restrict the availability of substrate and regulatory molecules may be of extreme importance to the biologically relevant proteolytic specificity, as biochemical analysis has shown that TTSPs have varying degrees of substrate promiscuity *in vitro* [40]. Inhibitors of TTSPs include members of the serpin family of SP inhibitors and the Kunitz-type inhibitors hepatocyte growth factor activator inhibitor-1 and 2 (HAI-1, HAI-2). Efficient initial activation is likely regulated by the association with inhibitors *in vivo* and may require both uncharacterized cofactors and specific extracellular conditions involving reactive oxygen species (ROS) and pH [41].

5.3 Physiology and Pathobiology

5.3.1 Hepsin/TMPRSS Subfamily

The TTSP enteropeptidase was discovered in the early 1900s in the laboratory of Ivan Pavlov, the renowned physiologist who was seminal to our understanding of gastrointestinal biology. Contemporary research has yielded a wealth of structural and functional information that has established this protease as a principal regulator of vertebrate food digestion. Enteropeptidase mRNA is expressed in the enterocytes and goblet cells that comprise the epithelial lining of the duodenum, and the tethered protein localizes to the apical surface where nutrient absorption occurs [42, 43]. Proenteropeptidase is believed to be activated by both trypsin and the SP duodenase, which is also synthesized in and secreted from the duodenal epithelium [44, 45]. Activated enteropeptidase is shed into the intestinal lumen where it converts pancreatic trypsinogen into active trypsin, resulting in the subsequent activation of several zymogens essential for food digestion and nutrient absorption. The delayed activation of trypsinogen at the duodenal lumen serves to protect pancreatic ducts from proteolytic damage, which under conditions of duodenopancreatic reflux is the primary cause of acute pancreatitis [46–48]. Accordingly, enteropeptidase exhibits remarkable specificity for trypsin activation that is engendered by the SP domain and the stem region and occurs at a highly conserved peptide (DDDDK) [49]. While excess trypsin activity can be devastating outside the small intestine, a reduction in intestinal enteropeptidase activity is the cause of congenital enteropeptidase deficiency (CED) and acquired enteropeptidase deficiency [50]. CED is a rare genetic disease caused by compound heterozygous mutations that result in proteolytically inactive constructs, while the acquired deficiency is thought to be caused by injury to the intestinal mucosa [12, 51]. Severe enzyme deficiency is primarily associated with CED and results in chronic diarrhea, edema, and growth retardation due to severe malnutrition [52]. Both pathologies respond well to enzyme replacement therapy, and patients with CED who are treated as infants do not require additional enzyme supplementation when they are adults, indicating that an initial amount of active trypsin may be sufficient for perpetuating trypsinogen activation in adults, following from trypsin's ability to activate its own zymogen [12, 53, 54].

Hepsin/TMPRSS1 was the first SP characterized to contain a transmembrane domain and was named based on its original identification in hepatocytes [55]. Subsequent analysis of hepsin mRNA has revealed a broad range of expression including the thymus, thyroid, pituitary gland, and pancreas [56, 57]. Early studies of hepsin proposed a role in hepatocyte development and blood clotting; however, hepsin-deficient mice fail to display any aberration in liver or coagulation phenotypes [58–61]. Recent analysis of hepsin-deficient mice has indicated that these animals exhibit a profound loss of hearing, evidenced by extremely elevated auditory-evoked brainstem response (ABR) thresholds [9]. The cochleae of hepsin-deficient mice display a deformed and hypertrophic tectorial membrane, as well as defects of nerve fiber compaction in the spiral ganglia and reduced expression of myelination genes. In addition, hepsin-deficient mice produce significantly less thyroxine hormone than wild-type littermates, which may be the cause of a concomitant reduction in the expression of large conductance voltage- and Ca^{2+}-activated potassium channels [9]. Interestingly, replacement therapy of thyroxine in hepsin-deficient mice does not ameliorate hearing loss or tectorial membrane malformation [62]. This raises the possibility that hepsin may regulate thyroxine responsiveness during development of the inner ear. Two other hepsin subfamily members, TMPRSS3 and TMPRSS5, are expressed in the spiral ganglia of the cochlea, with TMPRSS3 expression also found in the organ of Corti [63]. Several mutations in TMPRSS3 have been found in patients with nonsyndromic recessive hearing loss, which is believed to result from reduced TMPRSS3 proteolytic activity due to inefficient zymogen activation [11, 64]. TMPRSS3 can undergo autoactivation *in vitro;* however, it is unknown whether or not autoregulation or proteolytic cross activation between hepsin, TMPRSS3, and TMPRSS5 has a significant role in inner ear development or neuronal signaling [65]. Additional analysis should seek to determine the role of TMPRSS3 and TMPRSS5 in cochlear development and assess whether or not mutations in individual family members result in disparate or overlapping phenotypes.

5.3.2 Corin Subfamily

Corin is the only member of the eponymous subfamily, and it is unique among TTSPs in the occurrence of two frizzled domains that appear to be indispensible to its catalytic activity [66, 67]. Corin mRNA expression is highest in the myocardium of humans and rodents, with substantially lower expression found in the kidneys [68–70]. Corin is known to process two natriuretic peptides, atrial natriuretic peptide (ANP) and B-type natriuretic peptide (BNP), which are involved in regulating blood pressure, blood volume, and sodium homeostasis [21, 71]. Pro-ANP produced in cardiac myocytes is cleaved by corin during periods of stress, blood pressure spikes, and increased blood volume, allowing ANP to bind with receptors in the kidney and vasculature, which promotes salt excretion and vasodilation [23]. Corin does not require transmembrane tethering for its ability to activate pro-ANP, and a soluble form is found in human plasma [67]. Mutational analysis has revealed that both frizzled domains and the first four LDL repeats are required for efficient substrate recognition and cleavage [72, 73]. A human corin allele with two missense mutations in the second frizzled domain has been associated with chronic hypertension, and owing to its higher frequency in African Americans, it is proposed to contribute to an increased risk of heart disease in this population [66]. The mutations present in this corin variant significantly reduce zymogen activation, resulting in reduced processing of both pro-ANP and pro-BNP in human cell culture experiments [73].

As reduced corin activity results in the accumulation of inactive natriuretic peptides, plasma levels of corin, pro-ANP, and pro-BNP have become useful biomarkers for heart disease. Low corin levels and high levels of the proforms of ANP and BNP correlate with increased severity in patients with heart failure, and decreased corin zymogen activation may contribute to both disease onset and progression [23, 24, 69, 74]. Corin knockout mice develop to adulthood but display elevated levels of pro-ANP without any detectable amount of the biologically activated peptide [25]. These animals exhibit spontaneous hypertension that is exacerbated by a high salt diet, cardiac hypertrophy, and decreased cardiac function that manifests during senescence [25]. A separate study has revealed that corin deficiency may also contribute to a sodium retention phenotype that is observed in human proteinuric kidney disease. In this study, the loss of active ANP was correlated with a build-up of activated epithelial sodium channels (ENaCs) and other downstream effectors of sodium absorption that are hallmarks of disease pathology [26]. Abnormally increased corin activity has also been correlated with cardiac pathology, and precise regulation appears necessary for normal physiology [75]. There is strong impetus to delineate the regulatory mechanisms of corin activation as well as the downstream effectors of natriuretic peptide signaling. Recently, corin has been shown to modify pigmentation in mouse hair follicles [27], highlighting the necessity to explore the effects of corin expression in diverse tissues and identify novel substrates of this protease.

5.3.3
Matriptase Subfamily

The matriptase subfamily is the most widely studied group of TTSPs, with the founding member having a well-characterized role in epithelial development and carcinogenesis and the emerging role of matriptase-2 in iron metabolism. Matriptase expression localizes to the epithelial component of a wide variety of organs, as well as in monocytes and macrophages of the immune system [18, 76–78]. Expression of matriptase is dynamic during mouse development, with early embryonic expression detected in the olfactory placode, oral cavity, pharyngeal foregut, the midgut/hindgut region, and apical ectodermal regions of the limbs, followed by strong expression in the follicular and interfollicular epidermis and more subtle expression in the placenta [18, 78]. Postnatal expression is primarily in the epithelium of the gastrointestinal tract, transitional layer of the epidermis, hair follicles, and kidney tubules, with lower expression in the respiratory epithelium, the urinary tract, and the reproductive system of males and females [18, 79, 80]. Matriptase is capable of autoactivation *in vitro*, and its activation in primary human keratinocytes can be induced under slightly acidic conditions (pH 6.0) or upon exposure to ROS [41, 80–82]. Following from the ability to autoactivate, it has been proposed that matriptase is the initiator of key proteolytic signaling cascades that control epithelial development and differentiation. Matriptase function *in vivo* has been studied most extensively in the epidermis, where it has been shown to have pleiotropic effects on differentiation, homeostasis, and pathology.

The model for matriptase control of epidermal development and integrity has evolved to include multiple proteases and inhibitors, as well as the pH and calcium gradient present in the extracellular matrix (ECM) of skin [83]. Mutational loss of matriptase in mice results in a severely compromised epidermal barrier, causing death during the early neonatal period from terminal dehydration [14]. The epidermis of matriptase-deficient mice displays a variety of abnormalities; which include a reduction in lipid-matrix-associated vesicular bodies, disruption of junction complexes, and the loss of proteolytically processed profilaggrin [14–16, 84]. The processing of profilaggrin is crucial for keratinocyte differentiation and may involve proteolytic events that affect sodium and calcium influx through the plasma membrane [83]. An extracellular protease cascade begins with matriptase autoactivation at the plasma membrane, which may be contingent on its exposure to an acidic pH at the transition between the stratum granulosum and stratum corneum [41]. Matriptase activity is regulated by HAI-1, which binds to and inhibits free matriptase in the plasma membrane [85, 86]. Active matriptase cleaves the pro form of the glycosylphosphatidylinositol (GPI)-linked SP prostasin *in vivo*, resulting in its catalytic activation. This proteolytic step is critical to normal epidermal development, as matriptase-deficient mice and prostasin-deficient mice display identical terminal phenotypes [16, 87]. Prostasin and matriptase can activate ENaCs, which are required for normal epidermal development and may affect profilaggrin processing [83, 88, 89]. Importantly, processed filaggrin monomers are required for differentiation of the stratum corneum and are not generated when matriptase or prostasin are deficient [90].

In humans, a homozygous point mutation in the matriptase SP domain is the cause of autosomal recessive ichthyosis with hypotrichosis (ARIH), a congenital disorder characterized by thickened and scaling skin and sparse, fragile hair [13, 17]. Individuals with ARIH express a matriptase isoform that contains a missense mutation that results in a G827R substitution in the SP domain that lowers catalytic activity more than 100-fold and results in reduced prostasin activation and impaired profilaggrin processing in both mice and human epidermis [17, 91]. Furthermore, hypomorphic matriptase mice display hyperproliferation of basal keratinocytes and impaired desquamation in the stratum corneum and morphologically phenocopy human ARIH [17]. Excessive matriptase activity is associated with developmental pathology, as highlighted in mice that are deficient in the inhibitor HAI-1. In many adult and embryonic tissues, HAI-1 colocalizes with matriptase and may be involved in the initial activation of matriptase as well as be its primary inhibitor [78, 92, 93]. HAI-1-null mice do not survive embryogenesis because of loss of undifferentiated chorionic trophoblasts and failure to form the placental labyrinth [78]. Matriptase/HAI-1 double deficiency restores the cellular architecture of the placenta and is phenotypically indistinguishable from matriptase deficiency, indicating that HAI-1 inhibition of matriptase is required for murine placental development [78].

Chimeric HAI-1-deficient mice that maintain sufficient trophoblastic expression to survive to term have been generated; however, around day 7 these mice develop severe ichthyosis, and by day 16 they perish [77, 94, 95]. Histological examination of these HAI-1-deficient animals reveals hyperkeratosis of the forestomach, hyperkeratosis and acanthosis of the epidermis, and anomalous cuticular separation resulting in hypotrichosis. The TTSP inhibitor HAI-2 is also required for embryonic survival. HAI-2-null mice display defective neural tube closure and abnormal placental development [96]. HAI-2/matriptase double deficiency restores embryonic survival and placental development; however, it does not fully restore neural tube defects, indicating that proteases other than matriptase are regulated by HAI-2 [96]. Deregulated proteolytic activity involving matriptase has also been implicated in Netherton's syndrome, which is characterized by detachment and loss of the stratum corneum, exposing the living layers of the epidermis [97, 98]. The pathology of this syndrome is caused by the loss of the SP inhibitor LEKTI, which regulates the activity of kallikrein proteases at the granular and transitional layer of the epidermis [99]. The result of LEKI deficiency is unregulated kallikrein proteolysis that degrades corneodesmosomes in the lower layers of the epidermis, causing improper desquamation, which results in ectopic detachment of the stratum corneum. LEKTI-deficient mice display the key phenotypes associated with Netherton's syndrome, including aberrant proteolytic activity in the lower layers of the epidermis, which are all reversed when matriptase is deficient [100]. Taken together with the observation that matriptase can activate prokallikreins *in vitro*, this indicates that matriptase initiates ectopic kallikrein activity in the absence of LEKT1. Thus, precise regulation of matriptase as well as its downstream proteolytic targets is essential for the development of the epidermal strata and their barrier function.

Recent evidence suggests that matriptase is required not only for proper development of the epidermis but also for homeostasis of epithelial tissue in several disparate organs. Matriptase is highly expressed in the developing thymus and is required for the survival of nascent lymphocytes [14, 101]. T lymphocytes in matriptase-deficient mice display a fivefold increase in their apoptotic index, with the loss of ~50% of developing thymocytes [14]. The conditional deletion of matriptase from the gastrointestinal tract results in extreme enlargement of the colon, persistent diarrhea, growth retardation, and early mortality [15]. Examination of the gastric mucosa shows a dramatic disruption of colonic architecture, including a loss of mucinogen-producing goblet cells, edema, and pervasive inflammation [15]. Ablation of matriptase in the salivary gland results in the nearly complete loss of saliva production, however, without any apparent changes in glandular architecture. Matriptase is also reported to promote growth and morphogenesis of mammary epithelium in cell culture experiments [102]. Continual expression of matriptase is required for the maintenance of epithelial integrity in mice as adult mice with acute matriptase ablation display profound weight loss, and subsequently perish [15]. In these animals, it was shown that matriptase is required for the maintenance of intercellular tight junctions in both simple and stratified adult epithelium, thus providing a functional rationale for its requirement throughout the murine lifespan. Defects in the formation or function of tight junctions is strongly correlated with inappropriate expression of the claudin-2 protein, which may be the cause of pathologies associated with decreased matriptase activity in both the epidermis and intestine [15, 103]. Future work will seek to determine how matriptase activity regulates tight junction proteins and further illuminate the mechanisms by which targets of matriptase catalysis affect tissue growth and development.

Several matriptase substrates have been identified *in vitro* using cell-free or cell-culture-based methods. These include urokinase-type plasminogen activator (uPA), the G-protein-coupled protease-activated receptor-2 (PAR2), ENaCs, human acid-sensing ion channel 1 (ASIC1), macrophage-stimulating protein 1 (MSP-1), insulinlike growth factor binding protein-related protein-1 (IGFBP-rP1), and platelet-derived growth factor D (PDGF-D) [19, 76, 82, 89, 92, 104]. However, the physiological relevance of these substrates awaits further investigation. One of the first substrates to be reported, the pro form of hepatocyte growth factor (HGF) was recently identified as being critical *in vivo* during matriptase-mediated squamous cell carcinogenesis (Section 5.1.4.2).

Expression of matriptase-2 (TMPRSS6) is strongest in the liver, where it is coexpressed with the GPI-linked membrane protein hemojuvelin (HJV) at the surface of hepatocytes [22, 105]. Matriptase-2 regulation of HJV has recently been shown to play a significant role in regulating iron homeostasis of both mice and humans. Hepatocytes store the majority of cellular iron, where ferrous iron is detoxified and stored through complexion with the protein ferritin [106]. Matriptase-2 cleaves HJV at the hepatocyte plasma membrane, resulting in its deactivation as an initiator of BMP signaling [22]. When active, HJV acts as a cofactor for BMP/SMAD signaling, which increases transcription of the peptide hormone hepcidin, a critical regulator of blood iron levels [107–110]. Hepcidin protein binds to the iron transporter ferroportin, which it targets for degradation, causing decreased transport of iron into the bloodstream [111]. Under conditions of inactive matriptase-2, increased active HJV induces a BMP signaling cascade that upregulates hepcidin expression, resulting in degradation of ferroportin and a pathological reduction of iron transport to the blood. Accordingly, loss of function mutations of matriptase-2 in both mice and humans result in iron-refractory iron deficiency anemia (IRIDA), a condition that does not respond to oral administration of iron, presumably as a result of hepcidin-dependent cellular iron retention and impaired export [20, 112, 113].

Matriptase-2 is one of two enzymes known to proteolytically process HJV, with the second being the pro-protein convertase furin [114]. Both furin and matriptase-2 downregulate HJV signaling through peptide cleavage; however, unlike matriptase-2, the product of furin cleavage (sHJV) acts as an inhibitor of BMP signaling by binding to BMP ligand as a decoy receptor [115]. The matriptase-2 cleavage product does not act as a BMP decoy, and thus matriptase-2 inhibits HJV signaling solely through the reduction of active HJV enzyme at the cellular surface [22]. The importance of hepcidin is exemplified by the fact that its deregulation is the cause of a vast majority of pathologies in iron metabolism [116]. Stringent regulation of hepcidin expression is required to prevent both anemia and iron overload, which induces oxidative damage and has been linked to a variety of age-associated pathologies. Because furin does not compensate for the effects of matriptase-2 mutations, the two proteases appear to have evolved independent regulatory mechanisms that respond to nutritional iron absorption and/or blood–iron homeostasis differentially to regulate HJV signaling and hepcidin expression.

The last two members of the matriptase subfamily, matriptase-3 and polyserase-1, have not yet been associated with any physiological function. Matriptase-3 is broadly expressed in mouse tissues, with high expression in the brain, eye, skin, testis, and salivary gland and low expression in the heart, skeletal muscle, thymus, ovary, prostate, and uterus [117]. Human expression is similar to that of mice, with the strongest expression being found in the testis and ovary; and low expression in the brain, trachea, lung, and salivary gland [117]. Translated matriptase-3 protein localizes to the plasma membrane; is capable of cleaving gelatin, casein, and albumin; and is inhibited by serpin family protease inhibitors *in vitro* [117]. The *polyserase-1* gene is unique among TTSPs in encoding a protein with three SP domains, the C-terminal-most domain being inactive, and can generate a splice variant (serase-1B) that contains one SEA domain and the first of the three protease domains [31, 118]. Expression of full-length polyserase-1 mRNA is found in skeletal muscle, liver, brain, heart, and placenta, while the splice variant is detected in the liver, small intestine, pancreas, testis, and peripheral blood cells [31, 118]. It remains to be determined what relevance each of these two protein transcripts have to tissue physiology and whether or not they share biological substrates or mechanisms of regulation.

5.3.4
HAT/DESC1 Subfamily

The HAT/DESC subfamily of TTSPs is not well characterized, with expression data being limited to HAT and DESC1 and biological roles for these proteins inferred from available expression data and sparse *in vitro* analysis. All HAT/DESC genes are located in a single cluster in both mice and humans, with mice possessing seven genes and humans five genes that share enough sequence homology and conserved structural organization to be classified as TTSPs [30]. Human HAT protein was first isolated in the fluid secretions of patients with chronic airway diseases; however, mRNA expression has subsequently been identified in a wide range of tissues [119–121]. In respiratory epithelium, HAT protein localizes to the suprabasal layer of bronchial epithelium as well as the basal region of the associated cilia, where it has been proposed to play a role in mucus production [122, 123]. Other presumptive roles for HAT include fibrinogen processing to influence blood clotting, inducing the proliferation of bronchial fibroblasts, and proteolytic activation of influenza virus [124–126]. *In vitro*, HAT has been shown to activate PAR2, and a splice variant (AsP) has been identified that can cleave and activate pro-gamma-melanocyte-stimulating hormone [119, 120, 127]. However, participation of HAT in these and/or other processes *in vivo* remains to be established. HAT-like 2 and HAT-like 3 genes encode functional constructs in rodents; however, their human orthologs exist only as pseudogenes, suggesting that their function is either specific to rodents or is redundant with other proteases in higher primates [128]. HAT-like 4 and 5 do appear to encode functional proteases in humans; however, there is currently no data regarding expression or function of these TTSPs in either rodents or humans. DESC1 encodes a functional cell surface protease in mice, with high levels of expression in the epidermis and oral and reproductive epithelium [30, 129]. The function of DESC1, however, remains mysterious and has not been characterized in mice or humans. Future work needs to establish physiologically relevant roles for members of this TTSP subfamily, with due caution given to the interpretation of functional conservation between rodents and humans.

5.3.5 TTSPs in Cancer

Because TTSPs can modify epithelial development, differentiation, and barrier integrity, as well as degrade components of the ECM, their deregulation has been linked to both tumor formation and invasive progression. Importantly, several TTSPs have been identified to display aberrant expression in a variety of tumors and cancer cells, which often correlates with poor patient prognosis [4, 7, 130–132]. Following from the ability to modulate proteolysis at the cell surface, considerable attention has been focused on the role of TTSPs in cancer cell migration and invasion. Of particular interest are the proteins hepsin, TMPRSS2, and matriptase, which have been correlated with epithelial carcinogenesis and metastasis in diverse tissues. Deregulation of TTSP expression or proteolytic activity is known to have a strong correlation with several cellular phenotypes that occur during malignant transformation and has become a primary focus of cancer research.

Hepsin is reported as being highly upregulated in prostate cancer, and its overexpression is among the most consistent biomarkers for malignant transformation from benign prostate hyperplasia [130, 133–135]. In addition, high hepsin expression correlates with a high Gleason score for prostate tumors and poor clinical outcome with relapse following prostatectomy [60, 135–137]. Other than the aforementioned role in hearing, the physiological role of hepsin in tissues outside the developing cochlea remains unknown. There are, however, several potential targets of hepsin proteolysis that are implicated in carcinogenesis, including pro-HGF, pro-uPA, and the ECM protein laminin 322 [60, 138–140]. Activation of pro-HGF by hepsin may be the cause of invasive growth in ovarian cancer cells by increased phosphorylation and activation of c-Met hepatocyte growth factor receptor (HGFR), a process that may be regulated *in vivo* by the TTSP inhibitors HAI-1 and HAI-2 [138]. Hepsin activation of uPA has been proposed to initiate proteolysis at the tumor–stroma interface causing a disruption of the epithelial basement membrane and leading to cancer progression [139]. Hepsin cleavage of laminin-322 results in an increase of prostate cancer cell migration *in vitro*, which may translate to increased tumor cell motility *in vivo* as a result of ECM degradation [140].

Hepsin overexpression in the prostate of mice does not affect cell proliferation, but it does result in destabilization of the basement membrane, indicating that hepsin probably affects cancer cell migration and metastasis more so than tumor formation [8]. This view is fortified by the ability of hepsin overexpression to induce metastasis to remote organs in a murine model of nonmetastasizing prostate cancer, whereas inhibition of hepsin activity by antibodies leods to a decrease in invasiveness, but not growth of prostate and ovarian cancer cells [8, 141]. When and how hepsin expression and/or activity become altered in the human prostate during carcinogenesis and cancer progression is under investigation. Several hepsin SNPs have been associated with increased risk for prostate cancer; however, mapping has revealed that these were not located in the coding region, and a recent study reported no significant correlation between hepsin gene variation and prostate cancer [142, 143]. Intriguingly, *low* hepsin expression has been correlated with poor prognosis in patients with hepatocellular carcinoma, and there are conflicting reports of how hepsin levels influence the prognosis of patients with renal carcinomas [133, 144, 145]. Taken together, tissue-specific control may prove paramount to the use of therapeutic agents that target hepsin, such as antagonists and small molecule inhibitors. It remains unclear how hepsin is regulated at the transcriptional level as well as how its proteolytic activity is modified by inhibitors during cancer progression and metastasis *in vivo*.

It is widely accepted that the major implication of TMPRSS2 in cancer results from genetic fusion with members of the erythroblast-transformation-specific (ETS) transcription factors [146]. Fusions of TMPRSS2:ETS occur via complex chromosomal translocation events that combine the 5′ regulatory region of TMPRSS2 in frame with either the entire or partial coding sequence of ETS genes, resulting in androgen-induced expression of the transcription factor [147]. Four TMPRSS2-ETS somatic fusions have been identified, with fusions to ERG (ETS-related gene) being the most prevalent and found in a majority of patients with prostate cancer [148, 149]. Normal TMPRSS2 expression is localized to the prostate epithelium, and these fusions result in the overexpression of the ETS member proto-oncogene in the prostate [149, 150]. Furthermore, TMPRSS2:ERG fusion events appear to be induced in both malignant and nonmalignant prostate cells through androgen signaling, which can cause topoisomerase-II-beta-mediated DNA double-strand breaks at the regions of genetic fusion [151]. ETS transcription factors are known to regulate cell growth, and TMPRSS2:ETS fusions have been hypothesized to induce prostatic intraepithelial neoplasia (PIN) in mice, or evoke the transition from PIN to carcinoma in humans [150, 152–154]. There are several different isoforms of TMPRSS2:ERG fusions, which are proposed to affect pathology differently; however; at present, how these individual constructs and their expression levels correlate with clinical outcome is a matter of contention [155]. Regardless, TMPRSS2:ETS fusions are being assessed for both their potential as therapeutic targets and as noninvasive biomarkers for early detection of disease [156–158].

Overexpression, and in some cases, hyperactivation of the TTSP matriptase, has been described in a diverse and impressive array of epithelial cancers and has garnered significant attention from cancer biologists. Matriptase expression is known to be deregulated in breast, prostate, cervical, colorectal, gastric, and pancreatic carcinomas and in tumors of the lung, liver, and kidney, among others [3]. An increase in matriptase expression correlates with severity of tumors in the breast and prostate, and *de novo* expression is found in ovarian and cervical carcinoma and expression levels also correlate with histopathological grade [136, 159–161]. Furthermore, the ratio of matriptase/HAI-1 mRNA is increased in colorectal carcinoma, and the expression of HAI-1 and HAI-2 are reduced in prostate and endometrial carcinoma when compared with normal tissue [162–165]. Taken together, these observations have bolstered hypotheses that rampant matriptase proteolytic activity may be the cause of both early and late events during carcinogenesis and cancer progression.

Spatial deregulation of matriptase expression that expands from terminally differentiated epithelial cells to actively proliferating cells is implicated in the early events of carcinogenesis. In normal murine epidermis, matriptase expression is solely found in differentiated keratinocytes; however, treatment with carcinogens can induce ectopic expression in the basal layers of cells where carcinomas originate [18]. In mice, the ability of matriptase to initiate tumor formation is confirmed with transgenic expression in the basal layer of the epidermis, which results in spontaneous squamous cell carcinoma and dramatically increases carcinogen-induced tumorigenesis [18]. Importantly, the oncogenic properties of matriptase are completely abolished in dual transgenic mice that also express the inhibitor HAI-1 in the basal layer, providing strong evidence that matriptase proteolytic activity is the cause of these phenotypes. Stable knockdown of HAI-1 results in epithelial to mesenchymal transition (EMT) in pancreatic and lung cancer cells, and xenograft of matriptase overexpressing cells in mice increases the metastatic potential of gastric cancer cells, suggesting that matriptase may also promote the late stages of tumor dissemination [166, 167]. Recent *in vivo* analysis has shown that the oncogenic potential of matriptase in mouse epidermis requires *c-Met* and can be blocked by the pharmacological inhibition of the mammalian target of rapamycin (mTor) pathway [168]. Elevated matriptase increases processing of pro-HGF, resulting in the initiation of c-Met–mTor signaling pathway to induce proliferation and migration in primary epithelial cells, a pathway that may be relevant to a large proportion of squamous cell carcinomas. Accordingly, both matriptase and its cognate inhibitors are being actively pursued as candidates for pharmacological intervention and for usefulness as biomarkers of disease onset and progression.

Other TTSPs that are potentially significant to cancer initiation and progression include TMPRSS3, TMPRSS4, matriptase-2, and DESC-1; however, at present, information implicating these proteases is relatively sparse. A splice variant of TMPRSS3, TADG-12, is overexpressed in ovarian cancer; and expression levels appear to correlate with increased pathology [169]. TADG-12 is currently being investigated for its potential as a diagnostic marker and as a target for immunotherapy for ovarian cancer [170]. TMPRSS4 is upregulated in several epithelial cancers, where it has been proposed to affect disease progression through increased proteolytic activity via reduction of the tissue factor pathway inhibitor TFPI2 and/or through the induction of EMT [171, 172]. Matriptase-2 has been proposed to act as a tumor suppressor, as its increased expression results in a reduction of invasive potential in prostate and breast cancer cell lines and correlates with positive patient outcome [173, 174]. DESC-1 expression appears to be reduced in squamous cell carcinoma, perhaps causing a reduction in normal epithelial differentiation; whereas transgenic expression in mammalian kidney cells *in vitro* induces phenotypes associated with tumor growth [175, 176]. The association of these proteases with carcinogenesis is in a nascent stage, and further research is needed to characterize their proteolytic targets and downstream signaling nodes and to assess how these are affected during tumor growth and progression.

In summary, TTSPs have emerged as a unique family of proteases with profoundly diverse physiological functions. TTSPs are primarily expressed in epithelial tissue, and individual family members have been shown to contribute to the development and maintenance of organs with highly specialized epithelial components, such as the skin and inner ear. Other TTSPs have pleiotropic effects on physiology by contributing to the regulation of digestion, blood pressure, or iron metabolism. Research has sought to uncover the specific and shared substrates and inhibitors of these proteases and characterize their ability to initiate proteolytic cascades and affect signal transduction within the context of individual cell or tissue types. Several TTSPs are deregulated in carcinomas and may contribute to both tumor formation and metastasis through their ability to activate growth factors and modify components of the ECM. The loss of precise regulation of TTSP expression and/or catalytic activity may be the cause of several types of epithelial cancers. There is strong impetus to understand the events that lead to deregulation at the gene and protein level and how this precipitates various stages of tumorigenesis. Additionally, several TTSPs have yet to be assessed for their physiological function, and the mechanisms by which characterized members affect phenotypes are largely unknown. Progress in these areas will continue to add to our burgeoning knowledge of how proteases regulate diverse components of development, homeostasis, and pathology.

References

1. Bugge, T.H., Antalis, T.M., and Wu, Q. (2009) Type II transmembrane serine proteases. *J. Biol. Chem.*, **284** (35), 23177–23181.
2. Hooper, J.D., Clements, J.A., Quigley, J.P., and Antalis, T.M. (2001) Type II transmembrane serine proteases. Insights into an emerging class of cell surface proteolytic enzymes. *J. Biol. Chem.*, **276**, 857–860.
3. Szabo, R. and Bugge, T.H. (2008) Type II transmembrane serine proteases in development and disease. *Int. J. Biochem. Cell Biol.*, **40** (6–7), 1297–1316.
4. Webb, S.L., Sanders, A.J., Mason, M.D., and Jiang, W.G. (2011) Type II transmembrane serine protease (TTSP) deregulation in cancer. *Front. Biosci.*, **16**, 539–552.
5. Irving, P., Troxler, L., Heuer, T.S., Belvin, M., Kopczynski, C., Reichhart, J.M., Hoffmann, J.A., and Hetru, C. (2001) A genome-wide analysis of immune responses in Drosophila.

Proc. Natl. Acad. Sci. U.S.A., **98**, 15119–15124.

6. Choi, S.Y., Bertram, S., Glowacka, I., Park, Y.W., and Pohlmann, S. (2009) Type II transmembrane serine proteases in cancer and viral infections. *Trends Mol. Med.*, **15**, 303–312.
7. List, K. (2009) Matriptase: a culprit in cancer? *Future Oncol.*, **5** (1), 97–104.
8. Klezovitch, O., Chevillet, J., Mirosevich, J., Roberts, R.L., Matusik, R.J., and Vasioukhin, V. (2004) Hepsin promotes prostate cancer progression and metastasis. *Cancer Cell*, **6**, 185–195.
9. Guipponi, M., Tan, J., Cannon, P.Z., Donley, L., Crewther, P., Clarke, M., Wu, Q., Shepherd, R.K., and Scott, H.S. (2007) Mice deficient for the type II transmembrane serine protease, TMPRSS1/hepsin, exhibit profound hearing loss. *Am. J. Pathol.*, **171**, 608–616.
10. Kim, T.S., Heinlein, C., Hackman, R.C., and Nelson, P.S. (2006) Phenotypic analysis of mice lacking the Tmprss2-encoded protease. *Mol. Cell Biol.*, **26**, 965–975.
11. Wattenhofer, M., Sahin–Calapoglu, N., Andreasen, D., Kalay, E., Caylan, R., Braillard, B., Fowler-Jaeger, N., Reymond, A., Rossier, B.C., Karaguzel, A. *et al.* (2005) A novel TMPRSS3 missense mutation in a DFNB8/10 family prevents proteolytic activation of the protein. *Hum. Genet.*, **117**, 528–535.
12. Holzinger, A., Maier, E.M., Buck, C., Mayerhofer, P.U., Kappler, M., Haworth, J.C., Moroz, S.P., Hadorn, H.B., Sadler, J.E., and Roscher, A.A. (2002) Mutations in the proenteropeptidase gene are the molecular cause of congenital enteropeptidase deficiency. *Am. J. Hum. Genet.*, **70**, 20–25.
13. Basel-Vanagaite, L., Attia, R., Ishida-Yamamoto, A., Rainshtein, L., Ben Amitai, D., Lurie, R., Pasmanik-Chor, M., Indelman, M., Zvulunov, A., Saban, S. *et al.* (2007) Autosomal recessive ichthyosis with hypotrichosis caused by a mutation in ST14, encoding type II transmembrane serine protease matriptase. *Am. J. Hum. Genet.*, **80**, 467–477.
14. List, K., Haudenschild, C.C., Szabo, R., Chen, W., Wahl, S.M., Swaim, W., Engelholm, L.H., Behrendt, N., and Bugge, T.H. (2002) Matriptase/MT-SP1 is required for postnatal survival, epidermal barrier function, hair follicle development, and thymic homeostasis. *Oncogene*, **21**, 3765–3779.
15. List, K., Kosa, P., Szabo, R., Bey, A.L., Wang, C.B., Molinolo, A., and Bugge, T.H. (2009) Epithelial integrity is maintained by a matriptase-dependent proteolytic pathway. *Am. J. Pathol.*, **175**, 1453–1463.
16. Netzel-Arnett, S., Currie, B.M., Szabo, R., Lin, C.Y., Chen, L.M., Chai, K.X., Antalis, T.M., Bugge, T.H., and List, K. (2006) Evidence for a matriptase-prostasin proteolytic cascade regulating terminal epidermal differentiation. *J. Biol. Chem.*, **281**, 32941–32945.
17. List, K., Currie, B., Scharschmidt, T.C., Szabo, R., Shireman, J., Molinolo, A., Cravatt, B.F., Segre, J., and Bugge, T.H. (2007) Autosomal ichthyosis with hypotrichosis syndrome displays low matriptase proteolytic activity and is phenocopied in ST14 hypomorphic mice. *J. Biol. Chem.*, **282**, 36714–36723.
18. List, K., Szabo, R., Molinolo, A., Nielsen, B.S., and Bugge, T.H. (2006) Delineation of matriptase protein expression by enzymatic gene trapping suggests diverging roles in barrier function, hair formation, and squamous cell carcinogenesis. *Am. J. Pathol.*, **168**, 1513–1525.
19. Szabo, R., Rasmussen, A.L., Moyer, A.B., Kosa, P., Schafer, J.M., Molinolo, A.A., Gutkind, J.S., and Bugge, T.H. (2011) c-Met-induced epithelial carcinogenesis is initiated by the serine protease matriptase. *Oncogene*, **30**, 2003–2016.
20. Melis, M.A., Cau, M., Congiu, R., Sole, G., Barella, S., Cao, A., Westerman, M., Cazzola, M., and Galanello, R. (2008) A mutation in the TMPRSS6 gene, encoding a transmembrane serine protease that suppresses hepcidin production, in familial iron deficiency

anemia refractory to oral iron. *Haematologica*, **93**, 1473–1479.

21. Wu, Q., Xu-Cai, Y.O., Chen, S., and Wang, W. (2009) Corin: new insights into the natriuretic peptide system. *Kidney Int.*, **75**, 142–146.
22. Silvestri, L., Pagani, A., Nai, A., De Domenico, I., Kaplan, J., and Camaschella, C. (2008) The serine protease matriptase-2 (TMPRSS6) inhibits hepcidin activation by cleaving membrane hemojuvelin. *Cell Metab.*, **8**, 502–511.
23. Zhou, Y., Jiang, J., Cui, Y., and Wu, Q. (2009) Corin, atrial natriuretic peptide and hypertension. *Nephrol. Dial. Transplant.*, **24**, 1071–1073.
24. Dong, N., Chen, S., Yang, J., He, L., Liu, P., Zheng, D., Li, L., Zhou, Y., Ruan, C., Plow, E. *et al.* (2010) Plasma soluble corin in patients with heart failure. *Circ. Heart Fail.*, **3**, 207–211.
25. Chan, J.C., Knudson, O., Wu, F., Morser, J., Dole, W.P., and Wu, Q. (2005) Hypertension in mice lacking the proatrial natriuretic peptide convertase corin. *Proc. Natl. Acad. Sci. U.S.A.*, **102**, 785–790.
26. Polzin, D., Kaminski, H.J., Kastner, C., Wang, W., Kramer, S., Gambaryan, S., Russwurm, M., Peters, H., Wu, Q., Vandewalle, A. *et al.* (2010) Decreased renal corin expression contributes to sodium retention in proteinuric kidney diseases. *Kidney Int.*, **78**, 650–659.
27. Enshell-Seijffers, D., Lindon, C., and Morgan, B.A. (2008) The serine protease Corin is a novel modifier of the agouti pathway. *Development*, **135** (2), 217–225.
28. Viloria, C.G., Peinado, J.R., Astudillo, A., Garcia-Suarez, O., Gonzalex, O., Suarez, C., and Cal, S. (2007) Human DESC1 serine protease confers tumorigenic properties to MDCK cells and it is upregulated in tumours of different origin. *Br. J. Cancer*, **97** (2), 201–209.
29. Szabo, R., Wu, Q., Dickson, R.B., Netzel-Arnett, S., Antalis, T.M., and Bugge, T.H. (2003) Type II transmembrane serine proteases. *Thromb. Haemost.*, **90**, 185–193.
30. Hobson, J.P., Netzel-Arnett, S., Szabo, R., Rehault, S.M., Church, F.C., Strickland, D.K., Lawrence, D.A., Antalis, T.M., and Bugge, T.H. (2004) Mouse DESC1 is located within a cluster of seven DESC1-like genes and encodes a type II transmembrane serine protease that forms serpin inhibitory complexes. *J. Biol. Chem.*, **279**, 46981–46994.
31. Cal, S., Quesada, V., Garabaya, C., and Lopez-Otin, C. (2003) Polyserase-I, a human polyprotease with the ability to generate independent serine protease domains from a single translation product. *Proc. Natl. Acad. Sci. U.S.A.*, **100**, 9185–9190.
32. Cho, E.G., Kim, M.G., Kim, C., Kim, S.R., Seong, I.S., Chung, C., Schwartz, R.H., and Park, D. (2001) N-terminal processing is essential for release of epithin, a mouse type II membrane serine protease. *J. Biol. Chem.*, **276**, 44581–44589.
33. Matsushima, M., Ichinose, M., Yahagi, N., Kakei, N., Tsukada, S., Miki, K., Kurokawa, K., Tashiro, K., Shiokawa, K., Shinomiya, K. *et al.* (1994) Structural characterization of porcine enteropeptidase. *J. Biol. Chem.*, **269**, 19976–19982.
34. Afar, D.E., Vivanco, I., Hubert, R.S., Kuo, J., Chen, E., Saffran, D.C., Raitano, A.B., and Jakobovits, A. (2001) Catalytic cleavage of the androgen-regulated TMPRSS2 protease results in its secretion by prostate and prostate cancer epithelia. *Cancer Res.*, **61**, 1686–1692.
35. Ramsay, A.J., Hooper, J.D., Folgueras, A.R., Velasco, G., and Lopez-Otin, C. (2009) Matriptase-2 (TMPRSS6): a proteolytic regulator of iron homeostasis. *Haematologica*, **94**, 840–849.
36. Takeuchi, T., Shuman, M.A., and Craik, C.S. (1999) Reverse biochemistry: use of macromolecular protease inhibitors to dissect complex biological processes and identify a membrane-type serine protease in epithelial cancer and normal tissue. *Proc. Natl. Acad. Sci. U.S.A.*, **96** (20), 11054–11061.
37. Velasco, G., Cal, S., Quesada, V., Sanchez, L.M., and Lopez-Otin, C. (2002) Matriptase-2, a

membrane-bound mosaic serine proteinase predominantly expressed in human liver and showing degrading activity against extracellular matrix proteins. *J. Biol. Chem.*, **277**, 37637–37646.

38. Qiu, D., Owen, K., Gray, K., Bass, R., and Ellis, V. (2007) Roles and regulation of membrane-associated serine proteases. *Biochem. Soc. Trans.*, **35**, 583–587.
39. Rawlings, N.D. and Barrett, A.J. (1994) Families of serine peptidases. *Methods Enzymol.*, **244**, 19–61.
40. Beliveau, F., Desilets, A., and Leduc, R. (2009) Probing the substrate specificities of matriptase, matriptase-2, hepsin and DESC1 with internally quenched fluorescent peptides. *FEBS J.*, **276** (8), 2213–2226.
41. Chen, C.J., Wu, B.Y., Tsao, P.I., Chen, C.Y., Wu, M.H., Chan, Y.L., Lee, H.S., Johnson, M.D., Eckert, R.L., Chen, Y.W. *et al.* (2010) Increased matriptase zymogen activation in inflammatory skin disorders. *Am. J. Physiol. Cell Physiol.*, **300**, C406–C415.
42. Imamura, T. and Kitamoto, Y. (2003) Expression of enteropeptidase in differentiated enterocytes, goblet cells, and the tumor cells in human duodenum. *Am. J. Physiol. Gastrointest. Liver Physiol.*, **285** (6), G1235–G1241.
43. Yuan, X., Zheng, X., Lu, D., Rubin, D.C., Pung, C.Y., and Sadler, J.E. (1998) Structure of murine enterokinase (enteropeptidase) and expression in small intestine during development. *Am. J. Physiol.*, **274**, G342–G349.
44. Lu, D., Yuan, X., Zheng, X., and Sadler, J.E. (1997) Bovine proenteropeptidase is activated by trypsin, and the specificity of enteropeptidase depends on the heavy chain. *J. Biol. Chem.*, **272**, 31293–31300.
45. Zamolodchikova, T.S., Sokolova, E.A., Lu, D., and Sadler, J.E. (2000) Activation of recombinant proenteropeptidase by duodenase. *FEBS Lett.*, **466**, 295–299.
46. Fernandez–del Castillo, C. *et al.* (1994) Interstitial protease activation is the central event in progression to necrotizing pancreatitis. *Surgery*, **116** (3), 497–504.
47. Hartwig, W., Werner, J., Jimenez, R.E., Z'Graggen, K., Weimann, J., Lewandrowski, K.B., Warshaw, A.L., and Fernandez-del Castillo, C. (1999) Trypsin and activation of circulating trypsinogen contribute to pancreatitis-associated lung injury. *Am. J. Physiol.*, **277**, G1008–G1016.
48. Yang, K., Ding, Y.X., and Chin, W.C. (2007) K+-induced ion-exchanges trigger trypsin activation in pancreas acinar zymogen granules. *Arch. Biochem. Biophys.*, **459** (2), 256–263.
49. Zheng, X.L., Kitamoto, Y., and Sadler, J.E. (2009) Enteropeptidase, a type II transmembrane serine protease. *Front. Biosci. (Elite Ed.)*, **1**, 242–249.
50. Mann, N.S. and Mann, S.K. (1994) Enterokinase. *Proc. Soc. Exp. Biol. Med.*, **206** (2), 114–118.
51. Iyngkaran, N., Yadav, M., and Boey, C.G. (1995) Mucosal enterokinase activity in cow's milk protein sensitive enteropathy. *Singapore Med. J.*, **36**, 393–396.
52. Rutgeerts, L. and Eggermont, E. (1976) Human enterokinase. *Tijdschr. Gastroenterol.*, **19**, 231–246.
53. Ghishan, F.K., Lee, P.C., Lebenthal, E., Johnson, P., Bradley, C.A., and Greene, H.L. (1983) Isolated congenital enterokinase deficiency. Recent findings and review of the literature. *Gastroenterology*, **85**, 727–731.
54. Green, J.R., Bender, S.W., Posselt, H.G., and Lentze, M.J. (1984) Primary intestinal enteropeptidase deficiency. *J. Pediatr. Gastroenterol. Nutr.*, **3**, 630–633.
55. Leytus, S.P., Loeb, K.R., Hagen, F.S., Kurachi, K., and Davie, E.W. (1988) A novel trypsin-like serine protease (hepsin) with a putative transmembrane domain expressed by human liver and hepatoma cells. *Biochemistry*, **27**, 1067–1074.
56. Tsuji, A., Torres-Rosado, A., Arai, T., Le Beau, M.M., Lemons, R.S., Chou, S.H., and Kurachi, K. (1991) Hepsin, a cell membrane-associated protease. Characterization, tissue distribution,

and gene localization. *J. Biol. Chem.*, **266**, 16948–16953.

57. Tsuji, A., Torres-Rosado, A., Arai, T., Chou, S.H., and Kurachi, K. (1991) Characterization of hepsin, a membrane bound protease. *Biomed. Biochim. Acta*, **50**, 791–793.
58. Kazama, Y., Hamamoto, T., Foster, D.C., and Kisiel, W. (1995) Hepsin, a putative membrane-associated serine protease, activates human factor VII and initiates a pathway of blood coagulation on the cell surface leading to thrombin formation. *J. Biol. Chem.*, **270**, 66–72.
59. Kurachi, K., Torres-Rosado, A., and Tsuji, A. (1994) Hepsin. *Methods Enzymol.*, **244**, 100–114.
60. Wu, Q. and Parry, G. (2007) Hepsin and prostate cancer. *Front. Biosci.*, **12**, 5052–5059.
61. Zheng, X.L., Kitamoto, Y., and Sadler, J.E. (2009) Enteropeptidase, a type II transmembrane serine protease. *Front. Biosci. (Elite Ed.)*, **1**, 242–249.
62. Hanifa, S., Scott, H.S., Crewther, P., Guipponi, M., and Tan, J. (2010) Thyroxine treatments do not correct inner ear defects in tmprss1 mutant mice. *Neuroreport*, **21**, 897–901.
63. Guipponi, M., Toh, M.Y., Tan, J., Park, D., Hanson, K., Ballana, E., Kwong, D., Cannon, P.Z., Wu, Q., Gout, A. *et al.* (2008) An integrated genetic and functional analysis of the role of type II transmembrane serine proteases (TMPRSSs) in hearing loss. *Hum. Mutat.*, **29**, 130–141.
64. Lee, Y.J., Park, D., Kim, S.Y., and Park, W.J. (2003) Pathogenic mutations but not polymorphisms in congenital and childhood onset autosomal recessive deafness disrupt the proteolytic activity of TMPRSS3. *J. Med. Genet.*, **40**, 629–631.
65. Guipponi, M., Vuagniaux, G., Wattenhofer, M., Shibuya, K., Vazquez, M., Dougherty, L., Scamuffa, N., Guida, E., Okui, M., Rossier, C. *et al.* (2002) The transmembrane serine protease (TMPRSS3) mutated in deafness DFNB8/10 activates the epithelial sodium channel (ENaC) in vitro. *Hum. Mol. Genet.*, **11**, 2829–2836.
66. Dries, D.L., Victor, R.G., Rame, J.E., Cooper, R.S., Wu, X., Zhu, X., Leonard, D., Ho, S.I., Wu, Q., Post, W. *et al.* (2005) Corin gene minor allele defined by 2 missense mutations is common in blacks and associated with high blood pressure and hypertension. *Circulation*, **112**, 2403–2410.
67. Knappe, S., Wu, F., Masikat, M.R., Morser, J., and Wu, Q. (2003) Functional analysis of the transmembrane domain and activation cleavage of human corin: design and characterization of a soluble corin. *J. Biol. Chem.*, **278**, 52363–52370.
68. Hooper, J.D., Scarman, A.L., Clarke, B.E., Normyle, J.F., and Antalis, T.M. (2000) Localization of the mosaic transmembrane serine protease corin to heart myocytes. *Eur. J. Biochem.*, **267**, 6931–6937.
69. Ichiki, T., Huntley, B.K., Heublein, D.M., Sandberg, S.M., McKie, P.M., Martin, F.L., Jougasaki, M., and Burnett, J.C. Jr. (2011) Corin is present in the normal human heart, kidney, and blood, with pro-B-type natriuretic peptide processing in the circulation. *Clin. Chem.*, **57**, 40–47.
70. Yan, W., Sheng, N., Seto, M., Morser, J., and Wu, Q. (1999) Corin, a mosaic transmembrane serine protease encoded by a novel cDNA from human heart. *J. Biol. Chem.*, **274**, 14926–14935.
71. Vanderheyden, M., Vrints, C., Verstreken, S., Bartunek, J., Beunk, J., and Goethals, M. (2010) B-type natriuretic peptide as a marker of heart failure: new insights from biochemistry and clinical implications. *Biomark. Med.*, **4**, 315–320.
72. Knappe, S., Wu, F., Madlansacay, M.R., and Wu, Q. (2004) Identification of domain structures in the propeptide of corin essential for the processing of proatrial natriuretic peptide. *J. Biol. Chem.*, **279**, 34464–34471.
73. Wang, W., Liao, X., Fukuda, K., Knappe, S., Wu, F., Dries, D.L., Qin, J., and Wu, Q. (2008) Corin variant associated with hypertension and cardiac hypertrophy exhibits impaired zymogen activation and natriuretic

peptide processing activity. *Circ. Res.*, **103**, 502–508.

74. Chen, S., Sen, S., Young, D., Wang, W., Moravec, C.S., and Wu, Q. (2010) Protease corin expression and activity in failing hearts. *Am. J. Physiol. Heart Circ. Physiol.*, **299**, H1687–H1692.
75. Tran, K.L., Lu, X., Lei, M., Feng, Q., and Wu, Q. (2004) Upregulation of corin gene expression in hypertrophic cardiomyocytes and failing myocardium. *Am. J. Physiol. Heart Circ. Physiol.*, **287**, H1625–H1631.
76. Bhatt, A.S., Welm, A., Farady, C.J., Vasquez, M., Wilson, K., and Craik, C.S. (2007) Coordinate expression and functional profiling identify an extracellular proteolytic signaling pathway. *Proc. Natl. Acad. Sci. U.S.A.*, **104**, 5771–5776.
77. Fan, B., Brennan, J., Grant, D., Peale, F., Rangell, L., and Kirchhofer, D. (2007) Hepatocyte growth factor activator inhibitor-1 (HAI-1) is essential for the integrity of basement membranes in the developing placental labyrinth. *Dev. Biol.*, **303**, 222–230.
78. Szabo, R., Molinolo, A., List, K., and Bugge, T.H. (2007) Matriptase inhibition by hepatocyte growth factor activator inhibitor-1 is essential for placental development. *Oncogene*, **26**, 1546–1556.
79. Oberst, M.D., Singh, B., Ozdemirli, M., Dickson, R.B., Johnson, M.D., and Lin, C.Y. (2003) Characterization of matriptase expression in normal human tissues. *J. Histochem. Cytochem.*, **51**, 1017–1025.
80. Takeuchi, T., Harris, J.L., Huang, W., Yan, K.W., Coughlin, S.R., and Craik, C.S. (2000) Cellular localization of membrane-type serine protease 1 and identification of protease-activated receptor-2 and single-chain urokinase-type plasminogen activator as substrates. *J. Biol. Chem.*, **275**, 26333–26342.
81. Chen, Y.W., Wang, J.K., Chou, F.P., Chen, C.Y., Rorke, E.A., Chen, L.M., Chai, K.X., Eckert, R.L., Johnson, M.D., and Lin, C.Y. (2010) Regulation of the matriptase-prostasin cell surface proteolytic cascade by hepatocyte growth factor activator inhibitor-1 during epidermal differentiation. *J. Biol. Chem.*, **285**, 31755–31762.
82. Lee, S.L., Dickson, R.B., and Lin, C.Y. (2000) Activation of hepatocyte growth factor and urokinase/plasminogen activator by matriptase, an epithelial membrane serine protease. *J. Biol. Chem.*, **275**, 36720–36725.
83. Ovaere, P., Lippens, S., Vandenabeele, P., and Declercq, W. (2009) The emerging roles of serine protease cascades in the epidermis. *Trends Biochem. Sci.*, **34**, 453–463.
84. List, K., Szabo, R., Wertz, P.W., Segre, J., Haudenschild, C.C., Kim, S.Y., and Bugge, T.H. (2003) Loss of proteolytically processed filaggrin caused by epidermal deletion of Matriptase/MT-SP1. *J. Cell Biol.*, **163**, 901–910.
85. Lee, M.S., Tseng, I.C., Wang, Y., Kiyomiya, K., Johnson, M.D., Dickson, R.B., and Lin, C.Y. (2007) Autoactivation of matriptase in vitro: requirement for biomembrane and LDL receptor domain. *Am. J. Physiol. Cell Physiol.*, **293**, C95–105.
86. Oberst, M.D., Chen, L.Y., Kiyomiya, K., Williams, C.A., Lee, M.S., Johnson, M.D., Dickson, R.B., and Lin, C.Y. (2005) HAI-1 regulates activation and expression of matriptase, a membrane-bound serine protease. *Am. J. Physiol. Cell Physiol.*, **289**, C462–C470.
87. Leyvraz, C., Charles, R.P., Rubera, I., Guitard, M., Rotman, S., Breiden, B., Sandhoff, K., and Hummler, E. (2005) The epidermal barrier function is dependent on the serine protease CAP1/Prss8. *J. Cell Biol.*, **170**, 487–496.
88. Mauro, T., Guitard, M., Behne, M., Oda, Y., Crumrine, D., Komuves, L., Rassner, U., Elias, P.M., and Hummler, E. (2002) The ENaC channel is required for normal epidermal differentiation. *J. Invest. Dermatol.*, **118**, 589–594.
89. Vuagniaux, G., Vallet, V., Jaeger, N.F., Hummler, E., and Rossier, B.C. (2002) Synergistic activation of ENaC by three membrane-bound

channel-activating serine proteases (mCAP1, mCAP2, and mCAP3) and serum- and glucocorticoid-regulated kinase (Sgk1) in Xenopus Oocytes. *J. Gen. Physiol.*, **120**, 191–201.

90. Smith, F.J., Irvine, A.D., Terron-Kwiatkowski, A., Sandilands, A., Campbell, L.E., Zhao, Y., Liao, H., Evans, A.T., Goudie, D.R., Lewis-Jones, S. *et al.* (2006) Loss-of-function mutations in the gene encoding filaggrin cause ichthyosis vulgaris. *Nat. Genet.*, **38**, 337–342.

91. Alef, T., Torres, S., Hausser, I., Metze, D., Tursen, U., Lestringant, G.G., and Hennies, H.C. (2009) Ichthyosis, follicular atrophoderma, and hypotrichosis caused by mutations in ST14 is associated with impaired profilaggrin processing. *J. Invest. Dermatol.*, **129**, 862–869.

92. Bocheva, G., Rattenholl, A., Kempkes, C., Goerge, T., Lin, C.Y., D'Andrea, M.R., Stander, S., and Steinhoff, M. (2009) Role of matriptase and proteinase-activated receptor-2 in nonmelanoma skin cancer. *J. Invest. Dermatol.*, **129**, 1816–1823.

93. Szabo, R., Kosa, P., List, K., and Bugge, T.H. (2009) Loss of matriptase suppression underlies spint1 mutation-associated ichthyosis and postnatal lethality. *Am. J. Pathol.*, **174**, 2015–2022.

94. Nagaike, K., Kawaguchi, M., Takeda, N., Fukushima, T., Sawaguchi, A., Kohama, K., Setoyama, M., and Kataoka, H. (2008) Defect of hepatocyte growth factor activator inhibitor type 1/serine protease inhibitor, Kunitz type 1 (Hai-1/Spint1) leads to ichthyosis-like condition and abnormal hair development in mice. *Am. J. Pathol.*, **173**, 1464–1475.

95. Tanaka, H., Nagaike, K., Takeda, N., Itoh, H., Kohama, K., Fukushima, T., Miyata, S., Uchiyama, S., Uchinokura, S., Shimomura, T. *et al.* (2005) Hepatocyte growth factor activator inhibitor type 1 (HAI-1) is required for branching morphogenesis in the chorioallantoic placenta. *Mol. Cell Biol.*, **25**, 5687–5698.

96. Szabo, R., Hobson, J.P., Christoph, K., Kosa, P., List, K., and Bugge, T.H. (2009) Regulation of cell surface protease matriptase by HAI2 is essential for placental development, neural tube closure and embryonic survival in mice. *Development*, **136**, 2653–2663.

97. Smith, D.L., Smith, J.G., Wong, S.W., and deShazo, R.D. (1995) Netherton's syndrome. *Br. J. Dermatol.*, **133**, 153–154.

98. Smith, D.L., Smith, J.G., Wong, S.W., and deShazo, R.D. (1995) Netherton's syndrome: a syndrome of elevated IgE and characteristic skin and hair findings. *J. Allergy Clin. Immunol.*, **95**, 116–123.

99. Descargues, P., Deraison, C., Bonnart, C., Kreft, M., Kishibe, M., Ishida-Yamamoto, A., Elias, P., Barrandon, Y., Zambruno, G., Sonnenberg, A. *et al.* (2005) Spink5-deficient mice mimic Netherton syndrome through degradation of desmoglein 1 by epidermal protease hyperactivity. *Nat. Genet.*, **37**, 56–65.

100. Sales, K.U., Masedunskas, A., Bey, A.L., Rasmussen, A.L., Weigert, R., List, K., Szabo, R., Overbeek, P.A., and Bugge, T.H. (2010) Matriptase initiates activation of epidermal pro-kallikrein and disease onset in a mouse model of Netherton syndrome. *Nat. Genet.*, **42**, 676–683.

101. Kim, M.G., Chen, C., Lyu, M.S., Cho, E.G., Park, D., Kozak, C., and Schwartz, R.H. (1999) Cloning and chromosomal mapping of a gene isolated from thymic stromal cells encoding a new mouse type II membrane serine protease, epithin, containing four LDL receptor modules and two CUB domains. *Immunogenetics*, **49**, 420–428.

102. Lee, S.L., Huang, P.Y., Roller, P., Cho, E.G., Park, D., and Dickson, R.B. (2010) Matriptase/epithin participates in mammary epithelial cell growth and morphogenesis through HGF activation. *Mech. Dev.*, **127**, 82–95.

103. Buzza, M.S., Netzel-Arnett, S., Shea-Donohue, T., Zhao, A., Lin, C.Y., List, K., Szabo, R., Fasano, A., Bugge, T.H., and Antalis, T.M. (2010)

Membrane-anchored serine protease matriptase regulates epithelial barrier formation and permeability in the intestine. *Proc. Natl. Acad. Sci. U.S.A.*, **107**, 4200–4205.

104. Clark, E.B., Jovov, B., Rooj, A.K., Fuller, C.M., and Benos, D.J. (2010) Proteolytic cleavage of human acid-sensing ion channel 1 by the serine protease matriptase. *J. Biol. Chem.*, **285**, 27130–27143.

105. Ramsay, A.J., Reid, J.C., Velasco, G., Quigley, J.P., and Hooper, J.D. (2008) The type II transmembrane serine protease matriptase-2--identification, structural features, enzymology, expression pattern and potential roles. *Front. Biosci.*, **13**, 569–579.

106. Sibille, J.C., Kondo, H., and Aisen, P. (1989) Uptake of ferritin and iron bound to ferritin by rat hepatocytes: modulation by apotransferrin, iron chelators and chloroquine. *Biochim. Biophys. Acta*, **1010**, 204–209.

107. Andriopoulos, B. Jr., Corradini, E., Xia, Y., Faasse, S.A., Chen, S., Grgurevic, L., Knutson, M.D., Pietrangelo, A., Vukicevic, S., Lin, H.Y. *et al.* (2009) BMP6 is a key endogenous regulator of hepcidin expression and iron metabolism. *Nat. Genet.*, **41**, 482–487.

108. Celec, P. (2005) Hemojuvelin: a supposed role in iron metabolism one year after its discovery. *J. Mol. Med.*, **83** (7), 521–525.

109. Meynard, D., Kautz, L., Darnaud, V., Canonne-Hergaux, F., Coppin, H., and Roth, M.P. (2009) Lack of the bone morphogenetic protein BMP6 induces massive iron overload. *Nat. Genet.*, **41**, 478–481.

110. Niederkofler, V., Salie, R., and Arber, S. (2005) Hemojuvelin is essential for dietary iron sensing, and its mutation leads to severe iron overload. *J. Clin. Invest.*, **115** (8), 2180–2186.

111. Ramey, G., Deschemin, J.C., Durel, B., Canonne-Hergaux, F., Nicolas, G., and Vaulont, S. (2010) Hepcidin targets ferroportin for degradation in hepatocytes. *Haematologica*, **95**, 501–504.

112. Finberg, K.E., Heeney, M.M., Campagna, D.R., Aydinok, Y., Pearson, H.A., Hartman, K.R., Mayo, M.M., Samuel, S.M., Strouse, J.J., Markianos, K. *et al.* (2008) Mutations in TMPRSS6 cause iron-refractory iron deficiency anemia (IRIDA). *Nat. Genet.*, **40**, 569–571.

113. Folgueras, A.R., de Lara, F.M., Pendas, A.M., Garabaya, C., Rodriguez, F., Astudillo, A., Bernal, T., Cabanillas, R., Lopez-Otin, C., and Velasco, G. (2008) Membrane-bound serine protease matriptase-2 (Tmprss6) is an essential regulator of iron homeostasis. *Blood*, **112**, 2539–2545.

114. Silvestri, L., Pagani, A., and Camaschella, C. (2008) Furin-mediated release of soluble hemojuvelin: a new link between hypoxia and iron homeostasis. *Blood*, **111**, 924–931.

115. Henrich, S., Lindberg, I., Bode, W., and Than, M.E. (2005) Proprotein convertase models based on the crystal structures of furin and kexin: explanation of their specificity. *J. Mol. Biol.*, **345**, 211–227.

116. Lee, P.L. and Beutler, E. (2009) Regulation of hepcidin and iron-overload disease. *Annu. Rev. Pathol.*, **4**, 489–515.

117. Szabo, R., Netzel-Arnett, S., Hobson, J.P., Antalis, T.M., and Bugge, T.H. (2005) Matriptase-3 is a novel phylogenetically preserved membrane-anchored serine protease with broad serpin reactivity. *Biochem. J.*, **390**, 231–242.

118. Okumura, Y., Hayama, M., Takahashi, E., Fujiuchi, M., Shimabukuro, A., Yano, M., and Kido, H. (2006) Serase-1B, a new splice variant of polyserase-1/TMPRSS9, activates urokinase-type plasminogen activator and the proteolytic activation is negatively regulated by glycosaminoglycans. *Biochem. J.*, **400**, 551–561.

119. Hahner, S., Fassnacht, M., Hammer, F., Schammann, M., Weismann, D., Hansen, I.A., and Allolio, B. (2005) Evidence against a role of human airway trypsin-like protease--the human analogue of the growth-promoting rat adrenal secretory protease--in adrenal tumourigenesis. *Eur. J. Endocrinol.*, **152**, 143–153.

120. Iwakiri, K., Ghazizadeh, M., Jin, E., Fujiwara, M., Takemura, T., Takezaki, S., Kawana, S., Yasuoka, S., and

Kawanami, O. (2004) Human airway trypsin-like protease induces PAR-2-mediated IL-8 release in psoriasis vulgaris. *J. Invest. Dermatol.*, **122**, 937–944.
121. Yasuoka, S., Ohnishi, T., Kawano, S., Tsuchihashi, S., Ogawara, M., Masuda, K., Yamaoka, K., Takahashi, M., and Sano, T. (1997) Purification, characterization, and localization of a novel trypsin-like protease found in the human airway. *Am. J. Respir. Cell Mol. Biol.*, **16**, 300–308.
122. Chokki, M., Yamamura, S., Eguchi, H., Masegi, T., Horiuchi, H., Tanabe, H., Kamimura, T., and Yasuoka, S. (2004) Human airway trypsin-like protease increases mucin gene expression in airway epithelial cells. *Am. J. Respir. Cell Mol. Biol.*, **30**, 470–478.
123. Takahashi, M., Sano, T., Yamaoka, K., Kamimura, T., Umemoto, N., Nishitani, H., and Yasuoka, S. (2001) Localization of human airway trypsin-like protease in the airway: an immunohistochemical study. *Histochem. Cell Biol.*, **115**, 181–187.
124. Bottcher, E., Matrosovich, T., Beyerle, M., Klenk, H.D., Garten, W., and Matrosovich, M. (2006) Proteolytic activation of influenza viruses by serine proteases TMPRSS2 and HAT from human airway epithelium. *J. Virol.*, **80**, 9896–9898.
125. Matsushima, R., Takahashi, A., Nakaya, Y., Maezawa, H., Miki, M., Nakamura, Y., Ohgushi, F., and Yasuoka, S. (2006) Human airway trypsin-like protease stimulates human bronchial fibroblast proliferation in a protease-activated receptor-2-dependent pathway. *Am. J. Physiol. Lung Cell Mol. Physiol.*, **290**, L385–L395.
126. Yoshinaga, S., Nakahori, Y., and Yasuoka, S. (1998) Fibrinogenolytic activity of a novel trypsin-like enzyme found in human airway. *J. Med. Invest.*, **45** (1–4), 77–86.
127. Beaufort, N., Leduc, D., Eguchi, H., Mengele, K., Hellmann, D., Masegi, T., Kamimura, T., Yasuoka, S., Fend, F., Chignard, M. *et al.* (2007) The human airway trypsin-like protease modulates the urokinase receptor (uPAR, CD87) structure and functions. *Am. J. Physiol. Lung Cell Mol. Physiol.*, **292**, L1263–L1272.
128. Quesada, V., Ordonez, G.R., Sanchez, L.M., Puente, X.S., and Lopez-Otin, C. (2009) The Degradome database: mammalian proteases and diseases of proteolysis. *Nucleic Acids Res.*, **37**, D239–D243.
129. Lang, J.C. and Schuller, D.E. (2001) Differential expression of a novel serine protease homologue in squamous cell carcinoma of the head and neck. *Br. J. Cancer*, **84** (2), 237–243.
130. Magee, J.A., Araki, T., Patil, S., Ehrig, T., True, L., Humphrey, P.A., Catalona, W.J., Watson, M.A., and Milbrandt, J. (2001) Expression profiling reveals hepsin overexpression in prostate cancer. *Cancer Res.*, **61**, 5692–5696.
131. Netzel-Arnett, S., Hooper, J.D., Szabo, R., Madison, E.L., Quigley, J.P., Bugge, T.H., and Antalis, T.M. (2003) Membrane anchored serine proteases: a rapidly expanding group of cell surface proteolytic enzymes with potential roles in cancer. *Cancer Metastasis Rev.*, **22**, 237–258.
132. Wallrapp, C., Hahnel, S., Muller-Pillasch, F., Burghardt, B., Iwamura, T., Ruthenburger, M., Lerch, M.M., Adler, G., and Gress, T.M. (2000) A novel transmembrane serine protease (TMPRSS3) overexpressed in pancreatic cancer. *Cancer Res.*, **60**, 2602–2606.
133. Chen, Z., Fan, Z., McNeal, J.E., Nolley, R., Caldwell, M.C., Mahadevappa, M., Zhang, Z., Warrington, J.A., and Stamey, T.A. (2003) Hepsin and maspin are inversely expressed in laser capture microdissectioned prostate cancer. *J. Urol.*, **169**, 1316–1319.
134. Dhanasekaran, S.M., Barrette, T.R., Ghosh, D., Shah, R., Varambally, S., Kurachi, K., Pienta, K.J., Rubin, M.A., and Chinnaiyan, A.M. (2001) Delineation of prognostic biomarkers in prostate cancer. *Nature*, **412**, 822–826.
135. Stephan, C., Yousef, G.M., Scorilas, A., Jung, K., Jung, M., Kristiansen, G., Hauptmann, S., Kishi, T., Nakamura, T., Loening, S.A. *et al.* (2004) Hepsin is highly over expressed in and a new

candidate for a prognostic indicator in prostate cancer. *J. Urol.*, **171**, 187–191.
136. Saleem, M., Adhami, V.M., Zhong, W., Longley, B.J., Lin, C.Y., Dickson, R.B., Reagan-Shaw, S., Jarrard, D.F., and Mukhtar, H. (2006) A novel biomarker for staging human prostate adenocarcinoma: overexpression of matriptase with concomitant loss of its inhibitor, hepatocyte growth factor activator inhibitor-1. *Cancer Epidemiol. Biomarkers Prev.*, **15**, 217–227.
137. Stamey, T.A., Warrington, J.A., Caldwell, M.C., Chen, Z., Fan, Z., Mahadevappa, M., McNeal, J.E., Nolley, R., and Zhang, Z. (2001) Molecular genetic profiling of Gleason grade 4/5 prostate cancers compared to benign prostatic hyperplasia. *J. Urol.*, **166**, 2171–2177.
138. Herter, S., Piper, D.E., Aaron, W., Gabriele, T., Cutler, G., Cao, P., Bhatt, A.S., Choe, Y., Craik, C.S., Walker, N. *et al.* (2005) Hepatocyte growth factor is a preferred in vitro substrate for human hepsin, a membrane-anchored serine protease implicated in prostate and ovarian cancers. *Biochem. J.*, **390**, 125–136.
139. Moran, P., Li, W., Fan, B., Vij, R., Eigenbrot, C., and Kirchhofer, D. (2006) Pro-urokinase-type plasminogen activator is a substrate for hepsin. *J. Biol. Chem.*, **281**, 30439–30446.
140. Tripathi, M., Nandana, S., Yamashita, H., Ganesan, R., Kirchhofer, D., and Quaranta, V. (2008) Laminin-332 is a substrate for hepsin, a protease associated with prostate cancer progression. *J. Biol. Chem.*, **283**, 30576–30584.
141. Xuan, J.A., Schneider, D., Toy, P., Lin, R., Newton, A., Zhu, Y., Finster, S., Vogel, D., Mintzer, B., Dinter, H. *et al.* (2006) Antibodies neutralizing hepsin protease activity do not impact cell growth but inhibit invasion of prostate and ovarian tumor cells in culture. *Cancer Res.*, **66**, 3611–3619.
142. Holt, S.K., Kwon, E.M., Lin, D.W., Ostrander, E.A., and Stanford, J.L. (2010) Association of hepsin gene variants with prostate cancer risk and prognosis. *Prostate*, **70**, 1012–1019.
143. Pal, P., Xi, H., Kaushal, R., Sun, G., Jin, C.H., Jin, L., Suarez, B.K., Catalona, W.J., and Deka, R. (2006) Variants in the HEPSIN gene are associated with prostate cancer in men of European origin. *Hum. Genet.*, **120**, 187–192.
144. Betsunoh, H., Mukai, S., Akiyama, Y., Fukushima, T., Minamiguchi, N., Hasui, Y., Osada, Y., and Kataoka, H. (2007) Clinical relevance of hepsin and hepatocyte growth factor activator inhibitor type 2 expression in renal cell carcinoma. *Cancer Sci.*, **98**, 491–498.
145. Roemer, A., Schwettmann, L., Jung, M., Stephan, C., Roigas, J., Kristiansen, G., Loening, S.A., Lichtinghagen, R., and Jung, K. (2004) The membrane proteases adams and hepsin are differentially expressed in renal cell carcinoma are they potential tumor markers? *J. Urol.*, **172**, 2162–2166.
146. Narod, S.A., Seth, A., and Nam, R. (2008) Fusion in the ETS gene family and prostate cancer. *Br. J. Cancer*, **99**, 847–851.
147. Kumar-Sinha, C., Tomlins, S.A., and Chinnaiyan, A.M. (2008) Recurrent gene fusions in prostate cancer. *Nat. Rev. Cancer*, **8** (7), 497–511.
148. Soller, M.J., Isaksson, M., Elfving, P., Soller, W., Lundgren, R., and Panagopoulos, I. (2006) Confirmation of the high frequency of the TMPRSS2/ERG fusion gene in prostate cancer. *Genes Chromosomes Cancer*, **45**, 717–719.
149. Tomlins, S.A., Rhodes, D.R., Perner, S., Dhanasekaran, S.M., Mehra, R., Sun, X.W., Varambally, S., Cao, X., Tchinda, J., Kuefer, R. *et al.* (2005) Recurrent fusion of TMPRSS2 and ETS transcription factor genes in prostate cancer. *Science*, **310**, 644–648.
150. Tomlins, S.A., Laxman, B., Dhanasekaran, S.M., Helgeson, B.E., Cao, X., Morris, D.S., Menon, A., Jing, X., Cao, Q., Han, B. *et al.* (2007) Distinct classes of chromosomal rearrangements create oncogenic ETS gene fusions in prostate cancer. *Nature*, **448**, 595–599.
151. Bastus, N.C., Boyd, L.K., Mao, X., Stankiewicz, E., Kudahetti, S.C.,

Oliver, R.T., Berney, D.M., and Lu, Y.J. (2010) Androgen-induced TMPRSS2:ERG fusion in nonmalignant prostate epithelial cells. *Cancer Res.*, **70**, 9544–9548.

152. Carver, B.S., Tran, J., Chen, Z., Carracedo-Perez, A., Alimonti, A., Nardella, C., Gopalan, A., Scardino, P.T., Cordon-Cardo, C., Gerald, W. *et al.* (2009) ETS rearrangements and prostate cancer initiation. *Nature*, **457**, E1; discussion E2–3.

153. Klezovitch, O., Risk, M., Coleman, I., Lucas, J.M., Null, M., True, L.D., Nelson, P.S., and Vasioukhin, V. (2008) A causal role for ERG in neoplastic transformation of prostate epithelium. *Proc. Natl. Acad. Sci. U.S.A.*, **105**, 2105–2110.

154. Zong, Y., Xin, L., Goldstein, A.S., Lawson, D.A., Teitell, M.A., and Witte, O.N. (2009) ETS family transcription factors collaborate with alternative signaling pathways to induce carcinoma from adult murine prostate cells. *Proc. Natl. Acad. Sci. U.S.A.*, **106**, 12465–12470.

155. Huang, W. and Waknitz, M. (2009) ETS gene fusions and prostate cancer. *Am. J. Transl. Res.*, **1** (4), 341–351.

156. Han, B., Mehra, R., Dhanasekaran, S.M., Yu, J., Menon, A., Lonigro, R.J., Wang, X., Gong, Y., Wang, L., Shankar, S. *et al.* (2008) A fluorescence in situ hybridization screen for E26 transformation-specific aberrations: identification of DDX5-ETV4 fusion protein in prostate cancer. *Cancer Res.*, **68**, 7629–7637.

157. Hessels, D., Smit, F.P., Verhaegh, G.W., Witjes, J.A., Cornel, E.B., and Schalken, J.A. (2007) Detection of TMPRSS2-ERG fusion transcripts and prostate cancer antigen 3 in urinary sediments may improve diagnosis of prostate cancer. *Clin. Cancer Res.*, **13**, 5103–5108.

158. Laxman, B., Tomlins, S.A., Mehra, R., Morris, D.S., Wang, L., Helgeson, B.E., Shah, R.B., Rubin, M.A., Wei, J.T., and Chinnaiyan, A.M. (2006) Noninvasive detection of TMPRSS2:ERG fusion transcripts in the urine of men with prostate cancer. *Neoplasia*, **8**, 885–888.

159. Lee, J.W., Yong Song, S., Choi, J.J., Lee, S.J., Kim, B.G., Park, C.S., Lee, J.H., Lin, C.Y., Dickson, R.B., and Bae, D.S. (2005) Increased expression of matriptase is associated with histopathologic grades of cervical neoplasia. *Hum. Pathol.*, **36**, 626–633.

160. Tanimoto, H., Shigemasa, K., Tian, X., Gu, L., Beard, J.B., Sawasaki, T., and O'Brien, T.J. (2005) Transmembrane serine protease TADG-15 (ST14/Matriptase/MT-SP1): expression and prognostic value in ovarian cancer. *Br. J. Cancer*, **92**, 278–283.

161. Tsai, W.C., Chu, C.H., Yu, C.P., Sheu, L.F., Chen, A., Chiang, H., and Jin, J.S. (2008) Matriptase and survivin expression associated with tumor progression and malignant potential in breast cancer of Chinese women: tissue microarray analysis of immunostaining scores with clinicopathological parameters. *Dis. Markers*, **24**, 89–99.

162. Bergum, C. and List, K. (2010) Loss of the matriptase inhibitor HAI-2 during prostate cancer progression. *Prostate*, **70** (13), 1422–1428.

163. Nakamura, K., Hongo, A., Kodama, J., and Hiramatsu, Y. (2010) The role of hepatocyte growth factor activator inhibitor (HAI)-1 and HAI-2 in endometrial cancer. *Int. J. Cancer*, **128**, 2613–2624.

164. Oberst, M.D., Johnson, M.D., Dickson, R.B., Lin, C.Y., Singh, B., Stewart, M., Williams, A., al-Nafussi, A., Smyth, J.F., Gabra, H. *et al.* (2002) Expression of the serine protease matriptase and its inhibitor HAI-1 in epithelial ovarian cancer: correlation with clinical outcome and tumor clinicopathological parameters. *Clin. Cancer Res.*, **8**, 1101–1107.

165. Vogel, L.K., Saebo, M., Skjelbred, C.F., Abell, K., Pedersen, E.D., Vogel, U., and Kure, E.H. (2006) The ratio of Matriptase/HAI-1 mRNA is higher in colorectal cancer adenomas and carcinomas than corresponding tissue from control individuals. *BMC Cancer*, **6**, 176.

166. Tripathi, M., Potdar, A.A., Yamashita, H., Weidow, B., Cummings, P.T., Kirchhofer, D., and Quaranta, V. (2011)

Laminin-332 cleavage by matriptase alters motility parameters of prostate cancer cells. *Prostate*, **71**, 184–196.

167. Ustach, C.V., Huang, W., Conley-LaComb, M.K., Lin, C.Y., Che, M., Abrams, J., and Kim, H.R. (2010) A novel signaling axis of matriptase/PDGF-D/ss-PDGFR in human prostate cancer. *Cancer Res.*, **70**, 9631–9640.

168. Cheng, H., Fukushima, T., Takahashi, N., Tanaka, H., and Kataoka, H. (2009) Hepatocyte growth factor activator inhibitor type 1 regulates epithelial to mesenchymal transition through membrane-bound serine proteinases. *Cancer Res.*, **69**, 1828–1835.

169. Ihara, S., Miyoshi, E., Ko, J.H., Murata, K., Nakahara, S., Honke, K., Dickson, R.B., Lin, C.Y., and Taniguchi, N. (2002) Prometastatic effect of N-acetylglucosaminyltransferase V is due to modification and stabilization of active matriptase by adding beta 1–6 GlcNAc branching. *J. Biol. Chem.*, **277**, 16960–16967.

170. Sawasaki, T., Shigemasa, K., Gu, L., Beard, J.B., and O'Brien, T.J. (2004) The transmembrane protease serine (TMPRSS3/TADG-12) D variant: a potential candidate for diagnosis and therapeutic intervention in ovarian cancer. *Tumour Biol.*, **25**, 141–148.

171. Bellone, S., Anfossi, S., O'Brien, T.J., Cannon, M.J., Silasi, D.A., Azodi, M., Schwartz, P.E., Rutherford, T.J., Pecorelli, S., and Santin, A.D. (2009) Induction of human tumor-associated differentially expressed gene-12 (TADG-12/TMPRSS3)-specific cytotoxic T lymphocytes in human lymphocyte antigen-A2.1-positive healthy donors and patients with advanced ovarian cancer. *Cancer*, **115**, 800–811.

172. Dawelbait, G., Winter, C., Zhang, Y., Pilarsky, C., Grutzmann, R., Heinrich, J.C., and Schroeder, M. (2007) Structural templates predict novel protein interactions and targets from pancreas tumour gene expression data. *Bioinformatics*, **23**, i115–i124.

173. Jung, H., Lee, K.P., Park, S.J., Park, J.H., Jang, Y.S., Choi, S.Y., Jung, J.G., Jo, K., Park, D.Y., Yoon, J.H. *et al.* (2008) TMPRSS4 promotes invasion, migration and metastasis of human tumor cells by facilitating an epithelial-mesenchymal transition. *Oncogene*, **27**, 2635–2647.

174. Parr, C., Sanders, A.J., Davies, G., Martin, T., Lane, J., Mason, M.D., Mansel, R.E., and Jiang, W.G. (2007) Matriptase-2 inhibits breast tumor growth and invasion and correlates with favorable prognosis for breast cancer patients. *Clin. Cancer Res.*, **13**, 3568–3576.

175. Sanders, A.J., Parr, C., Martin, T.A., Lane, J., Mason, M.D., and Jiang, W.G. (2008) Genetic upregulation of matriptase-2 reduces the aggressiveness of prostate cancer cells in vitro and in vivo and affects FAK and paxillin localisation. *J. Cell Physiol.*, **216**, 780–789.

176. Sedghizadeh, P.P., Mallery, S.R., Thompson, S.J., Kresty, L., Beck, F.M., Parkinson, E.K., Biancamano, J., and Lang, J.C. (2006) Expression of the serine protease DESC1 correlates directly with normal keratinocyte differentiation and inversely with head and neck squamous cell carcinoma progression. *Head Neck*, **28**, 432–440.

6
Plasminogen Activators in Ischemic Stroke

Gerald Schielke and Daniel A. Lawrence

6.1
Introduction

Ischemic stroke is the leading cause of disability and the third most common cause of death in the world. Despite enormous efforts to identify interventions that protect brain tissue from ischemic damage, there is only one Food and Drug Administration (FDA)-approved treatment for stroke, the thrombolytic drug recombinant tissue-type plasminogen activator (rtPA) known clinically as *alteplase*. Unfortunately, treatment with rtPA is associated with increased risk of symptomatic intracranial hemorrhage (ICH); to be effective, treatment must be started within hours of the onset of symptoms and, therefore, only a small percentage of stroke patients are treated.

Plasminogen activators (PAs) are thrombolytic serine proteases that activate the zymogen plasminogen to the broadly acting enzyme plasmin, which in turn cleaves fibrin, the primary fibrillar protein component that forms the mesh work of a clot. Tissue-type plasminogen activator (tPA) and urokinase-type plasminogen activator (uPA), the two PAs in human tissue, have diverse functions in addition to fibrinolysis; these include roles in tissue remodeling, angiogenesis, inflammation, and tumor metastasis. PAs are regulated by multiple and complex mechanisms but a critical component in their regulation are specific inhibitors known as *ser*ine *p*rotease *in*hibitors (serpins). Plasminogen activator inhibitor-1 (PAI-1) is the major physiologic inhibitor of tPA and uPA in the plasma. In the brain, neuroserpin is expressed in neurons and is a primary regulator of PA activity. Protease nexin-1 (PN-1) is also expressed in the brain, although it is a less efficient tPA inhibitor and its physiological role is uncertain. There is a growing body of evidence that in stroke, brain tPA may be neurotoxic and contribute to the pathophysiological processes that lead to hemorrhage and ischemic cell death.

This review of the role of PAs in stroke encompasses both the clinical use of rtPA for thrombolysis in stroke management and preclinical studies addressing the role of endogenous PAs and their inhibitors in ischemic brain damage. A central premise of this review is that a more complete understanding of the role of the endogenous PA system in normal brain function and ischemic pathophysiology will

Matrix Proteases in Health and Disease, First Edition. Edited by Niels Behrendt.

lead to improvements in thrombolytic therapy for stroke, with greater effectiveness and reduced side effects. In addition, identifying mechanisms responsible for PA-induced brain damage may lead to alternative therapeutic strategies that protect ischemic brain tissue.

6.2 Rationale for Thrombolysis after Stroke

The use of PAs to treat stroke is based on the concept that reperfusion of ischemic brain tissue will reduce tissue death and therefore limit adverse functional outcomes. Early preclinical studies in rodent and primate models of ischemia identified critical thresholds for brain damage, which include both the extent of reduction in cerebral blood flow (CBF) and the duration of the ischemic event. A study in baboons demonstrated that as CBF is progressively reduced, brain electrical activity ceases (at CBF of about 20 ml 100 g min^{-1}), followed by depletion of ATP and disruption of ion homeostasis when CBF falls to 7 ml 100 g min^{-1} [1, 2]. In focal ischemic stroke, CBF is severely reduced near the site of vessel occlusion, but is partially maintained by collateral vessels in the periphery of the ischemic territory [3]. These observations led to the concept of the "*ischemic penumbra,*" defined as the rim of ischemic tissue where electrical activity is absent but ion gradients are intact. Studies have shown that unlike the severely ischemic core tissue that progresses to infarction, the tissue in the penumbra has the potential for recovery. However, this hypoperfused tissue is challenged by numerous factors that push it toward infarction, including exposure to excitotoxic neurotransmitters [4], energy demand from cortical spreading depressions and seizures [5, 6], blood–brain barrier (BBB) disruption [7], and inflammation [8]. It is now well established that restoration of CBF in this penumbral tissue can prevent the expansion of the ischemic core and reduce final infarct volume and functional deficits [9]. However, studies in animal models of focal ischemia have demonstrated that there is a short 2–4 h window of opportunity for salvaging this tissue by increasing blood flow to the penumbra [10–13].

Since 87% of strokes in humans are thromboembolic, thrombolysis presented a promising strategy for enhancing CBF in penumbral tissue and reducing infarct size. Early studies with urokinase and with streptokinase, a PA produced by the pathogenic streptococci, suggested that this strategy might be untenable because of severe hemorrhagic complications [14]. However, tPA, because of its fibrin specificity, was thought to offer potential advantages in this regard [15, 16]. The first study to demonstrate the effectiveness of tPA in limiting ischemic pathology was carried out in a rabbit embolic stroke model where intravenous (IV) tPA, 1 mg kg^{-1}, was given immediately after clots were injected into the cerebral circulation. tPA improved neurological outcomes at 24 h and did not increase the incidence of hemorrhage [16]. Since then, several autologous clot models have been developed in rats and mice, and the benefits and relative safety of IV tPA have been consistently demonstrated when the drug was administered within 2–4 h of the initiation of ischemia [11]. These promising results led to a series of clinical trials evaluating the efficacy and safety of rtPA and several other PAs as thrombolytics for the treatment of acute ischemic stroke.

6.2.1 Clinical Trials: Overview

The pivotal study that led to FDA approval of rtPA for the treatment of acute stroke was the NINDS stroke trial reported in 1995 [17]. This double-blind placebo-controlled trial assessed both changes in neurological deficit at 24 h and sustained favorable outcomes at three months after IV tPA (0.9 mg kg^{-1} body weight). rtPA treatment was initiated within 3 h of stroke onset, after computed tomography (CT) imaging confirmed the absence of hemorrhage. rtPA improved outcome at three months even though the rate of symptomatic intracerebral hemorrhage was increased by treatment from 0.1 to 6.4% [18]. The FDA approved the use of rtPA for acute stroke in 1996. Several other clinical trials have since been carried out that were important in defining treatment parameters for tPA in acute stroke and criterion for patient selection. ECASS was a large multicenter double-blind placebo-controlled trial of tPA (1.1 mg kg^{-1} body weight) given within 6 h of onset of symptoms [19]. This trial failed to show the effectiveness of tPA in its primary endpoint and the treated group had increased occurrence of large ICH. The ECASS II trial was similar but the dose was reduced to 0.9 mg kg^{-1} to match the NINDS study [20]. Although the inclusion criteria were more restrictive, the 6 h treatment window was maintained. The study found a nonsignificant trend for improved outcome and an increase in symptomatic ICH in the treated group (8.8 vs 3.4%). A study of rtPA treatment initiated between 3 and 5 h after stroke onset, the ATLANTIS trial, found no significant benefit of rtPA on the 90-day outcome measures [21]. However, a pooled analysis of the NINDS, ECASS, and ATLANTIS trials found an overall highly significant benefit of tPA treatment, which decreased with increasing time from onset of symptoms to treatment [22]. A significant improvement in three-month outcome was seen with treatment initiated as late as 4.5 h; however, the analysis showed that the greatest benefit is seen if the drug is given as soon as possible. Recent results from the ECASS III trial, where patients were treated as late as 4.5 h after symptom onset [23–25], confirmed the conclusion of the pooled analysis, demonstrating a benefit of rtPA on three-month outcome measures. A recent advisory statement from the American Heart Association/American Stroke Association recommended expanding the time window for rtPA to 4.5 h in selected patients [26].

This important series of trials:

1) Resulted in approval of the first and only drug to treat acute ischemic stroke. This was an important step in light of the numerous failed clinical studies of potential neuroprotective drugs.
2) Demonstrated a reperfusion time window similar to that seen in preclinical models of ischemia.

3) Expanded the time window from 3 to 4.5 h, thus increasing the number of stroke patients that will potentially benefit from thrombolysis.
4) Demonstrated that the earliest possible treatment will provide the most benefit.
5) Found that hemorrhagic complications are increased by rtPA treatment.

Despite its proven benefit and approval by regulatory agencies in most countries, the use of rtPA for thrombolysis in acute stroke is extremely low, with only about 3–8% of stroke patients treated. The reasons for this are complex but certainly the limited therapeutic time window and the fear of ICH have a major impact on treatment decisions. Currently, there is much interest in expanding the therapeutic time window using state-of-the-art imaging techniques to distinguish penumbral from irreversibly damaged tissue in individual patients and use this to select patients likely to benefit from thrombolytic therapy [27]. Initially, positron emission tomography (PET) imaging was used to identify potentially salvageable tissue [28, 29]. When PET imaging data was compared to magnetic resonance imaging (MRI) diffusion-weighted images (DWIs), a similar volume of infarcted tissue was predicted [30]. MRI can also measure tissue perfusion (perfusion-weighted imaging (PWI)), and the mismatch between the volume of brain tissue identified by DWI and PWI provides an estimate of the size of the penumbra. Although this concept is still under evaluation [31], two clinical rtPA trials using DWI/PWI have been reported. DEFUES was an open label study with a 3–6 h treatment window (mean treatment time 5.46 h) [32]. The study showed that early reperfusion was associated with clinical benefit and patients with a DWI/PWI mismatch benefited more, while those with no mismatch failed to show improved outcomes. EPITHET was a placebo-controlled randomized phase II trial testing the hypothesis that infarct growth would be less in DWI/PWI mismatched patients who were treated with rtPA in the 3–6 h window [33]. Secondly, reperfusion and functional outcome were hypothesized to be better in mismatched patients receiving treatment. These studies indicated that rtPA was associated with less infarct expansion (nonsignificant) and increased perfusion in mismatched patients. As seen in the DEFUSE trial, better clinical outcome was associated with increased perfusion. A recent reanalysis of the EPITHET data using coregistration in the analysis of the magnetic resonance images found greater regions of mismatched tissue and a significant effect of rtPA on infarct growth [34]. These findings offer hope that treating patients based on imaging of the volume of potentially viable penumbral tissue could result in better outcomes and possibly extend the therapeutic time window beyond 4.5 h in selected patients.

A series of trials with the PA desmoteplase were carried out on patients selected using DWI/PWI mismatch, with treatment starting 3–9 h after stroke onset. Desmoteplase, a recombinant form of PA originally isolated from vampire bat saliva, has the potential advantage over tPA of much higher fibrin specificity, longer plasma half-life, and potentially no enhancement of neurotoxicity (discussed below) [35, 36]. The initial phase II trials (DIAS and DEDAS) were promising, showing desmoteplase to be safe, with a low rate of ICH at the 125 μg kg^{-1} dose with increased reperfusion and improved clinical outcome [37, 38]. Unfortunately, the DIAS II phase III study, where patients were selected based on DWI/PWI mismatch, failed to show significant clinical benefit [39]. However, there was an unusually high response rate in the placebo group, which may have been responsible for the lack of significant benefit in the treatment group. Patients are currently being enrolled for a desmoteplase phase III study with selection based on vessel occlusion rather than DWI/PWI mismatch [40]. A meta-analysis of five DWI/PWI mismatch studies – DEFUSE, EPITHET, and the three desmoteplase trials – was recently reported [41]. This analysis found that delayed thrombolytic treatment in DWI/PWI-mismatched patients is associated with increased reperfusion but does not result in significant improvement in clinical outcome and concluded that treatment beyond the 4.5 h window cannot be recommended in routine care. However, prospective phase III trials of an extended treatment window using current refined imaging criteria are warranted.

Another approach to improving thrombolytic treatment for acute stroke is to identify novel PAs and thrombolytics with properties that make recanalization faster or reduce the adverse effects on hemorrhage and neurotoxicity compared to rtPA. To date, none of these strategies have advanced to regulatory approval. Preclinical studies are also evaluating novel thrombolytic strategies with the potential for reduced side effects and increased specificity of action (Table 6.1).

Despite its limitations, thrombolysis with rtPA is an important treatment option for acute ischemic stroke. It is underutilized (<10%) because of safety concerns, a short therapeutic time window, and insufficient hospital resources for rapid evaluation and treatment. In addition, rtPA only benefits a small portion of treated patients. Only one-third of the rtPA-treated stroke patients show substantially improved outcome when treated at 3 h after onset and this drops to one-sixth with treatment at 4.5 h [25, 59]. Thus, to improve thrombolytic treatment for stroke, in addition to getting more patients treated and reperfused earlier [60], it is important to understand the mechanisms responsible for ICH and the lack of efficacy in this substantial number of treated patients. Preclinical studies, discussed in the following section, have begun to unravel the complex actions of PAs and their endogenous inhibitors in normal and pathologic brain function.

6.3 Preclinical Studies

6.3.1 Localization of PAs, Neuroserpin, and Plasminogen in the Brain

The endogenous PAs, uPA, and tPA, are expressed in many different tissues; however, their best understood role *in vivo* is in blood where they convert plasminogen

Table 6.1 Thrombolytic Agents.

Agent	Mechanism of action	Status
rtPA	Plasminogen activator	FDA approved
Urokinase	Plasminogen activator	Intra-arterial trial (MELT) aborted and primary end point not reached [42]
rPro-urokinase	Plasminogen activator	Positive phase II trial for intra-arterial use Not approved by FDA [43, 44]
Reteplase	Recombinant tPA variant	Longer half-life than rtPA; under investigation for intra-arterial use [45, 46]
Tenecteplace	Recombinant tPA variant	Longer half-life and greater fibrin specificity than tPA More resistant to PAI-1 inhibition Phase IIB/III prematurely terminated Approved for myocardial infarction (MI) Further study indicated [47–49]
Alfimeprase	Direct fibrin degradation	Failed phase II [50, 51]
Desmoteplase	Plasminogen activator	Failed phase III Currently enrolling patients for DIAS III/IV trial (see text)
Plasmin	Direct fibrin degradation	Phase I, phase II trials, ClinicalTrials.gov Identifier: NCT01014975
BB10153	Plasminogen variant activated by thrombin	Under investigation [52]
Microplasmin	Recombinant truncated plasmin	Phase I safety study completed Preclinical studies suggest reduced ICH [53, 54]
Staphylokinase	Highly fibrin-selective PA with resistance to inhibition	Tested in MI Could be limited by immunogenicity [55]
RBC-coupled tPA	Increased plasma half-life and confining activity to vascular compartment	Preclinical proof of concept [56–58]

to plasmin in the process of fibrinolysis or thrombolysis [61]. Plasmin, a promiscuous protease, degrades fibrin and fibrinogen as well as coagulation factors, V and VIII. This process is tightly regulated by serine protease inhibitors of the serpin family, such as PAI-1 and α-2-antiplasmin. In the brain, as in other organs, a significant source of tPA is the endothelial cells (ECs) of the cerebral vasculature. tPA is secreted from the ECs in response to hypoxia and/or thrombus formation and is responsible for maintaining vessel patency and for the spontaneous recanalization seen in some strokes.

The PA/plasminogen system also plays an important role in the extracellular compartment of the brain as well as in the vascular lumen. In the developing nervous system, PAs are synthesized by glial and neuronal cells and have been associated with cell migration and remodeling [62–64]. In the adult uninjured brain, tPA is the primary PA present, as uPA expression in not detectable [63, 65, 66]. tPA mRNA transcripts have been detected in widespread brain regions including the hippocampus, hypothalamus, amygdala, cortex, and cerebellum and are primarily localized in neurons [63]. tPA protein and proteolytic activity are seen in these same regions; however, there are specific cell layers of the hippocampus (CA-1) and cerebellum (granule cells) that appear to express tPA messages but lack proteolytic activity [65]. This may be the result of regulation at the translational level or regional expression of the endogenous tPA inhibitors, neuroserpin, and/or PN-1 [67]. Neuroserpin, a member of the serpin gene family, is mainly expressed in the brain where it is thought to be the primary regulator of tPA activity [68]. This is unlike the related serpin PAI-1, which is the primary tPA inhibitor outside the central nervous system (CNS) but which has very low expression in the uninjured brain. Also, unlike PAI-1, which inhibits uPA and tPA equally well, neuroserpin inhibits tPA significantly faster than it does uPA [67]. In the normal adult brain, neuroserpin is primarily expressed in neurons and its regional localization is similar to that of tPA [68]. The serpin PN-1 is highly expressed during development and in specific brain structures in the adult. Its expression is increased in glial cells after neuronal injury; however, its importance in stroke is unclear [69–71].

Plasminogen is found in neurons of the hippocampus, cortex, hypothalamus, and cerebellum [72–74]. Hippocampal plasminogen immunoreactivity was observed in the pyramidal cell bodies, with especially strong staining in parvalbumin-positive GABAergic neurons. Plasminogen was also seen in layers II, III, and V of the cortex and in subpopulations of hypothalamic neurons. Thus, there is a brain-region-specific expression of the components of the tPA proteolytic cascade as well as the serpins that regulate their activity, suggesting an important role for this system in normal brain function. Indeed, tPA and neuroserpin have been shown to be involved in multiple CNS functions including learning- and memory-associated synaptic remodeling and long-term potentiation of neuronal activity (LTP) [75, 76], stress responses [77], anxiety-like behavior [78, 79], and regulation of CBF in response to increased neuronal activity [80]. CNS injury can also upregulate tPA and neuroserpin as well as uPA and PAI-1 [81–83].

6.4 The Association of Endogenous tPA with Excitotoxic and Ischemic Brain Injury

6.4.1 Excitotoxicity

Investigations begun in the 1980s demonstrated that the administration of rtPA improved CBF and reduced ischemic brain injury in models of thromboembolic stroke and this supported the clinical use of tPA in ischemic stroke [16]. However, by the early 1990s evidence began to emerge suggesting that the tPA/plasmin system might also have other functions in the CNS, unrelated to their well-understood role in fibrinolysis [76, 84, 85]. The availability of the tPA knock-out mouse [86] and other mouse genetic models by the mid-1990s accelerated these discoveries, and many studies began to suggest that under certain pathological conditions tPA in the CNS might promote neuronal injury [87, 88]. One of first studies to suggest a nonfibrinolytic role for tPA was published by Qian, and it demonstrated that a tPA message is markedly induced in the hippocampus in a kindling model of epilepsy, and in high-frequency brain stimulation producing LTP [84]. This work suggested that tPA might be involved in processes associated with learning and memory, and was soon followed by studies showing a role for tPA in motor learning [76] and LTP [75], both processes that involve synaptic remodeling. The observations by Qian also stimulated studies of tPA's role in CNS pathology. The rapid induction of tPA messages in the hippocampus, a brain region that is known to be selectively vulnerable to a variety of injurious conditions, led to studies of tPA in models of excitotoxicity. Excitotoxicity is a pathological process of neuronal cell death caused by excessive activation of glutamate receptors (NMDA, AMPA, and kainate receptors) and elevation of intracellular calcium levels [89]. In these studies, genetically modified mice lacking tPA (tPA$^{-/-}$) were found to be resistant to neuronal degeneration induced by hippocampal injection of the glutamate analog kainate [90]. These mice had decreased activation of microglia and were also less susceptible to Metrazol and kainate-induced seizures. In a subsequent study, kainate-induced hippocampal cell loss was restored in tPA$^{-/-}$ mice by tPA injected into the hippocampi and, in addition, wild-type mice infused with the tPA inhibitor PAI-1 were protected [91]. Later studies demonstrated that plasminogen null mice, but not fibrinogen null mice, were protected from kainate excitotoxic cell loss, implicating a nonfibrin substrate for plasmin in this process [92]. However, tPA, but not plasminogen, is required to see microglial activation in this model, suggesting both plasminogen-dependent and plasminogen-independent effects of tPA in the brain [74]. Excitotoxic damage in the cortex and striatum induced by another glutamate analog, NMDA, was also prevented by the coadministration of tPA inhibitor neuroserpin [93]. These actions may be specific to the NMDA receptor (NMDAR) since in primary neuronal cultures, neuroserpin protected against NMDA- but not AMPA-induced cell death. Overall, these studies demonstrate that endogenous tPA can play a significant role in the neuronal degeneration induced by excitotoxic levels of glutamate receptor activation.

6.4.2
Focal Ischemia

Excitotoxicity is an important mechanism contributing to ischemic brain injury [94]. Using animal models of focal ischemic stroke, the roles of both endogenous and exogenous tPA and other components of this proteolytic pathway have been studied. Since the early 1980s, multiple rodent models of focal ischemia have been developed [95]. They usually involve middle cerebral artery occlusion (MCAO), either permanently or temporarily with varying durations of reperfusion. Mechanical occlusion can be accomplished by ligation, electrocautery, or passage of a filament through the internal carotid artery to block the origin of the middle cerebral artery (MCA). Thromboembolic MCAO can be produced by injection of autologous clots or by inducing clot formation locally following an IV injection of a photosensitive dye (Rose Bengal) and activation by a laser focused on the MCA. The thromboembolic and intraluminal filament models allow reperfusion by tPA administration or filament withdrawal, respectively. CBF is often monitored by laser Doppler in the affected cortex to assure adequate ischemia and reperfusion. Comparing studies that use animal models of ischemia can be difficult because their properties vary greatly and the results can be influenced by small procedural differences. This is especially true in mice, where it is difficult to monitor and maintain the physiological status. The genetic background in genetically engineered mice is also a critical variable that must be correctly controlled for [96].

The role of endogenous tPA in ischemic stroke was first studied in mice using the intravascular filament model [97]. After 3 h of ischemia and 21 h of reperfusion, the stroke lesion size was approximately 50% smaller in $tPA^{-/-}$ mice compared to that in wild-type C57Bl6 controls. When rtPA was given by IV infusion 2 h after ischemia in $tPA^{-/-}$ or wild-type mice, lesion volumes were increased 33–50% in all groups. CBF in response to insertion and removal of the occluding filament was similar across groups. *In situ* zymography indicated that PA activity was induced by 2 h of ischemia in the brains of wild-type but not $tPA^{-/-}$ mice. These results clearly showed that endogenous and exogenous tPA can adversely affect the outcome of ischemic stroke and were provocative in light of the approval of rtPA for the treatment of stroke in humans less than two years earlier [98]. This controversy was further heightened when a very similar study using the same MCAO model as Wang (3 h of ischemia followed by 24 h reperfusion) reported essentially opposite results [99–101]. The reason for this difference was not clear; however, the lesion volume in the $tPA^{-/-}$ mice was only larger when compared to control mice on a "matched mixed 129/Sv and C57Bl/6 background," while the lesion was smaller when compared to wild-type C57BL/6 controls. The use of "matched mixed" backgrounds could have contributed to this difference since these mixed backgrounds were being bred separately and were, in essence, different strains. Consistent with this possibility, the lesion volume of the mixed "wild types" was surprisingly small for a 3 h MCAO, which is a relatively severe model. This controversy further highlights the importance of controlling for genetic backgrounds, and when mixed genetic backgrounds are studied, heterozygous mating and littermate controls should always be used [96]. Finally, it should be noted that CBF during and after ischemia was lower in the $tPA^{-/-}$ mice compared to that in control mice, suggesting the possibility that secondary thrombosis caused by microthrombi and lack of tPA-mediated lysis might have contributed to larger lesions in the $tPA^{-/-}$ mice in this study [100].

Since then, numerous studies have addressed this important issue, and have largely confirmed the observations of Wang. In the first of these follow-up studies, several of the components of the PA system were systematically evaluated in a permanent ligation MCAO model [102]. Five different genetic models were compared to wild-type control mice. This study confirmed that $tPA^{-/-}$ mice had significantly reduced lesion volumes, as did alpha-2-antiplasmin$^{-/-}$ mice. In contrast, PAI-1$^{-/-}$ and plasminogen null ($Plg^{-/-}$) mice had larger lesions sizes, while $uPA^{-/-}$ mice had lesions similar to wild-type mice. The PAI-1$^{-/-}$ and $tPA^{-/-}$ results are consistent with a role for tPA in ischemic brain damage, and the lack of a phenotype in the $uPA^{-/-}$ mice suggests that this effect is specific to tPA. Furthermore, in contrast to the excitotoxic hippocampal injury discussed above [74], the damaging effects of tPA in ischemic injury appear to be independent of plasminogen activation as the $Plg^{-/-}$ mice had the opposite phenotype of the $tPA^{-/-}$ mice. In a later study, this group also analyzed transgenic mice overexpressing PAI-1, and compared two types of MCAO, a permanent ligation model that does not allow reperfusion and thromboembolic MCAO model [103]. Following permanent MCAO ligation, PAI-1 overexpressing mice had smaller lesion volumes, suggesting that at least in models where reperfusion plays no role in ischemic damage, tPA inhibition is protective. In contrast, in the thromboembolic model the lesion volumes were larger in the PAI-1 overexpressing mice. Together, these data suggest that the role of tPA in cerebral ischemia is complex and may be context specific. For example, tPA activity within the CNS might exacerbate ischemic damage through interactions that do not involve plasminogen, but tPA in the blood stream, which can promote thrombolysis, might be beneficial in thromboembolic stroke. Consistent with this interpretation, recombinant PAI-1 injected directly into the cerebrospinal fluid reduced ischemic injury in both stroke models in wild-type mice, but had no effect in $tPA^{-/-}$ mice [103].

Other tPA inhibitors have also shown protection in different stroke models. Neuroserpin injection into the brain following permanent MCAO in rats reduced lesion volume by 64% with a corresponding reduction in both tPA and uPA activity in the cortex [82]. Likewise, neuroserpin overexpressing mice have significantly reduced brain tPA activity, and lesion volumes that were 30% smaller compared to wild-type controls after permanent MCAO [104]. Thus, the weight of the data from genetically modified mice and studies where inhibitors were administered directly into the brain supports the idea that endogenous brain tPA activity within the CNS can exacerbate ischemic brain damage. However, studies have also shown that intravascular tPA can be beneficial. For example, tPA delivered IV or intracarotid in thromboembolic models improves outcomes if given early [105–107]. In mechanical occlusion models, the findings with tPA administration are even more divergent and most likely depend on the specific application of the model [108, 109]. Finally, a direct comparison of the effect of tPA infusion on intraluminal thread and embolic stroke in rats demonstrated adverse effects on lesion size in the thread model and protection in the embolic model [110]. Together, these data confirm the importance of both intravascular plasminogen activation in clot lysis and the concept of neurotoxicity by tPA in the CNS during ischemia.

6.4.3 Global Ischemia

Two recent studies have also examined the role of tPA in models of transient global cerebral ischemia with reperfusion, and as in the early studies of focal ischemia the two studies appear to demonstrate opposite results. In this model, blood flow to the brain is interrupted by 20 min of bilateral common carotid artery occlusion, leading to delayed selective neuronal death in the hippocampus. In the first study, $tPA^{-/-}$ mice were reported to be protected from neuronal death in the hippocampal CA-1 region [111]. However, in the second study, the opposite effect of tPA deficiency was reported, with increasing neuronal cell death in the CA-1 layer of the hippocampus [112]. In the latter study, the authors proposed that tPA was neuroprotective, and used cell culture studies to suggest that tPA was involved in ischemic preconditioning, a process where a mild insult protects the brain from a subsequent lethal ischemic event. The reason for these discrepant results is unclear and additional studies are needed to clarify the role of endogenous tPA in global ischemia and confirm its potential role in preconditioning. Nonetheless, taken together with the studies discussed above, there is substantial evidence that both endogenous and exogenous tPA can exacerbate brain injury in many different rodent models of ischemia. The mechanisms responsible for the detrimental effects of tPA in the brain are not completely understood and are a topic of active research. Recent studies have focused on two critical actions of tPA within the CNS; modulation of the neuronal NMDAR and interactions with the neurovascular unit (NVU), causing disruption of the BBB. The following sections discuss these two potential mechanisms.

6.5 Mechanistic Studies of tPA in Excitotoxic and Ischemic Brain Injury

6.5.1 tPA and the NMDA Receptor

Cerebral ischemia results in a large increase in glutamate in the extracellular space [4], and it is well established that excessive activation of glutamate receptors causes neuronal cell injury (excitotoxicity), which is an important contributor to ischemic brain injury after stroke [113]. The NMDA subtype of glutamate receptor is thought to be largely responsible for excitotoxicity. In the normal brain, it functions in excitatory neurotransmission and is critical to LTP and synaptic plasticity [114], and tPA is known to play a role in both LTP and excitotoxic cell death (see above) leading to the hypothesis that there may be a causal link between tPA released with neuronal depolarization and regulation of NMDAR function. Numerous studies have explored this idea and have generated diverse data and several mechanistic models to explain the action of tPA in the modulation of NMDA signaling. The NMDAR is a tetramer composed of two NR1 and two NR2 (subtypes NR2A-D) subunits [115, 116]. The NR1 subunits express the receptor/channel properties, whereas the NR2 subunits have a modulatory function and can interact with the intracellular adaptor and scaffolding proteins. Glutamate activation of the receptor opens a nonselective ion channel, which allows calcium and sodium to enter the cell. In neuronal cultures, tPA was found to potentiate both NMDA-induced calcium entry and neuronal cell death [117]. This potentiation by tPA was blocked by PAI-1 but was not induced by plasmin treatment, and it was originally proposed that cleavage of the NR1 subunit was responsible for the potentiating effects of tPA on NMDAR function [117]. However, while the enhancing effect of tPA on NMDAR function has been consistently confirmed, several other studies have been unable to replicate direct tPA cleavage of NR1 subunit, and the exact mechanism remains controversial [118–130]. An alternative hypothesis that has been proposed is that rather than tPA cleaving a subunit of the NMDAR it is the interaction of a tPA–serpin complex that potentiates NMDAR function through engagement of the complex with LDL-receptor-related protein (LRP) [131]. This is consistent with data showing that tPA must be proteolytically active to enhance NMDA function because complex formation requires an active protease to bind the serpin [120, 131] and with data demonstrating that LRP mediates neuronal NMDAR signaling through an interaction that involves the intracellular LRP NPXY domain and the postsynaptic density, PSD protein-95 [132–137].

Regardless of the mechanism or receptor subunit, evidence demonstrates that tPA potentiates NMDAR function where it has effects in synaptic plasticity [75], neurovascular coupling [80], the response to acute stress [79], and in the pathological condition of excitotoxicity [90]. Thus, this potentiation could at least partially contribute to neuronal cell death resulting from ischemia and other CNS pathologies where tPA is released from depolarized cells, and may be a factor that limits the effectiveness of rtPA treatment in stroke patients due to entry of rtPA across compromised BBB.

6.5.2
tPA and the Blood–Brain Barrier

The BBB is a property of the cerebral vascular bed that restricts and regulates the movement of substance across the endothelial cells lining the vessels [138]. Regulation of the brain extracellular environment is critical to the maintenance of normal neuronal function [139], and the physical barrier that restricts passive movement of substances such as ions, macromolecules, and neurotransmitters is primarily due to the tight junctional complex between endothelial cells [140]. The junctional complex consists of adherens proteins, cadherins; and the tight junction proteins, claudins and occludin, that interact with scaffolding proteins (zonula occludens); and the actin cytoskeleton. In addition to the physical barrier, specific transporters, channels, and vesicular pathways regulate ion and water homeostasis, nutrient uptake, and removal of metabolic waste. The movement of cellular components such as neutrophils into the brain is also restricted by the BBB, although mononuclear cell can cross by diapedesis. The properties of the junctional complex can be modulated by interactions with the extracellular environment and surrounding cells. Because of the complex interactions of brain endothelial cells with surrounding components, the BBB is now considered part of the NVU, which consists of the endothelial cells, basal lamina, perivascular neurons, astrocytes, and pericytes. This complex is critical to the maintenance of tight junctions and the active regulation of the brain microvasculature. BBB damage is an important element of the pathophysiology of ischemic brain injury and leads to entry of macromolecules and inflammatory cells, and to brain edema [7]. Severe disruption of the BBB may ultimately lead to structural damage and hemorrhage. The molecular mechanisms that disrupt and dysregulate the BBB after ischemia are complex and not fully understood; however, recent evidence has begun to define a critical role for tPA. The pathway through which tPA affects the BBB involves the LRP, platelet-derived growth factor-CC (PDGF-CC), and the platelet-derived growth factor receptor α(PDGFRα). tPA may also influence the expression or activation of matrix metalloproteinases (MMPs) and their effects on extracellular matrix disruption, BBB opening, and hemorrhage.

6.5.3 tPA and the Blood–Brain Barrier – MMPs

MMPs are endopeptidases that are primarily secreted as inactive zymogens. When activated, MMPs can digest many components of the extracellular matrix and the tight junction proteins [141]. Their concentration in the brain is normally very low but increases in response to tissue injury and ischemia. Numerous studies have evaluated the role of MMPs in ischemic brain damage, where they are implicated in BBB damage through degradation of the basal lamina and junctional proteins [142–144]. MMP-9 and MMP-2 have received the most attention, although other members of the family have been implicated [145]. Several studies have found that tPA is associated with increases in the activity of MMP-9 in the rodent brain after stroke; however, the mechanism of this association is controversial [146, 147]. Ischemia-induced MMP-9 activity is markedly reduced in $tPA^{-/-}$ mice, and with neuroserpin injection following MCAO in rats [147–149]. Likewise, thrombolysis with tPA in thromboembolic stroke in rats is associated with increased brain MMP-9 activity between 6 and 24 h depending on the study [150, 151]. It has also been suggested that activation of pro-MMPs by tPA-generated plasmin is responsible for the increased activity [152]; however, this pathway seems unlikely since ischemia-induced brain MMP-9 activity appears normal in $Plg^{-/-}$ mice [147].

In addition, recent studies in MMP-9 null mice have shown that BBB disruption following permanent MCAO (6 h) or transient MCAO with rtPA administration (24 h) was not reduced by genetic deletion [147, 153]. Finally, although there is evidence that MMPs can be expressed by several cell types in the CNS, the cellular source responsible for increased activity in stroke is very likely invading inflammatory cells, since ischemia-induced brain MMP-9 activity was completely absent in chimeric mice lacking leukocyte MMP-9 but with otherwise normal MMP-9 expression [154]. Also, neutrophil depletion from the blood markedly reduces the amount of pro-MMP-9 in the brain after transient MCAO in rats [155], although in another study neutrophil depletion followed by a more severe MCAO model found no effect on the increase in active MMP-9 in the brain [156]. Given that the majority of MMP-9 activity in the brain in the first 24 h after stroke appears to come from infiltrating leukocytes, and that tPA in the CNS induces early loss of BBB integrity, it is possible that the BBB disruption induced by tPA/PDGF-CC (see below) promotes infiltration of leukocytes leading to increases in MMP-9 activity, which could further damage the BBB through MMP-mediated matrix degradation. Thus, the link between tPA and MMPs in stroke may be indirect through the tPA's ability to enhance leukocyte recruitment into the brain during the early response to cerebral ischemia. In this regard, it is interesting to note that tPA can also directly stimulate neutrophil degranulation and MMP-9 release [157]. The role of MMPs in the latter stages of stroke pathology, beyond 6–24 h, is clear. For example, analysis of human ischemic stroke tissue found elevated MMP-9 in the infarcted areas and in brain regions with hemorrhagic transformation [158]. Similarly, MMP-9-positive neutrophils are associated with microvessels showing basal lamina disruption and collagen IV degradation, and tissues with hemorrhage have the greatest number of MMP-9-positive neutrophils and disrupted microvessels. However, MMP-9 appears to be acting at later times following ischemic injury than tPA, and is associated with the entry of neutrophils that infiltrate the brain during the reperfusion phase.

6.5.4
tPA and the Blood–Brain Barrier – LRP

LRP is a member of the low-density lipoprotein (LDL) receptor family. It is a large endocytic receptor with multiple and complex functions including lipid metabolism and regulation of proteases and their inhibitors [159, 160]. The LRP extracellular domain has four clusters of ligand-binding repeats, which are sites responsible for specific binding of a large number of diverse ligands. Binding can result in internalization and degradation of the ligand and also trigger intracellular signaling. The LRP cytoplasmic domain interacts with intracellular adapter and scaffold proteins and is involved in signaling and intracellular trafficking. LRP is expressed in most tissues and has an important role in regulating plasma and tissue protease activity, partially through cellular uptake of serpin-bound serine proteases including the clearance of tPA from the plasma [161]. LRP expression in the brain is found in neurons, astrocytes, pericytes, and vascular smooth muscle, while endothelial cell expression is very low [162]. LRP binds tPA [163] and neuroserpin–tPA complexes [164] but not uncomplexed active or cleaved neuroserpin. Cleavage and shedding of the extracellular domain and release of the intracellular domain and its associated proteins have been described as an important step in some LRP signaling events [163, 165].

The role of tPA in early opening of the BBB in stroke was originally demonstrated in studies that showed accumulation of albumin-bound Evans Blue dye in the brain parenchyma 6 h after induction of permanent MCAO in mice [147]. In these studies, $tPA^{-/-}$ mice demonstrated greatly reduced BBB damage compared to wild-type controls, as did wild-type mice treated with intracerebral neuroserpin. In contrast, mice lacking uPA, Plg, and MMP-9 all showed BBB leakage similar to their wild-type controls, demonstrating that endogenous tPA is necessary for the early loss of BBB integrity after focal ischemia, and that this effect is independent of plasminogen and MMP-9. Increased tPA activity was also observed in the perivascular space surrounding cerebral vessels as early as 1 h after MCAO. Furthermore, direct injection of tPA into the cerebrospinal fluid induced opening of the BBB within 1 h in the absence of cerebral ischemia. This activity was also plasminogen independent, but did require active tPA, since chemically inactivated tPA [147], or an active-site mutant of tPA [166] did not induce BBB opening. This suggested the existence of a nonplasminogen substrate for tPA. The tPA-induced BBB opening was also LRP dependent since when tPA was coinjected with the LRP antagonist, the receptor-associated protein (RAP) or an anti-LRP antibody, the opening of the BBB was significantly attenuated. Thus, tPA is both necessary and sufficient to induce rapid BBB disruption most likely via a nonplasminogen substrate for tPA but requiring an interaction with LRP. It has been suggested that this process might involve shedding of LRP from the perivascular astrocytic endfeet [167]. However, the importance of LRP shedding in the BBB opening is not known. In another study, it was shown that cerebral ischemia induces expression of LRP in the brain, which is attenuated in $tPA^{-/-}$ mice, and that ischemia-induced activation of NF-κB and iNOS is tPA dependent via an LRP-mediated process, suggesting a linkage between tPA, LRP, and inflammatory pathways in BBB dysregulation after ischemia [168]. Finally, it has also been suggested that tPA can induce the expression of MMP-9 following cerebral ischemia in an LRP-dependent manner [146]. However, the role of this pathway in the early opening of the BBB is not clear since other studies have shown that MMP-9 null mice are not protected from early loss of barrier function 6 h after MCAO [147].

6.6
tPA and the Blood–Brain Barrier–PDGF-CC

The PDGFs are members of a family of genes that activate specific tyrosine kinase receptors in many different settings [169]. This family of genes includes not only the PDGFs A through D, but also the vascular endothelial growth factors (VEGFs) and placental growth factor (PGF). PDGFs are potent mitogens associated with processes such as angiogenesis. PDGFs are produced as homo- and heterodimers that can bind and activate two related PDGF receptors α and β(PDGFRα, PDGFRβ). PDGF-AA, PDGF-BB, and PDGF-AB are secreted in their active forms, whereas PDGF-CC and PDGF-DD contain N-terminal CUB domains that block receptor binding and must be proteolytically removed to activate the growth factor domain. In 2004, tPA was identified as an activator of PDGF-CC [170]. This action was highly specific since tPA could not cleave the closely related PDGF-DD, and since uPA could not activate PDGF-CC. Active PDGF-CC binds and signals through the PDGFRα. The observations that proteolytically active endogenous brain tPA increases focal ischemic lesion size and BBB opening independent of plasminogen suggested an alternate substrate [102, 147]. Thus, the observation that PDGF-CC was a substrate for tPA [170] led to studies examining whether PDGF-CC was involved in tPA-mediated BBB regulation [166, 171]. These studies demonstrated that even in the absence of ischemia, the direct injections of tPA or active PDGF-CC into the cerebrospinal fluid increased BBB opening as early as 1 h, but that injection of much higher doses of tPA in the blood did not affect the BBB of healthy mice. This suggested that tPA or active PDGF-CC in the brain could affect BBB permeability. Furthermore, tPA-induced BBB opening was greatly reduced by anti-PDGF-CC antibodies, suggesting that the PDGF-CC effect was downstream of tPA. Other PDGFRα agonists (PDGF-AA and PDGF-BB) could also induce opening of the BBB, strongly implicating PDGFRα in this process; however, neutralizing antibodies against PDGF-AB did not attenuate the tPA-induced opening confirming the role of PDGF-CC. Interestingly, the LRP antagonist RAP did not block active PDGF-CC-induced opening of the BBB as it did with tPA [147], suggesting that active PDGF-CC is downstream of both tPA and LRP, and raising the possibility that LRP may be involved in the activation of latent PDGF-CC by tPA since both tPA and PDGF-CC appear to bind to LRP, and since mouse embryonic fibroblasts genetically deficient in LRP do not process latent PDGF-CC in the presence of tPA [166]. Immunohistological and fluorescence studies of brain tissue from heterozygous reporter mice expressing GFP from the PDGFRα locus demonstrated that tPA, PDGF-CC, and PDGFRα are all localized to the cerebral arterioles and strongly associated with the perivascular astrocytes, precisely the location where transmission electron microscopy showed developing edema following injection of either tPA or active PDGF-CC. Interestingly, there was no major structural damage to the vasculature in this region, suggesting that activation of PDGFRα is a regulated process that controls BBB function. An important role for this pathway in stroke was supported by the *in vivo* results showing that 6 h after photothrombotic MCAO, PDGFRα is activated in the ischemic cortex of wild-type but not $tPA^{-/-}$ mice. Likewise, treatment with the PDGFRα antagonist, imatinib, or an anti-PDGF-CC antibody reduced BBB opening after MCAO. Finally, when mice subjected to photothrombotic stroke were reperfused with delayed rtPA thrombolysis 5 h after MCAO, they developed significant cortical hemorrhages that were markedly reduced by imatinib treatment. This suggests that blocking early PDGFRα-dependent BBB opening might ameliorate severe BBB damage and potentially hemorrhagic complications resulting from delayed rtPA treatment, possibly by preventing thrombolytic rtPA entry into the brain parenchyma where it could activate additional PDGF-CC further stimulating PDGFRα.

A recent study has questioned the role of PDGF-CC in BBB regulation, and reported that PDGF-CC is a potent neuroprotective factor that is critically required for neuronal survival [172]. They based their conclusions in part on the observations that PDGF-CC null mice have larger strokes than wild-type mice after permanent MCAO, and that PDGF-CC protected cultured neurons from induction of apoptosis. The differences between this study and that of Su *et al.* highlight the complexity of animal models of stroke; however, there are several potential explanations for the apparent differences. In the studies of MCAO in the PDGF-CC null mice, it is very likely that the significant congenital abnormalities in the brain and cerebral vasculature of PDGF-CC null mice could affect infarct size independent of any acute effects of PDGF-CC deficiency [173]. The latter conclusion that PDGF-CC is a potent neuroprotective factor is also problematic since it is based largely on studies of cultured neurons that express the PDGFRα; however, adult rodent neurons do not to express PDGFRα *in vivo* [173–177]. Thus, studies with neurons in culture may not be informative with regard to the effects of PDGF-CC *in vivo*. Finally, regarding BBB regulation there was at least one procedural difference between the two studies that might affect outcome. In the paper by Su *et al.*, BBB opening induced by active PDGF-CC was compared to opening induced by tPA, inactive tPA, or saline, whereas Tang compared PDGF-CC to injections of bovine serum albumin (BSA). Unfortunately, albumin is reported to have direct effects on cerebral function and may induce opening of tight junctions [178, 179]. Finally, a recent study of ICH in mice, currently available as an "Accepted Article" in the *Annals of Neurology*, has confirmed the observation that blocking PDGFRα signaling protects the BBB [180]. Additional studies are needed to clarify the mechanisms underlying the role of PDGF-CC in tPA-mediated effects on the BBB and hemorrhage; however, modulation of PDGFRα provides a new and promising approach to improving the outcomes of thrombolytic treatment of ischemic stroke.

6.7 Summary

The highly specific serine protease, rtPA, is a thrombolytic agent and the only approved treatment for ischemic stroke. Current guidelines recommend that treatment be initiated no later than 4.5 h after the onset of stroke symptoms because of the loss of effectiveness and increased occurrence of symptomatic ICH over time. This limitation greatly reduces the number of patients who are treated, and efforts to extend this narrow time window by selecting patients using advanced imaging to identify those with salvageable tissue are ongoing. Evidence has also emerged that the benefits of thrombolytic reperfusion with rtPA may be countered by the damaging actions of tPA in the brain parenchyma. Endogenous brain tPA and the serpins that regulate its activity have a significant role in normal CNS function where they promote neuronal plasticity, learning and memory, and help maintain local CBF appropriate to the needs of the tissue, and regulate normal BBB function. An important role for tPA in the pathophysiology of ischemic brain injury is now well established in animal models of stroke. Studies have focused on two areas related to the role of endogenous tPA in ischemic injury – NMDAR-induced excitotoxicity and BBB disruption leading to hemorrhagic transformation. Interestingly, although acting at different cell types, both these pathologic processes share some common properties related to the role of tPA. Excessive neuronal release of tPA into the extracellullar space or synaptic cleft is the initiator of a pathway that requires proteolytically active tPA, but not its classic substrate plasminogen, thus implicating an alternate tPA substrate(s). The substrates that have been proposed to be responsible for tPAs potentiation of NMDAR-dependent increases in intracellular calcium and excitotoxicity are the direct cleavage of the receptor itself or a serpin/tPA interaction, which, through binding to LRP, facilitates the LRP interaction with the NMDAR. The effects of tPA on the BBB are also LRP dependent and early BBB leakage after a stroke requires the proteolytic activation of latent PDGF-CC by tPA. This early barrier opening may allow entry of blood constituents, including rtPA in treated subjects, and leukocytes, which can cause further damage through MMP release and lead to hemorrhage. These normal and pathological CNS functions of tPA may be the underlying cause of the limited effectiveness, short therapeutic time window, and hemorrhagic side effects of rtPA treatment in acute stroke. Understanding the mechanisms of endogenous tPA action in normal and pathological CNS function will aid in the development of safer and more effective treatments for stroke. Clinical and preclinical efforts to improve thrombolytic therapy for stroke are ongoing and include identifying and developing novel fibrinolytic agents that do not overlap with brain tPA function, finding faster acting agents and combination therapies to counter adverse effects of rtPA, improving imaging for identification of salvageable brain tissue, developing mechanical clot removal techniques in combination with tPA, and improving the clinical procedures for implementing thrombolysis so that more patients are treated and treatment begins sooner. Thus, exploring the complex actions tPA in the brain continues to be an important path toward mitigating the devastating effect of ischemic stroke in man.

Acknowledgments

This work was supported by National Institutes of Health grants HL-55 374, HL-54 710, and HL-89 407 (to D.A. Lawrence). We thank E. J. Su for helpful discussions and reading of the manuscript.

References

1. Astrup, J., Siesjo, B.K., and Symon, L. (1981) Thresholds in cerebral ischemia – the ischemic penumbra. *Stroke*, **12** (6), 723–725.
2. Bell, B.A., Symon, L., and Branston, N.M. (1985) CBF and time thresholds for the formation of ischemic cerebral edema, and effect of reperfusion in baboons. *J. Neurosurg.*, **62** (1), 31–41.
3. Strong, A.J., Venables, G.S., and Gibson, G. (1983) The cortical ischaemic penumbra associated with occlusion of the middle cerebral artery in the cat: 1. Topography of changes in blood flow, potassium ion activity, and EEG. *J. Cereb. Blood Flow Metab.*, **3** (1), 86–96.
4. Benveniste, H., Drejer, J., Schousboe, A., and Diemer, N.H. (1984) Elevation of the extracellular concentrations of glutamate and aspartate in rat hippocampus during transient cerebral ischemia monitored by intracerebral microdialysis. *J. Neurochem.*, **43** (5), 1369–1374.
5. Shin, H.K., Dunn, A.K., Jones, P.B., Boas, D.A., Moskowitz, M.A., and Ayata, C. (2006) Vasoconstrictive neurovascular coupling during focal ischemic depolarizations. *J. Cereb. Blood Flow Metab.*, **26** (8), 1018–1030.
6. Back, T., Ginsberg, M.D., Dietrich, W.D., and Watson, B.D. (1996) Induction of spreading depression in the ischemic hemisphere following experimental middle cerebral artery occlusion: effect on infarct morphology. *J. Cereb. Blood Flow Metab.*, **16** (2), 202–213.
7. Betz, A.L. (1996) Alterations in cerebral endothelial cell function in ischemia. *Adv. Neurol.*, **71**, 301–311; discussion 311–313.
8. Barone, F.C. and Feuerstein, G.Z. (1999) Inflammatory mediators and stroke: new opportunities for novel therapeutics. *J. Cereb. Blood Flow Metab.*, **19** (8), 819–834.
9. Ginsberg, M.D. and Pulsinelli, W.A. (1994) The ischemic penumbra, injury thresholds, and the therapeutic window for acute stroke. *Ann. Neurol.*, **36** (4), 553–554.
10. Jones, T.H., Morawetz, R.B., Crowell, R.M., Marcoux, F.W., FitzGibbon, S.J., DeGirolami, U., and Ojemann, R.G. (1981) Thresholds of focal cerebral ischemia in awake monkeys. *J. Neurosurg.*, **54** (6), 773–782.
11. Overgaard, K., Sereghy, T., Pedersen, H., and Boysen, G. (1994) Effect of delayed thrombolysis with rt-PA in a rat embolic stroke model. *J. Cereb. Blood Flow Metab.*, **14** (3), 472–477.
12. Kaplan, B., Brint, S., Tanabe, J., Jacewicz, M., Wang, X.J., and Pulsinelli, W. (1991) Temporal thresholds for neocortical infarction in rats subjected to reversible focal cerebral ischemia. *Stroke*, **22** (8), 1032–1039.
13. Memezawa, H., Smith, M.L., and Siesjo, B.K. (1992) Penumbral tissues salvaged by reperfusion following middle cerebral artery occlusion in rats. *Stroke*, **23** (4), 552–559.
14. Hanaway, J., Torack, R., Fletcher, A.P., and Landau, W.M. (1976) Intracranial bleeding associated with urokinase therapy for acute ischemic hemispheral stroke. *Stroke*, **7** (2), 143–146.
15. Lyden, P.D., Madden, K.P., Clark, W.M., Sasse, K.C., and Zivin, J.A. (1990) Incidence of cerebral hemorrhage after treatment with tissue plasminogen activator or streptokinase following embolic stroke in rabbits [corrected]. *Stroke*, **21** (11), 1589–1593.
16. Zivin, J.A., Fisher, M., DeGirolami, U., Hemenway, C.C., and Stashak, J.A. (1985) Tissue plasminogen activator reduces neurological damage after cerebral embolism. *Science*, **230**, 1289–1292.
17. NINDA and Stroke rtPA Study Group (1995) Tissue plasminogen activator for acute ischemic stroke. The national institute of neurological disorders and stroke rt-PA stroke study group. *N. Engl. J. Med.*, **333** (24), 1581–1587.
18. NINDA and Stroke rtPA Study Group (1997) Intracerebral hemorrhage after intravenous t-PA therapy for ischemic stroke. *Stroke*, **28** (11), 2109–2118.

19. Hacke, W., Kaste, M., Fieschi, C., Toni, D., Lesaffre, E., von Kummer, R., Boysen, G., Bluhmki, E., Hoxter, G., Mahagne, M.H. *et al.* (1995) Intravenous thrombolysis with recombinant tissue plasminogen activator for acute hemispheric stroke. The European cooperative acute stroke study (ECASS). *JAMA*, **274** (13), 1017–1025.
20. Hacke, W., Kaste, M., Fieschi, C., von Kummer, R., Davalos, A., Meier, D., Larrue, V., Bluhmki, E., Davis, S., Donnan, G., Schneider, D., Diez-Tejedor, E., and Trouillas, P. (1998) Randomised double-blind placebo-controlled trial of thrombolytic therapy with intravenous alteplase in acute ischaemic stroke (ECASS II). Second European-Australasian acute stroke study investigators. *Lancet*, **352** (9136), 1245–1251.
21. Clark, W.M., Wissman, S., Albers, G.W., Jhamandas, J.H., Madden, K.P., and Hamilton, S. (1999) Recombinant tissue-type plasminogen activator (Alteplase) for ischemic stroke 3 to 5 hours after symptom onset. The ATLANTIS Study: a randomized controlled trial. Alteplase thrombolysis for acute noninterventional therapy in ischemic stroke. *J. Am. Med. Assoc.*, **282** (21), 2019–2026.
22. Hacke, W., Donnan, G., Fieschi, C., Kaste, M., von Kummer, R., Broderick, J.P., Brott, T., Frankel, M., Grotta, J.C., Haley, E.C. Jr., Kwiatkowski, T., Levine, S.R., Lewandowski, C., Lu, M., Lyden, P., Marler, J.R., Patel, S., Tilley, B.C., Albers, G., Bluhmki, E., Wilhelm, M., and Hamilton, S. (2004) Association of outcome with early stroke treatment: pooled analysis of ATLANTIS, ECASS, and NINDS rt-PA stroke trials. *Lancet*, **363** (9411), 768–774.
23. Cronin, C.A. (2010) Intravenous tissue plasminogen activator for stroke: a review of the ECASS III results in relation to prior clinical trials. *J. Emerg. Med.*, **38** (1), 99–105.
24. Hacke, W., Kaste, M., Bluhmki, E., Brozman, M., Davalos, A., Guidetti, D., Larrue, V., Lees, K.R., Medeghri, Z., Machnig, T., Schneider, D., von Kummer, R., Wahlgren, N., and Toni, D. (2008) Thrombolysis with alteplase 3 to 4.5 hours after acute ischemic stroke. *N. Engl. J. Med.*, **359** (13), 1317–1329.
25. Saver, J.L., Gornbein, J., Grotta, J., Liebeskind, D., Lutsep, H., Schwamm, L., Scott, P., and Starkman, S. (2009) Number needed to treat to benefit and to harm for intravenous tissue plasminogen activator therapy in the 3- to 4.5-hour window: joint outcome table analysis of the ECASS 3 trial. *Stroke*, **40** (7), 2433–2437.
26. Del Zoppo, G.J., Saver, J.L., Jauch, E.C., and Adams, H.P. Jr. (2009) Expansion of the time window for treatment of acute ischemic stroke with intravenous tissue plasminogen activator: a science advisory from the American heart association/American stroke association. *Stroke*, **40** (8), 2945–2948.
27. Donnan, G.A., Baron, J.C., Ma, H., and Davis, S.M. (2009) Penumbral selection of patients for trials of acute stroke therapy. *Lancet Neurol.*, **8** (3), 261–269.
28. Baron, J.C. (1999) Mapping the ischaemic penumbra with PET: implications for acute stroke treatment. *Cerebrovasc. Dis.*, **9** (4), 193–201.
29. Heiss, W.D., Kracht, L.W., Thiel, A., Grond, M., and Pawlik, G. (2001) Penumbral probability thresholds of cortical flumazenil binding and blood flow predicting tissue outcome in patients with cerebral ischaemia. *Brain*, **124** (Pt 1), 20–29.
30. Heiss, W.D., Sobesky, J., Smekal, U., Kracht, L.W., Lehnhardt, F.G., Thiel, A., Jacobs, A.H., and Lackner, K. (2004) Probability of cortical infarction predicted by flumazenil binding and diffusion-weighted imaging signal intensity: a comparative positron emission tomography/magnetic resonance imaging study in early ischemic stroke. *Stroke*, **35** (8), 1892–1898.
31. Parsons, M.W., Christensen, S., McElduff, P., Levi, C.R., Butcher, K.S., De Silva, D.A., Ebinger, M., Barber, P.A., Bladin, C., Donnan, G.A., and Davis, S.M. (2010) Pretreatment diffusion- and perfusion-MR lesion volumes have a crucial influence on

clinical response to stroke thrombolysis. *J. Cereb. Blood Flow Metab.*, **30** (6), 1214–1225.

32. Albers, G.W., Thijs, V.N., Wechsler, L., Kemp, S., Schlaug, G., Skalabrin, E., Bammer, R., Kakuda, W., Lansberg, M.G., Shuaib, A., Coplin, W., Hamilton, S., Moseley, M., and Marks, M.P. (2006) Magnetic resonance imaging profiles predict clinical response to early reperfusion: the diffusion and perfusion imaging evaluation for understanding stroke evolution (DEFUSE) study. *Ann. Neurol.*, **60** (5), 508–517.
33. Davis, S.M., Donnan, G.A., Parsons, M.W., Levi, C., Butcher, K.S., Peeters, A., Barber, P.A., Bladin, C., De Silva, D.A., Byrnes, G., Chalk, J.B., Fink, J.N., Kimber, T.E., Schultz, D., Hand, P.J., Frayne, J., Hankey, G., Muir, K., Gerraty, R., Tress, B.M., and Desmond, P.M. (2008) Effects of alteplase beyond 3 h after stroke in the Echoplanar Imaging Thrombolytic Evaluation Trial (EPITHET): a placebo-controlled randomised trial. *Lancet Neurol.*, **7** (4), 299–309.
34. Nagakane, Y., Christensen, S., Brekenfeld, C., Ma, H., Churilov, L., Parsons, M.W., Levi, C.R., Butcher, K.S., Peeters, A., Barber, P.A., Bladin, C.F., De Silva, D.A., Fink, J., Kimber, T.E., Schultz, D.W., Muir, K.W., Tress, B.M., Desmond, P.M., Davis, S.M., and Donnan, G.A. (2011) EPITHET: positive result after reanalysis using baseline diffusion-weighted imaging/perfusion-weighted imaging co-registration. *Stroke*, **42** (1), 59–64.
35. Paciaroni, M., Medeiros, E., and Bogousslavsky, J. (2009) Desmoteplase. *Expert Opin. Biol. Ther.*, **9** (6), 773–778.
36. Liberatore, G.T., Samson, A., Bladin, C., Schleuning, W.D., and Medcalf, R.L. (2003) Vampire bat salivary plasminogen activator (desmoteplase): a unique fibrinolytic enzyme that does not promote neurodegeneration. *Stroke*, **34** (2), 537–543.
37. Furlan, A.J., Eyding, D., Albers, G.W., Al-Rawi, Y., Lees, K.R., Rowley, H.A., Sachara, C., Soehngen, M., Warach, S., and Hacke, W. (2006) Dose escalation of desmoteplase for acute ischemic stroke (DEDAS): evidence of safety and efficacy 3 to 9 hours after stroke onset. *Stroke*, **37** (5), 1227–1231.
38. Hacke, W., Albers, G., Al-Rawi, Y., Bogousslavsky, J., Davalos, A., Eliasziw, M., Fischer, M., Furlan, A., Kaste, M., Lees, K.R., Soehngen, M., and Warach, S. (2005) The desmoteplase in acute ischemic stroke trial (DIAS): a phase II MRI-based 9-hour window acute stroke thrombolysis trial with intravenous desmoteplase. *Stroke*, **36** (1), 66–73.
39. Hacke, W., Furlan, A.J., Al-Rawi, Y., Davalos, A., Fiebach, J.B., Gruber, F., Kaste, M., Lipka, L.J., Pedraza, S., Ringleb, P.A., Rowley, H.A., Schneider, D., Schwamm, L.H., Leal, J.S., Sohngen, M., Teal, P.A., Wilhelm-Ogunbiyi, K., Wintermark, M., and Warach, S. (2009) Intravenous desmoteplase in patients with acute ischaemic stroke selected by MRI perfusion-diffusion weighted imaging or perfusion CT (DIAS-2): a prospective, randomised, double-blind, placebo-controlled study. *Lancet Neurol.*, **8** (2), 141–150.
40. Molina, C.A. (2010) Reperfusion therapies for acute ischemic stroke: current pharmacological and mechanical approaches. *Stroke*, **42** (Suppl. 1), S16–S19.
41. Mishra, N.K., Albers, G.W., Davis, S.M., Donnan, G.A., Furlan, A.J., Hacke, W., and Lees, K.R. (2010) Mismatch-based delayed thrombolysis: a meta-analysis. *Stroke*, **41** (1), e25–e33.
42. Ogawa, A., Mori, E., Minematsu, K., Taki, W., Takahashi, A., Nemoto, S., Miyamoto, S., Sasaki, M., and Inoue, T. (2007) Randomized trial of intraarterial infusion of urokinase within 6 hours of middle cerebral artery stroke: the middle cerebral artery embolism local fibrinolytic intervention trial (MELT) Japan. *Stroke*, **38** (10), 2633–2639.
43. Del Zoppo, G.J., Copeland, B.R., Waltz, T.A., Zyroff, J., Plow, E.F., and Harker, L.A. (1986) The beneficial effect of intracarotid urokinase on acute stroke in a baboon model. *Stroke*, **17** (4), 638–643.

44. Furlan, A., Higashida, R., Wechsler, L., Gent, M., Rowley, H., Kase, C., Pessin, M., Ahuja, A., Callahan, F., Clark, W.M., Silver, F., and Rivera, F. (1999) Intra-arterial prourokinase for acute ischemic stroke. The PROACT II study: a randomized controlled trial. Prolyse in acute cerebral thromboembolism. *J. Am. Med. Assoc.*, **282** (21), 2003–2011.
45. Martin, U., Kaufmann, B., and Neugebauer, G. (1999) Current clinical use of reteplase for thrombolysis. A pharmacokinetic-pharmacodynamic perspective. *Clin. Pharmacokinet.*, **36** (4), 265–276.
46. Qureshi, A.I., Siddiqui, A.M., Suri, M.F., Kim, S.H., Ali, Z., Yahia, A.M., Lopes, D.K., Boulos, A.S., Ringer, A.J., Saad, M., Guterman, L.R., and Hopkins, L.N. (2002) Aggressive mechanical clot disruption and low-dose intra-arterial third-generation thrombolytic agent for ischemic stroke: a prospective study. *Neurosurgery*, **51** (5), 1319–1327; discussion 1327–1329.
47. Davydov, L. and Cheng, J.W. (2001) Tenecteplase: a review. *Clin. Ther.*, **23** (7), 982–997; discussion 981.
48. Haley, E.C. Jr., Thompson, J.L., Grotta, J.C., Lyden, P.D., Hemmen, T.G., Brown, D.L., Fanale, C., Libman, R., Kwiatkowski, T.G., Llinas, R.H., Levine, S.R., Johnston, K.C., Buchsbaum, R., Levy, G., and Levin, B. (2010) Phase IIB/III trial of tenecteplase in acute ischemic stroke: results of a prematurely terminated randomized clinical trial. *Stroke*, **41** (4), 707–711.
49. Parsons, M.W., Miteff, F., Bateman, G.A., Spratt, N., Loiselle, A., Attia, J., and Levi, C.R. (2009) Acute ischemic stroke: imaging-guided tenecteplase treatment in an extended time window. *Neurology*, **72** (10), 915–921.
50. Deitcher, S.R., Funk, W.D., Buchanan, J., Liu, S., Levy, M.D., and Toombs, C.F. (2006) Alfimeprase: a novel recombinant direct-acting fibrinolytic. *Expert Opin. Biol. Ther.*, **6** (12), 1361–1369.
51. Killer, M., Ladurner, G., Kunz, A.B., and Kraus, J. (2010) Current endovascular treatment of acute stroke and future aspects. *Drug Discov. Today*, **15** (15–16), 640–647.
52. Gibson, C.M., Zorkun, C., Molhoek, P., Zmudka, K., Greenberg, M., Mueller, H., Wesdorp, J., Louwerenburg, H., Niederman, A., Westenburg, J., Bikkina, M., Batty, J., de Winter, J., Murphy, S.A., and McCabe, C.H. (2006) Dose escalation trial of the efficacy, safety, and pharmacokinetics of a novel fibrinolytic agent, BB-10153, in patients with ST elevation MI: results of the TIMI 31 trial. *J. Thromb. Thrombolysis*, **22** (1), 13–21.
53. Lapchak, P.A., Araujo, D.M., Pakola, S., Song, D., Wei, J., and Zivin, J.A. (2002) Microplasmin: a novel thrombolytic that improves behavioral outcome after embolic strokes in rabbits. *Stroke*, **33** (9), 2279–2284.
54. Thijs, V.N., Peeters, A., Vosko, M., Aichner, F., Schellinger, P.D., Schneider, D., Neumann-Haefelin, T., Rother, J., Davalos, A., Wahlgren, N., and Verhamme, P. (2009) Randomized, placebo-controlled, dose-ranging clinical trial of intravenous microplasmin in patients with acute ischemic stroke. *Stroke*, **40** (12), 3789–3795.
55. Collen, D. (1998) Staphylokinase: a potent, uniquely fibrin-selective thrombolytic agent. *Nat. Med.*, **4** (3), 279–284.
56. Armstead, W.M., Ganguly, K., Kiessling, J.W., Chen, X.H., Smith, D.H., Higazi, A.A., Cines, D.B., Bdeir, K., Zaitsev, S., and Muzykantov, V.R. (2009) Red blood cells-coupled tPA prevents impairment of cerebral vasodilatory responses and tissue injury in pediatric cerebral hypoxia/ischemia through inhibition of ERK MAPK activation. *J. Cereb. Blood Flow Metab.*, **29** (8), 1463–1474.
57. Ganguly, K., Krasik, T., Medinilla, S., Bdeir, K., Cines, D.B., Muzykantov, V.R., and Murciano, J.C. (2005) Blood clearance and activity of erythrocyte-coupled fibrinolytics. *J. Pharmacol. Exp. Ther.*, **312** (3), 1106–1113.
58. Murciano, J.C. and Muzykantov, V.R. (2003) Coupling of anti-thrombotic agents to red blood cells offers safer

and more effective management of thrombosis. *Discov. Med.*, **3** (18), 28–29.

59. Lees, K.R., Bluhmki, E., von Kummer, R., Brott, T.G., Toni, D., Grotta, J.C., Albers, G.W., Kaste, M., Marler, J.R., Hamilton, S.A., Tilley, B.C., Davis, S.M., Donnan, G.A., Hacke, W., Allen, K., Mau, J., Meier, D., Del Zoppo, G., De Silva, D.A., Butcher, K.S., Parsons, M.W., Barber, P.A., Levi, C., Bladin, C., and Byrnes, G. (2010) Time to treatment with intravenous alteplase and outcome in stroke: an updated pooled analysis of ECASS, ATLANTIS, NINDS, and EPITHET trials. *Lancet*, **375** (9727), 1695–1703.
60. Dirks, M., Niessen, L.W., van Wijngaarden, J.D., Koudstaal, P.J., Franke, C.L., van Oostenbrugge, R.J., Huijsman, R., Lingsma, H.F., Minkman, M.M., and Dippel, D.W. (2011) Promoting thrombolysis in acute ischemic stroke. *Stroke*, **42** (5), 1325–1330.
61. Plow, E.F., Herren, T., Redlitz, A., Miles, L.A., and Hoover-Plow, J.L. (1995) The cell biology of the plasminogen system. *FASEB J.*, **9** (10), 939–945.
62. Seeds, N.W., Verrall, S., Friedman, G., Hayden, S., Gadotti, D., Haffke, S., Christensen, K., Gardner, B., McGuire, P., and Krystosek, A. (1992) Plasminogen activators and plasminogen activator inhibitors in neural development. *Ann. N.Y. Acad. Sci.*, **667**, 32–40.
63. Ware, J.H., Dibenedetto, A.J., and Pittman, R.N. (1995) Localization of tissue plasminogen activator mRNA in adult rat brain. *Brain Res. Bull.*, **37** (3), 275–281.
64. Ware, J.H., DiBenedetto, A.J., and Pittman, R.N. (1995) Localization of tissue plasminogen activator mRNA in the developing rat cerebellum and effects of inhibiting tissue plasminogen activator on granule cell migration. *J. Neurobiol.*, **28** (1), 9–22.
65. Sappino, A.P., Madani, R., Huarte, J., Belin, D., Kiss, J.Z., Wohlwend, A., and Vassalli, J.D. (1993) Extracellular proteolysis in the adult murine brain. *J. Clin. Invest.*, **92** (2), 679–685.
66. Teesalu, T., Kulla, A., Simisker, A., Siren, V., Lawrence, D.A., Asser, T., and Vaheri, A. (2004) Tissue plasminogen activator and neuroserpin are widely expressed in the human central nervous system. *Thromb. Haemost.*, **92** (2), 358–368.
67. Hastings, G.A., Coleman, T.A., Haudenschild, C.C., Stefansson, S., Smith, E.P., Barthlow, R., Cherry, S., Sandkvist, M., and Lawrence, D.A. (1997) Neuroserpin, a brain-associated inhibitor of tissue plasminogen activator is localized primarily in neurons. Implications for the regulation of motor learning and neuronal survival. *J. Biol. Chem.*, **272** (52), 33062–33067.
68. Krueger, S.R., Ghisu, G.P., Cinelli, P., Gschwend, T.P., Osterwalder, T., Wolfer, D.P., and Sonderegger, P. (1997) Expression of neuroserpin, an inhibitor of tissue plasminogen activator, in the developing and adult nervous system of the mouse. *J. Neurosci.*, **17** (23), 8984–8996.
69. Hultman, K., Blomstrand, F., Nilsson, M., Wilhelmsson, U., Malmgren, K., Pekny, M., Kousted, T., Jern, C., and Tjarnlund-Wolf, A. (2010) Expression of plasminogen activator inhibitor-1 and protease nexin-1 in human astrocytes: Response to injury-related factors. *J. Neurosci. Res.*, **88** (11), 2441–2449.
70. Noel, M., Norris, E.H., and Strickland, S. (2011) Tissue plasminogen activator is required for the development of fetal alcohol syndrome in mice. *Proc. Natl. Acad. Sci. U.S.A.*, **108** (12), 5069–5074.
71. Reinhard, E., Suidan, H.S., Pavlik, A., and Monard, D. (1994) Glia-derived nexin/protease nexin-1 is expressed by a subset of neurons in the rat brain. *J. Neurosci. Res.*, **37** (2), 256–270.
72. Basham, M.E. and Seeds, N.W. (2001) Plasminogen expression in the neonatal and adult mouse brain. *J. Neurochem.*, **77** (1), 318–325.
73. Taniguchi, Y., Inoue, N., Morita, S., Nikaido, Y., Nakashima, T., Nagai, N., Okada, K., Matsuo, O., and Miyata, S. (2011) Localization of plasminogen in

mouse hippocampus, cerebral cortex, and hypothalamus. *Cell Tissue Res.*, **343** (2), 303–317.
74. Tsirka, S.E., Rogove, A.D., Bugge, T.H., Degen, J.L., and Strickland, S. (1997) An extracellular proteolytic cascade promotes neuronal degeneration in the mouse hippocampus. *J. Neurosci.*, **17** (2), 543–552.
75. Baranes, D., Lederfein, D., Huang, Y.Y., Chen, M., Bailey, C.H., and Kandel, E.R. (1998) Tissue plasminogen activator contributes to the late phase of LTP and to synaptic growth in the hippocampal mossy fiber pathway. *Neuron*, **21** (4), 813–825.
76. Seeds, N.W., Williams, B.L., and Bickford, P.C. (1995) Tissue plasminogen activator induction in Purkinje neurons after cerebellar motor learning. *Science*, **270** (5244), 1992–1994.
77. Pawlak, R., Rao, B.S., Melchor, J.P., Chattarji, S., McEwen, B., and Strickland, S. (2005) Tissue plasminogen activator and plasminogen mediate stress-induced decline of neuronal and cognitive functions in the mouse hippocampus. *Proc. Natl. Acad. Sci. U.S.A.*, **102** (50), 18201–18206.
78. Matys, T., Pawlak, R., Matys, E., Pavlides, C., McEwen, B.S., and Strickland, S. (2004) Tissue plasminogen activator promotes the effects of corticotropin-releasing factor on the amygdala and anxiety-like behavior. *Proc. Natl. Acad. Sci. U.S.A.*, **101** (46), 16345–16350.
79. Pawlak, R., Magarinos, A.M., Melchor, J., McEwen, B., and Strickland, S. (2003) Tissue plasminogen activator in the amygdala is critical for stress-induced anxiety-like behavior. *Nat. Neurosci.*, **6** (2), 168–174.
80. Park, L., Gallo, E.F., Anrather, J., Wang, G., Norris, E.H., Paul, J., Strickland, S., and Iadecola, C. (2008) Key role of tissue plasminogen activator in neurovascular coupling. *Proc. Natl. Acad. Sci. U.S.A.*, **105** (3), 1073–1078.
81. Salles, F.J. and Strickland, S. (2002) Localization and regulation of the tissue plasminogen activator-plasmin system in the hippocampus. *J. Neurosci.*, **22** (6), 2125–2134.
82. Yepes, M., Sandkvist, M., Wong, M.K., Coleman, T.A., Smith, E., Cohan, S.L., and Lawrence, D.A. (2000) Neuroserpin reduces cerebral infarct volume and protects neurons from ischemia-induced apoptosis. *Blood*, **96** (2), 569–576.
83. Hosomi, N., Lucero, J., Heo, J.H., Koziol, J.A., Copeland, B.R., and Del Zoppo, G.J. (2001) Rapid differential endogenous plasminogen activator expression after acute middle cerebral artery occlusion. *Stroke*, **32** (6), 1341–1348.
84. Qian, Z., Gilbert, M.E., Colicos, M.A., Kandel, E.R., and Kuhl, D. (1993) Tissue-plasminogen activator is induced as an immediate-early gene during seizure, kindling and long-term potentiation. *Nature*, **361** (6411), 453–457.
85. Sappino, A.P., Huarte, J., Belin, D., and Vassalli, J.D. (1989) Plasminogen activators in tissue remodeling and invasion: mRNA localization in mouse ovaries and implanting embryos. *J. Cell Biol.*, **109** (5), 2471–2479.
86. Carmeliet, P., Schoonjans, L., Kieckens, L., Ream, B., Degen, J., Bronson, R., De Vos, R., van den Oord, J.J., Collen, D., and Mulligan, R.C. (1994) Physiological consequences of loss of plasminogen activator gene function in mice. *Nature*, **368** (6470), 419–424.
87. Strickland, S. (2001) Tissue plasminogen activator in nervous system function and dysfunction. *Thromb. Haemost.*, **86** (1), 138–143.
88. Pawlak, R. and Strickland, S. (2002) Tissue plasminogen activator and seizures: a clot-buster's secret life. *J. Clin. Invest.*, **109** (12), 1529–1531.
89. Choi, D.W. (1985) Glutamate neurotoxicity in cortical cell culture is calcium dependent. *Neurosci. Lett.*, **58** (3), 293–297.
90. Tsirka, S.E., Gualandris, A., Amaral, D.G., and Strickland, S. (1995) Excitotoxin-induced neuronal degeneration and seizure are mediated by tissue plasminogen activator. *Nature*, **377** (6547), 340–344.

91. Tsirka, S.E., Rogove, A.D., and Strickland, S. (1996) Neuronal cell death and tPA. *Nature*, **384** (6605), 123–124.
92. Tsirka, S.E., Bugge, T.H., Degen, J.L., and Strickland, S. (1997) Neuronal death in the central nervous system demonstrates a non-fibrin substrate for plasmin. *Proc. Natl. Acad. Sci. U.S.A.*, **94** (18), 9779–9781.
93. Lebeurrier, N., Liot, G., Lopez-Atalaya, J.P., Orset, C., Fernandez-Monreal, M., Sonderegger, P., Ali, C., and Vivien, D. (2005) The brain-specific tissue-type plasminogen activator inhibitor, neuroserpin, protects neurons against excitotoxicity both in vitro and in vivo. *Mol. Cell. Neurosci.*, **30** (4), 552–558.
94. Hazell, A.S. (2007) Excitotoxic mechanisms in stroke: an update of concepts and treatment strategies. *Neurochem. Int.*, **50** (7–8), 941–953.
95. Mhairi Macrae, I. (1992) New models of focal cerebral ischaemia. *Br. J. Clin. Pharmacol.*, **34** (4), 302–308.
96. Sigmund, C.D. (2000) Viewpoint: are studies in genetically altered mice out of control? *Arterioscler. Thromb. Vasc. Biol.*, **20** (6), 1425–1429.
97. Wang, Y.F., Tsirka, S.E., Strickland, S., Stieg, P.E., Soriano, S.G., and Lipton, S.A. (1998) Tissue plasminogen activator (tPA) increases neuronal damage after focal cerebral ischemia in wild-type and tPA-deficient mice. *Nat. Med.*, **4** (2), 228–231.
98. Del Zoppo, G.J. (1998) tPA: a neuron buster, too? *Nat. Med.*, **4** (2), 148–150.
99. Ginsberg, M.D. (1999) On ischemic brain injury in genetically altered mice. *Arterioscler. Thromb. Vasc. Biol.*, **19** (11), 2581–2583.
100. Tabrizi, P., Wang, L., Seeds, N., McComb, J.G., Yamada, S., Griffin, J.H., Carmeliet, P., Weiss, M.H., and Zlokovic, B.V. (1999) Tissue plasminogen activator (tPA) deficiency exacerbates cerebrovascular fibrin deposition and brain injury in a murine stroke model: studies in tPA-deficient mice and wild-type mice on a matched genetic background. *Arterioscler. Thromb. Vasc. Biol.*, **19** (11), 2801–2806.
101. Kudryk, B.J. and Bini, A. (2000) Monoclonal antibody designated T2G1 reacts with human fibrin beta-chain but not with the corresponding chain from mouse fibrin. *Arterioscler. Thromb. Vasc. Biol.*, **20** (7), 1848–1849.
102. Nagai, N., De Mol, M., Lijnen, H.R., Carmeliet, P., and Collen, D. (1999) Role of plasminogen system components in focal cerebral ischemic infarction: a gene targeting and gene transfer study in mice. *Circulation*, **99** (18), 2440–2444.
103. Nagai, N., Suzuki, Y., Van Hoef, B., Lijnen, H.R., and Collen, D. (2005) Effects of plasminogen activator inhibitor-1 on ischemic brain injury in permanent and thrombotic middle cerebral artery occlusion models in mice. *J. Thromb. Haemost.*, **3** (7), 1379–1384.
104. Cinelli, P., Madani, R., Tsuzuki, N., Vallet, P., Arras, M., Zhao, C.N., Osterwalder, T., Rulicke, T., and Sonderegger, P. (2001) Neuroserpin, a neuroprotective factor in focal ischemic stroke. *Mol. Cell. Neurosci.*, **18** (5), 443–457.
105. Kilic, E., Hermann, D.M., and Hossmann, K.A. (1999) Recombinant tissue plasminogen activator reduces infarct size after reversible thread occlusion of middle cerebral artery in mice. *Neuroreport*, **10** (1), 107–111.
106. Zhang, R.L., Chopp, M., Zhang, Z.G., and Divine, G. (1998) Early (1 h) administration of tissue plasminogen activator reduces infarct volume without increasing hemorrhagic transformation after focal cerebral embolization in rats. *J. Neurol. Sci.*, **160** (1), 1–8.
107. Overgaard, K. (1994) Thrombolytic therapy in experimental embolic stroke. *Cerebrovasc. Brain Metab. Rev.*, **6** (3), 257–286.
108. Kilic, E., Bahr, M., and Hermann, D.M. (2001) Effects of recombinant tissue plasminogen activator after intraluminal thread occlusion in mice: role of hemodynamic alterations. *Stroke*, **32** (11), 2641–2647.
109. Meng, W., Wang, X., Asahi, M., Kano, T., Asahi, K., Ackerman, R.H., and Lo,

E.H. (1999) Effects of tissue type plasminogen activator in embolic versus mechanical models of focal cerebral ischemia in rats. *J. Cereb. Blood Flow Metab.*, **19** (12), 1316–1321.

110. Armstead, W.M., Nassar, T., Akkawi, S., Smith, D.H., Chen, X.H., Cines, D.B., and Higazi, A.A. (2006) Neutralizing the neurotoxic effects of exogenous and endogenous tPA. *Nat. Neurosci.*, **9** (9), 1150–1155.

111. Lee, S.R., Lok, J., Rosell, A., Kim, H.Y., Murata, Y., Atochin, D., Huang, P.L., Wang, X., Ayata, C., Moskowitz, M.A., and Lo, E.H. (2007) Reduction of hippocampal cell death and proteolytic responses in tissue plasminogen activator knockout mice after transient global cerebral ischemia. *Neuroscience*, **150** (1), 50–57.

112. Echeverry, R., Wu, J., Haile, W.B., Guzman, J., and Yepes, M. (2010) Tissue-type plasminogen activator is a neuroprotectant in the mouse hippocampus. *J. Clin. Invest.*, **120** (6), 2194–2205.

113. Lau, A. and Tymianski, M. (2010) Glutamate receptors, neurotoxicity and neurodegeneration. *Pflugers Arch.*, **460** (2), 525–542.

114. Rao, V.R. and Finkbeiner, S. (2007) NMDA and AMPA receptors: old channels, new tricks. *Trends Neurosci.*, **30** (6), 284–291.

115. Kohr, G. (2006) NMDA receptor function: subunit composition versus spatial distribution. *Cell Tissue Res.*, **326** (2), 439–446.

116. Paoletti, P. and Neyton, J. (2007) NMDA receptor subunits: function and pharmacology. *Curr. Opin. Pharmacol.*, **7** (1), 39–47.

117. Nicole, O., Docagne, F., Ali, C., Margaill, I., Carmeliet, P., MacKenzie, E.T., Vivien, D., and Buisson, A. (2001) The proteolytic activity of tissue-plasminogen activator enhances NMDA receptor-mediated signaling. *Nat. Med.*, **7** (1), 59–64.

118. Baron, A., Montagne, A., Casse, F., Launay, S., Maubert, E., Ali, C., and Vivien, D. (2010) NR2D-containing NMDA receptors mediate tissue plasminogen activator-promoted neuronal excitotoxicity. *Cell Death Differ.*, **17** (5), 860–871.

119. Benchenane, K., Castel, H., Boulouard, M., Bluthe, R., Fernandez-Monreal, M., Roussel, B.D., Lopez-Atalaya, J.P., Butt-Gueulle, S., Agin, V., Maubert, E., Dantzer, R., Touzani, O., Dauphin, F., Vivien, D., and Ali, C. (2007) Anti-NR1 N-terminal-domain vaccination unmasks the crucial action of tPA on NMDA-receptor-mediated toxicity and spatial memory. *J. Cell Sci.*, **120** (Pt 4), 578–585.

120. Liot, G., Benchenane, K., Leveille, F., Lopez-Atalaya, J.P., Fernandez-Monreal, M., Ruocco, A., Mackenzie, E.T., Buisson, A., Ali, C., and Vivien, D. (2004) 2,7-Bis-(4-amidinobenzylidene)-cycloheptan-1-one dihydrochloride, tPA stop, prevents tPA-enhanced excitotoxicity both in vitro and in vivo. *J. Cereb. Blood Flow Metab.*, **24** (10), 1153–1159.

121. Lopez-Atalaya, J.P., Roussel, B.D., Levrat, D., Parcq, J., Nicole, O., Hommet, Y., Benchenane, K., Castel, H., Leprince, J., To Van, D., Bureau, R., Rault, S., Vaudry, H., Petersen, K.U., Santos, J.S., Ali, C., and Vivien, D. (2008) Toward safer thrombolytic agents in stroke: molecular requirements for NMDA receptor-mediated neurotoxicity. *J. Cereb. Blood Flow Metab.*, **28** (6), 1212–1221.

122. Macrez, R., Bezin, L., Le Mauff, B., Ali, C., and Vivien, D. (2010) Functional occurrence of the interaction of tissue plasminogen activator with the NR1 subunit of N-Methyl-D-Aspartate receptors during stroke. *Stroke*, **41** (12), 2950–2955.

123. Matys, T. and Strickland, S. (2003) Tissue plasminogen activator and NMDA receptor cleavage. *Nat. Med.*, **9** (4), 371–372; author reply 372–373.

124. Norris, E.H. and Strickland, S. (2007) Modulation of NR2B-regulated contextual fear in the hippocampus by the tissue plasminogen activator system. *Proc. Natl. Acad. Sci. U.S.A.*, **104** (33), 13473–13478.

125. Pawlak, R., Melchor, J.P., Matys, T., Skrzypiec, A.E., and Strickland, S. (2005) Ethanol-withdrawal seizures

are controlled by tissue plasminogen activator via modulation of NR2B-containing NMDA receptors. *Proc. Natl. Acad. Sci. U.S.A.*, **102** (2), 443–448.

126. Samson, A.L. and Medcalf, R.L. (2006) Tissue-type plasminogen activator: a multifaceted modulator of neurotransmission and synaptic plasticity. *Neuron*, **50** (5), 673–678.

127. Yuan, H., Vance, K.M., Junge, C.E., Geballe, M.T., Snyder, J.P., Hepler, J.R., Yepes, M., Low, C.M., and Traynelis, S.F. (2009) The serine protease plasmin cleaves the amino-terminal domain of the NR2A subunit to relieve zinc inhibition of the N-methyl-D-aspartate receptors. *J. Biol. Chem.*, **284** (19), 12862–12873.

128. Kvajo, M., Albrecht, H., Meins, M., Hengst, U., Troncoso, E., Lefort, S., Kiss, J.Z., Petersen, C.C., and Monard, D. (2004) Regulation of brain proteolytic activity is necessary for the in vivo function of NMDA receptors. *J. Neurosci.*, **24** (43), 9734–9743.

129. Fernandez-Monreal, M., Lopez-Atalaya, J.P., Benchenane, K., Cacquevel, M., Dulin, F., Le Caer, J.P., Rossier, J., Jarrige, A.C., Mackenzie, E.T., Colloc'h, N., Ali, C., and Vivien, D. (2004) Arginine 260 of the amino-terminal domain of NR1 subunit is critical for tissue-type plasminogen activator-mediated enhancement of N-methyl-D-aspartate receptor signaling. *J. Biol. Chem.*, **279** (49), 50850–50856.

130. Fernandez-Monreal, M., Lopez-Atalaya, J.P., Benchenane, K., Leveille, F., Cacquevel, M., Plawinski, L., MacKenzie, E.T., Bu, G., Buisson, A., and Vivien, D. (2004) Is tissue-type plasminogen activator a neuromodulator? *Mol. Cell. Neurosci.*, **25** (4), 594–601.

131. Samson, A.L., Nevin, S.T., Croucher, D., Niego, B., Daniel, P.B., Weiss, T.W., Moreno, E., Monard, D., Lawrence, D.A., and Medcalf, R.L. (2008) Tissue-type plasminogen activator requires a co-receptor to enhance NMDA receptor function. *J. Neurochem.*, **107** (4), 1091–1101.

132. Bacskai, B.J., Xia, M.Q., Strickland, D.K., Rebeck, G.W., and Hyman, B.T. (2000) The endocytic receptor protein LRP also mediates neuronal calcium signaling via N-methyl-D-aspartate receptors. *Proc. Natl. Acad. Sci. U.S.A.*, **97** (21), 11551–11556.

133. Eugenin, E.A., King, J.E., Nath, A., Calderon, T.M., Zukin, R.S., Bennett, M.V., and Berman, J.W. (2007) HIV-tat induces formation of an LRP-PSD-95-NMDAR-nNOS complex that promotes apoptosis in neurons and astrocytes. *Proc. Natl. Acad. Sci. U.S.A.*, **104** (9), 3438–3443.

134. Martin, A.M., Kuhlmann, C., Trossbach, S., Jaeger, S., Waldron, E., Roebroek, A., Luhmann, H.J., Laatsch, A., Weggen, S., Lessmann, V., and Pietrzik, C.U. (2008) The functional role of the second NPXY motif of the LRP1 beta-chain in tissue-type plasminogen activator-mediated activation of N-methyl-D-aspartate receptors. *J. Biol. Chem.*, **283** (18), 12004–12013.

135. Qiu, Z., Strickland, D.K., Hyman, B.T., and Rebeck, G.W. (2002) alpha 2-Macroglobulin exposure reduces calcium responses to N-methyl-D-aspartate via low density lipoprotein receptor-related protein in cultured hippocampal neurons. *J. Biol. Chem.*, **277** (17), 14458–14466.

136. Zhou, L., Li, F., Xu, H.B., Luo, C.X., Wu, H.Y., Zhu, M.M., Lu, W., Ji, X., Zhou, Q.G., and Zhu, D.Y. (2010) Treatment of cerebral ischemia by disrupting ischemia-induced interaction of nNOS with PSD-95. *Nat. Med.*, **16** (12), 1439–1443.

137. Bard, L., Sainlos, M., Bouchet, D., Cousins, S., Mikasova, L., Breillat, C., Stephenson, F.A., Imperiali, B., Choquet, D., and Groc, L. (2010) Dynamic and specific interaction between synaptic NR2-NMDA receptor and PDZ proteins. *Proc. Natl. Acad. Sci. U.S.A.*, **107** (45), 19561–19566.

138. Abbott, N.J., Patabendige, A.A., Dolman, D.E., Yusof, S.R., and Begley, D.J. (2010) Structure and function of the blood-brain barrier. *Neurobiol. Dis.*, **37** (1), 13–25.

139. Betz, A.L. (1992) An overview of the multiple functions of the blood-brain barrier. *NIDA Res. Monogr.*, **120**, 54–72.
140. Sandoval, K.E. and Witt, K.A. (2008) Blood-brain barrier tight junction permeability and ischemic stroke. *Neurobiol. Dis.*, **32** (2), 200–219.
141. Morancho, A., Rosell, A., Garcia-Bonilla, L., and Montaner, J. (2010) Metalloproteinase and stroke infarct size: role for anti-inflammatory treatment? *Ann. N.Y. Acad. Sci.*, **1207**, 123–133.
142. Jin, R., Yang, G., and Li, G. (2010) Molecular insights and therapeutic targets for blood-brain barrier disruption in ischemic stroke: critical role of matrix metalloproteinases and tissue-type plasminogen activator. *Neurobiol. Dis.*, **38** (3), 376–385.
143. Rosell, A. and Lo, E.H. (2008) Multiphasic roles for matrix metalloproteinases after stroke. *Curr. Opin. Pharmacol.*, **8** (1), 82–89.
144. Rosenberg, G.A., Navratil, M., Barone, F., and Feuerstein, G. (1996) Proteolytic cascade enzymes increase in focal cerebral ischemia in rat. *J. Cereb. Blood Flow Metab.*, **16** (3), 360–366.
145. Suzuki, Y. (2010) Role of tissue-type plasminogen activator in ischemic stroke. *J. Pharmacol. Sci.*, **113** (3), 203–207.
146. Wang, X., Lee, S.R., Arai, K., Tsuji, K., Rebeck, G.W., and Lo, E.H. (2003) Lipoprotein receptor-mediated induction of matrix metalloproteinase by tissue plasminogen activator. *Nat. Med.*, **9** (10), 1313–1317.
147. Yepes, M., Sandkvist, M., Moore, E.G., Bugge, T.H., Strickland, D.K., and Lawrence, D.A. (2003) Tissue-type plasminogen activator induces opening of the blood-brain barrier via the LDL receptor-related protein. *J. Clin. Invest.*, **112** (10), 1533–1540.
148. Lee, S.R., Guo, S.Z., Scannevin, R.H., Magliaro, B.C., Rhodes, K.J., Wang, X., and Lo, E.H. (2007) Induction of matrix metalloproteinase, cytokines and chemokines in rat cortical astrocytes exposed to plasminogen activators. *Neurosci. Lett.*, **417** (1), 1–5.
149. Tsuji, K., Aoki, T., Tejima, E., Arai, K., Lee, S.R., Atochin, D.N., Huang, P.L., Wang, X., Montaner, J., and Lo, E.H. (2005) Tissue plasminogen activator promotes matrix metalloproteinase-9 upregulation after focal cerebral ischemia. *Stroke*, **36** (9), 1954–1959.
150. Aoki, T., Sumii, T., Mori, T., Wang, X., and Lo, E.H. (2002) Blood-brain barrier disruption and matrix metalloproteinase-9 expression during reperfusion injury: mechanical versus embolic focal ischemia in spontaneously hypertensive rats. *Stroke*, **33** (11), 2711–2717.
151. Kelly, M.A., Shuaib, A., and Todd, K.G. (2006) Matrix metalloproteinase activation and blood-brain barrier breakdown following thrombolysis. *Exp. Neurol.*, **200** (1), 38–49.
152. Del Zoppo, G.J. (2010) The neurovascular unit, matrix proteases, and innate inflammation. *Ann. N.Y. Acad. Sci.*, **1207**, 46–49.
153. Copin, J.C., Bengualid, D.J., Da Silva, R.F., Kargiotis, O., Schaller, K., and Gasche, Y. (2011) Recombinant tissue plasminogen activator induces blood-brain barrier breakdown by a matrix metalloproteinase-9-independent pathway after transient focal cerebral ischemia in mouse. *Eur. J. Neurosci.*, **34** (7), 1085–1092.
154. Gidday, J.M., Gasche, Y.G., Copin, J.C., Shah, A.R., Perez, R.S., Shapiro, S.D., Chan, P.H., and Park, T.S. (2005) Leukocyte-derived matrix metalloproteinase-9 mediates blood-brain barrier breakdown and is proinflammatory after transient focal cerebral ischemia. *Am. J. Physiol. Heart Circ. Physiol.*, **289** (2), H558–H568.
155. Justicia, C., Panes, J., Sole, S., Cervera, A., Deulofeu, R., Chamorro, A., and Planas, A.M. (2003) Neutrophil infiltration increases matrix metalloproteinase-9 in the ischemic brain after occlusion/reperfusion of the middle cerebral artery in rats. *J. Cereb. Blood Flow Metab.*, **23** (12), 1430–1440.
156. Harris, A.K., Ergul, A., Kozak, A., Machado, L.S., Johnson, M.H., and Fagan, S.C. (2005) Effect of neutrophil depletion on gelatinase expression,

edema formation and hemorrhagic transformation after focal ischemic stroke. *BMC Neurosci.*, **6**, 49.

157. Cuadrado, E., Ortega, L., Hernandez-Guillamon, M., Penalba, A., Fernandez-Cadenas, I., Rosell, A., and Montaner, J. (2008) Tissue plasminogen activator (t-PA) promotes neutrophil degranulation and MMP-9 release. *J. Leukoc. Biol.*, **84** (1), 207–214.

158. Rosell, A., Ortega-Aznar, A., Alvarez-Sabin, J., Fernandez-Cadenas, I., Ribo, M., Molina, C.A., Lo, E.H., and Montaner, J. (2006) Increased brain expression of matrix metalloproteinase-9 after ischemic and hemorrhagic human stroke. *Stroke*, **37** (6), 1399–1406.

159. Herz, J. and Strickland, D.K. (2001) LRP: a multifunctional scavenger and signaling receptor. *J. Clin. Invest.*, **108** (6), 779–784.

160. Lillis, A.P., Van Duyn, L.B., Murphy-Ullrich, J.E., and Strickland, D.K. (2008) LDL receptor-related protein 1: unique tissue-specific functions revealed by selective gene knockout studies. *Physiol. Rev.*, **88** (3), 887–918.

161. Biessen, E.A., van Teijlingen, M., Vietsch, H., Barrett-Bergshoeff, M.M., Bijsterbosch, M.K., Rijken, D.C., van Berkel, T.J., and Kuiper, J. (1997) Antagonists of the mannose receptor and the LDL receptor-related protein dramatically delay the clearance of tissue plasminogen activator. *Circulation*, **95** (1), 46–52.

162. Lillis, A.P., Mikhailenko, I., and Strickland, D.K. (2005) Beyond endocytosis: LRP function in cell migration, proliferation and vascular permeability. *J. Thromb. Haemost.*, **3** (8), 1884–1893.

163. Bu, G., Williams, S., Strickland, D.K., and Schwartz, A.L. (1992) Low density lipoprotein receptor-related protein/alpha 2-macroglobulin receptor is an hepatic receptor for tissue-type plasminogen activator. *Proc. Natl. Acad. Sci. U.S.A.*, **89** (16), 7427–7431.

164. Makarova, A., Mikhailenko, I., Bugge, T.H., List, K., Lawrence, D.A., and Strickland, D.K. (2003) The low density lipoprotein receptor-related protein modulates protease activity in the brain by mediating the cellular internalization of both neuroserpin and neuroserpin-tissue-type plasminogen activator complexes. *J. Biol. Chem.*, **278** (50), 50250–50258.

165. May, P., Reddy, Y.K., and Herz, J. (2002) Proteolytic processing of low density lipoprotein receptor-related protein mediates regulated release of its intracellular domain. *J. Biol. Chem.*, **277** (21), 18736–18743.

166. Su, E.J., Fredriksson, L., Geyer, M., Folestad, E., Cale, J., Andrae, J., Gao, Y., Pietras, K., Mann, K., Yepes, M., Strickland, D.K., Betsholtz, C., Eriksson, U., and Lawrence, D.A. (2008) Activation of PDGF-CC by tissue plasminogen activator impairs blood-brain barrier integrity during ischemic stroke. *Nat. Med.*, **14** (7), 731–737.

167. Polavarapu, R., Gongora, M.C., Yi, H., Ranganthan, S., Lawrence, D.A., Strickland, D., and Yepes, M. (2007) Tissue-type plasminogen activator-mediated shedding of astrocytic low-density lipoprotein receptor-related protein increases the permeability of the neurovascular unit. *Blood*, **109** (8), 3270–3278.

168. Zhang, X., Polavarapu, R., She, H., Mao, Z., and Yepes, M. (2007) Tissue-type plasminogen activator and the low-density lipoprotein receptor-related protein mediate cerebral ischemia-induced nuclear factor-κB pathway activation. *Am. J. Pathol.*, **171** (4), 1281–1290.

169. Fredriksson, L., Li, H., and Eriksson, U. (2004) The PDGF family: four gene products form five dimeric isoforms. *Cytokine Growth Factor Rev.*, **15** (4), 197–204.

170. Fredriksson, L., Li, H., Fieber, C., Li, X., and Eriksson, U. (2004) Tissue plasminogen activator is a potent activator of PDGF-CC. *EMBO J.*, **23** (19), 3793–3802.

171. Su, E.J., Fredriksson, L., Schielke, G.P., Eriksson, U., and Lawrence, D.A. (2009) Tissue plasminogen activator-mediated PDGF signaling and neurovascular coupling in stroke.

J. Thromb. Haemost., **7** (Suppl. 1), 155–158.

172. Tang, Z., Arjunan, P., Lee, C., Li, Y., Kumar, A., Hou, X., Wang, B., Wardega, P., Zhang, F., Dong, L., Zhang, Y., Zhang, S.Z., Ding, H., Fariss, R.N., Becker, K.G., Lennartsson, J., Nagai, N., Cao, Y., and Li, X. (2010) Survival effect of PDGF-CC rescues neurons from apoptosis in both brain and retina by regulating GSK3beta phosphorylation. *J. Exp. Med.*, **207** (4), 867–880.

173. Fredriksson, L., Nilsson, I., Su, E.J., Andrea, J., Ding, H., Betsholtz, C., Eriksson, U., and Lawrence, D.A. (2012) Platelet-derived growth factor C deficiency in C57BL/6 mice leads to abnormal cerebral vascularization, loss of neuroependymal integrity, and ventricular abnormalities. *Am. J. Pathol, http://dx.doi.org/10.1016/j.ajpath.2011.12.006*

174. Andrae, J., Gallini, R., and Betsholtz, C. (2008) Role of platelet-derived growth factors in physiology and medicine. *Genes Dev.*, **22** (10), 1276–1312.

175. He, Y., Cai, W., Wang, L., and Chen, P. (2009) A developmental study on the expression of PDGFalphaR immunoreactive cells in the brain of postnatal rats. *Neurosci. Res.*, **65** (3), 272–279.

176. Jackson, E.L., Garcia-Verdugo, J.M., Gil-Perotin, S., Roy, M., Quinones-Hinojosa, A., VandenBerg, S., and Alvarez-Buylla, A. (2006) PDGFR alpha-positive B cells are neural stem cells in the adult SVZ that form glioma-like growths in response to increased PDGF signaling. *Neuron*, **51** (2), 187–199.

177. Pringle, N.P., Mudhar, H.S., Collarini, E.J., and Richardson, W.D. (1992) PDGF receptors in the rat CNS: during late neurogenesis, PDGF alpha-receptor expression appears to be restricted to glial cells of the oligodendrocyte lineage. *Development*, **115** (2), 535–551.

178. Chang, C., Wang, X., and Caldwell, R.B. (1997) Serum opens tight junctions and reduces ZO-1 protein in retinal epithelial cells. *J. Neurochem.*, **69** (2), 859–867.

179. Ivens, S., Gabriel, S., Greenberg, G., Friedman, A., and Shelef, I. (2010) Blood-brain barrier breakdown as a novel mechanism underlying cerebral hyperperfusion syndrome. *J. Neurol.*, **257** (4), 615–620.

180. Ma, Q., Huang, B., Khatibi, N., Rolland, W.II., Suzuki, H., Zhang, J.H., and Tang, J. (2011) PDGFR-α inhibition preserves blood-brain barrier after intracerebral hemorrhage. *Ann. Neurol*, **70**, 920–931.

7
Bacterial Abuse of Mammalian Extracellular Proteases during Tissue Invasion and Infection

Claudia Weber, Heiko Herwald, and Sven Hammerschmidt

7.1
Introduction

Proteases are essential enzymes for all mammalian organisms as they conduct hydrolytic cleavage of peptide bonds in proteins and peptides, leading either to irreversible degradation or activation of their substrate structures. In terms of specificity, some proteases target a unique protein in a highly selective and efficient way, whereas others are promiscuous in their degradative activity. Proteases differ in their spatial localization, being either intracellular or extracellular [1]. Extracellular proteases play an important role in the blood coagulation system, extracellular matrix (ECM) and cell surface remodeling, blood pressure control, and the proper function of the immune system and are therefore strictly regulated by several mechanisms. The performance of extracellular proteases is controlled by gene expression, posttranslational modifications such as glycosylation, metal binding, degradation, activation of their inactive zymogens by proteolysis, blockade by endogenous inhibitors, as well as targeting to specific intra- and extracellular compartments [1, 2]. Nevertheless, some microorganisms have gained the potential to not only interfere with this rigorous regulation but also subvert proteases and their activators and inhibitors for their own survival and prosperity. The recruited host proteases endow microorganisms with resistance against factors of the innate and acquired immune systems, for example, through degradation of circulating antibodies, complement proteins, and/or antimicrobial peptides, or promote their invasion into the blood and dissemination. Apart from viruses, fungi, and parasites, a wide variety of pathogenic bacteria are able to abuse host-derived extracellular proteases for initiation or maintenance of an infection. This report provides an overview of the sophisticated strategies that bacteria evolved to interact with extracellular proteases and key elements of proteolytic cascades of various physiological systems of the host.

Matrix Proteases in Health and Disease, First Edition. Edited by Niels Behrendt.

7.2
Tissue and Cell Surface Remodeling Proteases

Tissue microenvironment comprises ECM and ECM-deposited factors such as chemokines, cytokines, and growth factors and is remodeled dynamically to adjust to the current physiological needs by proteases of the metzincin superfamily. This family is characterized by the presence of a catalytic Zn (II) atom in the active center complexed by three histidine residues, which are crucial for the proteolytic activity. In addition, the members of the metzincin superfamily comprise an invariant common structure in their metalloproteinase domain, the so-called met turn that is characterized by a conserved methionine residue downstream of the third zinc-binding histidine in a β-turn motif [3]. Medically important members of this family are matrix metalloproteinases (MMPs), a disintegrin and metalloproteinases (ADAMs), and a disintegrin and metalloproteinases with thrombospondin motifs (ADAMTSs), whose ubiquitous expression in various tissues has been frequently reported.

7.2.1
Matrix Metalloproteinases (MMPs)

The MMP family, also called matrixins, contains 25 members in mammals. On the basis of their structural characteristics, they can be divided into membrane-anchored (MT-MMP, membrane-type matrix metalloproteinase) and secreted-type MMPs [4]. The latter are secreted as a proenzyme and require proteolytic cleavage for activation [5]. MMPs degrade at least one kind of ECM protein, although more recent studies also imply target substrates such as other proteases, protease inhibitors, clotting factors, adhesion molecules, chemo- and cytokines, growth factors, hormones, as well as their receptors and antimicrobial peptides. The tight control required to orchestrate the degradation of the extraordinary diversity of MMP substrates is established by the necessity of MMPs for induction, secretion, and activation in order to achieve full activity and the existence of natural inhibitors termed tissue inhibitors of metalloproteinases (TIMPs). Inappropriate regulation of MMPs causes many pathological complications such as cancer metastasis, inflammatory tissue damage, and microbial invasion and dissemination. A wide variety of gram-positive and gram-negative bacteria modulate the expression and function of MMPs. A number of MMPs are upregulated during sepsis and endotoxemia, either directly or indirectly by alterations of the expression of cytokines, growth factors, hormones, and cell–ECM interactions [6]. For example, tumor necrosis factor-α (TNF-α), a proinflammatory cytokine expressed on lipopolysaccharide (LPS) treatment, can induce MMP-9 expression in the human monocytic cell line HL60 [7]. LPS can also elicit the production of MMPs in a vast number of primary cells from neutrophils, monocytes/macrophages, and mast cells through to epithelial, endothelial, microglial cells, and fibroblasts [7–12]. After intraperitoneal injection of LPS in mice, transcription of MMP-13, MMP-9, MT1-MMP, MMP-12, and MMP10 was elevated in the liver of mice [13]. A dramatic elevation of MMP-3, MMP-9, MMP-14, and MMP-28 expression on oral infection of mice with *Salmonella enterica* was noted in Peyer's patches and the mesenteric lymph nodes, while *Yersinia enterocolitica* increases MMP-3, MMP-8, and MMP-9 in these immune organs [14]. Mice infected intracorneally with *Pseudomonas aeruginosa* showed an elevated expression of membrane-type MMPs MT-MMP-4, MT-MMP-5, and MT-MMP-6 in the cornea [15]. These results indicate that secreted as well as membrane-type MMPs can be induced by a vast variety of different bacterial species and their products.

Bacteria have developed strategies to take advantage of functional host proteases spurred to action by inflammatory processes during infections. However, bacteria have also the capacity to activate the induced pro-MMPs by their own endogenously produced virulence proteases. The transmembrane aspartic protease PgtE of *S. enterica* is an omptin that activates the macrophage-derived pro-MMP-9 [16]. Elastase, an enzyme produced by *P. aeruginosa*, the most common pathogen associated with cystic fibrosis, was shown to cleave pro-MMP1, -8, and -9 into their active forms [17]. Moreover, a thiol proteinase of *Streptococcus pyogenes*, a virulence factor known as SpeB, proteolytically processes inactive precursors of MMP-2 and MMP-9 to active enzymes that, in turn, shed Fas ligand (FasL) and TNF-α from the cellular membrane. This led to the production of soluble (s)FasL and TNF-α and induction of apoptosis in the host cells via binding to the corresponding receptors Fas and TNF receptor 1 [18]. The induction of apoptosis is probably a strategy employed for bacterial dissemination by breaking the endothelial barrier separating the tissue from the blood vessels. Indeed, half of the invasive strains of *Streptococcus pneumoniae*, the causative agent of pneumonia and meningitis, produce the zinc metalloproteinase ZmpC, which is able to proteolytically activate pro-MMP-9 [19]. Another mechanism to enter the blood stream is the destruction of tight junctions. *Neisseria meningitidis* infections enhance MMP-8, which degrades occludin, one of the major constituents of tight junctions [20]. Bacteria have also the ability to inhibit selected functions of MMPs. For example, the staphylococcal superantigen-like protein 5 (SSL5) binds and inhibits neutrophil MMP-9, thereby suppressing transmigration of neutrophils in response to *N*-formyl-methionyl-leucyl-phenylalanine (fMLP) across basement membranes [21].

MMP activity is regulated by a group of endogenous proteins called *tissue inhibitors of metalloproteinases* (TIMP-1 to TIMP-4) that bind to catalytic and alternate sites of the activated MMPs. There is evidence that bacteria are capable of modulating these inhibitors. Culture supernatants of *Porphyromonas gingivalis* added to human gingival fibroblasts increased the expression of MMP-1, MMP-2, MMP-3, and MMP-14; and decreased TIMP-1 production in parallel [22]. Sonicated extracts from *P. gingivalis* incubated with human periodontal ligament cells degrade TIMP-1 and TIMP-2, leading to decreased inhibitory activity of TIMP-1 against interstitial collagenase (MMP-1) and MMP-2 in this experimental setup [23]. Apart from inhibiting MMPs, TIMP-3 is enabled to cross inhibit other metzincin family members, namely ADAM17 and ADAMTS1, -4, and -5 [24].

7.2.2
A Disintegrin and Metalloproteinases (ADAMs)

The human genome contains 25 ADAM encoding genes. Out of these, 4 are pseudogenes and 13 exhibit proteinase activity [25]. In contrast to MMPs, the metalloproteinase domain in ADAMs is followed by a disintegrin, a cysteine-rich and an epidermal growth factor (EGF)-like domain. The disintegrin domain comprises a structure referred to as *disintegrin loop* whose role in binding to integrins is controversially discussed [26]. The C-terminus of membrane-type ADAMs is formed by a transmembrane and a cytoplasmatic domain, which includes several sites for phosphorylation and interaction with signaling molecules. Similar to the MMP family, there are ADAMs such as ADAM9, ADAM12, and ADAM28 that lack transmembrane and cytoplasmatic domains and therefore occur as secreted proteases [27].

One of the major functions of proteolytic ADAMs is the shedding of ectodomains of membrane proteins such as membrane-anchored cytokines, chemokines, growth factors, hormones and their receptors, cell–cell communication proteins such as cadherins, and Notch/Delta-like as well as ECM protein receptors such as the hyaluronic acid receptor (CD44) [25]. Strikingly, ECM proteins are not secure from being degraded by ADAMs, as evident in the cleavage of fibronectin by ADAM12 [28]. The prototypic example for an ADAM sheddase is ADAM17, also called the TNF-α converting enzyme (TACE), which orchestrates immune and inflammatory responses via activation of pro-TNF-α and is also essential for activation of heparin-binding EGF-like growth factor (HB-EGF) during development [29]. When ADAM17 was found to cleave the precursor form of TNF-α, the full impact of ADAM proteinases on fundamental aspects of mammalian physiology such as immunology, endocrinology, and development became apparent. Dysregulation of ADAMs has been implicated from then on in cancer and immunological, neurological, and cardiovascular diseases as well as in infections [30, 31].

It has become obvious in the recent years that bacteria are physiologic activators of the expression of ADAM family proteases, for example, ADAM17. TNF-α is induced on infection and sepsis and initially occurs as a type II transmembrane protein. Then, TNF-α is processed by ADAM17 into its soluble, biologically active form. TACE is upregulated in polymorphonuclear leukocytes (PMNs) of patients with severe sepsis and peritonitis in both blood and peritoneal fluid [32]. Stimulation of human monocytes with LPS increases the activity of TACE *in vitro* concomitantly with a decrease of membrane-localized TNF-α [33]. Both the gram-positive *Staphylococcus aureus* and the gram-negative *P. aeruginosa* activate IL-6 and TACE transcription, respectively, in epithelial cells [34]. A great number of strains of *Vibrio cholerae* produce a cytolysin (VCC, *Vibrio cholerae* cytolysin) that forms oligomeric transmembrane pores in mammalian cells, thereby lysing the target cells. The monomeric VCC binds to eukaryotic cells and is cleaved at the N-terminus to generate the active pore-forming toxin. Murine fibroblasts lacking ADAM17 bind comparable amounts of protoxin; however, oligomerization and cleavage of pro-VCC is markedly decreased in TACE-deficient cells, whereas TACE overexpressing HEK-293 cells revealed a clearly enhanced cleaving and oligomerization potency [35]. A similar strategy to facilitate entry and transcytosis through host tissue is employed by *S. aureus*, although this mechanism does not take advantage of the proteolytic activity of ADAMs. The cytotoxin α-hemolysin (Hla) is secreted as a water-soluble monomer that undergoes a series of conformational changes to form heptameric pores on susceptible host cell membranes triggering alterations in ion gradients, activation of stress pathways, loss of membrane integrity, and cell death. It was assumed for a long time that clustered phosphocholine head groups of the membrane lipids serve as a high affinity binding site for the toxin [36]. However, this model fails to explain the species specificity, the time course, and the saturability of Hla binding. Moreover, the findings that protease treatment abrogates toxin binding point to a ligand–proteinaceous receptor interaction [37]. This inconsistence was solved recently when ADAM10 was identified as the membrane receptor for Hla. ADAM10 aggregates in discrete punctae inside A549 cells on treatment with wild-type Hla. ADAM10 then colocalizes with caveolin-1-enriched lipid rafts, which is dependent on the ability of the toxin to form stable oligomers. Genetic ablation of ADAM10 by RNAi diminished the dephosporylation of focal adhesion kinase (FAK), Src, and p130cas, which represent important signaling molecules in focal adhesions. In addition, phagocytosis is decreased in FAK-deficient neutrophils [38]. The Hla–ADAM10 complex obviously inhibits this signaling cascade and leads to disruption of focal adhesions, thereby facilitating detachment of cells from structured tissues compromising tissue integrity. This could be a strategy of *S. aureus* to abuse host-derived proteases to perturb cellular barriers and escape phagocytosis or endocytosis and eradication by immune cells [39].

7.2.3
A Disintegrin and Metalloproteinase with Thrombospondin Motif (ADAMTS)

Another family of metzincin proteases with homology to the ADAMs is the ADAMTS family. The 19 mammalian members lack a transmembrane, a cytoplasmatic, and an EGF domain characteristic of the ADAM family. However, they contain a spacer region and one or more thrombospondin (TSP)-like motif(s), which are thought to be attached to the cell membrane. Cleavage of the TSP-like motif regulates both the proteinase activity and the localization of the enzyme [40, 41]. Taking into consideration the complex multidomain structure of ADAMTS proteinases, it is not surprising that they are implicated in many physiological and pathological processes such as osteoarthritis, cancer, chronic asthma, and connective tissue disorders and thrombotic thrombocytopenic purpura (TTP) [42–48].

Remarkably little is known about the role and regulation of ADAMTSs during infections with bacteria. ADAMTS4 is induced in human chondrocytes, susceptible strains of mice, and patients infected with *Borrelia burgdorferi*. Moreover, activated ADAMTS4 was increased in synovial fluid samples from patients with active lyme arthritis and elevated in joints of *B. burgdorferi*-infected mice [49]. Nevertheless, as the 18 other members of this ADAMTS family are involved in crucial physiological processes, it can be hypothesized that the interaction of other bacteria with proteinases of this family will be discovered.

7.3
Proteases of the Blood Coagulation and the Fibrinolytic System

The blood coagulation and fibrinolytic systems are the two antagonistic cascades in hemostasis. These two cascades are in a meticulous equilibrium in nonpathological conditions and so guarantee proper blood supply to all organs by protecting the integrity of the blood transport pathways. Blood coagulation is a complex process. Damage to the walls of blood vessels is covered by platelet- and fibrin-containing plugs that on one hand stop bleeding while on the other hand preserving normal blood flow in intact parts of the vasculature. Moreover, clotting factors initiate repair of the injured vein after break down of the clots by proteases of the fibrinolytic system. These two systems play an important role in different homeostatic aspects of normal physiology such as vessel and tissue regeneration, wound repair, inflammation, and defense against infections. Therefore, perturbations of the delicate equilibrium, either transient or genetically determined, mostly result in life-threatening pathological defects such as vascular diseases, hemophilia, and thrombotic complications such as myocardial infarction, pulmonary embolism, and cerebral apoplexy as well.

7.3.1
Proteases of the Blood Coagulation System

Coagulation is a sophisticated event consisting of a first immediate cellular reaction and a second system based on the action of multiple proteins called *clotting factors*, referred to as *primary and secondary hemostasis*, respectively. Coagulation leads to the formation of a blood clot (thrombus) that contains a network of polymerized fibrin with aggregates of platelets.

The cellular primary reaction is triggered instantly by injury to the endothelium of blood vessels exposing subendothelial layers containing endothelium-derived von Willebrand factor (vWF) and collagen. Circulating platelets bind to collagen and vWF and change their shape when activated from smooth disks to spiny spheres and are now optimally equipped to form sticky clots, which are further stabilized by fibrin from the secondary stage of hemostasis. The coagulation cascade of the secondary hemostasis comprises two converging pathways leading to thrombin and fibrin generation: (i) the intrinsic pathway also known as *contact activation pathway* and (ii) the extrinsic pathway often referred to as tissue factor (TF) pathway, which was shown to be more important [50, 51]. However, both pathways consist of a series of proteolytic reactions in which a zymogen of a serine protease is activated to catalyze the next reaction in the cascade.

The extrinsic pathway is started in response to vascular injury when the transmembrane protein TF, expressed in the adventitia and subendothelial cells such as fibroblasts and macrophages, comes into contact with its ligand, the circulating factor VII (FVII) [50]. The interaction of TF with FVII and factor X (FX) leads to generation of FXa from FX, which forms the prothrombinase complex together with activated factor V (FVa). This complex is responsible for the cleavage of a small amount of prothrombin to thrombin that amplifies the proteolytic cascade by activation of more FV and factor VIII (FVIII) that, bound to factor IXa (FIXa), cleaves FX. Finally, thrombin converts fibrinogen to fibrin that is further covalently cross-linked to form fibrin meshworks by the transamidase activity of the thrombin-activated factor XIII (FXIII) [51].

The intrinsic pathway starts with the autocatalytic activation of a component of the contact system, here factor XII (FXII, Hageman factor). The cue is the contact of FXII with polyanions such as polyphosphates secreted by activated platelets or nucleic acids from damaged cells [52, 53].

Bacteria feature a vast array of surface-located polyanions that may lead to the activation of the intrinsic part of the coagulation system. Surfaces of gram-negative bacteria are decorated with negatively charged LPS, whereas gram-positive bacteria possess negatively charged teichoic and lipoteichoic acids (LTAs). In fact, bronchial instillation of *Escherichia coli* LPSs or LTAs of *S. aureus* was able to activate bronchoalveolar coagulation. However, LPS and LTA also concurrently inhibit anticoagulant mechanisms measurable by reductions in antithrombin, activated protein C, plasminogen (PLG) activator activity, and elevated levels of soluble thrombomodulin and PLG activator inhibitor type 1 (PAI-1) [54].

There are different classes of endogenous bacterial proteins exploited by bacteria to usurp the coagulation system of their hosts. On the one hand, pathogens are able to use secreted proteins for recruitment of host proteases; on the other hand, they utilize surface-located proteins for binding and/or proteolytic activation of coagulation proteins. Staphylocoagulase (Coa) is a 60–70 kDa protein secreted by *S. aureus* that activates human prothrombin and directly initiates blood clotting by forming an equimolar complex with prothrombin and inducing a conformational change [55, 56]. The clot generation by fibrinogen activation plays an important role in the pathogenesis of abscess formation in multiple organs, which is considered to be a strategy of staphylococci to replicate in an environment protected from the host immune cells by a pseudocapsule. When the abscess ruptures to release the bacteria into the blood stream, a new round of abscess formation is initiated, eventually resulting in a lethal outcome for the infected person. Together with another staphylococcal clotting factor, the von-Willebrand-factor-binding protein (vWbp), which is colocalized with Coa, thrombin, and fibrin in the eosinophilic pseudocapsule, Coa was found to be absolutely pivotal for abscess formation and persistence in infected tissues. Abscess formation was completely abrogated when mice were infected with a Coa/vWbp double mutant, and antibodies blocking the association of prothrombin with these two staphylococcal factors prevented abscess formation in the infected mice [57]. Coa and vWbp are not the only virulence determinants by which *S. aureus* targets thrombin, although the mechanism of the alternative activation of thrombin is rather indirect. *S. aureus* α-toxin, the major and long known staphylococcal cytolysin, promotes the activation of prothrombin by elevation of the exocytotic release of factor V from α-granules [58]. The gram-negative bacterium *Vibrio vulnificus* is a causative agent of foodborne diseases as it secretes two zinc metalloproteases causing serious hemorrhagic complications. Vibrio extracellular proteases (vEPs) and vEP-MO6 cleave various coagulation-associated proteins such as prothrombin, fibrinogen, and FX and the fibrinolysis-associated protein PLG. The two antagonistic hemostasis systems are most probably not simultaneously activated. A first wave of prothrombin activation could lead to a rapid buildup of a fibrin mesh that might protect the bacteria from the host's immune surveillance followed by a second wave of fibrin cleavage, either directly or mediated by activation of PLG, that enables the bacteria to use the blood flow for dissemination [59, 60].

Another class of bacterial proteins has been reported to bind and hyperactivate thrombin, namely, porins from *Salmonella typhimurium*. Porins are proteins embedded in the outer membrane of gram-negative bacteria that control cell permeability by forming cross-membrane channels. Purified porins were shown to enhance thrombin activity, presumably by a nonproteolytic mechanism involving conformational changes in the active site of thrombin, and, moreover, inhibit the antithrombin (AT) activity exerted by AT III or α_2-macroglobulin(α_2M) by an unknown mechanism [61]. There are also bacterial species reported that bind endogenous proteases to their surface such as *P. gingivalis*, the gram-negative anaerobic key pathogen implicated in chronic periodontitis, a destructive inflammatory disease of the gingiva (gums) [62]. *P. gingivalis* expresses two cysteine proteases, the lysine-specific gingipain (gingipain-K) and the arginine-specific gingipain (gingipain R), which are located on the surface of *Porphyromonas* and in some strains are released [63, 64]. Gingipains have been associated with dose- and time-dependent activation of prothrombin and FX, a zymogen component of the coagulation pathway [65, 66]. The same gingipains were able to activate the coagulation FIX, the factor directly upstream of FX, that is converted to its active form either by the FVII-TF complex during extrinsic coagulation or by activated factor XI (FXI) in the context of the intrinsic pathway [67]. These findings, together with the fact that gingipains can also spur prekallikrein to action, suggest a role of gingipains as a causative agent in the pathogenesis of disseminated intravascular coagulation in sepsis and highlight the emerging relationship between periodontitis and cardiovascular disease [68].

7.3.2 Proteases of the Fibrinolytic System

Once the coagulation is triggered at the site of vascular injury it is crucial that there are mechanisms to impair the expansion of the hemostatic plug into healthy vasculature to ensure maintenance of normal blood flow and to initiate and facilitate the process of wound healing. The fibrinolytic system, which takes over this function, is composed primarily of three serine proteases present as zymogens in the blood. Among these, the glycoprotein PLG, which is produced in the liver, is the principal mediator. PLG is cleaved to dual-chained active PLG by two main effectors, proteases themselves, namely tissue-type plasminogen activator (tPA), released from activated endothelial cells in response to venous occlusion and urokinase-type plasminogen activator (uPA) secreted as pro-urokinase and activated by plasmin and contact system factors in the presence of the receptor for urokinase-type plasminogen activator (uPAR) [51]. tPA, bound to fibrin clots, activates PLG, likewise bound to the fibrin meshwork that, in turn, cleaves and breaks down fibrin, thereby resolving the plug [69]. Apart from degrading fibrin, plasmin is known to use the ECM proteins, fibronectin, laminin, and vitronectin, as well as pro-MMPs as substrates in the process of tissue renewing and wound healing. The fibrinolysis cascade is highly regulated by inhibitors such as the serpins PAI1–3 and the thrombin-activatable fibrinolysis inhibitor (TAFI) acting on tPA and uPA as well as the serpin α_2-antiplasmin, antithrombin, and α_2M, which inhibit plasmin [70]. Factors of the fibrinolytic cascade play an important role in wound healing and counteraction of blood coagulation. Furthermore, they are crucial for cell migration, tissue remodeling, angiogenesis, and embryogenesis [71]. Hence, pathologic variances in the fibrinolytic cascade entail the development of hemostatic abnormalities, ranging from insignificant transient changes to severe life-threatening diseases such as disseminated intravascular coagulation, myocardial infarction, and hyperfibrinolysis [72–74]. Some of the noncongenial conditions mentioned above can be triggered indirectly by infections with microorganisms and the resulting inflammation and toxic shock [72]. Strikingly, bacterial pathogens have evolved a panel of mechanisms to subvert the fibrinolytic cascades for colonization, invasion, and dissemination by binding, activating, or inhibiting host-derived proteases, activators, or inhibitors of the fibrinolysis system.

Group A, C, and G streptococci produce streptokinase, the prototype of a complexing bacterial PLG activator that has been structurally analyzed by X-ray diffraction in complex with the catalytic unit of plasmin referred to as *microplasmin* [75]. Loops of the α-and γ-domains of the streptokinase interact with microplasmin. The formation of this stoichiometric complex induces nonproteolytic conformational changes in PLG or plasmin molecules, resulting in hydrolytic activation of other PLG molecules and protection of plasmin against enzymatic inactivation by the serpin α_2-antiplasmin [76]. This protection against α_2-antiplasmin was not seen when the nonhomologous *S. aureus* enzyme staphylokinase was complexed with plasmin, unless fibrin or bacterial cell surfaces were present. Moreover, binding of PLG to fibrin or bacterial surfaces even enhances the staphylokinase-induced PLG activation [77]. Similar to streptokinase and staphylokinase, proteins secreted by other streptococcal species such as PauA and PauB from *Streptococcus uberis* [78], Esk from *Streptococcus equisimilis* [79], and PadA from *Streptococcus dysgalactiae* bind and activate PLG [80]. However, the activated PLG remains susceptible to α_2-antiplasmin. Recently, a plasmin(ogen)-binding protein secreted by the group B streptococcus *Streptococcus agalactiae* was described, which has moderate sequence identity to streptokinase albeit lacking its intrinsic PLG activation potency. However, it is able to enhance the activation of PLG mediated by uPA and tPA [81].

Prominent members of the group of proteolytic bacterial PLG activators are the omptins Pla of *Yersinia pestis* and PgtE of *Salmonella enterica*, which convert

PLG by peptide bond hydrolysis into plasmin such as tPA and uPA. Similar to the streptokinase–plasmin complex from *S. pyogenes*, plasmin is protected from α_2-antiplasmin action in this complex, which is thought to be the main effect of PgtE rather than strong activation of PLG via PgtE [82]. Pla seems to be highly important for dissemination from subcutaneous and intradermal infection sites to the lymph nodes and blood circulation, as Pla-deficient strains were avirulent when applied subcutaneously but as virulent as the Pla-sufficient ones when injected intravenously [83].

Most bacterial species only recruit PLG to their surface without being able to activate it. Nevertheless, the immobilization of PLG on cell surfaces such as the bacterial envelope causes conformational opening of PLG that facilitates its activation by the host-derived PLG activators tPA and uPA [84]. Often, the acquisition of PLG is accompanied by a protection against the inhibitor α_2-antiplasmin [85]. tPA is the major activator of fibrinolysis and possesses fibrin-binding activity, while uPA is involved both in fibrinolysis and cell adhesion and migration, tissue remodeling, and apoptosis [86, 87]. PLG-binding proteins of bacteria belong to different protein classes: (i) nonclassical surface proteins, (ii) choline-binding proteins, (iii) lipoproteins, and (iv) covalently anchored proteins containing the sortase motif LPxTG.

Nonclassical PLG-binding proteins include several glycolytic enzymes such as glyceraldehyde-3-phosphate dehydrogenase (GAPDH) and enolase from different bacterial species. The GAPDH from *S. pyogenes* (Plr or SDH) binds PLG via its C-terminal lysine residue, while the two lysine residues at the C-terminus of the enolases of *S. pyogenes* and *S. pneumoniae* have no direct role in PLG binding. The essential nine-residue PLG-binding motif of the pneumococcal enolase, localized between amino acids 248 and 256, has been shown to be highly conserved among several enolases from various bacterial species [88–90]. Strikingly, this motif is a key factor for the plasmin-mediated degradation of ECM proteins and for transmigration of pneumococci through fibrin clots [91]. Although these glycolytic enzymes lack typical signal peptides and membrane anchorage motifs required for surface export and cell wall anchorage, respectively, they were detected on microbial cell surfaces [88, 92–94].

Another group of proteins comprising a member that holds PLG-binding activity are the cholin-binding proteins as in the case of the choline-binding protein E (CbpE) of *S. pneumoniae*, also referred to as phosphorylcholine esterase (Pce). The biological relevance of the CbpE–PLG interaction is supported by the finding that *cbpE*-deficient pneumococcal mutants display a reduced level of PLG binding and activation and show decreased ability to cross the ECM in an *in vitro* model [95]. More importantly, the CbpE-bound plasmin(ogen) fulfills a dual function in enhancing migration of pneumococci through cell barriers formed by endothelial cells and dissemination by facilitating adherence to endothelial and epithelial cells, while proteolytically active plasmin degrades intercellular junctions by cleavage of the main component of endothelial adherence junctions cadherin [96].

S. pyogenes produces a high-affinity plasmin(ogen) receptor, the M-like protein PLG-binding group A streptococcal M protein (PAM) that belongs to the LPxTG-anchored surface proteins [97, 98]. The concerted action of PAM and streptokinase in these streptococci improves significantly their ability to overcome microvascular occlusions and, hence, enhances streptococcal virulence [99]. Further examples of LPxTG-anchored proteins with the capability of binding PLG are the plasmin- and fibronectin-binding protein A (PfbA) and the pneumococcal adherence and virulence factor B (PavB) from *S. pneumoniae*. A strain deficient in PfbA was less competent than the wild-type strain in its ability to adhere to and invade human lung and laryngeal epithelial cells and exhibited a decreased antiphagocytic activity in human peripheral blood [100]. The interaction of PavB with PLG depends on the number of SSURE (Streptococcal SUrface REpeats) present in the core unit of PavB and contributes to nasopharyngeal colonization of mice. The observed delay of this mutant in transmigration to the lungs is due to the reduced ability in PLG binding [101].

The lipoproteins outer surface protein A and C (Osp A and C) and OspEF-related proteins P, A, and C (erpP, A, and C) bind PLG. These proteins endow the spirochete *B. burgdorferi* with the ability to penetrate human skin layers and disseminate in the blood to cause, for example, Lyme disease after being transmitted by ticks [102, 103]. OspA and OspC enable the bacterium to degrade components of the ECM and to penetrate specialized endothelial layers such as the blood–brain barrier [104, 105]. Moreover, experiments with PLG-knockout mice demonstrated an essential role of PLG in transmigration of the spirochetes to the salivary gland of the ticks after feeding [106]. Erp proteins are constitutively expressed through all stages of infection, pointing to a role in acute as well as long-term infection. Beyond its role in dissemination, PLG binding by *B. burgdorferi* is probably involved in immune avoidance as the acquired proteolytic activity degrades specific antibodies and components of the complement system. The finding that ErpP also binds Factor H, the key regulator of the alternative pathway of the complement system, suggests a sophisticated escape mechanism of this bacterial species for persistence in the mammalian host [103]. A similar phenomenon was detected for the complement regulator-acquiring surface protein 1 (CRASP-1), a multifunctional protein of *B. burgdorferi* that binds not only PLG and Factor H but also different collagens, fibronectin, laminin, and bone morphogenic protein 2 (BMP-2), a pleiotrophic human adhesion protein of the transforming growth factor-β protein family [107].

In addition, *B. burgdorferi* and other bacteria not only abuse one component of the fibrinolytic system but also subvert other factors of this system for infection, namely uPA, its receptor uPAR and PAIs, the serpins that counteract the fibrinolytic activity of plasmin. The spirochete is able to induce the expression of uPA in its zymogen form in monocytes and provides an appropriate surface for subsequent interactions between (pro)uPA and plasmin(ogen) in an inflammatory site [108]. Moreover, *B. burgdorferi* can stimulate upregulation of the membrane receptor for urokinase uPAR by CD14 and TLR2 activation in human monocytes that together with its ligand uPA can, in turn, activate more bacteria-bound PLG, thereby enhancing the proteolytic capacity of the spirochete [109, 110]. The induction of PAIs seems counterproductive for the benefit of the bacterium at first glance, but studies showed that the transmigration of *B. burgdorferi* is not impaired by PAI-2, the PAI secreted by monocytes and macrophages. Contrarily, the invasion of monocytic cells across a reconstituted basement membrane (matrigel) is significantly diminished by PAI-2 [111]. Another inhibitor of fibrinolysis, which is targeted by bacteria, is the TAFI that removes C-terminal lysine residues from fibrin, thereby attenuating accelerated plasmin formation. TAFI is inactivated by omptins Pla and PgtE of *Y. pestis* and *S. enterica*, respectively, by C-terminal cleavage resulting in increased fibrinolysis and enhanced spread of the pathogenic bacteria during infection [112]. Taken together, the findings mentioned in this paragraph highlight the importance of the blood coagulation and the PLG system for the bacterial infection, dissemination, and persistence.

7.4 Contact System

The contact system consists of two serine proteases (plasma kallikrein (PK) and FXII, respectively) and one nonenzymatic cofactor (high-molecular-weight kininogen, or HK). Its activation on negatively charged or cellular surfaces leads to the release of bradykinin (a nonapeptide) from the HK precursor by the action of PK and/or the activation of FXI by activated FXII [113]. While bradykinin is considered an important vasoactive mediator evoking the classic signs of inflammation: *dolor, rubor, calor,* and *tumor* (pain, redness, heat, and swelling), activation of FXI triggers an induction of the intrinsic pathway of coagulation and eventually causes the formation of a fibrin network [114, 115]. A systemic activation of the contact system is often seen in patients suffering from severe infectious diseases such as sepsis and septic shock, and a consumption of contact factors in these patients is an indication of a fatal outcome of the disease [116]. Notably, kinin levels in plasma samples from patients with sepsis can reach pathologic concentrations [117]. It is therefore currently believed that a massive kinin release contributes to complications such as hypovolemic hypotension, which is a hallmark in severe infections. In addition to these complications, rat and murine sepsis models have shown that systemic contact activation can also lead to serious hemostatic dysfunction, as the infected animals develop severe pulmonary bleedings, which are not seen when they are treated with contact system inhibitors [52, 118, 119].

To prevent an overwhelming and uncontrolled activation of the contact system, activated PK, FXII, and FXI are efficiently targeted by serine proteinase inhibitors. C1 esterase inhibitor (C1-INH) is considered the most important physiological regulator, but other inhibitors such as α_2M, antithrombin III, and protein C inhibitor are also known to block the activity of contact factors [113]. Clinical sepsis studies have shown that application of C1-INH attenuates contact activation in patients and is combined with a beneficial effect on hypotension and renal dysfunction [120]. More recently, a double-blind placebo-controlled study was conducted, with healthy volunteers receiving an intravenous injection of *E. coli* LPS. It was found that C1-INH application exerts anti-inflammatory effects in the absence of classic complement activation [121], suggesting that C1-INH is an interesting drug candidate for treating severe infectious diseases.

7.4.1 Mechanisms of Bacteria-Induced Contact Activation

Under normal hemostatic conditions, activation of the contact system follows the waterfall principle requiring an exact proteolytic processing of all contact factors involved, a mechanism also referred to as *limited proteolysis*. However, many bacterial and fungal proteases are also able to activate the contact system by employing different modes of activation. This can lead to an activation of FXII and PK by a less specific cleavage mechanism or a direct processing HK, which can yield the release of active kinins. Microbial proteases of species such as *Candida albicans, Clostridium histolyticum, Bacillus stearothermophilus, Bacillus subtilis, P. gingivalis, Serratia marcescens*, and *V. vulnificus* have been found to activate FXII and/or PK, which in turn results in cleavage of HK by PK and the subsequent release of BK [122, 123]. Other proteases such as from *P. gingivalis, S. aureus, S. pyogenes*, and *Streptomyces caespitosus* act directly on HK to release active kinins [122, 123]. Interestingly, unlike the factors of the contact system, microbial proteases do not necessarily belong to the family of serine proteases, indicating that convergent evolution might have been the driving force that triggers the microbially induced activation of the contact system. As the release of kinins will evoke an increase in the vascular permeability, it has been speculated that this mechanism will supply the microorganisms with nutrients and allow their dissemination within their host [123].

The activation of the contact system on microbial surfaces seems to follow different principles and may also have other functions. It was already in the early 1980s when it was reported that surface factors from gram-negative and gram-positive bacteria, such as endotoxin, peptidoglycan, and teichoic acid, can cause an activation of the contact system [124, 125]. Within the last three decades many bacterial and fungal species that are able to recruit and activate contact factors at their surface have been identified [122, 123]. These findings led Opal and Esmon to conclude that pattern recognition molecules of the innate immune system function in a manner that is remarkably similar to that of contact factors [126]. Indeed, evidence is accumulating that the contact system is part of the innate immune system. It has been, for instance, reported that activation and processing of the contact system on microbial surfaces prompts the generation of kininogen-derived antimicrobial peptides [127, 128]. Other studies have shown that mice lacking kinin receptors, B1 and B2 receptors, have an impaired immune response. Thus, it was found that B1 receptor knockout animals have a diminished ability to recruit neutrophils to the inflamed site and animals lacking the B2 receptor suffer from increased hepatic bacterial burden and concomitant dramatic weight loss when infected with *Listeria monocytogenes*. Kinin receptors are therefore an interesting target for drug development. However, a SIRS (systemic inflammatory response syndrome) and sepsis study conducted in 1993 dampens these expectations. In a multicenter randomized, double-blind placebo-controlled trial, patients were treated with deltibant (a B2R antagonist) for 28 days or until death. The results show that deltibant had no significant effect on risk-adjusted 28-day survival, but there was a nonsignificant trend toward improvement in the risk-adjusted 7-day survival analysis and a statistically significant improvement in the subset of patients with gram-negative infections [128]. Taking into consideration that both kinin receptors seem to have an important function in severe infectious diseases, it is tempting to speculate that an approach involving a combination of a B1 and B2 receptor antagonist may be more promising. In contrast to kinin antagonists, inhibitors acting on contact factors have only been tested in animal models so far. It has been, for instance, reported that blocking FXII activity with a monoclonal antibody prevents irreversible hypotension and prolonged survival in the infected animals. Other studies using a rat *Salmonella* sepsis model have shown that application of an irreversible PK inhibitor leads to less severe lung lesions in the infected animals [118, 129], while a contact system inhibitor targeting HK is able to increase survival in a murine streptococcal sepsis model when given in combination with an antibiotic [130].

7.5 Conclusion and Future Prospectives

A vast amount of data has accumulated during the recent years emphasizing the elaborate mechanisms that bacteria have evolved to usurp proteolytic activity of the host for their own benefit and survival. In doing so, pathogens exploit different endogenous and host-derived macromolecules and interfere at various physiological and cell biological levels. On the bacterial side, LPSs and LTAs play an important role, so do miscellaneous kinds of secreted or surface-located bacterial and host-derived proteins on both the bacterial and mammalian side. Diverse physiological systems of the mammalian body are targeted by microbes on that account, sometimes redundantly, starting with blood coagulation and fibrinolysis right through to the contact system, the immune system, and tissue homeostasis. Some bacteria are known to merely elevate the expression of host proteases relying on the host's own physiologic reaction to activate them, while others take a more hands-on role in the activation process (Figure 7.1). Interestingly, recent studies suggest that the exploitation of host-derived proteases is not exclusively restricted to pathogenic bacteria but rather extends to commensal and probiotic bacteria

Figure 7.1 Schematic depiction of the modes of recruitment of host protease activity by bacteria. Bacteria trigger the expression (mRNA and protein) of host proteases that are activated (as indicated by a shift of shape from round to crescentic) by other host proteases in the course of concurrent inflammation [1] or bind already activated proteases [2]. Bacteria secrete and bind back virulence factors that can activate host proteases by complex formation [3] or proteolytic processing [4]. Bacteria express adhesins belonging to different protein types on their surface by means of which they recruit inactive host proteases to their cells envelope that get activated by other host proteases [5]. The corresponding host proteases fulfill a plethora of different tasks in physiology in a wide range of the hosts' homeostasis systems such as coagulation, fibrinolysis, contact system, and immune system targeting different aspects and factors in these pathways. These ways of subverting host proteases eventually lead to dissemination and persistence of bacterial infections.

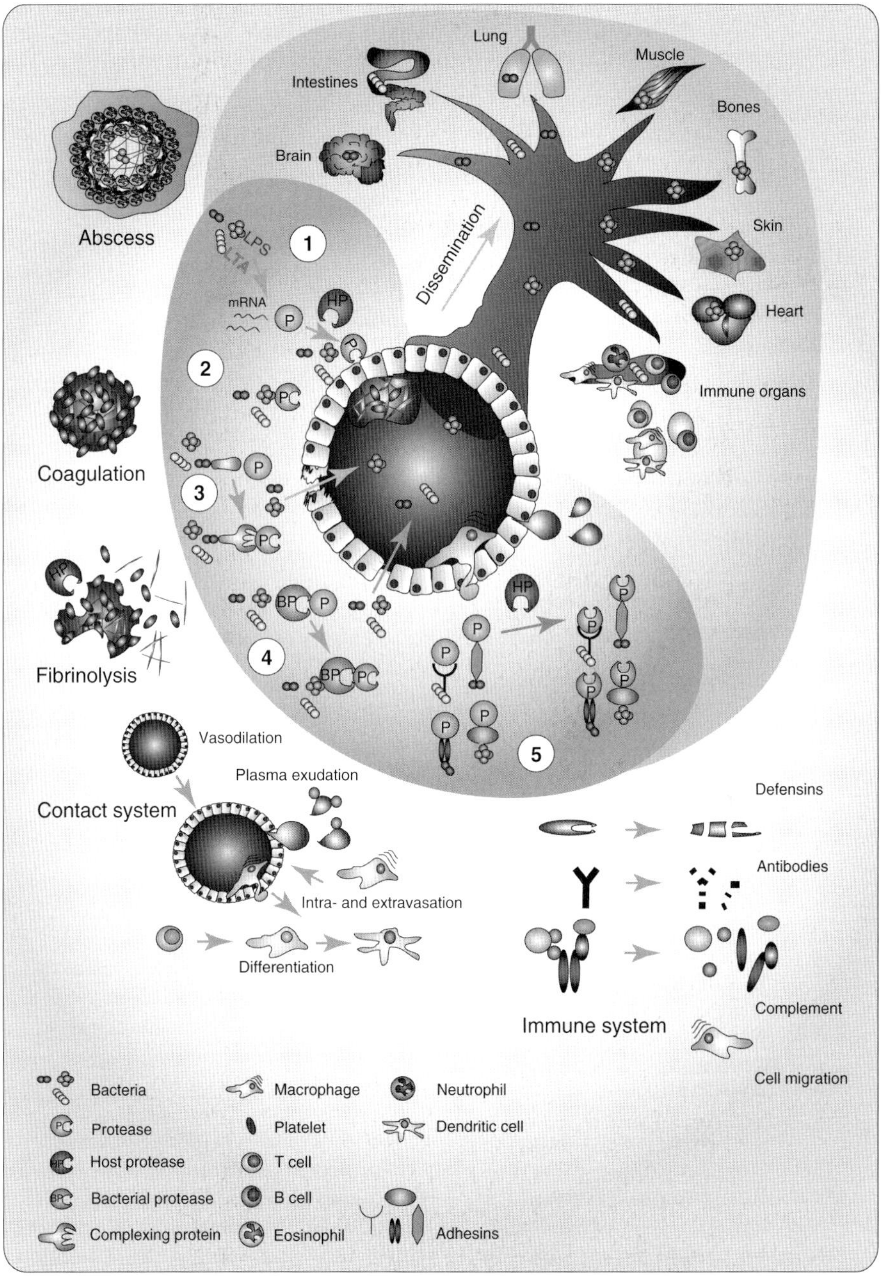

Lung
Muscle
Intestines
Bones
Brain
Skin
Abscess
1
LPS
LTA
Dissemination
mRNA
P
HP
Heart
PC
2
Immune organs
Coagulation
3
Fibrinolysis
BP
4
5
Vasodilation
Plasma exudation
Contact system
Defensins
Antibodies
Intra- and extravasation
Differentiation
Immune system
Complement
Cell migration
Bacteria
Macrophage
Neutrophil
Protease
Platelet
Dendritic cell
Host protease
T cell
Bacterial protease
B cell
Complexing protein
Eosinophil
Adhesins

[130, 131]. However, the beneficial effects of this interaction are not known yet, although the assumption is justified that, on the one hand, the binding to surface-located proteases might promote colonization with commensals and, on the other hand, the recruitment of secreted proteases might interfere with the exploitation of proteases by pathogens. Taken together, identification of bacterial factors manipulating host-derived proteolytic systems, their host receptor molecules, and the induced signal transduction cascades are of pivotal importance to come up with novel strategies to combat infectious diseases.

Acknowledgments

Research in the laboratory of S.H. is supported by grants from the Deutsche Forschungsgemeinschaft (DFG **HA 3125/2-1**, **HA 3125/3-1**, and **HA 3125/4-1**, **GRK840**, TRR34 project C10), Excellence Initiative Mecklenburg-Vorpommern (**FKZ UG 08010**), and the EU FP7 *CAREPNEUMO* (**EU-CP223111**, European Union). The research of H.H. is supported by the Swedish Research Council (project 7480). We thank Krystin Krauel for helpful discussion. Our apologies in advance to the authors of primary articles whom we have failed to cite owing to space restrictions.

References

1. Lopez-Otin, C. and Bond, J.S. (2008) Proteases: multifunctional enzymes in life and disease. *J. Biol. Chem.*, **283**, 30433–30437.
2. Overall, C.M. and Lopez-Otin, C. (2002) Strategies for MMP inhibition in cancer: innovations for the post-trial era. *Nat. Rev. Cancer*, **2**, 657–672.
3. Shiomi, T., Lemaitre, V., D'Armiento, J., and Okada, Y. (2010) Matrix metalloproteinases, a disintegrin and metalloproteinases, and a disintegrin and metalloproteinases with thrombospondin motifs in non-neoplastic diseases. *Pathol. Int.*, **60**, 477–496.
4. Itoh, Y. and Seiki, M. (2006) MT1-MMP: a potent modifier of pericellular microenvironment. *J. Cell Physiol.*, **206**, 1–8.
5. Opdenakker, G., Van den Steen, P.E., Dubois, B., Nelissen, I., Van Coillie, E., Masure, S., Proost, P., and Van Damme, J. (2001) Gelatinase B functions as regulator and effector in leukocyte biology. *J. Leukoc. Biol.*, **69**, 851–859.
6. Albert, J., Radomski, A., Soop, A., Sollevi, A., Frostell, C., and Radomski, M.W. (2003) Differential release of matrix metalloproteinase-9 and nitric oxide following infusion of endotoxin to human volunteers. *Acta Anaesthesiol. Scand.*, **47**, 407–410.
7. Kim, T. (2000) Lipopolysaccharide activates matrix metalloproteinase-2 in endothelial cells through an NF-kB-dependent pathway. *Biochem. Biophys. Res. Commun.*, **269**, 401–405.
8. Pugin, J., Widmer, M.C., Kossodo, S., Liang, C.M., Preas, H.L. II, and Suffredini, A.F. (1999) Human neutrophils secrete gelatinase B in vitro and in vivo in response to endotoxin and proinflammatory mediators. *Am. J. Respir. Cell Mol. Biol.*, **20**, 458–464.
9. Gottschall, P.E., Yu, X., and Bing, B. (1995) Increased production of gelatinase B (matrix metalloproteinase-9)

and interleukin-6 by activated rat microglia in culture. *J. Neurosci. Res.*, **42**, 335–342.
10. Lai, W.C., Zhou, M., Shankavaram, U., Peng, G., and Wahl, L.M. (2003) Differential regulation of lipopolysaccharide-induced monocyte matrix metalloproteinase (MMP)-1 and MMP-9 by p38 and extracellular signal-regulated kinase 1/2 mitogen-activated protein kinases. *J. Immunol.*, **170**, 6244–6249.
11. Tanaka, A., Yamane, Y., and Matsuda, H. (2001) Mast cell MMP-9 production enhanced by bacterial lipopolysaccharide. *J. Vet. Med. Sci.*, **63**, 811–813.
12. Warner, R.L., Bhagavathula, N., Nerusu, K.C., Lateef, H., Younkin, E., Johnson, K.J., and Varani, J. (2004) Matrix metalloproteinases in acute inflammation: induction of MMP-3 and MMP-9 in fibroblasts and epithelial cells following exposure to pro-inflammatory mediators in vitro. *Exp. Mol. Pathol.*, **76**, 189–195.
13. Pagenstecher, A., Stalder, A.K., Kincaid, C.L., Volk, B., and Campbell, I.L. (2000) Regulation of matrix metalloproteinases and their inhibitor genes in lipopolysaccharide-induced endotoxemia in mice. *Am. J. Pathol.*, **157**, 197–210.
14. Handley, S.A. and Miller, V.L. (2007) General and specific host responses to bacterial infection in Peyer's patches: a role for stromelysin-1 (matrix metalloproteinase-3) during Salmonella enterica infection. *Mol. Microbiol.*, **64**, 94–110.
15. Dong, Z., Katar, M., Alousi, S., and Berk, R.S. (2001) Expression of membrane-type matrix metalloproteinases 4, 5, and 6 in mouse corneas infected with P. aeruginosa. *Invest. Ophthalmol. Vis. Sci.*, **42**, 3223–3227.
16. Ramu, P., Lobo, L.A., Kukkonen, M., Bjur, E., Suomalainen, M., Raukola, H., Miettinen, M., Julkunen, I., Holst, O., Rhen, M., Korhonen, T.K., and Lahteenmaki, K. (2008) Activation of pro-matrix metalloproteinase-9 and degradation of gelatin by the surface protease PgtE of Salmonella enterica serovar Typhimurium. *Int. J. Med. Microbiol.*, **298**, 263–278.
17. Okamoto, T., Akaike, T., Suga, M., Tanase, S., Horie, H., Miyajima, S., Ando, M., Ichinose, Y., and Maeda, H. (1997) Activation of human matrix metalloproteinases by various bacterial proteinases. *J. Biol. Chem.*, **272**, 6059–6066.
18. Tamura, F., Nakagawa, R., Akuta, T., Okamoto, S., Hamada, S., Maeda, H., Kawabata, S., and Akaike, T. (2004) Proapoptotic effect of proteolytic activation of matrix metalloproteinases by Streptococcus pyogenes thiol proteinase (Streptococcus pyrogenic exotoxin B). *Infect. Immun.*, **72**, 4836–4847.
19. Oggioni, M.R., Memmi, G., Maggi, T., Chiavolini, D., Iannelli, F., and Pozzi, G. (2003) Pneumococcal zinc metalloproteinase ZmpC cleaves human matrix metalloproteinase 9 and is a virulence factor in experimental pneumonia. *Mol. Microbiol.*, **49**, 795–805.
20. Schubert-Unkmeir, A., Konrad, C., Slanina, H., Czapek, F., Hebling, S., and Frosch, M. (2010) Neisseria meningitidis induces brain microvascular endothelial cell detachment from the matrix and cleavage of occludin: a role for MMP-8. *PLoS Pathog.*, **6**, e1000874.
21. Itoh, S., Hamada, E., Kamoshida, G., Takeshita, K., Oku, T., and Tsuji, T. (2010) Staphylococcal superantigen-like protein 5 inhibits matrix metalloproteinase 9 from human neutrophils. *Infect. Immun.*, **78**, 3298–3305.
22. Zhou, J. and Windsor, L.J. (2006) Porphyromonas gingivalis affects host collagen degradation by affecting expression, activation, and inhibition of matrix metalloproteinases. *J. Periodontal Res.*, **41**, 47–54.
23. Sato, Y., Kishi, J., Suzuki, K., Nakamura, H., and Hayakawa, T. (2009) Sonic extracts from a bacterium related to periapical disease activate gelatinase A and inactivate tissue inhibitor of metalloproteinases TIMP-1 and TIMP-2. *Int. Endod. J.*, **42**, 1104–1111.

24. Gomis-Ruth, F.X. (2003) Structural aspects of the metzincin clan of metalloendopeptidases. *Mol. Biotechnol.*, **24**, 157–202.
25. Edwards, D.R., Handsley, M.M., and Pennington, C.J. (2008) The ADAM metalloproteinases. *Mol. Aspects Med.*, **29**, 258–289.
26. Huang, J., Bridges, L.C., and White, J.M. (2005) Selective modulation of integrin-mediated cell migration by distinct ADAM family members. *Mol. Biol. Cell*, **16**, 4982–4991.
27. Seals, D.F. and Courtneidge, S.A. (2003) The ADAMs family of metalloproteases: multidomain proteins with multiple functions. *Genes Dev.*, **17**, 7–30.
28. Roy, R., Wewer, U.M., Zurakowski, D., Pories, S.E., and Moses, M.A. (2004) ADAM 12 cleaves extracellular matrix proteins and correlates with cancer status and stage. *J. Biol. Chem.*, **279**, 51323–51330.
29. Sahin, U., Weskamp, G., Kelly, K., Zhou, H.M., Higashiyama, S., Peschon, J., Hartmann, D., Saftig, P., and Blobel, C.P. (2004) Distinct roles for ADAM10 and ADAM17 in ectodomain shedding of six EGFR ligands. *J. Cell Biol.*, **164**, 769–779.
30. Mochizuki, S. and Okada, Y. (2007) ADAMs in cancer cell proliferation and progression. *Cancer Sci.*, **98**, 621–628.
31. Tousseyn, T., Jorissen, E., Reiss, K., and Hartmann, D. (2006) (Make) stick and cut loose–disintegrin metalloproteases in development and disease. *Birth Defects Res. C Embryo Today*, **78**, 24–46.
32. Kermarrec, N., Selloum, S., Plantefeve, G., Chosidow, D., Paoletti, X., Lopez, A., Mantz, J., Desmonts, J.M., Gougerot-Pocidalo, M.A., and Chollet-Martin, S. (2005) Regulation of peritoneal and systemic neutrophil-derived tumor necrosis factor-alpha release in patients with severe peritonitis: role of tumor necrosis factor-alpha converting enzyme cleavage. *Crit. Care Med.*, **33**, 1359–1364.
33. Robertshaw, H.J. and Brennan, F.M. (2005) Release of tumour necrosis factor alpha (TNFalpha) by TNFalpha cleaving enzyme (TACE) in response to septic stimuli in vitro. *Br. J. Anaesth.*, **94**, 222–228.
34. Gomez, M.I., Sokol, S.H., Muir, A.B., Soong, G., Bastien, J., and Prince, A.S. (2005) Bacterial induction of TNF-alpha converting enzyme expression and IL-6 receptor alpha shedding regulates airway inflammatory signaling. *J. Immunol.*, **175**, 1930–1936.
35. Valeva, A., Walev, I., Weis, S., Boukhallouk, F., Wassenaar, T.M., Endres, K., Fahrenholz, F., Bhakdi, S., and Zitzer, A. (2004) A cellular metalloproteinase activates vibrio cholerae pro-cytolysin. *J. Biol. Chem.*, **279**, 25143–25148.
36. Valeva, A., Hellmann, N., Walev, I., Strand, D., Plate, M., Boukhallouk, F., Brack, A., Hanada, K., Decker, H., and Bhakdi, S. (2006) Evidence that clustered phosphocholine head groups serve as sites for binding and assembly of an oligomeric protein pore. *J. Biol. Chem.*, **281**, 26014–26021.
37. Hildebrand, A., Pohl, M., and Bhakdi, S. (1991) Staphylococcus aureus alpha-toxin. Dual mechanism of binding to target cells. *J. Biol. Chem.*, **266**, 17195–17200.
38. Kasorn, A., Alcaide, P., Jia, Y., Subramanian, K.K., Sarraj, B., Li, Y., Loison, F., Hattori, H., Silberstein, L.E., Luscinskas, W.F., and Luo, H.R. (2009) Focal adhesion kinase regulates pathogen-killing capability and life span of neutrophils via mediating both adhesion-dependent and -independent cellular signals. *J. Immunol.*, **183**, 1032–1043.
39. Wilke, G.A. and Bubeck Wardenburg, J. (2010) Role of a disintegrin and metalloprotease 10 in Staphylococcus aureus alpha-hemolysin-mediated cellular injury. *Proc. Natl. Acad. Sci. U S A*, **107**, 13473–13478.
40. Porter, S., Clark, I.M., Kevorkian, L., and Edwards, D.R. (2005) The ADAMTS metalloproteinases. *Biochem. J.*, **386**, 15–27.
41. Novak, U. (2004) ADAM proteins in the brain. *J. Clin. Neurosci.*, **11**, 227–235.

42. Majumdar, M.K., Askew, R., Schelling, S., Stedman, N., Blanchet, T., Hopkins, B., Morris, E.A., and Glasson, S.S. (2007) Double-knockout of ADAMTS-4 and ADAMTS-5 in mice results in physiologically normal animals and prevents the progression of osteoarthritis. *Arthritis Rheum.*, **56**, 3670–3674.
43. Colige, A., Sieron, A.L., Li, S.W., Schwarze, U., Petty, E., Wertelecki, W., Wilcox, W., Krakow, D., Cohn, D.H., Reardon, W., Byers, P.H., Lapiere, C.M., Prockop, D.J., and Nusgens, B.V. (1999) Human Ehlers-Danlos syndrome type VII C and bovine dermatosparaxis are caused by mutations in the procollagen I N-proteinase gene. *Am. J. Hum. Genet.*, **65**, 308–317.
44. Di Valentin, E., Crahay, C., Garbacki, N., Hennuy, B., Gueders, M., Noel, A., Foidart, J.M., Grooten, J., Colige, A., Piette, J., and Cataldo, D. (2009) New asthma biomarkers: lessons from murine models of acute and chronic asthma. *Am. J. Physiol. Lung Cell Mol. Physiol.*, **296**, L185–L197.
45. Dubail, J., Kesteloot, F., Deroanne, C., Motte, P., Lambert, V., Rakic, J.M., Lapiere, C., Nusgens, B., and Colige, A. (2010) ADAMTS-2 functions as anti-angiogenic and anti-tumoral molecule independently of its catalytic activity. *Cell Mol. Life Sci.*, **67**, 4213–4232.
46. Furlan, M., Robles, R., and Lammle, B. (1996) Partial purification and characterization of a protease from human plasma cleaving von Willebrand factor to fragments produced by in vivo proteolysis. *Blood*, **87**, 4223–4234.
47. Iruela-Arispe, M.L., Carpizo, D., and Luque, A. (2003) ADAMTS1: a matrix metalloprotease with angioinhibitory properties. *Ann. N. Y. Acad. Sci.*, **995**, 183–190.
48. Levy, G.G., Nichols, W.C., Lian, E.C., Foroud, T., McClintick, J.N., McGee, B.M., Yang, A.Y., Siemieniak, D.R., Stark, K.R., Gruppo, R., Sarode, R., Shurin, S.B., Chandrasekaran, V., Stabler, S.P., Sabio, H., Bouhassira, E.E., Upshaw, J.D. Jr., Ginsburg, D., and Tsai, H.M. (2001) Mutations in a member of the ADAMTS gene family cause thrombotic thrombocytopenic purpura. *Nature*, **413**, 488–494.
49. Behera, A.K., Hildebrand, E., Szafranski, J., Hung, H.H., Grodzinsky, A.J., Lafyatis, R., Koch, A.E., Kalish, R., Perides, G., Steere, A.C., and Hu, L.T. (2006) Role of aggrecanase 1 in lyme arthritis. *Arthritis Rheum.*, **54**, 3319–3329.
50. Monroe, D.M. and Key, N.S. (2007) The tissue factor-factor VIIa complex: procoagulant activity, regulation, and multitasking. *J. Thromb. Haemost.*, **5**, 1097–1105.
51. Adams, R.L. and Bird, R.J. (2009) Review article: coagulation cascade and therapeutics update: relevance to nephrology. Part 1: overview of coagulation, thrombophilias and history of anticoagulants. *Nephrology (Carlton)*, **14**, 462–470.
52. Oehmcke, S. and Herwald, H. (2010) Contact system activation in severe infectious diseases. *J. Mol. Med.*, **88**, 121–126.
53. Frick, I.M., Bjorck, L., and Herwald, H. (2007) The dual role of the contact system in bacterial infectious disease. *Thromb. Haemost.*, **98**, 497–502.
54. Hoogerwerf, J.J., de Vos, A.F., Levi, M., Bresser, P., van der Zee, J.S., Draing, C., von Aulock, S., and van der Poll, T. (2009) Activation of coagulation and inhibition of fibrinolysis in the human lung on bronchial instillation of lipoteichoic acid and lipopolysaccharide. *Crit. Care Med.*, **37**, 619–625.
55. Friedrich, R., Panizzi, P., Fuentes-Prior, P., Richter, K., Verhamme, I., Anderson, P.J., Kawabata, S., Huber, R., Bode, W., and Bock, P.E. (2003) Staphylocoagulase is a prototype for the mechanism of cofactor-induced zymogen activation. *Nature*, **425**, 535–539.
56. Hemker, H.C., Bas, B.M., and Muller, A.D. (1975) Activation of a pro-enzyme by a stoichiometric reaction with another protein. The reaction between prothrombin and staphylocoagulase. *Biochim. Biophys. Acta*, **379**, 180–188.
57. Cheng, A.G., McAdow, M., Kim, H.K., Bae, T., Missiakas, D.M., and

Schneewind, O. (2010) Contribution of coagulases towards Staphylococcus aureus disease and protective immunity. *PLoS Pathog.*, **6**, e1001036.

58. Arvand, M., Bhakdi, S., Dahlback, B., and Preissner, K.T. (1990) Staphylococcus aureus alpha-toxin attack on human platelets promotes assembly of the prothrombinase complex. *J. Biol. Chem.*, **265**, 14377–14381.
59. Chang, A.K., Kim, H.Y., Park, J.E., Acharya, P., Park, I.S., Yoon, S.M., You, H.J., Hahm, K.S., Park, J.K., and Lee, J.S. (2005) Vibrio vulnificus secretes a broad-specificity metalloprotease capable of interfering with blood homeostasis through prothrombin activation and fibrinolysis. *J. Bacteriol.*, **187**, 6909–6916.
60. Kwon, J.Y., Chang, A.K., Park, J.E., Shin, S.Y., Yoon, S.M., and Lee, J.S. (2007) Vibrio extracellular protease with prothrombin activation and fibrinolytic activities. *Int. J. Mol. Med.*, **19**, 157–163.
61. Di Micco, B., Di Micco, P., Lepretti, M., Stiuso, P., Donnarumma, G., Iovene, M.R., Capasso, R., and Tufano, M.A. (2005) Hyperproduction of fibrin and inefficacy of antithrombin III and alpha2 macroglobulin in the presence of bacterial porins. *Int. J. Exp. Pathol.*, **86**, 241–245.
62. Socransky, S.S., Haffajee, A.D., Cugini, M.A., Smith, C., and Kent, R.L. Jr. (1998) Microbial complexes in subgingival plaque. *J. Clin. Periodontol.*, **25**, 134–144.
63. Roper, J.M., Raux, E., Brindley, A.A., Schubert, H.L., Gharbia, S.E., Shah, H.N., and Warren, M.J. (2000) The enigma of cobalamin (Vitamin B12) biosynthesis in Porphyromonas gingivalis. Identification and characterization of a functional corrin pathway. *J. Biol. Chem.*, **275**, 40316–40323.
64. DeCarlo, A.A., Paramaesvaran, M., Yun, P.L., Collyer, C., and Hunter, N. (1999) Porphyrin-mediated binding to hemoglobin by the HA2 domain of cysteine proteinases (gingipains) and hemagglutinins from the periodontal pathogen Porphyromonas gingivalis. *J. Bacteriol.*, **181**, 3784–3791.
65. Imamura, T., Potempa, J., Tanase, S., and Travis, J. (1997) Activation of blood coagulation factor X by arginine-specific cysteine proteinases (gingipain-Rs) from Porphyromonas gingivalis. *J. Biol. Chem.*, **272**, 16062–16067.
66. Imamura, T., Banbula, A., Pereira, P.J., Travis, J., and Potempa, J. (2001) Activation of human prothrombin by arginine-specific cysteine proteinases (Gingipains R) from porphyromonas gingivalis. *J. Biol. Chem.*, **276**, 18984–18991.
67. Imamura, T., Tanase, S., Hamamoto, T., Potempa, J., and Travis, J. (2001) Activation of blood coagulation factor IX by gingipains R, arginine-specific cysteine proteinases from Porphyromonas gingivalis. *Biochem. J.*, **353**, 325–331.
68. Imamura, T., Pike, R.N., Potempa, J., and Travis, J. (1994) Pathogenesis of periodontitis: a major arginine-specific cysteine proteinase from Porphyromonas gingivalis induces vascular permeability enhancement through activation of the kallikrein/kinin pathway. *J. Clin. Invest.*, **94**, 361–367.
69. Gale, A.J. (2010) Current understanding of hemostasis. *Toxicol. Pathol.*, **39**, 273–280.
70. Rau, J.C., Beaulieu, L.M., Huntington, J.A., and Church, F.C. (2007) Serpins in thrombosis, hemostasis and fibrinolysis. *J. Thromb. Haemost.*, **5** (Suppl. 1), 102–115.
71. Castellino, F.J. and Ploplis, V.A. (2005) Structure and function of the plasminogen/plasmin system. *Thromb. Haemost.*, **93**, 647–654.
72. Levi, M., Schultz, M., and van der Poll, T. (2010) Disseminated intravascular coagulation in infectious disease. *Semin. Thromb. Hemost.*, **36**, 367–377.
73. Gorog, D.A. (2010) Prognostic value of plasma fibrinolysis activation markers in cardiovascular disease. *J. Am. Coll. Cardiol.*, **55**, 2701–2709.
74. Carpenter, S.L. and Mathew, P. (2008) Alpha2-antiplasmin and its deficiency: fibrinolysis out of balance. *Haemophilia*, **14**, 1250–1254.

75. Wang, X., Lin, X., Loy, J.A., Tang, J., and Zhang, X.C. (1998) Crystal structure of the catalytic domain of human plasmin complexed with streptokinase. *Science*, **281**, 1662–1665.
76. Wang, H., Lottenberg, R., and Boyle, M.D. (1995) Analysis of the interaction of group A streptococci with fibrinogen, streptokinase and plasminogen. *Microb. Pathog.*, **18**, 153–166.
77. Molkanen, T., Tyynela, J., Helin, J., Kalkkinen, N., and Kuusela, P. (2002) Enhanced activation of bound plasminogen on Staphylococcus aureus by staphylokinase. *FEBS Lett.*, **517**, 72–78.
78. Johnsen, L.B., Rasmussen, L.K., Petersen, T.E., Etzerodt, M., and Fedosov, S.N. (2000) Kinetic and structural characterization of a two-domain streptokinase: dissection of domain functionality. *Biochemistry*, **39**, 6440–6448.
79. Nowicki, S.T., Minning-Wenz, D., Johnston, K.H., and Lottenberg, R. (1994) Characterization of a novel streptokinase produced by Streptococcus equisimilis of non-human origin. *Thromb. Haemost.*, **72**, 595–603.
80. Leigh, J.A., Hodgkinson, S.M., and Lincoln, R.A. (1998) The interaction of *Streptococcus dysgalactiae* with plasmin and plasminogen. *Vet. Microbiol.*, **61**, 121–135.
81. Wiles, K.G., Panizzi, P., Kroh, H.K., and Bock, P.E. (2010) Skizzle is a novel plasminogen- and plasmin-binding protein from Streptococcus agalactiae that targets proteins of human fibrinolysis to promote plasmin generation. *J. Biol. Chem.*, **285**, 21153–21164.
82. Kukkonen, M. and Korhonen, T.K. (2004) The omptin family of enterobacterial surface proteases/adhesins: from housekeeping in Escherichia coli to systemic spread of Yersinia pestis. *Int. J. Med. Microbiol.*, **294**, 7–14.
83. Sodeinde, O.A., Subrahmanyam, Y.V., Stark, K., Quan, T., Bao, Y., and Goguen, J.D. (1992) A surface protease and the invasive character of plague. *Science*, **258**, 1004–1007.
84. Plow, E.F., Freaney, D.E., Plescia, J., and Miles, L.A. (1986) The plasminogen system and cell surfaces: evidence for plasminogen and urokinase receptors on the same cell type. *J. Cell Biol.*, **103**, 2411–2420.
85. Rouy, D. and Angles-Cano, E. (1990) The mechanism of activation of plasminogen at the fibrin surface by tissue-type plasminogen activator in a plasma milieu in vitro. Role of alpha 2-antiplasmin. *Biochem. J.*, **271**, 51–57.
86. Rossignol, P., Luttun, A., Martin-Ventura, J.L., Lupu, F., Carmeliet, P., Collen, D., Angles-Cano, E., and Lijnen, H.R. (2006) Plasminogen activation: a mediator of vascular smooth muscle cell apoptosis in atherosclerotic plaques. *J. Thromb. Haemost.*, **4**, 664–670.
87. Crippa, M.P. (2007) Urokinase-type plasminogen activator. *Int. J. Biochem. Cell. Biol.*, **39**, 690–694.
88. Bergmann, S., Rohde, M., Chhatwal, G.S., and Hammerschmidt, S. (2001) alpha-Enolase of Streptococcus pneumoniae is a plasmin(ogen)-binding protein displayed on the bacterial cell surface. *Mol. Microbiol.*, **40**, 1273–1287.
89. Bergmann, S., Rohde, M., Chhatwal, G.S., and Hammerschmidt, S. (2004) Characterization of plasmin(ogen) binding to Streptococcus pneumoniae. *Indian J. Med. Res.*, **119** (Suppl.), 29–32.
90. Ehinger, S., Schubert, W.D., Bergmann, S., Hammerschmidt, S., and Heinz, D.W. (2004) Plasmin(ogen)-binding alpha-enolase from Streptococcus pneumoniae: crystal structure and evaluation of plasmin(ogen)-binding sites. *J. Mol. Biol.*, **343**, 997–1005.
91. Bergmann, S., Rohde, M., Preissner, K.T., and Hammerschmidt, S. (2005) The nine residue plasminogen-binding motif of the pneumococcal enolase is the major cofactor of plasmin-mediated degradation of extracellular matrix, dissolution of fibrin and transmigration. *Thromb. Haemost.*, **94**, 304–311.
92. Bergmann, S., Rohde, M., and Hammerschmidt, S. (2004)

Glyceraldehyde-3-phosphate dehydrogenase of Streptococcus pneumoniae is a surface-displayed plasminogen-binding protein. *Infect. Immun.*, **72**, 2416–2419.

93. Jin, H., Song, Y.P., Boel, G., Kochar, J., and Pancholi, V. (2005) Group A streptococcal surface GAPDH, SDH, recognizes uPAR/CD87 as its receptor on the human pharyngeal cell and mediates bacterial adherence to host cells. *J. Mol. Biol.*, **350**, 27–41.

94. Schaumburg, J., Diekmann, O., Hagendorff, P., Bergmann, S., Rohde, M., Hammerschmidt, S., Jansch, L., Wehland, J., and Karst, U. (2004) The cell wall subproteome of Listeria monocytogenes. *Proteomics*, **4**, 2991–3006.

95. Attali, C., Frolet, C., Durmort, C., Offant, J., Vernet, T., and Di Guilmi, A.M. (2008) Streptococcus pneumoniae choline-binding protein E interaction with plasminogen/plasmin stimulates migration across the extracellular matrix. *Infect. Immun.*, **76**, 466–476.

96. Attali, C., Durmort, C., Vernet, T., and Di Guilmi, A.M. (2008) The interaction of Streptococcus pneumoniae with plasmin mediates transmigration across endothelial and epithelial monolayers by intercellular junction cleavage. *Infect. Immun.*, **76**, 5350–5356.

97. Berge, A. and Sjobring, U. (1993) PAM, a novel plasminogen-binding protein from Streptococcus pyogenes. *J. Biol. Chem.*, **268**, 25417–25424.

98. Piard, J.C., Hautefort, I., Fischetti, V.A., Ehrlich, S.D., Fons, M., and Gruss, A. (1997) Cell wall anchoring of the Streptococcus pyogenes M6 protein in various lactic acid bacteria. *J. Bacteriol.*, **179**, 3068–3072.

99. Sun, H., Ringdahl, U., Homeister, J.W., Fay, W.P., Engleberg, N.C., Yang, A.Y., Rozek, L.S., Wang, X., Sjobring, U., and Ginsburg, D. (2004) Plasminogen is a critical host pathogenicity factor for group A streptococcal infection. *Science*, **305**, 1283–1286.

100. Yamaguchi, M., Terao, Y., Mori, Y., Hamada, S., and Kawabata, S. (2008) PfbA, a novel plasmin- and fibronectin-binding protein of Streptococcus pneumoniae, contributes to fibronectin-dependent adhesion and antiphagocytosis. *J. Biol. Chem.*, **283**, 36272–36279.

101. Jensch, I., Gamez, G., Rothe, M., Ebert, S., Fulde, M., Somplatzki, D., Bergmann, S., Petruschka, L., Rohde, M., Nau, R., and Hammerschmidt, S. (2010) PavB is a surface-exposed adhesin of Streptococcus pneumoniae contributing to nasopharyngeal colonization and airways infections. *Mol. Microbiol.*, **77**, 22–43.

102. Hu, L.T., Perides, G., Noring, R., and Klempner, M.S. (1995) Binding of human plasminogen to Borrelia burgdorferi. *Infect. Immun.*, **63**, 3491–3496.

103. Brissette, C.A., Haupt, K., Barthel, D., Cooley, A.E., Bowman, A., Skerka, C., Wallich, R., Zipfel, P.F., Kraiczy, P., and Stevenson, B. (2009) Borrelia burgdorferi infection-associated surface proteins ErpP, ErpA, and ErpC bind human plasminogen. *Infect. Immun.*, **77**, 300–306.

104. Coleman, J.L., Roemer, E.J., and Benach, J.L. (1999) Plasmin-coated borrelia Burgdorferi degrades soluble and insoluble components of the mammalian extracellular matrix. *Infect. Immun.*, **67**, 3929–3936.

105. Grab, D.J., Perides, G., Dumler, J.S., Kim, K.J., Park, J., Kim, Y.V., Nikolskaia, O., Choi, K.S., Stins, M.F., and Kim, K.S. (2005) Borrelia burgdorferi, host-derived proteases, and the blood-brain barrier. *Infect. Immun.*, **73**, 1014–1022.

106. Coleman, J.L., Gebbia, J.A., Piesman, J., Degen, J.L., Bugge, T.H., and Benach, J.L. (1997) Plasminogen is required for efficient dissemination of B. burgdorferi in ticks and for enhancement of spirochetemia in mice. *Cell*, **89**, 1111–1119.

107. Hallstrom, T., Haupt, K., Kraiczy, P., Hortschansky, P., Wallich, R., Skerka, C., and Zipfel, P.F. (2010) Complement regulator-acquiring surface protein 1 of Borrelia burgdorferi binds to human bone morphogenic protein 2, several extracellular matrix proteins, and plasminogen. *J. Infect. Dis.*, **202**, 490–498.

108. Fuchs, H., Simon, M.M., Wallich, R., Bechtel, M., and Kramer, M.D. (1996) Borrelia burgdorferi induces secretion of pro-urokinase-type plasminogen activator by human monocytes. *Infect. Immun.*, **64**, 4307–4312.
109. Coleman, J.L., Gebbia, J.A., and Benach, J.L. (2001) Borrelia burgdorferi and other bacterial products induce expression and release of the urokinase receptor (CD87). *J. Immunol.*, **166**, 473–480.
110. Coleman, J.L. and Benach, J.L. (2003) The urokinase receptor can be induced by Borrelia burgdorferi through receptors of the innate immune system. *Infect. Immun.*, **71**, 5556–5564.
111. Haile, W.B., Coleman, J.L., and Benach, J.L. (2006) Reciprocal upregulation of urokinase plasminogen activator and its inhibitor, PAI-2, by Borrelia burgdorferi affects bacterial penetration and host-inflammatory response. *Cell Microbiol.*, **8**, 1349–1360.
112. Valls Seron, M., Haiko, J., PG, D.E.G., Korhonen, T.K., and Meijers, J.C. (2010) Thrombin-activatable fibrinolysis inhibitor is degraded by Salmonella enterica and Yersinia pestis. *J. Thromb. Haemost.*, **8**, 2232–2240.
113. Colman, R.W. and Schmaier, A.H. (1997) Contact system: a vascular biology modulator with anticoagulant, profibrinolytic, antiadhesive, and proinflammatory attributes. *Blood*, **90**, 3819–3843.
114. Leeb-Lundberg, L.M., Marceau, F., Muller-Esterl, W., Pettibone, D.J., and Zuraw, B.L. (2005) International union of pharmacology. XLV. Classification of the kinin receptor family: from molecular mechanisms to pathophysiological consequences. *Pharmacol. Rev.*, **57**, 27–77.
115. Emsley, J., McEwan, P.A., and Gailani, D. (2010) Structure and function of factor XI. *Blood*, **115**, 2569–2577.
116. Pixley, R.A. and Colman, R.W. (1997) in *The Kallikrein-Kinin System in Sepsis Syndrome. Handbook of Immunopharmacology – The Kinin System* (ed. S.G. Farmer), Academic Press, New York, pp. 173–186.
117. Mattsson, E., Herwald, H., Cramer, H., Persson, K., Sjöbring, U., and Björck, L. (2001) Staphylococcus aureus induces release of bradykinin in human plasma. *Infect. Immun.*, **69**, 3877–3882.
118. Persson, K., Mörgelin, M., Lindbom, L., Alm, P., Björck, L., and Herwald, H. (2000) Severe lung lesions caused by *Salmonella* are prevented by inhibition of the contact system. *J. Exp. Med.*, **192**, 1415–1424.
119. Oehmcke, S. and Herwald, H. (2009) Contact system activation in severe infectious diseases. *J. Mol. Med.*, **88**, 121–126.
120. Hack, C.E., Ogilvie, A.C., Eisele, B., Eerenberg, A.J., Wagstaff, J., and Thijs, L.G. (1993) C1-inhibitor substitution therapy in septic shock and in the vascular leak syndrome induced by high doses of interleukin-2. *Intensive Care Med.*, **19** (Suppl. 1), S19–S28.
121. Dorresteijn, M.J., Visser, T., Cox, L.A., Bouw, M.P., Pillay, J., Koenderman, A.H., Strengers, P.F., Leenen, L.P., van der Hoeven, J.G., Koenderman, L., and Pickkers, P. (2010) C1-esterase inhibitor attenuates the inflammatory response during human endotoxemia. *Crit. Care Med.*, **38**, 2139–2145.
122. Tapper, H. and Herwald, H. (2000) Modulation of hemostatic mechanisms in bacterial infectious diseases. *Blood*, **96**, 2329–2337.
123. Potempa, J. and Pike, R.N. (2009) Corruption of innate immunity by bacterial proteases. *J. Innate Immun.*, **1**, 70–87.
124. Kalter, E.S., van Dijk, W.C., Timmerman, A., Verhoef, J., and Bouma, B.N. (1983) Activation of purified human plasma prekallikrein triggered by cell wall fractions of *Escherichia coli and Staphylococcus aureus. J. Infect. Dis.*, **148**, 682–691.
125. Kalter, E.S., Daha, M.R., ten Cate, J.W., Verhoef, J., and Bouma, B.N. (1985) Activation and inhibition of Hageman factor-dependent pathways and the complement system in uncomplicated bacteremia or bacterial shock. *J. Infect. Dis.*, **151**, 1019–1027.
126. Opal, S.M. and Esmon, C.T. (2003) Bench-to-bedside review: functional

relationships between coagulation and the innate immune response and their respective roles in the pathogenesis of sepsis. *Crit. Care*, **7**, 23–38.

127. Frick, I.M., Åkesson, P., Herwald, H., Mörgelin, M., Malmsten, M., Nägler, D.K., and Björck, L. (2006) The contact system–a novel branch of innate immunity generating antibacterial peptides. *EMBO J.*, **25**, 5569–5578.

128. Nordahl, E.A., Rydengård, V., Mörgelin, M., and Schmidtchen, A. (2005) Domain 5 of high molecular weight kininogen is antibacterial. *J. Biol. Chem.*, **280**, 34832–34839.

129. Pixley, R.A., De La Cadena, R., Page, J.D., Kaufman, N., Wyshock, E.G., Chang, A., Taylor, F.B. Jr., and Colman, R.W. (1993) The contact system contributes to hypotension but not disseminated intravascular coagulation in lethal bacteremia. *In vivo* use of a monoclonal anti-factor XII antibody to block contact activation in baboons. *J. Clin. Invest.*, **91**, 61–68.

130. Candela, M., Centanni, M., Fiori, J., Biagi, E., Turroni, S., Orrico, C., Bergmann, S., Hammerschmidt, S., and Brigidi, P. (2010) DnaK from Bifidobacterium animalis subsp. lactis is a surface-exposed human plasminogen receptor upregulated in response to bile salts. *Microbiology*, **156**, 1609–1618.

131. Hurmalainen, V., Edelman, S., Antikainen, J., Baumann, M., Lahteenmaki, K., and Korhonen, T.K. (2007) Extracellular proteins of Lactobacillus crispatus enhance activation of human plasminogen. *Microbiology*, **153**, 1112–1122.

8
Experimental Approaches for Understanding the Role of Matrix Metalloproteinases in Cancer Invasion

Elena Deryugina

8.1
Introduction: Functional Roles of MMPs in Physiological Processes Involving the Induction and Sustaining of Cancer Invasion

Cancer invasion is one of the hallmarks of malignant progression that is clinically manifested in the penetration of primary tumor cells into adjacent territories beyond the normal constrains of the tissue of origin [1, 2]. In cancer patients, the depth of cancer invasion serves as a diagnostic factor for primary tumor staging as well as a risk factor for lymph node involvement and a prognostic factor for the development of metastatic disease and long-term survival [3]. Many experimental model systems have directly linked the extent of tumor invasion to the levels of local and distant metastases, thereby pointing to the disruption of specific processes exploited by cancer cells during metastatic invasion as one of the most efficient ways to control cancer spread.

Mechanistically, the invasion of cancer cells requires the remodeling of extracellular matrix (ECM) proteins and modifications of cell–matrix and cell–cell contacts to facilitate the escape and directional locomotion of individual tumor cells or cohorts of tumor cells from the primary tumor to the neighboring tissues [4]. Matrix metalloproteinases (MMPs), capable of ECM degradation, proteolytic release and activation of ECM-sequestered growth factors and cytokines, and modifications of cell surface molecules, constitute a family of proteolytic enzymes that have been long implicated in cancer progression and metastasis [5]. Nevertheless, despite several decades of intensive research, the specific mechanisms underlying functions of MMPs in distinct cancer-related processes *in vivo* are still not fully elucidated. While a few MMPs can exhibit clear cancer-protective functions [6], a large volume of experimental evidence points to MMPs as enzymes that promote tumor growth, tumor angiogenesis, and metastatic invasion, and in some models, even epithelial cell transformation. However, critical processes such as angiogenic switching, followed by angiogenesis-dependent tumor cell intravasation and tumor-orchestrated preparation of premetastatic niches, are mainly facilitated by the MMPs originating from bone-marrow-derived myeloid cells recruited to the corresponding original or secondary tumor sites.

Matrix Proteases in Health and Disease, First Edition. Edited by Niels Behrendt.

Cancer invasion is initiated when tumor cells have acquired the ability to leave the primary tumor, either individually or in cohorts, and have switched on the machineries for directional migration within the three-dimensional milieu of the surrounding stroma. These machineries involve the modulation of cytoskeleton structure and creation of contractile forces, the turnover of substrate adhesions and reconfiguration of actin microfilaments and microtubules, and the generation of podosomes and invadopodia, which localize protease-mediated degradation of the ECM [7]. Depending on the tissue of origin, primary tumor cells exploit different programs to initiate persistent migration and invasion. Thus, carcinoma cells must undergo epithelial–mesenchymal transition (EMT) [8] to breach the epithelial basement membrane (BM), while mesenchymal fibrosarcoma cells or glioma cells appear intrinsically equipped with high invasive potential to penetrate into connective tissues.

The onset of invasion is manifested by the escape of leading cells from the primary tumor and the generation of ECM tracks and paths, which assist the ensuing invasion of trailing cells. Although passive shedding mechanisms of tumor dissemination have been proposed [9], the actively invading cancer cells are believed to be those cells that reach the abluminal surface of angiogenic or coopted preexisting blood vessels and enter the circulation. This notion is supported by clinical data demonstrating that the presence of tumor cells in the circulation is strictly dependent on the invasive stage of cancer [10]. However, the invading cancer cells constitute a heterogeneous population and only few complete the intravasation step, survive the shear forces of blood circulation, overcome proapoptotic signals during extravasation and colonization at the secondary organs, and successfully form the expanding metastases. This intrinsic heterogeneity of primary cancer cells makes it extremely difficult and problematic to attribute any specific molecule, including individual MMPs, with a unique, deterministic role during distinct phases of tumor development [11]. In experimental settings, the demonstration of functional importance of MMPs in cancer invasion and tumor progression can also be complicated by the notion that many MMPs not only exhibit overlapping and opposing functions but also frequently reveal their specific roles only in conjunction with additional cell surface or intracellular molecules or in the presence of certain ECM proteins. Different model systems designed to study distinct stages of cancer progression arm scientists with experimental approaches to unravel the specific functions of individual MMPs in cancer invasion and illuminate their unexpected and contrasting properties. Such experimental systems and approaches will be a critical focus in this chapter.

8.2
EMT: a Prerequisite of MMP-Mediated Cancer Invasion or a Coordinated Response to Growth-Factor-Induced MMPs?

It has become generally accepted that during carcinoma progression, aggressive tumor cells lose their epithelial features such as cell polarity and intercellular adherens junctions and adopt a mesenchymal phenotype associated with the loss of contact inhibition and acquisition of enhanced directional motility. This so-called EMT, originally identified as a reversible mechanism of mesenchymal differentiation during embryonic development [12], is now regarded as the major mechanism underlying a shift to cancer invasion and malignancy [13–18]. Originating from the tumor-associated stroma, EMT-inducing signaling molecules such as HGF, EGF, PDGF, and TGF-β trigger activation of several transcription factors in tumor cells, including Twist, Snail, Slug, ZEB1, and FOXC2. These factors, in turn, induce multiple intracellular signaling networks, enhancing cell invasion and metastasis. In carcinoma cells *in vitro*, the ongoing EMT is manifested by the progressive loss of various epithelial markers, including E-cadherin, some cytokeratins, and α-catenin, and the gain of mesenchymal markers such as N-cadherin, vimentin, fibronectin, α-smooth muscle actin, and nuclear β-catenin [18].

Despite a certain level of controversy regarding the presence or expression levels of EMT markers *in vivo* [19, 20], EMT is commonly considered as a prerequisite for the onset of cell migration, invasion, and metastatic dissemination in epithelial tumors [13, 15–18, 21]. Correspondingly, the resulting invasive cancer cells are usually viewed as those that have actually undergone EMT, and the induction of cancer invasion is frequently regarded as a direct consequence of EMT.

In addition to major transcriptional factors and protein markers of EMT, several MMPs have also been associated with EMT-induced cancer invasion. At the gene expression level, the upregulation of *MMP1, MMP2, MMP3, MMP7, MMP13,* and *MMP14* has been observed during EMT in several carcinoma cell lines [21, 22]. The involvement of MMPs in EMT is also indicated by the sensitivity of factor-induced EMT to MMP inhibitors [23]. However, the views on the functional roles of MMPs during EMT are not coherent with regard to the positioning of MMP induction in EMT signaling cascades, that is, downstream or upstream of early EMT-inducing transcription factors. A conventional point of view is that the induction of MMP expression follows the onset of the EMT program and leads to MMP-mediated ECM degradation; however, ample evidence points to a scenario where specific MMPs act as inducers of EMT cascades.

8.2.1 MMP-Induced EMT

The strongest evidence of an EMT-inducing role of MMPs comes from the models employing exposure of tumor cells to exogenous MMPs *in vitro* or transgenic expression of MMPs *in vivo*. Thus, the exposure of normal mouse mammary epithelial SCp2 cells to recombinant MMP-3 resulted in the loss of cell–cell interactions and acquisition of a scattered morphology, downregulation of epithelial cytokeratins, and upregulation of the mesenchymal marker vimentin, all being manifestations of the classical EMT phenotype [24]. If SCp2 cells were stably transfected with an autoactivating MMP-3 construct and orthotopically explanted into mammary fat pads, they formed highly invasive tumors with mesenchymal-like cells at the tumor periphery. Furthermore, the expression of autoactivating *MMP3* transgene targeted to mammary epithelium in mice resulted in the development of spontaneous malignant mammary lesions in transgenic animals [24]. Importantly, these tumors were virtually absent in the bitransgenic animals that coexpressed MMP-3 and TIMP-1, indicating that the enzymatic activity of MMP-3 was required to promote mammary neoplasias with EMT-like characteristics [24, 25]. Since in normal mammary gland, MMP-3 is mostly produced by stromal fibroblasts, these findings also exemplify the mechanisms by which stroma-produced MMPs can alter tissue microenvironment, leading to a malignant transformation of normal epithelial cells. Unexpectedly, however, the genetic depletion of MMP-3 results in accelerated progression of carcinogen-induced epithelial tumors to more invasive epithelioid and metastatic spindle cell phenotypes in knockout mice [26]. These apparently contrasting scenarios, involving MMP-3, have not yet been resolved.

EMT can also be induced by MMP-9 and MMP-28. The SCp2 mammary cells stably transfected with an autoactivated MMP-9 construct demonstrated a robust EMT associated with increased cell scattering, loss of cytokeratin, and acquisition of vimentin in the motile cells [27]. Stable and irreversible EMT was shown to be triggered by the overexpression of proteolytically active MMP-28 in the lung A549 adenocarcinoma cells [28]. At the same time, MMP-28-stimulated EMT was accompanied by the loss of cell surface E-cadherin, increased levels of bioactive TGF-β, upregulation of MT1-MMP and MMP-9, and increased cell invasion of collagen matrices. Interestingly, the onset of EMT was sensitive to a general MMP inhibitor GM6001, but once TGF-β-dependent EMT had occurred, the cell mesenchymal-like phenotype became MMP independent [28]. In a similar manner, once initiated, the progression of invasive mammary carcinomas in transgenic MMP-3 mice became independent of continuous MMP-3 expression [24].

Degradation or shedding of E-cadherin by several MMPs, including MMP-3, MMP-7, and MMP-9, has been mechanistically attributed to the induction of EMT downstream of growth-factor-induced EMT cascades, but upstream of EMT-inducing signaling pathways. Thus, *in vitro* treatment of SCp2 mouse mammary epithelial cells with exogenous MMP-3 caused the loss of intact E-cadherin and upregulated Snail1, leading to increased cell motility and invasiveness [29]. Analogously, proteolytic disruption of E-cadherin by MMP-3 or MMP-9 in normal renal tubular epithelial cells directly mediated EMT downstream of TGF-β1 but upstream of Slug, a repressor of E-cadherin promoter [23]. Shedding of E-cadherin ectodomain by the autoactivating mutant of MMP-7 (aMat) caused the decrease in E-cadherin at cell–cell junctions and promoted migration of aMat-transfected lung A549 adenocarcinoma cells in a wound-healing model *in vitro* [30]. Furthermore, treatment of lung tumor 16HBE cells with soluble E-cadherin fragments induced expression of several MMPs, including MMP-2, MMP-9, and MT1-MMP, and promoted cell invasion into chick embryo heart explants *ex vivo* and collagen gels *in vitro* [31]. Conversely, an overexpression of E-cadherin in highly invasive bronchial BZR tumor cells resulted in the decrease of β-catenin transcriptional activity, impaired invasion, and a concomitant decrease of several MMPs, including MMP-1, MMP-3, MMP-9, and MT1-MMP [32].

The acquisition of EMT markers and induction of mesenchymal-like behavior characteristics was observed in human ovarian carcinoma OVCA 433 cells transfected with the internalization-defective MT1-MMP mutant, MT1-Y/F, resulting in the sustained cell surface activity of MT1-MMP [33]. Compared to their wild-type counterparts, MT1-Y/F transfectants completely lost E-cadherin expression, while they significantly increased the expression of N-cadherin and vimentin. This mesenchymal marker makeover was accompanied by a corresponding increase in cell motility and invasion [33]. A novel mechanism, whereby MT1-MMP induces EMT features, was recently demonstrated for prostate cancer cells, in which a combined expression of MT1-MMP and a variant of FGFR4 stimulated matrix degradation and invasive tumor cell growth. Reciprocally, downregulation of MT1-MMP and FGFR4 by RNA interference induced expression of E-cadherin, suppressed N-cadherin, and blocked tumor cell invasion [34].

8.2.2
EMT-Induced MMPs

A conventional view of the functional role of MMPs in EMT is that the completion of EMT is manifested by proteolytic degradation of BMs and remodeling of the ECM by the EMT-induced MMPs [18]. An important aspect of this MMP-mediated matrix remodeling during EMT is related to the increased production and deposition of matrix proteins by tumor and also stromal cells during EMT-induced cancer invasion. MMP-mediated matrix degradation and remodeling associated with EMT has been initially demonstrated in a 2D *in vitro* model, where human rectal adenocarcinoma L-10 cells were stimulated with EMT-inducing HGF [35]. A dependency of cohort migration on MMP activity in this model system was attributed to MMP-2 and MT1-MMP, which were immunolocalized predominantly to the leading edges of the front cells, with the trailing cells being MMP-negative. In addition, *in situ* zymography demonstrated that during cohort migration, the gelatin matrix was degraded and reorganized by the leading cells [35], suggesting that matrix reorganization is essential for EMT-induced cell invasion.

MT1-MMP and MT2-MMP were demonstrated to cooperatively function as proinvasive factors for breast carcinoma MCF-7 undergoing Snail-induced EMT [36]. In the chick embryo CAM model, the expression and, importantly, the proteolytic activity of these two membrane-anchored MMPs were obligatory for the EMT invasion program induced by Snail1, while a number of secreted MMPs, including MMP-1, MMP-2, MMP-3, MMP-7, MMP-9, and MMP-13, were dispensable. Remarkably, the Snail1-induced program was completely recapitulated by the direct expression of MT1-MMP or MT2-MMP in MCF-7 cells, suggesting that in this model system, these two membrane-tethered MMPs could also trigger an EMT program [36]. However, the expression of MT1-MMP did not confer EMT-associated invasive and metastatic abilities to cancer cells isolated from spontaneously developed murine mammary tumors that had undergone EMT [37], thereby diminishing EMT-inducing capabilities of MT1-MMP. The cadherin switch from E-cadherin to N-cadherin can offer an alternative mechanism for MMP-mediated invasion of

MCF-7 cells undergoing EMT. Thus, complex formation between N-cadherin and FGF receptor has been shown to upregulate *MMP9* expression via MAPK/ERK signaling pathway and induce MMP-9-mediated invasion of N-cadherin-expressing MCF-7 cells in response to exposure to FGF *in vitro* [38].

Furthermore, an EMT program can be triggered by reciprocal activation of cancer cells and cancer-associated fibroblasts (CAFs). Soluble factors produced by tumor IL-6-activated CAFs caused an EMT program manifested by inhibition of E-cadherin and induction of vimentin, Snail, and Twist in PC-3 prostate cancer. In turn, this fibroblast-triggered EMT induced MMP-dependent enhanced invasion, tumor growth, and spontaneous metastasis of prostate carcinoma xenografts [39].

Altogether, it remains unresolved if EMT triggers MMP expression, allowing for the onset of MMP-mediated cell invasion of the ECM, or certain growth factors trigger MMP expression and MMP-mediated ECM remodeling, allowing for manifestation of EMT features by the invading cancer cells. However, whether or not MMPs directly induce the EMT or are induced during EMT or both, the resulting proteolytic activity of MMPs can directly contribute to the earliest invasion-associated step of the metastatic cascade, that is, escape of cancer cells from the primary tumor.

8.3
Escape from the Primary Tumor: MMP-Mediated Invasion of Basement Membranes

The invasion-dependent escape is manifested as persistent locomotion of individual cells or chains of cells away from the primary tumor as well as collective, cohort migration of multicellular clusters, strands, or sheets. In epithelial cancers, the progression of carcinoma *in situ* to invasive carcinoma involves the breaching of the epithelial BM, which is composed of type IV collagen, laminin complexes, and different proteoglycans, followed by invasion into interstitial ECM, enriched in types I and III collagen and fibronectin [40]. The initiation of invasive behavior in other cancer types may not involve breaching of the defined BM demarcation line, but involve penetration into fibrotic or condensed ECM enriched in BM proteins. With an exception of the amoeboid, protease-independent mode of directional cell invasion observed in the presence of multiple protease inhibitors [41], most of experimental data consistently indicate that both the invasive escape of cancer cells and the ensuing remodeling of stromal matrix by the leader cells critically depend on the proteolytic activity of MMPs.

8.3.1
In vitro Models of BM Invasion: Matrigel Invasion in Transwells

One of the most popular *in vitro* models of individual tumor cell invasion is Matrigel invasion in Transwells [42, 43]. Since Matrigel is enriched in laminin, type IV collagen, heparan sulfate proteoglycans, and growth factors, Matrigel invasion is commonly used as an *in vitro* surrogate model for breaching the BM. The tumor cells are placed as single cell suspensions into the upper chamber and are allowed to penetrate across the porous membrane occluded by Matrigel into the bottom chamber, where they are counted.

That the activity of MMPs is involved in Matrigel invasion has been convincingly demonstrated by modulation of the invasive ability of tumor cells by either overexpression or downregulation of TIMP-1. Thus, inhibition of TIMP-1 expression by antisense RNA resulted in the acquisition of invasive behavior in noninvasive Swiss 3T3 cells [44], whereas overexpression of inducible TIMP-1 significantly suppressed the Matrigel invasion ability of mouse B16F10 melanoma cells [45]. In addition, Matrigel invasion of tumor cells, induced by CAFs, was also sensitive to MMP inhibition, although the identity of invasion-mediating MMPs was not determined [39].

MT1-MMP expression and MT1-MMP-mediated MMP-2 activation were shown to be critical for invasion of Matrigel by human HT-1080 fibrosarcoma [46]. Both inhibition with BB-94 and downregulation of MT1-MMP with dsRNA reduced Matrigel invasion of transfected HT-1080 cells [47]. The Matrigel model was used to confirm that both PEX-mediated dimerization of MT1-MMP and dimerization-dependent MMP-2 activation were essential for HT-1080 cell invasion [48]. MT1-MMP-transfected breast carcinoma MCF-7 cells displayed enhanced Matrigel invasiveness, independent, however, of MMP-2 transfection [49]. The direct correlation of Matrigel invasion with MT1-MMP expression was also demonstrated for human melanoma BLM cells by overexpressing MT1-MMP or by knocking it down with shRNA retroviruses [50].

The expression levels and the activity of secreted gelatinases, MMP-2 and MMP-9, have also been positively linked to the extent of Matrigel invasion by tumor sublines isolated form spontaneous murine mammary tumors [37]. Matrigel invasion of HT-1080 human fibrosarcoma, producing both MMP-2 and MMP-9, can be significantly inhibited by peptides binding to the C-terminal domains of the respective MMPs [51]. The p38 MAPK-mediated stabilization of MMP-2/MMP-9 mRNA and downstream activation of MMP-2/MMP-9 proenzymes were essential for Matrigel invasion of human bladder carcinoma HTB9 and HTB5 cells [52]. Matrigel invasion of human glioblastoma SNB19 cells was inhibited by downregulation of MMP-9 with specific dsRNA constructs or the MMP-9 hemopexin domain [53, 54]. In a similar manner, downregulation of MMP-9 with shRNA reduced Matrigel invasion of prostate cancer cells [55]. Conversely, overexpression of MMP-9 and its activation by a plasmin/MMP-3 cascade induced Matrigel invasion of breast cancer MDA-MB-231 cells, an effect sensitive to TIMP-1 and activation-blocking mAb 7-11C [56], indicating that the activation step of MMP-9 proenzyme may be a limiting factor in the induction of cell invasion through the BM.

Several collagenases have been demonstrated to facilitate Matrigel invasion. Thus, transfection of MMP-1-null prostate cancer cells, LNCaP and RWPE2, with MMP-1 cDNA resulted in the increased invasiveness of Matrigel [57]. Furthermore, inhibition of MMP-1 activity with specific MMP-1 inhibitor FN-439 or immunodepletion of MMP-1 with a specific antibody, significantly decreased Matrigel invasion of MMP-1-expressing PC-3 and DU145 prostate carcinoma cells [57]. The endogenous expression of another collagenase, MMP-13, was essential for Matrigel invasion of sphere-forming cells isolated from human U251 glioblastoma cell line, and downregulation of MMP-13 expression by shRNA substantially suppressed the invasion potential of these cancer-stem-cell-like cells [58].

Matrilysins and stromelysins were also demonstrated to promote Matrigel invasion. Thus, Matrigel invasion, induced by overexpression of MMP-3 in aggressive fibromatosis tumor cells, was effectively diminished to control levels by GM6001 or TIMP-1 [59]. Correspondingly, downregulation of MMP-3 expression either by RNAi or IFN-γ significantly decreased Matrigel invasion of TNF-α-stimulated T98G human glioma cells [60]. Inhibition of MMP-7 expression and activity in metastatic colon carcinoma SW620 cells diminished their ability to cross Matrigel barriers [61]. The use of the Matrigel invasion model indicates that the optimal extent of MMP activation is critical to eliciting maximal invasion efficiency. Thus, human fibrosarcoma, breast carcinoma, and glioma cells, expressing intermediate levels of MT1-MMP, were the most effective in the invasion of Matrigel [62]. Furthermore, a gradual increase of TIMP-2 concentrations during Matrigel invasion of MT1-MMP-overexpressing HT-1080 cells demonstrated that at a certain level, MMP inhibition resulted in the increase of tumor cell invasion, likely due to a decrease of excessive matrix degradation [62]. Finally, the efficiency of Matrigel invasion correlates well not only with the expression and activity of individual MMPs but also with tumorigenic, angiogenic, and/or metastatic abilities of MMP-expressing tumor cells [37, 39, 57], indicating that this surrogate model for invasion of BM can adequately reflect the *in vivo* aggressiveness of cancer cells.

8.3.2
Ex Vivo Models of BM Invasion: Transmigration through the Intact BM

A few studies have failed to demonstrate the inhibition of Matrigel invasion by broad-range MMP inhibitors or natural MMP inhibitors, TIMPs [63, 64]. Although contradicting the published findings on Matrigel invasion from other laboratories, these studies have raised an important notion that MMP activity can sometimes be dispensable for invasion of artificial matrices, whereas the traversing of native BM may be critically dependent on MMP-mediated proteolysis [65].

Transmigration of tumor cells through the BM recovered from the intact peritoneum offers an *ex vivo* invasion model involving native, properly organized matrix. This model demonstrated that the ability of various tumor cells to traverse peritoneal BM depended exclusively on the expression and catalytic activity of the three membrane-type MMPs, namely, MT1-, MT2-, and MT3-MMP, but not on MT4- or MT6-MMP or soluble MMP-2 and MMP-9 [63]. In agreement, the capacity of human ovarian carcinoma cells to invade through the mesothelial cell monolayer into the underlying collagen-rich submesothelial matrix and establish metastatic foci in the peritoneal wall was linked to expression and activity of MT1-MMP [33, 66, 67].

A more extensive utilization of the above-described models of *ex vivo* BM invasion and the development of new models would help to resolve the apparently controversial issue of whether soluble MMPs, for example, gelatinases MMP-2 and MMP-9, capable of efficient degradation of at least one of the major proteinaceous components of BM, that is, type IV collagen, play a role in assisting tumor cells to breach BM and enter into the surrounding interstitial stroma. Furthermore, many secreted MMPs, including MMP-1 [68, 69], MMP-2 [70–75], and MMP-9 [76, 77], dock at specific cell surface molecules to exert their proteolytic activity, thereby diminishing the significance of functional discrimination between the membrane-tethered MT-MMPs and secreted MMPs.

8.3.3
In Vivo Models of BM Invasion: Invasion of the CAM in Live Chick Embryos

An *in vivo* model of breaching through the BM barrier is represented by the invasion of the CAM in live chick embryos. Although tumor invasion of the CAM was interpreted in several studies as the ability of cancer cells to breach the type I collagen barrier [78, 79], this *in vivo* model is actually extremely well suited to study the initial breaching of BM since intact CAM is topped with a tight layer of ectoderm and underlined with a dense capillary plexus [80, 81]. Both tissue structures express high levels of type IV collagen [36, 82] and therefore, present a proper BM-like barrier for the tumor cells grafted onto the CAM. Thus, the ability of Snail1-induced breast carcinoma MCF-7 cells to invade the type-IV-collagen-positive barrier of the CAM and penetrate into the CAM mesoderm was shown to depend exclusively on the expression of MT1-MMP and MT2-MMP, but not on secreted MMPs such as MMP-1, MMP-2, MMP-3, MMP-7, MMP-9, or MMP-13 [36]. Since the chick embryo can sustain the growth of primary tumors without species-specific barriers and also appears to support the development of tumors of different tissue origin, it appears that invasion of the CAM may offer a highly efficient model to study BM invasion *in vivo*.

8.4
Invasive Front Formation: Evidence for MMP Involvement *In Vivo*

The idea that escape from the primary tumor critically depends on the proteolytic activity of MMPs is supported by the general notion that the highest levels of MMP expression and MMP functional activity in aggressive tumors are localized to the invasive fronts, where MMPs can be expressed differentially either in tumor cells or stromal cells. Elevated expression of MT1-MMP, partially in association with FGFR4, was observed at the tumor–stroma border and tumor invasive front in breast adenocarcinomas [34]. The analysis of melanoma and fibrosarcoma xenografts demonstrated that MMP-2, MMP-9, and MT1-MMP were predominantly expressed at the tumor–stroma border, and furthermore, that functionally active MMP-2 was restricted to the invasive front [83, 84]. MMP-2 and MMP-9 were more strongly expressed at the invasive edges of malignant gliomas, both in cancer patients and in mice with orthotopic cranial implants [85, 86]. In colorectal cancer patients, the immunoreactive MMP-9 at peripheral tumor borders was localized to deeply invading tumor nests, whereas superficial tumor areas were only faintly positive [87]. Extensive analyses of invasive breast carcinomas demonstrated that elevated MMP-9 at the invasive front could be associated with the stromal fibroblasts [88] or mononuclear inflammatory cells [89]. Peritumoral neutrophils were the major source of MMP-9 at the tumor-invading edge in patients with hepatocellular carcinomas [90]. Although the clinical relevance of elevated expression of MMPs at invasive fronts to patient survival remains controversial for many cancer types [87, 89], the findings from several experimental model systems involving genetic overexpression or knockout of individual MMPs strongly support the notion that the penetration of leading cells into surrounding stroma is dependent on MMPs and their proteolytic activity.

8.4.1
MMP-Dependent Invasion in Spontaneous Tumors Developing in Transgenic Mice

Genetically modified transgenic mice prone to spontaneous, oncogene-driven carcinogenesis provide strong evidence that transformation of benign premalignant adenomas into malignant carcinomas is associated with the development of invasive fronts that localize high levels of MMP expression and MMP activity. The MMPs expressed in oncogene-driven cancer models are believed to mirror those identified in human cancers both in terms of expression levels and histological localization.

The induced expression of several MMPs, demonstrated in transgenic mice during transition to invasive malignant carcinomas, is frequently confined to the stroma surrounding the invasive foci. Thus, in a multistage (human papillomavirus) HPV16 transgenic model, squamous cell carcinomas progress with the appearance of an invasive front, which is formed where infiltrating mast cells degranulate and release MMP-9 that is activated by specific serine proteases [91, 92]. Transgenic RIP1-Tag2 mice expressing SV40T antigen under the control of the insulin promoter provide an experimental model of multistage carcinogenesis of pancreatic islets [93]. The progression of *in situ* carcinoma lesions into invasive carcinomas in these mice is associated with accumulation of MMP-9-positive leukocytes, and the invasive tumor formation is abrogated in MMP-9 knockout RIP1-Tag2 mice or in RIP1-Tag2 mice treated with a broad-spectrum MMP inhibitor BB-94 [94]. In mouse models of human cervical carcinogenesis driven by HPV oncogenes, the progressive transition from low-grade dysplasia to high-grade dysplasias to invasive carcinomas was paralleled by the concordant increase in gelatinase activity and accumulation of MMP-9-positive tumor-associated macrophages [95]. In MMTV transgenic models of oncogene-driven breast cancer progression, the transition from preinvasive to invasive breast carcinomas demonstrated a clear localization of stromal MMP-13 to the invasive front of developing tumors, closely resembling the pattern observed in human invasive ductal carcinoma [96].

In a recently described model of prostatic neuroendocrine cancer, transgenic mice expressing SV40 large T antigen under control of cryptdin-2 gene develop spontaneous tumors where stromal MMP-2, epithelial MMP-7, and macrophage

MMP-9 increase concurrently with the transition to invasive metastatic carcinoma [97]. While blocking of overall MMP activities with a broad-range MMP inhibitor reduced tumor burden, specific genetic depletion of MMP-2 decreased lung metastasis, and blood vessel density, whereas depletion of MMP-7 reduced endothelial coverage and decreased vessel size. Surprisingly, mice lacking MMP-9 had increased perivascular invasion, indicating the protective role of inflammatory cell MMP-9 in this model system [97].

Transgenic mouse models, where expression of a transgene is targeted to epithelial cells, allow for the elucidation of invasive mechanisms governed by overexpression of individual MMPs in tumor cells. Thus, *de novo* overexpression of MT1-MMP in epithelial cells of mammary tumors spontaneously developing in MT1-MMP transgenic mice was especially pronounced in the invading adenocarcinoma cells and associated with increased rates of lung metastases, suggesting that overexpression of MT1-MMP can contribute to breast cancer progression [98].

8.4.2
MMP-Dependent Invasion of Tumor Grafts in MMP-Competent Mice

Studies employing surgically implanted tumor xenografts have provided additional evidence that formation of an invasive front and ensuing cancer metastasis depend on the expression and activity of MMPs, in particular, on MT1-MMP and MT1-MMP-activated MMP-2. The conditional expression of MT1-MMP cDNA in canine MDCK cells, which normally do not form tumors when injected into nude mice, results in the formation of tumor xenografts that actively invade the adjacent muscular layers [99]. In a melanoma model of spontaneous metastasis, the expression and activation of MMP-2, colocalized with MT1-MMP, was demonstrated at the invading tumor front in subcutaneous human melanoma xenografts developing in nude mice after implantation of highly aggressive MV3 and BLM cell lines [83, 100]. Importantly, subcutaneous tumors developing from the BML cells, in which MT1-MMP was downregulated by shRNA interference, lost their capacity of thoracic and abdominal invasion and concomitantly, lost their ability to develop lung metastases [50]. Comprehensive analyses of fresh human cutaneous melanocytic lesions, comprising all stages of melanocytic tumor progression, also strongly suggest that coordinated expression of MT1-MMP and activated MMP-2 is required for melanoma invasion and development of metastases [100].

The proinvasive role of tumor MMP-9 was demonstrated in several intracranial glioma models, where glioma cells are orthotopically implanted directly into the brain. Thus, intracranial xenografts of human U87MG/AEG-1 glioma cells overexpressing MMP-9 as a result of ectopic expression of the astrocyte elevated gene-1, *AEG-1*, were highly invasive compared with the parental cells [85]. Conversely, specific downregulation of MMP-9, in conjunction with either cathepsin B [53] or uPAR [101, 102] or both, uPA and uPAR [103], significantly inhibited invasion of intracranial xenografts developing from human SNB19 or U87MG glioma cells.

On the other hand, tumor deficiency in select MMPs can lead to increased cancer invasiveness. Thus, MMP-2 knockout mouse glioma cells implanted into wild-type hosts demonstrated a twofold increase in perivascular invasion into the brain parenchyma compared with their MMP-2-producing counterparts [104], emphasizing the putative protective functions of tumor-derived MMPs.

Overall, the findings from graft models provide strong evidence for a role of MMPs in tumor invasion, although this role could be either proinvasive or anti-invasive, indicating a complex interplay between MMP-expressing cells during this stage of cancer progression.

8.4.3
Invasion of MMP-Competent Tumor Grafts in MMP-Deficient Mice

The significance of stromal MT1-MMP in tumor invasion and metastatic spread was elegantly demonstrated when MT1-MMP-deficient mice were crossed with MMTV-PyMT transgenic mice, producing animals lacking MT1-MMP but, surprisingly, developing mammary gland hyperplasias faster than their wild-type PyMT littermates [105]. Despite the accelerated growth of MT1-MMP-deficient PyMT mammary tumors, orthotopic transplantations into the cleared mammary fat pads of syngeneic recipients demonstrated a remarkable reduction of lung metastasis in MT1-MMP-deficient recipients and confirmed that MT1-MMP-mediated proteolysis of collagenous stroma by stromal cells is important for breast cancer metastasis [105].

MT1-MMP-dependent tumor invasion *in vivo*, however, also requires stromal-derived MMP-2, as demonstrated by a study in which epithelial cells isolated from MT1-MMP/p53 double-knockout mice were additionally transduced with *v-src* to increase their tumorigenic potential [106]. These immortalized epithelial cells were then transfected with a doxycycline-inducible MT1-MMP and transplanted subcutaneously into MMP-2 knockout mice or their wild-type littermates. While expression of MT1-MMP promoted tumor growth in *MMP2*+/+ hosts, the presence of functional MT1-MMP was not sufficient for tumor development in MMP-2 knockout recipients and did not induce invasive degradation of type IV collagen at cell–collagen interface unless MMP-2 was supplied via transfection or coimplantation of MMP-2-positive fibroblasts. These data indicate that the cooperation between stroma-derived MMP-2 and tumor-derived MT1-MMP is required for tumor invasion and expansion via MMP-2-mediated remodeling of tumor-associated BM [106].

A specific role for stromal MMP-2 was shown in a model where transformed mouse astrocytes, wild type or MMP-2 deficient, were implanted intracranially into RAG/MMP-2 double-knockout recipients. Surprisingly, the complete absence of MMP-2 when the gene was knocked out in both stromal and implanted tumor cells, led to the development of more invasive tumors because of increased density of new blood vessels, possibly providing tracks for glioma cell perivascular migration [104].

Whether or not stromal MMP-9 is required for prostate cancer invasion was analyzed in the RAG-1/MMP-9 double-knockout mice subcutaneously implanted with human PC-3 carcinoma. The lack of host MMP-9 did not affect the tumor incidence, tumor growth kinetics, and microvascular density in subcutaneous xenografts, thereby indicating that, at least in this model, stromal MMP-9 was neither necessary nor sufficient for subcutaneous PC-3 tumor growth [107].

Collectively, the findings from different *in vivo* models indicate that MMPs, via distinct mechanisms, can either positively or negatively regulate the formation of the invasive front in primary tumors. Nevertheless, advancing of malignant cells beyond the invasive front would require proteolysis-dependent modifications of the surrounding collagenous stroma by the leading cancer cells.

8.5 Invasion at the Leading Edge: MMP-Mediated Proteolysis of Collagenous Stroma

After the breaching of BM, the invading tumor cells encounter the stromal ECM rich in interstitial collagens, constituting the bulk of fibrillar proteins in the connective tissue *in vivo*. Collagen turnover involves constant MMP-mediated remodeling, which is deregulated in cancer, often resulting in the increased expression and elevated deposition of collagen with altered organization and stiffening [108, 109]. Persistent invasion through the collagenous stroma is an essential step in the metastatic dissemination of cancer cells, which has been demonstrated in several *in vitro* and *in vivo* model systems. These models, discussed below, have shown that invading tumor cells employ diversified molecular programs depending on the physical and chemical context of the confronting ECM.

8.5.1 Collagen Invasion in Transwells

A modification of the Boyden chamber/Transwell assay, where the separating porous filter is coated with native type I collagen, has repeatedly confirmed that penetration of collagenous matrices depends on the activity of MMPs. By employing native collagen in this model, Sodek *et al.* demonstrated that invasion of human ovarian carcinoma cell lines, including HEY, ES-2, and OVCA429, depended on overall MMP activity since cell invasion was almost completely abrogated by a broad-range MMP inhibitor, GM6001 [64]. The requirement for specific proteolytic activity in collagen invasion was demonstrated for a number of soluble as well as membrane-bound MMPs expressed by various tumor cell lines. Thus, MMP-1 activity was implicated in the invasion of human melanoma A2058 cells [110]. By inhibition with natural and synthetic MMP inhibitors or RNA silencing, collagen invasion of human glioma U251 and U178 cells was linked to proteolytic activity of MMP-12 [111]. Efficient collagen invasion of human ovarian OVCA 433 carcinoma required expression of wild-type MT1-MMP [33]. Interestingly, more than a twofold increase in collagen invasion was observed if these cells were transfected with a proteolytically active phosphorylation mutant of MT1-MMP, which is persistently retained on the cell surface because of reduced internalization [33].

A combination approach where tumor cells are mixed with stromal cells allows for elucidation of complex interplays governing MMP-mediated invasion. Thus, a novel proteolytic cascade involving activation of tumor-secreted pro-MMP-1 by fibroblast-produced MMP-3 was elucidated in Transwell assays involving melanoma A2058 cells mixed with stromal fibroblasts [110]. In admixtures of human squamous carcinoma cells and fibroblasts isolated from MMP-deficient mice, fibroblast-stimulated collagen invasion of tumor cells was completely dependent on the expression in fibroblasts of MT1-MMP, partially dependent on the expression of MMP-2 and independent of MMP-9 [112]. These *in vitro* results, excluding the role of fibroblast MMP-9, apparently contradict the *in vivo* data on a functional role of stromal MMP-9 in tumor development in mice from the same study [112], thus illustrating that caution should be exercised in extrapolation of *in vitro* data to *in vivo* conditions.

8.5.2
Invasion of Collagen Matrices by Overlaid Tumor Cells

When individual tumor cells are placed atop thick matrices generated from native type I collagen, the cells invade the 3D gels in an MMP-dependent manner. Thus, collagen gel invasion by cancer cells of mesenchymal origin, such as HT-1080 fibrosarcoma, or of epithelial origin, such as squamous cell SCC-1 carcinoma, was exclusively dependent on the activity of MT1-MMP and proceeded independently of the activity of MT2-MMP or MT3-MMP; soluble MMP-1, MMP-2, MMP-8, MMP-9, and MMP-13; or a soluble construct of MT1-MMP [78, 113]. Similar dependence on MT1-MMP of collagen invasion by overlaid tumor cells was demonstrated for human ovarian carcinoma DOV13 cell cultures, where both the expression and functional activity of MT1-MMP were induced by 3D, but not 2D, collagen, in a β1 integrin/EGFR1-dependent manner [114].

In the 2D/3D collagen invasion model, the invading MT1-MMP-positive tumor cells remodel the collagen substratum, creating a network of tunnels lined with collagen degradation products [78]. If genetically modified r/r collagen, resistant to the initial 3/4-1/4 collagenase cleavage [115], was used in this model, the tunnels were not formed and the penetration of cells into gels was completely abrogated, indicating the necessity of native collagen remodeling for cell invasion into the collagenous stroma [79]. The invasion pattern of native collagen matrices was consistent with both individual and collective cell invasion. However, the invasion of less-dense 3D collagen networks, created by pepsinized collagen, proceeds in an MMP-independent manner and without tunnel formation [64, 78], emphasizing that native, nonpepsinized collagen should be employed as a true surrogate ECM barrier *in vitro*.

A modification of the 2D/3D collagen invasion model, simulating organotypic conditions, is represented by cultivation of tumor cells atop 3D collagen gels polymerized with CAFs. The invasion of a highly malignant variant of the keratinocyte cell line, HaCaT-ras, into the CAF-containing collagen gels was dependent on MMP activity as it was efficiently inhibited by an MMP inhibitor Ro28-2653 [116].

In combination with MMP-deficient or MMP-overexpressing CAFs, this model presents a rigorous system to study the MMP-mediated interplay between cancer cells and stromal fibroblasts.

8.5.3 Models of 3D Collagen Invasion

The collagen spheroid model represents an *in vitro* assay well suited to study the initial invasion of tumor cells under 3D conditions. In this model system, tumor cell aggregates are incorporated into 3D native collagen gels, where tumor cells, especially those of nonepithelial tissue origin, rapidly scatter and within a few days, form the so-called "starburst" structures. The efficiency of cell invasion is estimated by applying different parameters such as length of invasion and number of invading cells within the starburst. Several modifications of this collagen spheroid model have been introduced, providing valuable information about the mechanisms involved in early invasion events.

MMP dependency of tumor escape from spheroids is critically dependent on the biochemical properties of collagen matrices. Thus, although 3D invasion of ovarian carcinoma OVCA429 cells from spheroids into native collagen was efficiently blocked by the MMP inhibitor GM6001, invasion into 3D collagen I reconstituted from pepsin-extracted collagen or Matrigel proceeded despite inhibitor-based MMP blockage [64]. When MT1-MMP-overexpressing HT-1080 cells were embedded into collagen matrices of variable densities, MMP dependence, although partial, was observed only in high-density matrices. In contrast, the cells invaded collagen gels of low and medium densities in an MMP-independent manner and displayed non-proteolytic, amoeboidlike migration in the presence of multiple protease inhibitors [117]. Importantly, the dependency of collagen invasion on MMP activity was attributed only to a collective migration of HT-1080 fibrosarcoma as well as breast carcinoma MDA-MB-231 cells, while individual cell invasion was largely MMP independent [117]. Selective immunohistochemical staining demonstrated that collagenolysis occurred at the spheroid–collagen interface and that the invading cells realigned the collagen fibrils, creating collagenolytic tracks [117].

When glioma U251 spheroids overexpressing MT1-MMP were embedded into 3D collagen to mimic cell escape from solid tumors, cystlike structures were formed in which interconnected tumor cells modify and expand the matrix by pushing collagen fibrils ahead of sphere boundaries in a TIMP-2-dependent manner [118]. These findings provide an example where MMP-mediated invasion results in directional matrix invasion by individual tumor cells and cell chains concomitant with a coordinated proteolysis-dependent expansion of an initial tumor mass as a coherent cell ensemble.

An additional approach to study the invasion of tumor cells from a defined interface is when tumor cells are first attached to gelatin-coated microcarrier beads, which then are embedded into native collagen, supporting 3D cell scattering. It was demonstrated in this model that MT1-MMP overexpressed in noninvasive

HeLa cells conferred them with the ability to proteolytically invade the surrounding collagen in a GM6001-sensitive mode [43].

Another modification providing a defined demarcation line for initial invasion involves incorporation of tumor cells into a collagen droplet, which is then embedded into a 3D collagen gel. Individual proteins, for example, fibronectin or fibrinogen, can be incorporated into the surrounding collagen gel, thus enriching the ECM confronted by invading tumor cells. By employing this model, we have recently demonstrated that MMP-mediated (GM6001-sensitive) invasion of a high-disseminating variant of prostate carcinoma PC-3 cells, PC-hi/diss [119], observed in 100% collagen gels is switched in fibrin-enriched collagen gels to a serine-protease-driven invasion, relatively insensitive to MMP inhibitors [120]. Thus, the specific composition of ECM surrounding the primary tumor can determine whether proteolysis-dependent invasion of tumor cells would be sustained by the MMP-dependent or MMP-independent mechanisms.

8.5.4
Invasion of Collagenous Stroma *In Vivo*

Among *in vivo* experimental systems, the chick embryo CAM model, where intramesodermally injected tumor cells form microtumors and initiate interstitial invasion, allows for probing the mechanisms of early tumor invasion in live animals. The labeling of tumor cells and chick embryo vasculature with contrasting markers facilitates microscopic visualization of tumor cells escaping the microtumors [121]. High levels of mesoderm invasion and its dependence on MMPs were recently confirmed in this model system for the cells from highly disseminating variant of HT-1080 fibrosarcoma, HT-hi/diss (E. Deryugina, unpublished results). In control embryos, cell scattering from *in vivo* microtumors proceeds via both individual and collective cell invasion with many disseminating tumor cells invading along CAM blood vessels, a cell behavior we refer to as *vasculotropism*. However, in GM6001-treated embryos, tumor cell invasion was significantly inhibited and the treatment mostly affected the collective type of HT-1080 invasion. In contrast, the invasive escape of PC-hi/diss prostate carcinoma cells appears to be relatively independent of MMPs, but critically dependent on the activity of serine proteases, including uPA-generated plasmin [120]. The use of opposing proteolytic machineries employed by mesenchymal tumor cells versus carcinoma cells, suggests that depending on the tissue of origin, cancer cells can exploit different classes of proteases to escape from the primary tumor and initiate invasion into the surrounding stroma.

The MMTV-Neu mice provided a valuable model system to study the relationship between stiffness of collagenous matrix and cancer. It has been shown that the transition to invasive cancer is associated with an incremental stiffening of mammary gland caused by LOX-mediated (lysyl oxidase) collagen cross-linking [109]. In the absence of MMPs, LOX-mediated stiffening diminishes tumor cell invasion in 3D matrices [122]. Furthermore, invasive carcinoma cells of different tissue origin can produce an isoform of type I collagen that is resistant to all collagenolytic MMPs and is deposited as insoluble fibers used by tumor cells for building MMP-resistant invasion paths [123].

8.5.5 Dynamic Imaging of ECM Proteolysis during Path-Making *In vitro* and *In Vivo*

Advances in microscopic imaging technology resulted in a new appreciation of the complexity of cancer cell invasion *in vivo* [124–128]. Confocal/multiphoton microscopy was used to show that pericellular degradation of ECM by human colon and breast carcinomas within type IV collagen was significantly increased by tumor-associated fibroblasts embedded into adjacent layer of type I collagen [129]. Specific MMP-based imaging probes have also been introduced for *in vivo* imaging of MMP activity in live tumor-bearing animals [130].

Dynamic live imaging of individual cancer cells invading 3D fibrillar collagen matrices has provided a mechanism explaining how invasive cancer cells coordinate collagen remodeling to accommodate a switch from individual to collective invasion [131]. The detailed proteolytic mapping of collagen remodeling revealed the attachment of collagen fibrils at the most anterior edge of the invading cell, followed by active MT1-MMP- and MMP-2-mediated collagenolysis at more posterior zones, and completed by realignment of cleaved collagen fibers at the rear, accommodating MMP-dependent, collective multicellular invasion along paths of least mechanical resistance [132, 133]. A key addition to the concept of path-making tumor cells was demonstrated when imaging of collectively invading carcinoma cells mixed with stromal fibroblasts revealed that the leading cell was always a fibroblast and that carcinoma cells actually moved behind the fibroblast, which generated the tracks in the ECM [134]. The latter observations raise several questions, including whether those tumor cells that follow leading fibroblasts can migrate in an MMP-independent manner and if so, whether these cells are competent to enter the circulation at the points of vascular intravasation or need assistance from other proteolytically active, MMP-expressing accessory cells.

8.6 Tumor Angiogenesis and Cancer Invasion: MMP-Mediated Interrelationships

Histological examinations of tissue samples from cancer patients indicate that increased tumor invasion can be associated with increased tumor angiogenesis. The most notable are gliomas characterized by a high proliferation rate, high levels of angiogenesis, and marked local invasion [135]. In experimental model systems, inhibition or induction of angiogenesis is usually accompanied by a corresponding reduction or increase of tumor growth and invasion, thereby providing positive correlations between the levels of tumor angiogenesis and extent of cancer invasion and metastasis. However, these overall correlations do not indicate causal mechanisms as to whether ensuing invasion of primary tumor cells is fueled by newly formed blood vessels or whether invading and expanding tumor cells fuel tumor angiogenesis.

8.6.1
Angiogenic Switch: MMP-9-Induced Neovascularization as a Prerequisite for Blood-Vessel-Dependent Cancer Invasion

The angiogenic switch, leading to the formation of a network of newly formed blood vessels, has been long recognized as one of the rate-limiting events in tumor progression [1, 136]. Clinical evidence for the association of tumor angiogenesis with cancer progression came from studies demonstrating a significant correlation between the density of microvessels in tissue sections of invasive breast carcinomas and the occurrence of metastases [137, 138]. In addition, vessel density was reported to be a prognostic indicator of lymph node metastases in prostate and gastric cancers [139, 140], strengthening the concept that neovascularization of developing tumors is a prerequisite of cancer invasion and metastasis.

Genetic models of cancer progression have strongly linked various MMPs to the initiation of tumor angiogenesis and angiogenesis-dependent cancer invasion. Surprisingly, the indicated MMPs were mostly stromal and not tumor-derived MMPs. Furthermore, a majority of studies have pointed to MMP-9 originating from stromal cells as a critical factor for initiation of tumor angiogenesis. In particular, MMP-9 supplied by inflammatory cells, including neutrophils, macrophages, and mast cells, was shown to trigger initial angiogenic events [92, 94, 141].

The fact that bone-marrow-derived inflammatory cells and their MMP-9 are critically important for tumor angiogenesis has been demonstrated in several mouse models involving mice deficient in a specific inflammatory cell type or mice deficient in MMP-9, which were then reconstituted with wild-type bone marrow. In mast-cell-deficient HPV16 transgenic mice, in which premalignant angiogenesis is ablated and the incidence of invasive tumors is decreased, both tumor angiogenesis and invasion are restored after transplantation of wild-type bone marrow, apparently replenishing MMP-9-delivering mast cells [91]. Bone marrow rescue of MMP-9-deficient HPV16 transgenic mice indicated that restorative MMP-9 could also be delivered by infiltrating neutrophils and tumor-associated macrophages [92].

The lack of MMP-9 in immunodeficient mice transplanted with tumor cells invariably results in impaired tumor angiogenesis, which can be rescued by bone marrow transplantation, leading to the delivery of stromal MMP-9 by tumor-infiltrating leukocytes. Thus, implantation of human ovarian cancer cells into the peritoneal cavity of nude mice lacking the *MMP9* gene significantly reduced microvessel density and decreased macrophage infiltration into the lesions, both of which were restored by transplantation of MMP-9-positive spleen cells [142]. Immature myeloid Gr+CD11b+ immune suppressor cells, isolated from the spleens of wild-type donors and presumably able to deliver myeloid cell MMP-9, can restore impaired angiogenesis and induce vessel maturation when injected subcutaneously with the Lewis lung cancer cells into MMP-9 knockout hosts [143].

A series of bone marrow transplantation experiments carried out in wild-type and MMP-9 knockout mice implanted with human neuroblastoma tumors demonstrated that bone-marrow-derived MMP-9 was critical for the recruitment of leukocytes from bone marrow into the tumor stroma and for the integration of bone-marrow-derived endothelial cells into the tumor vasculature [144]. Furthermore, expression of MMP-9 by bone marrow-derived cells (BMDCs) in the stroma was required for the formation of a mature vasculature and coverage of endothelial cells with pericytes within developing tumors [145].

Although the above-described models have demonstrated that stromal MMP-9 is functionally important for tumor angiogenesis, the fact that MMP-9-delivering leukocytes are critical for the *induction* of angiogenic switch has been proved in mouse models of oncogene-driven multistage carcinogenesis in MMP-9-deficient mice. The involvement of inflammatory cell MMP-9 in the triggering of angiogenesis was demonstrated in RIP1-Tag2 transgenic mice prone to develop spontaneous pancreatic islet cancer concomitant with the infiltration of premalignant lesions with MMP-9-positive leukocytes [94]. The MMP-9-deficient RIP1-Tag2 mice demonstrate significant reduction of angiogenesis-dependent cancer progression, which was rescued by transplantation of wild-type bone marrow, thus directly linking influxing MMP-9-positive inflammatory cells with angiogenic switching and tumor invasion. This mechanism appeared to involve MMP-9-dependent release of matrix-sequestered VEGF [94]. A similar MMP-9-dependent angiogenic switch induced by released VEGF was demonstrated in a mouse model of cervical cancer in which HPV-driven invasive carcinogenesis was significantly inhibited by a bisphosphonate MMP-9 inhibitor, zoledronic acid [95].

Although the origin of the MMP-9 implicated in angiogenesis and tumor invasion in the pancreatic islet cancer model was initially traced to tumor-associated macrophages [94, 95], the triggering of angiogenesis and subsequent cancer invasion were later specifically linked to inflammatory neutrophils and neutrophil MMP-9 [141]. Furthermore, enhanced neutrophil influx and the persistent activity of neutrophil MMP-9 compensated for the lack of tumor-associated macrophages in CCR2-deficient K14-HPV/E(2) transgenic mice, which have impaired monocyte infiltration [146].

Several aspects of the neutrophil MMP-9-induced angiogenic switch appear important for cancer invasion. First, while monocytes need time and priming to differentiate into macrophages capable of MMP-9 production [147], inflammatory neutrophils are exclusively well equipped to immediately trigger the angiogenic response by releasing their secretory granules containing presynthesized pro-MMP-9 [148, 149]. Second, neutrophil pro-MMP-9 is uniquely produced as TIMP-free proenzyme readily available for activation and release of proangiogenic factors, bFGF and VEGF [150, 151]. Third, pro-MMP-9 delivered by inflammatory neutrophils regulates tumor angiogenesis and tumor cell dissemination in a coordinated manner [120]. Thus, tumor angiogenesis in human prostate carcinoma and fibrosarcoma xenografts in mouse and chick embryo models was significantly inhibited when neutrophil influx was specifically blocked by neutralization of IL-8, a potent neutrophil chemoattractant. However, the inhibitory effects of anti-IL-8 treatment on tumor angiogenesis were completely reverted by the delivery of purified neutrophil pro-MMP-9, but not if it was precomplexed with TIMP-1 [120]. Furthermore, neutrophil MMP-9 can significantly modulate the potency of certain inflammatory cytokines in a positive feedback manner [152, 153], thereby providing a countercurrent mechanism for inflammation-driven cancer invasion and angiogenesis [154].

8.6.2
Mutual Reliance of MMP-Mediated Angiogenesis and Cancer Invasion

Positive correlations between the levels of tumor invasion and neovascularization suggest the existence of specific mechanism that may sustain this mutual reliance. Tumor cells are often found in close association with angiogenic vessels at the leading edge of primary tumors, indicating that angiogenic vessels might provide mechanical support and instructive cues for disseminating tumor cells. On the other hand, an expanding tumor produces a number of potent factors, which can induce neovascularization inside the primary tumor as well as in its close proximity. During gastrointestinal tumor progression, endothelial cells induce CXCL6-mediated neutrophil chemotaxis, which contributes to tumor cell invasion and metastasis by attracting and activating neutrophils loaded with multiple proteases, including proangiogenic MMP-9 [155]. In patients with hepatocellular carcinomas, proinflammatory IL-17^+ cells recruit MMP-9-delivering neutrophils, promoting angiogenesis at the invading tumor edge [90].

Histological examinations of primary tumors formed on the CAM of chick embryos clearly indicate close proximity of highly disseminating HT-1080 cells with the blood vessels at the invasive front [156]. Similarly, HT-1080 cells leaving intramesodermal microtumors are often found tightly associated with blood vessels, engaging them for collective as well as individual cell invasion [121]. Close vascular associations are disrupted by treatment with the MMP inhibitor GM6001, which significantly reduces velocity of HT-1080 cell invasion. These observations suggest that along with the functional activation of ECM-engaging molecules such as $\alpha v\beta 3$ integrin by MT1-MMP [74, 157], MMPs can also be involved in the invasion-assisting, proteolytic modifications of the abluminal surface of blood vessels.

The involvement of MMPs in sustaining angiogenesis-mediated tumor invasion is evidenced by experimental model systems in which differential overexpression or downregulation of individual MMPs results in concomitant changes of tumor angiogenesis and tumor growth and invasion. Thus, overexpression of MT1-MMP in human breast carcinoma MCF-7 and glioma U251 cells led to the upregulation of VEGF expression and coordinated increase of tumor growth, invasion, and angiogenesis in immunodeficient mice [49, 118]. Treatment of developing xenografts with MMP inhibitor AG3340 caused concurrent inhibition of MT1-MMP and VEGF-A expression in HaCaT-ras cancer cells and concomitantly, regression of immature neovasculature [116]. A recent study further elucidated complex mechanisms of MT1-MMP-mediated enhancement of VEGF expression and demonstrated that MT1-MMP regulates the translocation of VEGFR-2 to the cell surface, where VEGFR-2 complexes with MT1-MMP via its hemopexin domain and signals VEGF-A transcription via docked Src [158]. Interestingly, the activity of soluble MMP-2 or MMP-7 efficiently substituted in this model the catalytically dead MT1-MMP mutant [158], pointing to the lack of strict dependency of VEGF-A expression on the proteolytic activity of MT1-MMP.

Mutual dependence of MMP-mediated tumor development and MMP-mediated tumor angiogenesis is illustrated by the negative effects of a membrane-anchored glycoprotein, RECK, on MT1-MMP, MMP-2, and MMP-9 expression. Thus, overexpression of RECK in HT-1080 fibrosarcoma resulted in the coordinated suppression of angiogenesis, due to deficient MMP-mediated collagen remodeling by tumor cells, and massive tumor cell death, due to vascular deficiency [159]. In an orthotopic transplantation model in which prostate carcinoma PC-3 cells were implanted into immunodeficient mice, inhibition of MMP-1 with a specific inhibitor resulted in the simultaneous inhibition of tumor growth and tumor angiogenesis, concurrent with a reduction of lung metastases [57].

A few studies have demonstrated that MMPs can negatively, but still coordinately, regulate tumor growth and angiogenesis. Thus, in an orthotopic mouse model of glioblastoma multiforme, genetic ablation of MMP-2 in both stromal and glioma cells resulted in a significant and coordinated increase of tumor vascularity and perivascular invasion, indicating that MMP-2 can act as a negative regulator of tumor angiogenesis and angiogenesis-dependent tumor invasion [104]. A direct treatment of breast carcinoma MCF-7 xenografts with MMP-9 adenovirus resulted in a coordinated decrease in tumor growth rates and levels of angiogenesis because of increased levels of MMP-9-generated endostatin [160]. Stromal MMP-19 is another MMP that appears to control cancer progression since host MMP-19 deficiency is accompanied by increase in early angiogenic response and enhanced tumor invasion [161].

8.6.3 Apparent Distinction between MMP-Mediated Tumor Angiogenesis and Cancer Invasion

Despite evidence indicating the functional connections between MMP-mediated angiogenesis and angiogenesis-related invasion, failure of MMP inhibitor trials [162] in conjunction with only modest benefits of anti-VEGF therapy in cancer patients [163] and tumor refractoriness to antiangiogenic treatments in numerous experimental models [164–170] suggest that tumor angiogenesis may not be required to sustain late stages of cancer progression. Moreover, genetic or pharmacological inhibition of tumor neovascularization appears to create a condition of increased tumor invasiveness and evoke the appearance of more aggressive tumors, indicating that ongoing cancer invasion can progress independent of VEGF-mediated angiogenesis [171].

Thus, in the RIP1-Tag2 mouse model of pancreatic islet cancer, where MMP-9-triggers VEGF-dependent angiogenic switching [94], the anti-VEGFR therapy initiated at early stages of tumor development causes only temporal vascular inhibition and tumor stasis, followed by regrowth of tumors with more invasive and aggressive phenotypes [170]. Heightened invasiveness was also elicited by targeting the VEGF pathway in a glioblastoma model [170]. VEGFR2 blockade in glioma and other tumor types involved normalization of tumor vessels via increase in pericyte coverage, which is required for blood vessel maturation and induction of MMP-mediated degradation of pathologically thick BM [164, 172]. Furthermore, inhibition of MMPs with a broad-ranged inhibitor or targeting of type IV collagen cryptic sites did not slow the rates of revascularization, indicating that MMPs may not be involved in the aggressive regrowth of tumor vessels [165]. Finally, genetic ablation of MMP-9 in a transgenic model of prostatic neudocrine cancer significantly enhanced the numbers of invasive foci and perivascular tumor cell invasion, but concomitantly, resulted in a significantly inhibited angiogenesis [97], thereby exemplifying that increased tumor invasion can occur in the absence of increased angiogenesis.

The paradoxical consequences of antiangiogenic therapies on tumor invasion and metastasis point to the possibility that both the induction of tumor angiogenesis at the early stages of cancer progression and the suppression of ongoing tumor angiogenesis at the late stages of cancer development can exert proinvasive and prometastatic effects. These effects may have distinct physiological bases but appear to lead to the formation of angiogenic or coopted neovasculature that is more permissive for tumor cell intravasation and dissemination.

8.7
Cancer Cell Intravasation: MMP-Dependent Vascular Invasion

It is generally accepted that invasion of cancer cells, which escaped from the primary tumor, culminates in intravasation, that is, entry of tumor cells into blood or lymphatic vessels. Therefore, it is plausible to suggest that tumor cells that can invade blood vessels, possibly by MMP-mediated processes, are those aggressive cancer cells that would enter the circulation and seed distant metastases. Supporting this notion, the invasion of small blood vessels outside of the main tumor has been validated as an important prognostic factor predicting distant metastasis in stage I breast cancer [173]. Furthermore, imaging of intravasation events *in vivo* in the primary tumors derived from metastatic and nonmetastatic rat mammary adenocarcinoma cell lines indicated that metastatic cells showed greater orientation toward blood vessels and that nonmetastatic cells disintegrated on entering into blood vessels [174]. However, microscopic visualization of rare events of tumor cell entry into the vasculature is not an efficient way to quantitatively analyze the involvement of specific molecules in the intravasation process.

A unique model that allows for mechanistic and quantitative studies of intravasation is the CAM model of spontaneous metastasis. The model is based on grafting of tumor cells on the CAM of 9–11-day-old chick embryos, serving as natural immunodeficient hosts. The developing tumors usually induce angiogenesis, visualized as blood vessel ingrowths originating from the ectoderm capillary plexus. Owing to the nature of blood circulation in the embryo, spontaneously intravasated tumor cells are trapped back in the capillary network of the CAM, which therefore serves as a repository of intravasated cells. Indeed, within three to five days following tumor cell implantation, intravasated cells can be visualized microscopically in the portions of the CAM, distal to the site of primary tumor formation and, importantly, quantified by PCR. Primary tumors on the CAM are formed in close proximity to the ectoderm capillary plexus, which provides efficient nutrient supply and gas exchange to developing tumors. Therefore, inhibition of tumor angiogenesis in the chick embryo model does not result in a dramatic decrease in tumor growth, allowing for a unique possibility to correlate directly the rates of intravasation with the levels of angiogenesis.

By using the CAM intravasation model, several groups have been able to demonstrate that intravasation of tumor cells involves proteases, including serine proteases such as uPA [175, 176] and several MMPs [36, 79, 84, 156, 176]. However, it is still unresolved whether the intravasation event *per se*, namely, the physical crossing by cancer cells of a blood vessel barrier, is MMP dependent or involves MMP-independent locomotion. Furthermore, the origin and nature of MMPs involved in tumor cell intravasation remain controversial. Thus, tumor MMP-9 has been implicated in intravasation of several human tumor cell lines [176], but was excluded by siRNA silencing in intravasation studies employing human HT-1080 fibrosarcoma [156]. Instead, downregulation of tumor MMP-9 surprisingly resulted in a several-fold increase of HT-1080 cell intravasation. Our studies also demonstrated a lack of intravasation-promoting capacity for MT1-MMP [156], apparently in contrast to the findings linking the ability of HT-1080 cells to invade the BM of the CAM exclusively with MT1-MMP-mediated proteolysis [36].

To mechanistically investigate the tissue origin of intravasation-promoting MMPs, we turned to host MMPs produced by inflammatory cells influxing early-stage primary tumors. Comparative analysis of intravasation rates manifested by high- and low-disseminating variants of HT-1080 fibrosarcoma and PC-3 prostate carcinoma demonstrated that the levels of tumor cell intravasation were proportional to the levels of angiogenesis in primary tumors, which, in turn, was directly linked to tumor infiltration by MMP-9-positive neutrophils. Anti IL-8 treatment, specifically blocking the influx of MMP-9-positive neutrophils into primary hi/diss tumors, resulted in a significant inhibition of tumor angiogenesis and coordinately, tumor cell intravasation, indicating that inflammatory-neutrophil-derived MMP-9 was a critical determinant of tumor angiogenesis and intravasation. This notion was validated in rescue experiments in which delivery of TIMP-free neutrophil pro-MMP-9 to IL-8-deprived HT-hi/diss tumors concomitantly restored inhibited angiogenesis and the levels of intravasation. However, when the TIMP-complexed MMP-9 produced by most other cell types was used in rescue expreriments, this form failed to restore inhibited angiogenesis and intravasation [120].

A sophisticated intravital technique has been recently introduced for multiple evaluations of the same tumor in living mice by employing cancer cells expressing photoswitchable proteins [177]. These proteins constitute a relatively new group of GFP-like fluorophores, which irreversibly change their green fluorescence to the red spectrum after exposure to blue light, allowing for the marking of individual cells and then, for continuous tracking of marked cells. By selectively photoswitching the fluorophore in a small group of metastatic breast cancer MTLn3 cells within a developing tumor, limited migration of marked tumor cells was demonstrated in the regions of mammary gland devoid of blood vessels. In contrast, in the vascular microenvironment, photoswitched carcinoma cells infiltrated surrounding areas and lined up along the blood vessels. The subsequent appearance of photoswitched tumor cells in the lungs suggested that the vascular microenvironment promoted invasion and intravasation of cancer cells [177]. Future applications of this exquisite technique to MMP-competent or MMP-deficient tumors developing in wild-type or MMP knockout mice might facilitate studies of specific *in vivo* roles of individual MMPs during intravasation in the mammalian setting.

8.8 Cancer Cell Extravasation: MMP-Dependent Invasion of the Endothelial Barrier and Subendothelial Stroma

Despite a vast literature on tumor cell extravasation, the issue whether or not this step of the metastatic cascade depends on MMPs is not resolved and is still controversial. The main impediment appears to lay in a difficulty to discriminate between two interconnected but distinct extravasation-related processes, that is, the *transmigration* of tumor cells through the vascular endothelial lining and endothelial BM and the *survival* of newly extravasated tumor cells in the underlying extravascular stroma. Both these processes are required for a successful establishment of metastatic foci at the secondary site. Specific MMPs, which can originate from tumor cells, endothelial cells, stromal fibroblasts, or inflammatory leukocytes, might assist in both the breaching of endothelial BM and degradation of subendothelial stroma and ultimately, determine the survival rates of extravasated tumor cells.

8.8.1 Transmigration across Endothelial Monolayers *In Vitro*

A Transwell model, in which tumor cells are allowed to penetrate and migrate through an endothelial monolayer pregrown on top of a porous filter toward chemoattractants placed into the lower chamber, is frequently used as a surrogate model to study extravasation *in vitro*. In a modification of this basic approach, the endothelial monolayers could be generated on porous membranes additionally covered with BM ECM [178]. Extravasation can be also analyzed in Transwell chambers under shear flow conditions, partially recapitulating circulation forces *in vivo* [179]. In the different variations of these models, MMP-overexpressing or MMP-deficient tumor cells can be employed to study the functional role of tumor MMPs in the process of extravasation. Alternatively, endothelial cells from MMP transgenic or knockout mice, if available, can be presented as the monolayers to analyze the role of stromal MMPs. In addition, exogenous MMPs or MMP inhibitors can be readily incorporated into the assays, thereby diversifying conditions for tumor cell extravasation [180].

Despite apparent simplicity of the Transwell extravasation model, conclusive findings regarding the role of specific MMPs in crossing over the endothelial layer are rather scarce. Individual downregulation of MMP-1 or MMP-2 by shRNA interference did not change the ability of the highly metastatic variant of MDA-MB-231 carcinoma, LM2, to cross over an endothelial cell barrier *in vitro*. However, the combined shRNA ablation of MMP-1 and MMP-2 along with ablation of COX2 and epiregulin significantly inhibited Transwells extravasation of LM2 cells [181]. MMP-9-dependent degradation of BM underlying the endothelial monolayer was demonstrated for murine 5T2MM myeloma cells, in which MMP-9 production was induced in response to the interactions with the endothelial cells [182]. MMP-3- and MMP-10-mediated degradation of the ECM localized underneath the endothelial lining was shown to promote transmigration of B16F10 melanoma cells *in vitro* and provided a mechanism by which these two MMPs might disrupt vascular integrity *in vivo* [180].

8.8.2
Tumor Cell Extravasation *In Vivo*

In a few *in vivo* studies, tumor cell extravasation has been investigated as a specific step of spontaneous metastasis. By utilizing dual-channel *in vivo* imaging in the orthotopic model of human prostate cancer, it was demonstrated that after escaping from the primary tumor, PC-3 carcinoma cells reside initially in the blood vessels of the lung and liver, where they exhibit active MMPs [183]. Although the nature of individual MMPs was not identified in this study, the findings suggested that MMP activity can be exploited by intravascular tumor cells in order to cross vascular barriers and generate micrometastases at the secondary sites. In a spontaneous lung metastasis model, orthotopically implanted breast MDA-MB-231 carcinoma cells were shown to employ MMP-3 and MMP-10 to disrupt the ECM barrier of pulmonary vasculature and enhance lung extravasation and colonization [180].

In the most frequently used *in vivo* experimental model of extravasation, tumor cells are introduced directly into the circulation, thereby bypassing the initial steps of the metastatic cascade such as tumor escape and intravasation. In both avian and mammalian model systems, the live imaging of extravasating cells indicates the involvement of invasive processes and survival mechanisms, which together determine the fate of the extravasated tumor cells [121, 184–188].

Despite an apparent requirement for MMP activity to cross the endothelial BM and invade the underlying stroma, the role of tumor-derived MMPs during extravasation *in vivo* also remains poorly validated. An involvement of collagen-degrading MMPs in the extravasation process was indicated in the experiments where mAb HUIV26, recognizing proteolytically exposed cryptic epitopes in the triple helical structure of native collagen, inhibited lung colonization by B16F10 melanoma cells in the chick embryo model [189]. However, a putative inhibition of overall MMP activity by overexpression of TIMP-1 in mouse B16F10 melanoma did not affect cell extravasation from the CAM vasculature, although decreased proliferation of extravasated cells [190].

Conflicting results have also been demonstrated for the role of tumor MMPs during extravasation in mouse models. Thus, overexpression of MT1-MMP in a mouse lung carcinoma caused a threefold increase in the survival rate and in the number of lung nodules in syngeneic mice [70], whereas overexpression or downregulation of MT1-MMP in human melanoma BML cells did not affect early stages of lung colonization in immunodeficient mice [50]. Individual downregulation of *MMP1* or *MMP2* expression by specific shRNA did not change the extent of lung extravasation *in vivo* of the high metastatic variant of MDA-MB-231 breast carcinoma. However, when MMP-1 and MMP-2 were downregulated simultaneously and in conjunction with COX2 and epiregulin, pulmonary extravasation was significantly reduced, indicating that tumor-derived MMP-1 and MMP-2 can assist the metastatic cancer cells in breaching the endothelium in the secondary site [181].

The functional role of specific stromal-derived MMPs in the experimental metastasis model is demonstrated in studies in which tumor cells are inoculated into MMP-deficient mice. It is interesting that genetic depletion of MMP-2 did not affect the levels of lung colonization by intravenously inoculated Lewis lung carcinoma, whereas the lack of MMP-7 surprisingly caused a substantial increase in lung metastases [191]. However, when Lewis lung carcinoma or B16-BL6 melanoma cells were inoculated into MMP-9 deficient mice, there was a significant reduction of tumor burden, indicating a role of this stromal MMP in extravasation [191, 192]. Host rescue transplantations confirmed that tumor colonization was dependent on the expression of MMP-9 in BMDCs, most likely in neutrophils rather than in alveolar macrophages, reduction of which in the lungs did not have any significant effect on tumor cell survival [191]. A specific role for neutrophils during early stages of tumor cell extravasation is further supported by the findings demonstrating that inflammatory neutrophils increase lung retention of intravenously inoculated human melanoma 1205 Lu cells and significantly facilitate development of lung metastases by reducing tumor cell clearing [179]. In the brain microenvironment, host MMP-9 produced by activated astrocytes in the vicinity of growing metastases, may specifically contribute to colonization of melanoma MDA-MB-435 cells by assisting tumor invasion and providing initial proangiogenic stimuli [193].

Collectively, the above-discussed studies employing *in vivo* experimental metastasis models indicate that individual tumor-derived MMPs can influence the extravasation ability of some tumor cells. However, it is stromal MMPs, in particular myeloid-cell-derived MMP-9, which appear to determine the fate of extravasating cancer cells during spontaneous metastasis. Once delivered to the secondary site or *de novo* induced within the secondary site, specific MMPs can modify local microenvironments, facilitating colonization of spontaneously intravasated tumor cells.

8.9
Metastatic Site: Involvement of MMPs in the Preparation, Colonization, and Invasion of Distal Organ Stroma

The mechanisms underlying organ-specific metastasis exhibited by many solid cancers have indicated that primary tumor cells can produce the instructive cues that modulate the normal physiology of preferred distant sites and alter the microenvironment to assist the arrival of metastatic cells. Primary tumor cells can induce the egress of distinct types of BMDCs and govern their migration to future metastatic sites. In conjunction with local resident stromal cells, infiltrating BMDCs participate in the construction of permissive premetastatic niches and ECM scaffolds for extravasating tumor cells. Being inhabited by cancer cells, these specialized distant microenvironments become instructive niches or locales, which promote tumor cell survival, induce cell proliferation, and eventually, support neovascularization and invasive expansion of established micrometastases [194, 195].

The putative involvement of MMPs at the metastatic site is indicated in several of the above-mentioned processes. First, specific MMPs contribute to the molecular signature of primary tumor cells, conferring them with the ability to complete all steps of the metastatic cascade, including the ability to establish metastatic foci in the preferred sites. Second, MMPs can be induced in the distal sites in response to factors emanating from the primary tumor. These MMPs can be *de novo* expressed by the interstitial stromal cells or delivered to premetastatic sites by the infiltrating BMDCs. Furthermore, the initial egress of BMDCs can also be induced in an MMP-dependent manner. Finally, following the establishment of micrometastases, MMPs can be specifically expressed at the tumor–stroma border, where a new invasive front is created, once again facilitating the proteolytic remodeling of the stroma and invasive expansion of metastatic cells.

8.9.1
MMPs as Determinants of Organ-Specific Metastases

A view on metastasis as an evolutionary process culminating in the appearance of tumor cells that have acquired the capacity of preferential organ colonization raises a question regarding the nature of metastasis predisposition factors and the identity of genes that mediate tropism for a specific distant tissue [195, 196]. For example, genomic analyses of the *in-vivo*-selected organotropic variants of the MD-MBA-231 breast cancer cells, exhibiting propensity to metastasize to the lungs, the bone or the brain, led to the identification of organ-specific gene expression signatures, which include genes encoding for distinct MMPs [197, 198].

Using genetic and pharmacological approaches in a mouse xenograft model of human breast cancer, it was demonstrated that *MMP1* and *MMP2* were among the signature genes that confer primary breast cancer cells with the ability to seed pulmonary metastasis by facilitating assembly of new tumor blood vessels, release of tumor cells into the circulation at the primary site, and breaching of blood capillaries in the lungs [181, 199]. Bone-specific metastasis by breast cancer cells is mediated by two distinct tumor MMPs, namely, MMP-1 and *a disintegrin and metalloproteinase with thrombospondin motifs*, ADAMTS1, which proteolytically release the membrane-bound EGF-like growth factors and orchestrate a paracrine signaling cascade to modulate the bone microenvironment in favor of osteoclastogenesis and osteoclast-mediated bone remodeling [200]. Surprisingly, no *MMP* gene was identified in the signature defining the ability of breast cancer cells to metastasize into brain [201]. However, in breast cancer patients, three MMP genes overexpressed in primary tumors, namely, *MMP10,*

MMP13, and *MMP16*, were found in the signature associated with lymph node metastases [202].

Overall, it appears that in certain combinations, individual MMPs can constitute important determinants of organ-specific metastases in breast cancer. However, it is unclear whether MMPs, individually or in combination, would be part of "organ-specific" signatures in cancers other than breast carcinoma. In addition, distinct MMPs and MMP pathways that determine organ specificity of metastases may have contrasting or overlapping functional activities in the primary tumor and secondary sites. In other words, sets of tumorigenic and metastatic genes can share identical MMPs, exhibiting diversified roles during different phases of cancer progression.

8.9.2 MMP-Dependent Preparation of the PreMetastatic Microenvironment

According to the concept of premetastatic niches, the primary tumor, almost in anthropomorphic manner, "orchestrates" specific changes in distant organs and "instructs" resident stromal cells and infiltrating BMDCs to create specialized domains to facilitate organ-specific metastasis [194, 203]. Since the stroma of secondary organs is envisioned to be proteolytically modified to accommodate the arriving tumor cells, the role of MMPs was anticipated. Correspondingly, functional contributions of MMPs were indeed demonstrated in several niche-involving processes and pathways described below. Complex networks of interacting paracrine factors exchanged between the primary tumor, the bone marrow, and the stromal cells of distant sites include select MMPs, among which MMP-9 derived mainly from the mobilized BMDCs, appears to play the most essential role.

MMP-mediated modifications of the ECM at premetastatic sites was initially demonstrated in a model of spontaneous lung metastasis, where intradermally implanted Lewis lung carcinoma cells, which are prone to lung metastasis, induced both the mobilization of immature VEGFR1+$\alpha4\beta1$+ myeloid cells from the bone marrow and the expression in premetastatic lungs of fibronectin, an $\alpha4\beta1$ ligand [204]. Through a VEGFR-mediated pathway, the interactions of BMDCs with the lung tissue also induce the expression of MMP-9, which can proteolytically alter lung ECM, making it more receptive for tumor cell implantation and growth. The MMP-9-mediated preparation of lung niches was confirmed in MMP-9 knockout mice, which exhibited reduced BMDC cluster formation and metastatic spread [204]. Functional contribution of BMDC-derived MMP-9 in the preparation of lungs for tumor cell metastasis was also demonstrated in mice bearing mammary 4T1 adenocarcinomas, where Gr1+CD11b+ myeloid cells were shown to infiltrate premetastatic lungs before tumor cell arrival. These immature BMDCs produced MMP-9 that facilitated proteolytic vascular remodeling and increased tumor metastasis. Importantly, genetic depletion of MMP-9 normalized the aberrant vasculature in the premetastatic lung and diminished lung metastasis [205].

An orthotopic model of human breast cancer implicated another gelatinase, MMP-2, in preparation of the lung microenvironment for tumor cell arrival. Thus,

LOX, secreted by hypoxic MDA-MB-231 carcinoma cells, was shown to accumulate in premetastatic lungs of xenograft-bearing mice. By enhancing fibronectin deposition and cross-linking collagen IV in the lungs, LOX facilitates the recruitment of CD11b+ BMDCs, their adhesion to collagen fibrils, and production of MMP-2 [206]. This myeloid-cell-derived MMP-2, in turn, cleaves collagen IV and releases collagen IV peptides, which enhance *de novo* infiltration of BMDCs and metastasizing tumor cells. Genetic ablation of MMP-2 in knockout mice or inhibition of MMPs with a gelatinase inhibitor dramatically changed the pattern of CD11b + cell accumulation in the lung and prevented cell invasion into the surrounding stroma, indicating that the activity of inflammatory cell MMP-2 is required for premetastatic niche preparation [207]. That the host MMP-2 is involved in the formation of premetastatic niches in the lung complements the earlier identification of *MMP2* as a tumor progression gene associated with lung-specific breast cancer metastasis [181].

Highlighting the role of MMP-mediated modifications of the ECM at premetastatic sites are the findings demonstrating that MMPs proteolytically modify the vasculature to facilitate extravasation of metastasizing tumor cells. In a lung metastasis model, primary Lewis lung carcinoma tumors specifically induced MMP-9 in premetastatic pulmonary endothelial cells, thereby significantly promoting ensuing lung metastasis in wild type, but not in MMP-9 knockout mice [208]. In a melanoma metastasis model, the intradermal B16F10 tumors increased the permeability of premetastatic pulmonary vasculature, in part due to degradation of endothelial BM by tumor-induced endothelial cell MMP-3 and MMP-10, leading to enhanced tumor cell extravasation [180]. If pulmonary vascular permeability was attenuated by a combined lentivirus-based RNA interference, both myeloid cell infiltration and tumor cell extravasation were inhibited, therefore strongly implicating two stromal MMPs, MMP-3 and MMP-10, in the preparation of the premetastatic microenvironment [180].

An important aspect of the premetastatic niche physiology is the possibility that arriving tumor cells are presented to a microenvironment, which can prevent tumor cell proliferation and induce tumor cell dormancy. Furthermore, therapeutic treatments can result in the formation of the so-called chemoresistant niches that promote the survival of cancer cells and serve as a reservoir for eventual tumor relapse. Thus, in a mouse model of Burkitt's lymphoma, doxorubicin treatment induced stromal cells in the thymus to express and release of TIMP-1 and IL-6, which in turn protected lymphoma cells from the drug-induced apoptosis [209]. This study provides an example of how anticancer therapy can paradoxically elicit tumor cell survival programs that, in addition to other mechanisms, may involve TIMP inhibition of antimetastatic MMPs.

Despite the extensive investigation, many aspects of functional contribution of MMPs in the establishment of the metastatic site remain poorly addressed and await experimental resolution. It is unknown whether niche-residing metastatic tumor cells would eventually be surrounded by newly deposited BM or ECM, penetration of which would require the proteolytic activity of MMPs. It is also not known whether the tumor cells in the metastatic niche would exploit the same invasion mechanisms that were utilized at the primary tumor site, namely, MMP-mediated breaching of the BM or a fibrotic matrix in order to initiate their path-making and expansive invasion into the surrounding stroma. It could be that those MMPs, which are involved in the preparation of premetastatic niches and the initial expansion of established metastatic foci, should be targeted by anticancer therapy rather than the MMPs at the primary site.

8.9.3
Invasive Expansion of Cancer Cells at the Metastatic Site

In general, the functional involvement of MMPs in the invasive expansion of metastatic deposits has been long indicated by the elevated expression and activity of specific MMPs in tumor and stromal cells at the secondary sites. Furthermore, the differential expression of MMPs between primary tumor and metastases suggests a functional importance of specific MMPs in the invasion at distal sites. In view of a concept of mesenchymal to epithelial transition (MET) [21, 210, 211], maybe this invasive expansion of metastases would involve another round of MMP-induced EMT or EMT-induced MMPs. These and other mechanisms in which MMPs are exploited by tumor cells to facilitate or control tumor cell progression can be explored in order to illuminate the role of MMPs at this final stage of the metastatic cascade.

Modifications of experimental metastasis models, in which tumor cells are inoculated directly into preferred metastatic sites, are used to investigate the role of MMPs in metastatic invasion *in vivo*. Overexpression or downregulation of individual MMPs in tumor cells and knocking out MMPs in mice furthers the discrimination between the contributions of tumor-derived and stroma-derived MMPs.

Specific involvement of MMPs and their proteolytic activity has been demonstrated in a distant metastasis model of breast and lung cancer, where tumor cells are inoculated directly into the brain. By using this model, rat mammary adenocarcinoma ENU1564 cells were shown to form invasive brain metastases, which exhibited increased expression of MMP-2 and *de novo* expression of MMP-3 and MMP-9 and sensitivity of metastatic invasion to the MMP inhibitor PD166793 [212]. MMP-2 and MMP-9 were implicated in the invasive expansion of Lewis lung carcinoma, especially at early stages after cell inoculations into brain, where lung carcinoma deposits invaded perivascular spaces and brain parenchyma, leaving fragments of collagen fibrils in the gaps between neoplastic cells [213].

Functional contribution of MT1-MMP in growth and invasion of cancer cells at the metastatic site has been shown in experimental models of lung and bone metastases. Kinetic analysis of metastatic growth of human BML melanoma cells injected directly into the lungs of SCID mice has demonstrated that being dispensable during early phases of melanoma development and initial lung colonization, tumor cell-expressed MT1-MMP was required for subsequent expansion of metastases, as evidenced by a significant reduction of metastatic burden if MT1-MMP was knocked down [50].

A bone metastasis model appears to be specifically suited to study functional contribution of MMPs in the invasive expansion of metastases. To study intraosseous metastases, cancer cells with a propensity of bone metastasis, for example, prostate cancer cells, are surgically implanted into the tibia, followed by monitoring of tumor growth over time by noninvasive imaging in live animals and by immunohistochemistry in decalcified bones [214]. In an intratibial model of bone metastasis, overexpression of MT1-MMP in prostate cancer cells was shown to increase tumor growth and osteolytic invasion of skeletal metastatic deposits via MT1-MMP-mediated resorption of bone matrix and shedding of RANK ligand [215], causing increased cell migration via RANK/Src signaling cascade [216].

In addition to MT1-MMP, stromal MMPs have also been implicated in RANKL processing in prostate cancer bone metastases. In a rodent model of human prostate cancer in the bone, stromal MMP-7 is upregulated in the bone-resorbing osteoclasts, specifically at the tumor-bone interface, where MMP-7 processes RANK ligand to a soluble form that promotes osteoclast activation. Importantly, genetic ablation of host MMP-7 in RAG-2 immunodeficient mice significantly reduced osteolysis induced by rat prostate adenocarcinoma [217], validating a functional role for stromal MMP-7 in metastatic invasion.

Specific contributions of various MMPs to invasive expansion of cancer cells at the metastatic site suggest that individual MMPs can be attractive therapeutic targets for the control of metastatic disease. Originating in tumor cells or induced in the stromal cells, these metastasis-expanding MMPs can exert their clear proteolytic functions at the "second" invasive front.

8.10 Perspectives: MMPs in the Early Metastatic Dissemination and Awakening of Dormant Metastases

Strong correlations have been established between the levels of cancer invasion and the presence of metastases in local lymph nodes and distant organs. However, the actual extent of primary cancer invasion at the time when the very first metastases are seeded is not known. Recent multiparameter analyses have suggested that clinically relevant distant metastases can be initiated long before identification of the primary tumor [218], indicating that treatment of putative and present metastases can be more relevant than metastasis prevention [219]. Correspondingly, invasion observed in resected tumors may mainly reflect the heightened motility of aggressive tumor cells at relatively late stages of cancer progression and indicate a high probability that metastatic events have already occurred. If so, the tumor size and extent of tumor invasion can be the best prognostic factors for distant metastases [173], but mechanistically, may not be directly linked to the initiation of overt metastases.

Whether clinically relevant cancer dissemination occurs at very early stages of primary tumor development, not accessible for timely evaluation, or at later stages when detectable, primary tumors present with local invasion, it is plausible that the metastasis-initiating cells would employ MMPs to penetrate the BM and directionally invade the adjacent stroma to reach the points of intravasation into blood or lymphatic vessels at the original site or to accommodate the invasive expansion of established micrometastases at the secondary sites. At present, there is no data that would directly link MMPs with the very early metastasis-initiating cells or events.

Demonstrating that MMPs are involved in the initiation of metastases at very early stages of tumor development would require addressing whether or not the dissemination of the earliest metastasis-initiating cells, possibly occurring before evident cell invasion, depends on MMPs. If the deposition of clinically relevant metastases occurs at the level of less than 1 million of primary tumor cells [218], are these tumors already angiogenic, and if so, does this switch require MMP activity? Are clinically relevant metastases initiated by cancer stem/stem-like cells in an MMP-dependent manner? Do the early metastasis-initiating cells employ *de novo* induced MMPs or do they intrinsically express distinct MMPs to escape from the primary tumor and reach the premetastatic niche?

Finally, in view that metastatic disease can progress long after resection of solid primary tumors or initiate after long-term remissions of leukemias, it appears that investigation of the mechanisms underlying MMP-dependent awakening of dormant metastases and residual tumor cells can constitute a main focus of upcoming research in the MMP field. It is possible that individual tumor and/or stromal MMPs can be involved in Src-mediated awakening of dormant metastases, recently demonstrated for latent bone metastases in breast cancer [220]. This possibility might switch MMP inhibition therapy from the *targeting* of MMP activities in primary tumors and overt metastases to the *regulation* of those specific MMPs that are involved in the initiation and invasive expansion of latent metastases seeded most probably at the very earliest stages of cancer progression. The armamentarium of already available experimental models and development of novel approaches will aid both in identification of such MMPs and their validation as clinically relevant pharmacological targets.

References

1. Hanahan, D. and Weinberg, R.A. (2000) The hallmarks of cancer. *Cell*, **100** (1), 57–70.
2. Weinberg, R.A. (2008) Mechanisms of malignant progression. *Carcinogenesis*, **29** (6), 1092–1095.
3. Mareel, M. and Leroy, A. (2003) Clinical, cellular, and molecular aspects of cancer invasion. *Physiol. Rev.*, **83** (2), 337–376.
4. Woessner, J.F. Jr. (1991) Matrix metalloproteinases and their inhibitors in connective tissue remodeling. *FASEB J.*, **5** (8), 2145–2154.
5. Kessenbrock, K., Plaks, V., and Werb, Z. (2010) Matrix metalloproteinases: regulators of the tumor microenvironment. *Cell*, **141** (1), 52–67.
6. Martin, M.D. and Matrisian, L.M. (2007) The other side of MMPs: protective roles in tumor progression. *Cancer Metastasis Rev.*, **26** (3–4), 717–724.
7. Gimona, M., Buccione, R., Courtneidge, S.A., and Linder, S. (2008) Assembly and biological role of podosomes and invadopodia. *Curr. Opin. Cell Biol.*, **20** (2), 235–241.

8. Lee, J.M., Dedhar, S., Kalluri, R., and Thompson, E.W. (2006) The epithelial-mesenchymal transition: new insights in signaling, development, and disease. *J. Cell Biol.*, **172** (7), 973–981.
9. Bockhorn, M., Jain, R.K., and Munn, L.L. (2007) Active versus passive mechanisms in metastasis: do cancer cells crawl into vessels, or are they pushed? *Lancet Oncol.*, **8** (5), 444–448.
10. Ferro, P., Franceschini, M.C., Bacigalupo, B., Dessanti, P., Falco, E., Fontana, V., Gianquinto, D., Pistillo, M.P., Fedeli, F., and Roncella, S. (2010) Detection of circulating tumour cells in breast cancer patients using human mammaglobin RT-PCR: association with clinical prognostic factors. *Anticancer Res.*, **30** (6), 2377–2382.
11. Deryugina, E.I. and Quigley, J.P. (2006) Matrix metalloproteinases and tumor metastasis. *Cancer Metastasis Rev.*, **25** (1), 9–34.
12. Hay, E.D. (1995) An overview of epithelio-mesenchymal transformation. *Acta Anat. (Basel)*, **154** (1), 8–20.
13. Thiery, J.P. (2003) Epithelial-mesenchymal transitions in development and pathologies. *Curr. Opin. Cell Biol.*, **15** (6), 740–746.
14. Thiery, J.P. and Sleeman, J.P. (2006) Complex networks orchestrate epithelial-mesenchymal transitions. *Nat. Rev. Mol. Cell Biol.*, **7** (2), 131–142.
15. Zavadil, J., Haley, J., Kalluri, R., Muthuswamy, S.K., and Thompson, E. (2008) Epithelial-mesenchymal transition. *Cancer Res.*, **68** (23), 9574–9577.
16. Thiery, J.P., Acloque, H., Huang, R.Y., and Nieto, M.A. (2009) Epithelial-mesenchymal transitions in development and disease. *Cell*, **139** (5), 871–890.
17. Yilmaz, M. and Christofori, G. (2009) EMT, the cytoskeleton, and cancer cell invasion. *Cancer Metastasis Rev.*, **28** (1–2), 15–33.
18. Kalluri, R. and Weinberg, R.A. (2009) The basics of epithelial-mesenchymal transition. *J. Clin. Invest.*, **119** (6), 1420–1428.
19. Tarin, D., Thompson, E.W., and Newgreen, D.F. (2005) The fallacy of epithelial mesenchymal transition in neoplasia. *Cancer Res.*, **65** (14), 5996–6000; discussion 6000–5991.
20. Christiansen, J.J. and Rajasekaran, A.K. (2006) Reassessing epithelial to mesenchymal transition as a prerequisite for carcinoma invasion and metastasis. *Cancer Res.*, **66** (17), 8319–8326.
21. Hugo, H., Ackland, M.L., Blick, T., Lawrence, M.G., Clements, J.A., Williams, E.D., and Thompson, E.W. (2007) Epithelial--mesenchymal and mesenchymal--epithelial transitions in carcinoma progression. *J. Cell Physiol.*, **213** (2), 374–383.
22. Billottet, C., Tuefferd, M., Gentien, D., Rapinat, A., Thiery, J.P., Broet, P., and Jouanneau, J. (2008) Modulation of several waves of gene expression during FGF-1 induced epithelial-mesenchymal transition of carcinoma cells. *J. Cell Biochem.*, **104** (3), 826–839.
23. Zheng, G., Lyons, J.G., Tan, T.K., Wang, Y., Hsu, T.T., Min, D., Succar, L., Rangan, G.K., Hu, M., Henderson, B.R., Alexander, S.I., and Harris, D.C. (2009) Disruption of E-cadherin by matrix metalloproteinase directly mediates epithelial-mesenchymal transition downstream of transforming growth factor-beta1 in renal tubular epithelial cells. *Am.J. Pathol.*, **175** (2), 580–591.
24. Sternlicht, M.D., Lochter, A., Sympson, C.J., Huey, B., Rougier, J.P., Gray, J.W., Pinkel, D., Bissell, M.J., and Werb, Z. (1999) The stromal proteinase MMP3/stromelysin-1 promotes mammary carcinogenesis. *Cell*, **98** (2), 137–146.
25. Sternlicht, M.D., Bissell, M.J., and Werb, Z. (2000) The matrix metalloproteinase stromelysin-1 acts as a natural mammary tumor promoter. *Oncogene*, **19** (8), 1102–1113.
26. McCawley, L.J., Crawford, H.C., King, L.E. Jr., Mudgett, J., and Matrisian, L.M. (2004) A protective role for matrix metalloproteinase-3 in squamous cell carcinoma. *Cancer Res.*, **64** (19), 6965–6972.
27. Orlichenko, L.S. and Radisky, D.C. (2008) Matrix metalloproteinases stimulate epithelial-mesenchymal transition

during tumor development. *Clin. Exp. Metastasis*, **25** (6), 593–600.

28. Illman, S.A., Lehti, K., Keski-Oja, J., and Lohi, J. (2006) Epilysin (MMP-28) induces TGF-beta mediated epithelial to mesenchymal transition in lung carcinoma cells. *J. Cell Sci.*, **119** (Pt 18), 3856–3865.
29. Przybylo, J.A. and Radisky, D.C. (2007) Matrix metalloproteinase-induced epithelial-mesenchymal transition: tumor progression at Snail's pace. *Int. J. Biochem. Cell Biol.*, **39** (6), 1082–1088.
30. McGuire, J.K., Li, Q., and Parks, W.C. (2003) Matrilysin (matrix metalloproteinase-7) mediates E-cadherin ectodomain shedding in injured lung epithelium. *Am. J. Pathol.*, **162** (6), 1831–1843.
31. Nawrocki-Raby, B., Gilles, C., Polette, M., Bruyneel, E., Laronze, J.Y., Bonnet, N., Foidart, J.M., Mareel, M., and Birembaut, P. (2003) Upregulation of MMPs by soluble E-cadherin in human lung tumor cells. *Int. J. Cancer*, **105** (6), 790–795.
32. Nawrocki-Raby, B., Gilles, C., Polette, M., Martinella-Catusse, C., Bonnet, N., Puchelle, E., Foidart, J.M., Van Roy, F., and Birembaut, P. (2003) E-Cadherin mediates MMP down-regulation in highly invasive bronchial tumor cells. *Am. J. Pathol.*, **163** (2), 653–661.
33. Moss, N.M., Liu, Y., Johnson, J.J., Debiase, P., Jones, J., Hudson, L.G., Munshi, H.G., and Stack, M.S. (2009) Epidermal growth factor receptor-mediated membrane type 1 matrix metalloproteinase endocytosis regulates the transition between invasive versus expansive growth of ovarian carcinoma cells in three-dimensional collagen. *Mol. Cancer Res.*, **7** (6), 809–820.
34. Sugiyama, N., Varjosalo, M., Meller, P., Lohi, J., Hyytiainen, M., Kilpinen, S., Kallioniemi, O., Ingvarsen, S., Engelholm, L.H., Taipale, J., Alitalo, K., Keski-Oja, J., and Lehti, K. (2010) Fibroblast growth factor receptor 4 regulates tumor invasion by coupling fibroblast growth factor signaling to extracellular matrix degradation. *Cancer Res.*, **70** (20), 7851–7861.
35. Nabeshima, K., Inoue, T., Shimao, Y., Okada, Y., Itoh, Y., Seiki, M., and Koono, M. (2000) Front-cell-specific expression of membrane-type 1 matrix metalloproteinase and gelatinase A during cohort migration of colon carcinoma cells induced by hepatocyte growth factor/scatter factor. *Cancer Res.*, **60** (13), 3364–3369.
36. Ota, I., Li, X.Y., Hu, Y., and Weiss, S.J. (2009) Induction of a MT1-MMP and MT2-MMP-dependent basement membrane transmigration program in cancer cells by Snail1. *Proc. Natl. Acad. Sci. U.S.A.*, **106** (48), 20318–20323.
37. Tester, A.M., Ruangpanit, N., Anderson, R.L., and Thompson, E.W. (2001) MMP-9 secretion and MMP-2 activation distinguish invasive and metastatic sublines of a mouse mammary carcinoma system showing epithelial-mesenchymal transition traits. *Clin. Exp. Metastasis*, **18** (7), 553–560.
38. Hazan, R.B., Qiao, R., Keren, R., Badano, I., and Suyama, K. (2004) Cadherin switch in tumor progression. *Ann. N. Y. Acad. Sci.*, **1014**, 155–163.
39. Giannoni, E., Bianchini, F., Masieri, L., Serni, S., Torre, E., Calorini, L., and Chiarugi, P. (2010) Reciprocal activation of prostate cancer cells and cancer-associated fibroblasts stimulates epithelial-mesenchymal transition and cancer stemness. *Cancer Res.*, **70** (17), 6945–6956.
40. Rowe, R.G. and Weiss, S.J. (2008) Breaching the basement membrane: who, when and how? *Trends Cell Biol.*, **18** (11), 560–574.
41. Wolf, K., Mazo, I., Leung, H., Engelke, K., von Andrian, U.H., Deryugina, E.I., Strongin, A.Y., Brocker, E.B., and Friedl, P. (2003) Compensation mechanism in tumor cell migration: mesenchymal-amoeboid transition after blocking of pericellular proteolysis. *J. Cell Biol.*, **160** (2), 267–277.
42. Albini, A., Iwamoto, Y., Kleinman, H.K., Martin, G.R., Aaronson, S.A., Kozlowski, J.M., and McEwan, R.N.

(1987) A rapid in vitro assay for quantitating the invasive potential of tumor cells. *Cancer Res.*, **47** (12), 3239–3245.

43. Palmisano, R. and Itoh, Y. (2010) Analysis of MMP-dependent cell migration and invasion. *Methods Mol. Biol.*, **622**, 379–392.
44. Denhardt, D.T., Khokha, R., Yagel, S., Overall, C.M., and Parhar, R.S. (1992) Oncogenic consequences of down-modulating TIMP expression in 3T3 cells with antisense RNA. *Matrix Suppl.*, **1**, 281–285.
45. Khokha, R., Zimmer, M.J., Graham, C.H., Lala, P.K., and Waterhouse, P. (1992) Suppression of invasion by inducible expression of tissue inhibitor of metalloproteinase-1 (TIMP-1) in B16-F10 melanoma cells. *J. Natl. Cancer Inst.*, **84** (13), 1017–1022.
46. Sato, H., Takino, T., Okada, Y., Cao, J., Shinagawa, A., Yamamoto, E., and Seiki, M. (1994) A matrix metalloproteinase expressed on the surface of invasive tumour cells. *Nature*, **370** (6484), 61–65.
47. Ueda, J., Kajita, M., Suenaga, N., Fujii, K., and Seiki, M. (2003) Sequence-specific silencing of MT1-MMP expression suppresses tumor cell migration and invasion: importance of MT1-MMP as a therapeutic target for invasive tumors. *Oncogene*, **22** (54), 8716–8722.
48. Itoh, Y., Takamura, A., Ito, N., Maru, Y., Sato, H., Suenaga, N., Aoki, T., and Seiki, M. (2001) Homophilic complex formation of MT1-MMP facilitates proMMP-2 activation on the cell surface and promotes tumor cell invasion. *EMBO J.*, **20** (17), 4782–4793.
49. Sounni, N.E., Devy, L., Hajitou, A., Frankenne, F., Munaut, C., Gilles, C., Deroanne, C., Thompson, E.W., Foidart, J.M., and Noel, A. (2002) MT1-MMP expression promotes tumor growth and angiogenesis through an up-regulation of vascular endothelial growth factor expression. *FASEB J.*, **16** (6), 555–564.
50. Bartolome, R.A., Ferreiro, S., Miquilena-Colina, M.E., Martinez-Prats, L., Soto-Montenegro, M.L., Garcia-Bernal, D., Vaquero, J.J., Agami, R., Delgado, R., Desco, M., Sanchez-Mateos, P., and Teixido, J. (2009) The chemokine receptor CXCR4 and the metalloproteinase MT1-MMP are mutually required during melanoma metastasis to lungs. *Am. J. Pathol.*, **174** (2), 602–612.
51. Bjorklund, M., Heikkila, P., and Koivunen, E. (2004) Peptide inhibition of catalytic and noncatalytic activities of matrix metalloproteinase-9 blocks tumor cell migration and invasion. *J. Biol. Chem.*, **279** (28), 29589–29597.
52. Kumar, B., Koul, S., Petersen, J., Khandrika, L., Hwa, J.S., Meacham, R.B., Wilson, S., and Koul, H.K. (2010) p38 mitogen-activated protein kinase-driven MAPKAPK2 regulates invasion of bladder cancer by modulation of MMP-2 and MMP-9 activity. *Cancer Res.*, **70** (2), 832–841.
53. Lakka, S.S., Gondi, C.S., Yanamandra, N., Olivero, W.C., Dinh, D.H., Gujrati, M., and Rao, J.S. (2004) Inhibition of cathepsin B and MMP-9 gene expression in glioblastoma cell line via RNA interference reduces tumor cell invasion, tumor growth and angiogenesis. *Oncogene*, **23** (27), 4681–4689.
54. Ezhilarasan, R., Jadhav, U., Mohanam, I., Rao, J.S., Gujrati, M., and Mohanam, S. (2009) The hemopexin domain of MMP-9 inhibits angiogenesis and retards the growth of intracranial glioblastoma xenograft in nude mice. *Int. J. Cancer*, **124** (2), 306–315.
55. Xu, D., McKee, C.M., Cao, Y., Ding, Y., Kessler, B.M., and Muschel, R.J. (2010) Matrix metalloproteinase-9 regulates tumor cell invasion through cleavage of protease nexin-1. *Cancer Res.*, **70** (17), 6988–6998.
56. Ramos-DeSimone, N., Hahn-Dantona, E., Sipley, J., Nagase, H., French, D.L., and Quigley, J.P. (1999) Activation of matrix metalloproteinase-9 (MMP-9) via a converging plasmin/stromelysin-1 cascade enhances tumor cell invasion. *J. Biol. Chem.*, **274** (19), 13066–13076.

57. Pulukuri, S.M. and Rao, J.S. (2008) Matrix metalloproteinase-1 promotes prostate tumor growth and metastasis. *Int. J. Oncol.*, **32** (4), 757–765.

58. Inoue, A., Takahashi, H., Harada, H., Kohno, S., Ohue, S., Kobayashi, K., Yano, H., Tanaka, J., and Ohnishi, T. (2010) Cancer stem-like cells of glioblastoma characteristically express MMP-13 and display highly invasive activity. *Int. J. Oncol.*, **37** (5), 1121–1131.

59. Kong, Y., Poon, R., Nadesan, P., Di Muccio, T., Fodde, R., Khokha, R., and Alman, B.A. (2004) Matrix metalloproteinase activity modulates tumor size, cell motility, and cell invasiveness in murine aggressive fibromatosis. *Cancer Res.*, **64** (16), 5795–5803.

60. Cheng, S.M., Xing, B., Li, J.C., Cheung, B.K., and Lau, A.S. (2007) Interferon-gamma regulation of TNFalpha-induced matrix metalloproteinase 3 expression and migration of human glioma T98G cells. *Int. J. Cancer*, **121** (6), 1190–1196.

61. Dunn, K.M., Lee, P.K., Wilson, C.M., Iida, J., Wasiluk, K.R., Hugger, M., and McCarthy, J.B. (2009) Inhibition of hyaluronan synthases decreases matrix metalloproteinase-7 (MMP-7) expression and activity. *Surgery*, **145** (3), 322–329.

62. Deryugina, E.I., Luo, G.X., Reisfeld, R.A., Bourdon, M.A., and Strongin, A. (1997) Tumor cell invasion through matrigel is regulated by activated matrix metalloproteinase-2. *Anticancer Res.*, **17** (5A), 3201–3210.

63. Hotary, K., Li, X.Y., Allen, E., Stevens, S.L., and Weiss, S.J. (2006) A cancer cell metalloprotease triad regulates the basement membrane transmigration program. *Genes Dev.*, **20** (19), 2673–2686.

64. Sodek, K.L., Brown, T.J., and Ringuette, M.J. (2008) Collagen I but not Matrigel matrices provide an MMP-dependent barrier to ovarian cancer cell penetration. *BMC Cancer*, **8**, 223.

65. Sabeh, F., Shimizu-Hirota, R., and Weiss, S.J. (2009) Protease-dependent versus -independent cancer cell invasion programs: three-dimensional amoeboid movement revisited. *J. Cell Biol.*, **185** (1), 11–19.

66. Adley, B.P., Gleason, K.J., Yang, X.J., and Stack, M.S. (2009) Expression of membrane type 1 matrix metalloproteinase (MMP-14) in epithelial ovarian cancer: high level expression in clear cell carcinoma. *Gynecol. Oncol.*, **112** (2), 319–324.

67. Moss, N.M., Barbolina, M.V., Liu, Y., Sun, L., Munshi, H.G., and Stack, M.S. (2009) Ovarian cancer cell detachment and multicellular aggregate formation are regulated by membrane type 1 matrix metalloproteinase: a potential role in I.p. metastatic dissemination. *Cancer Res.*, **69** (17), 7121–7129.

68. Boire, A., Covic, L., Agarwal, A., Jacques, S., Sherifi, S., and Kuliopulos, A. (2005) PAR1 is a matrix metalloprotease-1 receptor that promotes invasion and tumorigenesis of breast cancer cells. *Cell*, **120** (3), 303–313.

69. Yang, E., Boire, A., Agarwal, A., Nguyen, N., O'Callaghan, K., Tu, P., Kuliopulos, A., and Covic, L. (2009) Blockade of PAR1 signaling with cell-penetrating pepducins inhibits Akt survival pathways in breast cancer cells and suppresses tumor survival and metastasis. *Cancer Res.*, **69** (15), 6223–6231.

70. Tsunezuka, Y., Kinoh, H., Takino, T., Watanabe, Y., Okada, Y., Shinagawa, A., Sato, H., and Seiki, M. (1996) Expression of membrane-type matrix metalloproteinase 1 (MT1-MMP) in tumor cells enhances pulmonary metastasis in an experimental metastasis assay. *Cancer Res.*, **56** (24), 5678–5683.

71. Brooks, P.C., Stromblad, S., Sanders, L.C., von Schalscha, T.L., Aimes, R.T., Stetler-Stevenson, W.G., Quigley, J.P., and Cheresh, D.A. (1996) Localization of matrix metalloproteinase MMP-2 to the surface of invasive cells by interaction with integrin alpha v beta 3. *Cell*, **85** (5), 683–693.

72. Deryugina, E.I., Bourdon, M.A., Reisfeld, R.A., and Strongin, A. (1998)

Remodeling of collagen matrix by human tumor cells requires activation and cell surface association of matrix metalloproteinase-2. *Cancer Res.*, **58** (16), 3743–3750.

73. Hofmann, U.B., Westphal, J.R., Waas, E.T., Becker, J.C., Ruiter, D.J., and van Muijen, G.N. (2000) Coexpression of integrin alpha(v)beta3 and matrix metalloproteinase-2 (MMP-2) coincides with MMP-2 activation: correlation with melanoma progression. *J. Invest. Dermatol.*, **115** (4), 625–632.
74. Deryugina, E.I., Bourdon, M.A., Jungwirth, K., Smith, J.W., and Strongin, A.Y. (2000) Functional activation of integrin alpha V beta 3 in tumor cells expressing membrane-type 1 matrix metalloproteinase. *Int. J. Cancer*, **86** (1), 15–23.
75. Deryugina, E.I., Ratnikov, B., Monosov, E., Postnova, T.I., DiScipio, R., Smith, J.W., and Strongin, A.Y. (2001) MT1-MMP initiates activation of pro-MMP-2 and integrin alphavbeta3 promotes maturation of MMP-2 in breast carcinoma cells. *Exp. Cell Res.*, **263** (2), 209–223.
76. Yu, Q. and Stamenkovic, I. (1999) Localization of matrix metalloproteinase 9 to the cell surface provides a mechanism for CD44-mediated tumor invasion. *Genes Dev.*, **13** (1), 35–48.
77. Redondo-Munoz, J., Ugarte-Berzal, E., Garcia-Marco, J.A., del Cerro, M.H., Van den Steen, P.E., Opdenakker, G., Terol, M.J., and Garcia-Pardo, A. (2008) Alpha4beta1 integrin and 190-kDa CD44v constitute a cell surface docking complex for gelatinase B/MMP-9 in chronic leukemic but not in normal B cells. *Blood*, **112** (1), 169–178.
78. Sabeh, F., Ota, I., Holmbeck, K., Birkedal-Hansen, H., Soloway, P., Balbin, M., Lopez-Otin, C., Shapiro, S., Inada, M., Krane, S., Allen, E., Chung, D., and Weiss, S.J. (2004) Tumor cell traffic through the extracellular matrix is controlled by the membrane-anchored collagenase MT1-MMP. *J. Cell Biol.*, **167** (4), 769–781.
79. Li, X.Y., Ota, I., Yana, I., Sabeh, F., and Weiss, S.J. (2008) Molecular dissection of the structural machinery underlying the tissue-invasive activity of membrane type-1 matrix metalloproteinase. *Mol. Biol. Cell*, **19** (8), 3221–3233.
80. Romanoff, A.L. (1960) *The Avian Embryo*, The Macmillan Co., New York.
81. Armstrong, P.B., Quigley, J.P., and Sidebottom, E. (1982) Transepithelial invasion and intramesenchymal infiltration of the chick embryo chorioallantois by tumor cell lines. *Cancer Res.*, **42** (5), 1826–1837.
82. Rowe, R.G., Li, X.Y., Hu, Y., Saunders, T.L., Virtanen, I., Garcia de Herreros, A., Becker, K.F., Ingvarsen, S., Engelholm, L.H., Bommer, G.T., Fearon, E.R., and Weiss, S.J. (2009) Mesenchymal cells reactivate Snail1 expression to drive three-dimensional invasion programs. *J. Cell Biol.*, **184** (3), 399–408.
83. Hofmann, U.B., Eggert, A.A., Blass, K., Brocker, E.B., and Becker, J.C. (2003) Expression of matrix metalloproteinases in the microenvironment of spontaneous and experimental melanoma metastases reflects the requirements for tumor formation. *Cancer Res.*, **63** (23), 8221–8225.
84. Partridge, J.J., Madsen, M.A., Ardi, V.C., Papagiannakopoulos, T., Kupriyanova, T.A., Quigley, J.P., and Deryugina, E.I. (2007) Functional analysis of matrix metalloproteinases and tissue inhibitors of metalloproteinases differentially expressed by variants of human HT-1080 fibrosarcoma exhibiting high and low levels of intravasation and metastasis. *J. Biol. Chem.*, **282** (49), 35964–35977.
85. Liu, L., Wu, J., Ying, Z., Chen, B., Han, A., Liang, Y., Song, L., Yuan, J., Li, J., and Li, M. (2010) Astrocyte elevated gene-1 upregulates matrix metalloproteinase-9 and induces human glioma invasion. *Cancer Res.*, **70** (9), 3750–3759.
86. Emdad, L., Sarkar, D., Lee, S.G., Su, Z.Z., Yoo, B.K., Dash, R., Yacoub, A., Fuller, C.E., Shah, K., Dent, P., Bruce, J.N., and Fisher, P.B. (2010) Astrocyte elevated gene-1: a novel target for human glioma therapy. *Mol. Cancer Ther.*, **9** (1), 79–88.

87. Jensen, S.A., Vainer, B., Bartels, A., Brunner, N., and Sorensen, J.B. (2010) Expression of matrix metalloproteinase 9 (MMP-9) and tissue inhibitor of metalloproteinases 1 (TIMP-1) by colorectal cancer cells and adjacent stroma cells--associations with histopathology and patients outcome. *Eur. J. Cancer*, **46** (18), 3233–3242.
88. Del Casar, J.M., Gonzalez, L.O., Alvarez, E., Junquera, S., Marin, L., Gonzalez, L., Bongera, M., Vazquez, J., and Vizoso, F.J. (2009) Comparative analysis and clinical value of the expression of metalloproteases and their inhibitors by intratumor stromal fibroblasts and those at the invasive front of breast carcinomas. *Breast Cancer Res. Treat.*, **116** (1), 39–52.
89. Gonzalez, L.O., Gonzalez-Reyes, S., Marin, L., Gonzalez, L., Gonzalez, J.M., Lamelas, M.L., Merino, A.M., Rodriguez, E., Pidal, I., del Casar, J.M., Andicoechea, A., and Vizoso, F. (2010) Comparative analysis and clinical value of the expression of metalloproteases and their inhibitors by intratumour stromal mononuclear inflammatory cells and those at the invasive front of breast carcinomas. *Histopathology*, **57** (6), 862–876.
90. Kuang, D.M., Zhao, Q., Wu, Y., Peng, C., Wang, J., Xu, Z., Yin, X.Y., and Zheng, L. (2011) Peritumoral neutrophils link inflammatory response to disease progression by fostering angiogenesis in hepatocellular carcinoma. *J. Hepatol.*, **54**, 948–955.
91. Coussens, L.M., Raymond, W.W., Bergers, G., Laig-Webster, M., Behrendtsen, O., Werb, Z., Caughey, G.H., and Hanahan, D. (1999) Inflammatory mast cells up-regulate angiogenesis during squamous epithelial carcinogenesis. *Genes Dev.*, **13** (11), 1382–1397.
92. Coussens, L.M., Tinkle, C.L., Hanahan, D., and Werb, Z. (2000) MMP-9 supplied by bone marrow-derived cells contributes to skin carcinogenesis. *Cell*, **103** (3), 481–490.
93. Hanahan, D. (1985) Heritable formation of pancreatic beta-cell tumours in transgenic mice expressing recombinant insulin/simian virus 40 oncogenes. *Nature*, **315** (6015), 115–122.
94. Bergers, G., Brekken, R., McMahon, G., Vu, T.H., Itoh, T., Tamaki, K., Tanzawa, K., Thorpe, P., Itohara, S., Werb, Z., and Hanahan, D. (2000) Matrix metalloproteinase-9 triggers the angiogenic switch during carcinogenesis. *Nat. Cell Biol.*, **2** (10), 737–744.
95. Giraudo, E., Inoue, M., and Hanahan, D. (2004) An amino-bisphosphonate targets MMP-9-expressing macrophages and angiogenesis to impair cervical carcinogenesis. *J. Clin. Invest.*, **114** (5), 623–633.
96. Almholt, K., Green, K.A., Juncker-Jensen, A., Nielsen, B.S., Lund, L.R., and Romer, J. (2007) Extracellular proteolysis in transgenic mouse models of breast cancer. *J. Mammary Gland Biol. Neoplasia*, **12** (1), 83–97.
97. Littlepage, L.E., Sternlicht, M.D., Rougier, N., Phillips, J., Gallo, E., Yu, Y., Williams, K., Brenot, A., Gordon, J.I., and Werb, Z. (2010) Matrix metalloproteinases contribute distinct roles in neuroendocrine prostate carcinogenesis, metastasis, and angiogenesis progression. *Cancer Res.*, **70** (6), 2224–2234.
98. Ha, H.Y., Moon, H.B., Nam, M.S., Lee, J.W., Ryoo, Z.Y., Lee, T.H., Lee, K.K., So, B.J., Sato, H., Seiki, M., and Yu, D.Y. (2001) Overexpression of membrane-type matrix metalloproteinase-1 gene induces mammary gland abnormalities and adenocarcinoma in transgenic mice. *Cancer Res.*, **61** (3), 984–990.
99. Soulie, P., Carrozzino, F., Pepper, M.S., Strongin, A.Y., Poupon, M.F., and Montesano, R. (2005) Membrane-type-1 matrix metalloproteinase confers tumorigenicity on nonmalignant epithelial cells. *Oncogene*, **24** (10), 1689–1697.
100. Hofmann, U.B., Westphal, J.R., Zendman, A.J., Becker, J.C., Ruiter, D.J., and van Muijen, G.N. (2000) Expression and activation of matrix metalloproteinase-2 (MMP-2) and its

co-localization with membrane-type 1 matrix metalloproteinase (MT1-MMP) correlate with melanoma progression. *J. Pathol.*, **191** (3), 245–256.
101. Lakka, S.S., Gondi, C.S., Yanamandra, N., Dinh, D.H., Olivero, W.C., Gujrati, M., and Rao, J.S. (2003) Synergistic down-regulation of urokinase plasminogen activator receptor and matrix metalloproteinase-9 in SNB19 glioblastoma cells efficiently inhibits glioma cell invasion, angiogenesis, and tumor growth. *Cancer Res.*, **63** (10), 2454–2461.
102. Lakka, S.S., Gondi, C.S., Dinh, D.H., Olivero, W.C., Gujrati, M., Rao, V.H., Sioka, C., and Rao, J.S. (2005) Specific interference of urokinase-type plasminogen activator receptor and matrix metalloproteinase-9 gene expression induced by double-stranded RNA results in decreased invasion, tumor growth, and angiogenesis in gliomas. *J. Biol. Chem.*, **280** (23), 21882–21892.
103. Gondi, C.S., Lakka, S.S., Dinh, D.H., Olivero, W.C., Gujrati, M., and Rao, J.S. (2004) Downregulation of uPA, uPAR and MMP-9 using small, interfering, hairpin RNA (siRNA) inhibits glioma cell invasion, angiogenesis and tumor growth. *Neuron Glia Biol.*, **1** (2), 165–176.
104. Du, R., Petritsch, C., Lu, K., Liu, P., Haller, A., Ganss, R., Song, H., Vandenberg, S., and Bergers, G. (2008) Matrix metalloproteinase-2 regulates vascular patterning and growth affecting tumor cell survival and invasion in GBM. *Neuro Oncol.*, **10** (3), 254–264.
105. Szabova, L., Chrysovergis, K., Yamada, S.S., and Holmbeck, K. (2008) MT1-MMP is required for efficient tumor dissemination in experimental metastatic disease. *Oncogene*, **27** (23), 3274–3281.
106. Taniwaki, K., Fukamachi, H., Komori, K., Ohtake, Y., Nonaka, T., Sakamoto, T., Shiomi, T., Okada, Y., Itoh, T., Itohara, S., Seiki, M., and Yana, I. (2007) Stroma-derived matrix metalloproteinase (MMP)-2 promotes membrane type 1-MMP-dependent tumor growth in mice. *Cancer Res.*, **67** (9), 4311–4319.
107. Nabha, S.M., Bonfil, R.D., Yamamoto, H.A., Belizi, A., Wiesner, C., Dong, Z., and Cher, M.L. (2006) Host matrix metalloproteinase-9 contributes to tumor vascularization without affecting tumor growth in a model of prostate cancer bone metastasis. *Clin. Exp. Metastasis*, **23** (7–8), 335–344.
108. Page-McCaw, A., Ewald, A.J., and Werb, Z. (2007) Matrix metalloproteinases and the regulation of tissue remodelling. *Nat. Rev. Mol. Cell Biol.*, **8** (3), 221–233.
109. Levental, K.R., Yu, H., Kass, L., Lakins, J.N., Egeblad, M., Erler, J.T., Fong, S.F., Csiszar, K., Giaccia, A., Weninger, W., Yamauchi, M., Gasser, D.L., and Weaver, V.M. (2009) Matrix crosslinking forces tumor progression by enhancing integrin signaling. *Cell*, **139** (5), 891–906.
110. Benbow, U., Schoenermark, M.P., Mitchell, T.I., Rutter, J.L., Shimokawa, K., Nagase, H., and Brinckerhoff, C.E. (1999) A novel host/tumor cell interaction activates matrix metalloproteinase 1 and mediates invasion through type I collagen. *J. Biol. Chem.*, **274** (36), 25371–25378.
111. Sarkar, S., Nuttall, R.K., Liu, S., Edwards, D.R., and Yong, V.W. (2006) Tenascin-C stimulates glioma cell invasion through matrix metalloproteinase-12. *Cancer Res.*, **66** (24), 11771–11780.
112. Zhang, W., Matrisian, L.M., Holmbeck, K., Vick, C.C., and Rosenthal, E.L. (2006) Fibroblast-derived MT1-MMP promotes tumor progression in vitro and in vivo. *BMC Cancer*, **6**, 52.
113. Fisher, K.E., Pop, A., Koh, W., Anthis, N.J., Saunders, W.B., and Davis, G.E. (2006) Tumor cell invasion of collagen matrices requires coordinate lipid agonist-induced G-protein and membrane-type matrix metalloproteinase-1-dependent signaling. *Mol. Cancer*, **5**, 69.
114. Barbolina, M.V., Adley, B.P., Ariztia, E.V., Liu, Y., and Stack, M.S. (2007) Microenvironmental regulation of membrane type 1 matrix metalloproteinase activity in ovarian carcinoma

cells via collagen-induced EGR1 expression. *J. Biol. Chem.*, **282** (7), 4924–4931.

115. Wu, H., Byrne, M.H., Stacey, A., Goldring, M.B., Birkhead, J.R., Jaenisch, R., and Krane, S.M. (1990) Generation of collagenase-resistant collagen by site-directed mutagenesis of murine pro alpha 1(I) collagen gene. *Proc. Natl. Acad. Sci. U.S.A.*, **87** (15), 5888–5892.
116. Woenne, E.C., Lederle, W., Zwick, S., Palmowski, M., Krell, H., Semmler, W., Mueller, M.M., and Kiessling, F. (2010) MMP inhibition blocks fibroblast-dependent skin cancer invasion, reduces vascularization and alters VEGF-A and PDGF-BB expression. *Anticancer Res.*, **30** (3), 703–711.
117. Wolf, K., Wu, Y.I., Liu, Y., Geiger, J., Tam, E., Overall, C., Stack, M.S., and Friedl, P. (2007) Multi-step pericellular proteolysis controls the transition from individual to collective cancer cell invasion. *Nat. Cell Biol.*, **9** (8), 893–904.
118. Deryugina, E.I., Soroceanu, L., and Strongin, A.Y. (2002) Up-regulation of vascular endothelial growth factor by membrane-type 1 matrix metalloproteinase stimulates human glioma xenograft growth and angiogenesis. *Cancer Res.*, **62** (2), 580–588.
119. Conn, E.M., Botkjaer, K.A., Kupriyanova, T.A., Andreasen, P.A., Deryugina, E.I., and Quigley, J.P. (2009) Comparative analysis of metastasis variants derived from human prostate carcinoma cells. Roles in intravasation of VEGF-mediated angiogenesis and uPA-mediated invasion. *Am. J. Pathol.*, **175** (4), 1638–1652.
120. Bekes, E.M., Schweighofer, B., Kupriyanova, T.A., Ardi, V.C., Zajac, E., Quigley, J.P., and Deryugina, E.I. (2011) Tumor-recruited neutrophils and neutrophil TIMP-free MMP-9 regulate coordinately the levels of tumor angiogenesis and efficiency of malignant cell intravasation. *Am. J. Pathology*, **179** (3), 1455–1470.
121. Deryugina, E.I. and Quigley, J.P. (2008) Chick embryo chorioallantoic membrane model systems to study and visualize human tumor cell metastasis. *Histochem. Cell Biol.*, **130** (6), 1119–1130.
122. Zaman, M.H., Trapani, L.M., Sieminski, A.L., Mackellar, D., Gong, H., Kamm, R.D., Wells, A., Lauffenburger, D.A., and Matsudaira, P. (2006) Migration of tumor cells in 3D matrices is governed by matrix stiffness along with cell-matrix adhesion and proteolysis. *Proc. Natl. Acad. Sci. U.S.A.*, **103** (29), 10889–10894.
123. Makareeva, E., Han, S., Vera, J.C., Sackett, D.L., Holmbeck, K., Phillips, C.L., Visse, R., Nagase, H., and Leikin, S. (2010) Carcinomas contain a matrix metalloproteinase-resistant isoform of type I collagen exerting selective support to invasion. *Cancer Res.*, **70** (11), 4366–4374.
124. Condeelis, J. and Segall, J.E. (2003) Intravital imaging of cell movement in tumours. *Nat. Rev. Cancer*, **3** (12), 921–930.
125. Condeelis, J., Singer, R.H., and Segall, J.E. (2005) The great escape: when cancer cells hijack the genes for chemotaxis and motility. *Annu. Rev. Cell Dev. Biol.*, **21**, 695–718.
126. Wang, W., Goswami, S., Sahai, E., Wyckoff, J.B., Segall, J.E., and Condeelis, J.S. (2005) Tumor cells caught in the act of invading: their strategy for enhanced cell motility. *Trends Cell Biol.*, **15** (3), 138–145.
127. Carragher, N.O. (2009) Profiling distinct mechanisms of tumour invasion for drug discovery: imaging adhesion, signalling and matrix turnover. *Clin. Exp. Metastasis*, **26** (4), 381–397.
128. Provenzano, P.P., Eliceiri, K.W., and Keely, P.J. (2009) Multiphoton microscopy and fluorescence lifetime imaging microscopy (FLIM) to monitor metastasis and the tumor microenvironment. *Clin. Exp. Metastasis*, **26** (4), 357–370.
129. Sameni, M., Cavallo-Medved, D., Dosescu, J., Jedeszko, C., Moin, K., Mullins, S.R., Olive, M.B., Rudy,

D., and Sloane, B.F. (2009) Imaging and quantifying the dynamics of tumor-associated proteolysis. *Clin. Exp. Metastasis*, **26** (4), 299–309.

130. Scherer, R.L., McIntyre, J.O., and Matrisian, L.M. (2008) Imaging matrix metalloproteinases in cancer. *Cancer Metastasis Rev.*, **27** (4), 679–690.

131. Wolf, K. and Friedl, P. (2009) Mapping proteolytic cancer cell-extracellular matrix interfaces. *Clin. Exp. Metastasis*, **26** (4), 289–298.

132. Friedl, P. and Wolf, K. (2008) Tube travel: the role of proteases in individual and collective cancer cell invasion. *Cancer Res.*, **68** (18), 7247–7249.

133. Friedl, P. and Wolf, K. (2010) Plasticity of cell migration: a multiscale tuning model. *J. Cell Biol.*, **188** (1), 11–19.

134. Gaggioli, C., Hooper, S., Hidalgo-Carcedo, C., Grosse, R., Marshall, J.F., Harrington, K., and Sahai, E. (2007) Fibroblast-led collective invasion of carcinoma cells with differing roles for RhoGTPases in leading and following cells. *Nat. Cell Biol.*, **9** (12), 1392–1400.

135. Tate, M.C. and Aghi, M.K. (2009) Biology of angiogenesis and invasion in glioma. *Neurotherapeutics*, **6** (3), 447–457.

136. Hanahan, D. and Folkman, J. (1996) Patterns and emerging mechanisms of the angiogenic switch during tumorigenesis. *Cell*, **86** (3), 353–364.

137. Weidner, N., Semple, J.P., Welch, W.R., and Folkman, J. (1991) Tumor angiogenesis and metastasis--correlation in invasive breast carcinoma. *N. Engl. J. Med.*, **324** (1), 1–8.

138. Benoy, I.H., Salgado, R., Elst, H., Van Dam, P., Weyler, J., Van Marck, E., Scharpe, S., Vermeulen, P.B., and Dirix, L.Y. (2005) Relative microvessel area of the primary tumour, and not lymph node status, predicts the presence of bone marrow micrometastases detected by reverse transcriptase polymerase chain reaction in patients with clinically non-metastatic breast cancer. *Breast Cancer Res.*, **7** (2), R210–R219.

139. Weidner, N., Carroll, P.R., Flax, J., Blumenfeld, W., and Folkman, J. (1993) Tumor angiogenesis correlates with metastasis in invasive prostate carcinoma. *Am. J. Pathol.*, **143** (2), 401–409.

140. Yano, H., Kinuta, M., Tateishi, H., Nakano, Y., Matsui, S., Monden, T., Okamura, J., Sakai, M., and Okamoto, S. (1999) Mast cell infiltration around gastric cancer cells correlates with tumor angiogenesis and metastasis. *Gastric Cancer*, **2** (1), 26–32.

141. Nozawa, H., Chiu, C., and Hanahan, D. (2006) Infiltrating neutrophils mediate the initial angiogenic switch in a mouse model of multistage carcinogenesis. *Proc. Natl. Acad. Sci. U.S.A.*, **103** (33), 12493–12498.

142. Huang, S., Van Arsdall, M., Tedjarati, S., McCarty, M., Wu, W., Langley, R., and Fidler, I.J. (2002) Contributions of stromal metalloproteinase-9 to angiogenesis and growth of human ovarian carcinoma in mice. *J. Natl. Cancer Inst.*, **94** (15), 1134–1142.

143. Yang, L., DeBusk, L.M., Fukuda, K., Fingleton, B., Green-Jarvis, B., Shyr, Y., Matrisian, L.M., Carbone, D.P., and Lin, P.C. (2004) Expansion of myeloid immune suppressor Gr+CD11b+ cells in tumor-bearing host directly promotes tumor angiogenesis. *Cancer Cell*, **6** (4), 409–421.

144. Chantrain, C.F., Shimada, H., Jodele, S., Groshen, S., Ye, W., Shalinsky, D.R., Werb, Z., Coussens, L.M., and DeClerck, Y.A. (2004) Stromal matrix metalloproteinase-9 regulates the vascular architecture in neuroblastoma by promoting pericyte recruitment. *Cancer Res.*, **64** (5), 1675–1686.

145. Jodele, S., Chantrain, C.F., Blavier, L., Lutzko, C., Crooks, G.M., Shimada, H., Coussens, L.M., and Declerck, Y.A. (2005) The contribution of bone marrow-derived cells to the tumor vasculature in neuroblastoma is matrix metalloproteinase-9 dependent. *Cancer Res.*, **65** (8), 3200–3208.

146. Pahler, J.C., Tazzyman, S., Erez, N., Chen, Y.Y., Murdoch, C., Nozawa, H., Lewis, C.E., and Hanahan, D. (2008)

Plasticity in tumor-promoting inflammation: impairment of macrophage recruitment evokes a compensatory neutrophil response. *Neoplasia*, **10** (4), 329–340.

147. Xie, B., Laouar, A., and Huberman, E. (1998) Fibronectin-mediated cell adhesion is required for induction of 92-kDa type IV collagenase/gelatinase (MMP-9) gene expression during macrophage differentiation. The signaling role of protein kinase C-beta. *J. Biol. Chem.*, **273** (19), 11576–11582.
148. Witko-Sarsat, V., Rieu, P., Descamps-Latscha, B., Lesavre, P., and Halbwachs-Mecarelli, L. (2000) Neutrophils: molecules, functions and pathophysiological aspects. *Lab Invest.*, **80** (5), 617–653.
149. Opdenakker, G., Van den Steen, P.E., Dubois, B., Nelissen, I., Van Coillie, E., Masure, S., Proost, P., and Van Damme, J. (2001) Gelatinase B functions as regulator and effector in leukocyte biology. *J. Leukoc. Biol.*, **69** (6), 851–859.
150. Ardi, V.C., Kupriyanova, T.A., Deryugina, E.I., and Quigley, J.P. (2007) Human neutrophils uniquely release TIMP-free MMP-9 to provide a potent catalytic stimulator of angiogenesis. *Proc. Natl. Acad. Sci. U.S.A.*, **104** (51), 20262–20267.
151. Ardi, V.C., Van den Steen, P.E., Opdenakker, G., Schweighofer, B., Deryugina, E.I., and Quigley, J.P. (2009) Neutrophil MMP-9 proenzyme, unencumbered by TIMP-1, undergoes efficient activation in vivo and catalytically induces angiogenesis via a basic fibroblast growth factor (FGF-2)/FGFR-2 pathway. *J. Biol. Chem.*, **284** (38), 25854–25866.
152. Van den Steen, P.E., Proost, P., Wuyts, A., Van Damme, J., and Opdenakker, G. (2000) Neutrophil gelatinase B potentiates interleukin-8 tenfold by aminoterminal processing, whereas it degrades CTAP-III, PF-4, and GRO-alpha and leaves RANTES and MCP-2 intact. *Blood*, **96** (8), 2673–2681.
153. Van Den Steen, P.E., Wuyts, A., Husson, S.J., Proost, P., Van Damme, J., and Opdenakker, G. (2003) Gelatinase B/MMP-9 and neutrophil collagenase/MMP-8 process the chemokines human GCP-2/CXCL6, ENA-78/CXCL5 and mouse GCP-2/LIX and modulate their physiological activities. *Eur. J. Biochem.*, **270** (18), 3739–3749.
154. Opdenakker, G. and Van Damme, J. (2004) The countercurrent principle in invasion and metastasis of cancer cells. Recent insights on the roles of chemokines. *Int. J. Dev. Biol.*, **48** (5–6), 519–527.
155. Gijsbers, K., Gouwy, M., Struyf, S., Wuyts, A., Proost, P., Opdenakker, G., Penninckx, F., Ectors, N., Geboes, K., and Van Damme, J. (2005) GCP-2/CXCL6 synergizes with other endothelial cell-derived chemokines in neutrophil mobilization and is associated with angiogenesis in gastrointestinal tumors. *Exp. Cell Res.*, **303** (2), 331–342.
156. Deryugina, E.I., Zijlstra, A., Partridge, J.J., Kupriyanova, T.A., Madsen, M.A., Papagiannakopoulos, T., and Quigley, J.P. (2005) Unexpected effect of matrix metalloproteinase down-regulation on vascular intravasation and metastasis of human fibrosarcoma cells selected in vivo for high rates of dissemination. *Cancer Res.*, **65** (23), 10959–10969.
157. Deryugina, E.I., Ratnikov, B.I., Postnova, T.I., Rozanov, D.V., and Strongin, A.Y. (2002) Processing of integrin alpha(v) subunit by membrane type 1 matrix metalloproteinase stimulates migration of breast carcinoma cells on vitronectin and enhances tyrosine phosphorylation of focal adhesion kinase. *J. Biol. Chem.*, **277** (12), 9749–9756.
158. Eisenach, P.A., Roghi, C., Fogarasi, M., Murphy, G., and English, W.R. (2010) MT1-MMP regulates VEGF-A expression through a complex with VEGFR-2 and Src. *J. Cell Sci.*, **123** (Pt 23), 4182–4193.
159. Oh, J., Takahashi, R., Kondo, S., Mizoguchi, A., Adachi, E., Sasahara, R.M., Nishimura, S., Imamura, Y., Kitayama, H., Alexander, D.B., Ide, C., Horan, T.P., Arakawa, T., Yoshida, H.,

Nishikawa, S., Itoh, Y., Seiki, M., Itohara, S., Takahashi, C., and Noda, M. (2001) The membrane-anchored MMP inhibitor RECK is a key regulator of extracellular matrix integrity and angiogenesis. *Cell*, **107** (6), 789–800.

160. Bendrik, C., Robertson, J., Gauldie, J., and Dabrosin, C. (2008) Gene transfer of matrix metalloproteinase-9 induces tumor regression of breast cancer in vivo. *Cancer Res.*, **68** (9), 3405–3412.

161. Jost, M., Folgueras, A.R., Frerart, F., Pendas, A.M., Blacher, S., Houard, X., Berndt, S., Munaut, C., Cataldo, D., Alvarez, J., Melen-Lamalle, L., Foidart, J.M., Lopez-Otin, C., and Noel, A. (2006) Earlier onset of tumoral angiogenesis in matrix metalloproteinase-19-deficient mice. *Cancer Res.*, **66** (10), 5234–5241.

162. Coussens, L.M., Fingleton, B., and Matrisian, L.M. (2002) Matrix metalloproteinase inhibitors and cancer: trials and tribulations. *Science*, **295** (5564), 2387–2392.

163. Fukumura, D. and Jain, R.K. (2007) Tumor microvasculature and microenvironment: targets for anti-angiogenesis and normalization. *Microvasc. Res.*, **74** (2–3), 72–84.

164. Tong, R.T., Boucher, Y., Kozin, S.V., Winkler, F., Hicklin, D.J., and Jain, R.K. (2004) Vascular normalization by vascular endothelial growth factor receptor 2 blockade induces a pressure gradient across the vasculature and improves drug penetration in tumors. *Cancer Res.*, **64** (11), 3731–3736.

165. Mancuso, M.R., Davis, R., Norberg, S.M., O'Brien, S., Sennino, B., Nakahara, T., Yao, V.J., Inai, T., Brooks, P., Freimark, B., Shalinsky, D.R., Hu-Lowe, D.D., and McDonald, D.M. (2006) Rapid vascular regrowth in tumors after reversal of VEGF inhibition. *J. Clin. Invest.*, **116** (10), 2610–2621.

166. Shojaei, F., Wu, X., Malik, A.K., Zhong, C., Baldwin, M.E., Schanz, S., Fuh, G., Gerber, H.P., and Ferrara, N. (2007) Tumor refractoriness to anti-VEGF treatment is mediated by CD11b+Gr1+ myeloid cells. *Nat. Biotechnol.*, **25** (8), 911–920.

167. Shojaei, F. and Ferrara, N. (2008) Refractoriness to antivascular endothelial growth factor treatment: role of myeloid cells. *Cancer Res.*, **68** (14), 5501–5504.

168. Claes, A., Wesseling, P., Jeuken, J., Maass, C., Heerschap, A., and Leenders, W.P. (2008) Antiangiogenic compounds interfere with chemotherapy of brain tumors due to vessel normalization. *Mol. Cancer Ther.*, **7** (1), 71–78.

169. Ebos, J.M., Lee, C.R., Cruz-Munoz, W., Bjarnason, G.A., Christensen, J.G., and Kerbel, R.S. (2009) Accelerated metastasis after short-term treatment with a potent inhibitor of tumor angiogenesis. *Cancer Cell*, **15** (3), 232–239.

170. Paez-Ribes, M., Allen, E., Hudock, J., Takeda, T., Okuyama, H., Vinals, F., Inoue, M., Bergers, G., Hanahan, D., and Casanovas, O. (2009) Antiangiogenic therapy elicits malignant progression of tumors to increased local invasion and distant metastasis. *Cancer Cell*, **15** (3), 220–231.

171. Bergers, G. and Hanahan, D. (2008) Modes of resistance to anti-angiogenic therapy. *Nat. Rev. Cancer*, **8** (8), 592–603.

172. Winkler, F., Kozin, S.V., Tong, R.T., Chae, S.S., Booth, M.F., Garkavtsev, I., Xu, L., Hicklin, D.J., Fukumura, D., di Tomaso, E., Munn, L.L., and Jain, R.K. (2004) Kinetics of vascular normalization by VEGFR2 blockade governs brain tumor response to radiation: role of oxygenation, angiopoietin-1, and matrix metalloproteinases. *Cancer Cell*, **6** (6), 553–563.

173. Westenend, P.J., Meurs, C.J., and Damhuis, R.A. (2005) Tumour size and vascular invasion predict distant metastasis in stage I breast cancer. Grade distinguishes early and late metastasis. *J. Clin. Pathol.*, **58** (2), 196–201.

174. Wyckoff, J.B., Jones, J.G., Condeelis, J.S., and Segall, J.E. (2000) A critical step in metastasis: in vivo analysis of intravasation at the primary tumor. *Cancer Res.*, **60** (9), 2504–2511.

175. Ossowski, L. (1988) In vivo invasion of modified chorioallantoic membrane by tumor cells: the role of cell surface-bound urokinase. *J. Cell Biol.*, **107** (6, Pt 1), 2437–2445.

176. Kim, J., Yu, W., Kovalski, K., and Ossowski, L. (1998) Requirement for specific proteases in cancer cell intravasation as revealed by a novel semiquantitative PCR-based assay. *Cell*, **94** (3), 353–362.

177. Kedrin, D., Gligorijevic, B., Wyckoff, J., Verkhusha, V.V., Condeelis, J., Segall, J.E., and van Rheenen, J. (2008) Intravital imaging of metastatic behavior through a mammary imaging window. *Nat. Methods*, **5** (12), 1019–1021.

178. Brandt, B., Heyder, C., Gloria-Maercker, E., Hatzmann, W., Rotger, A., Kemming, D., Zanker, K.S., Entschladen, F., and Dittmar, T. (2005) 3D-extravasation model -- selection of highly motile and metastatic cancer cells. *Semin. Cancer Biol.*, **15** (5), 387–395.

179. Huh, S.J., Liang, S., Sharma, A., Dong, C., and Robertson, G.P. (2010) Transiently entrapped circulating tumor cells interact with neutrophils to facilitate lung metastasis development. *Cancer Res.*, **70** (14), 6071–6082.

180. Huang, Y., Song, N., Ding, Y., Yuan, S., Li, X., Cai, H., Shi, H., and Luo, Y. (2009) Pulmonary vascular destabilization in the premetastatic phase facilitates lung metastasis. *Cancer Res.*, **69** (19), 7529–7537.

181. Gupta, G.P., Nguyen, D.X., Chiang, A.C., Bos, P.D., Kim, J.Y., Nadal, C., Gomis, R.R., Manova-Todorova, K., and Massague, J. (2007) Mediators of vascular remodelling co-opted for sequential steps in lung metastasis. *Nature*, **446** (7137), 765–770.

182. Vande Broek, I., Vanderkerken, K., Van Camp, B., and Van Riet, I. (2008) Extravasation and homing mechanisms in multiple myeloma. *Clin. Exp. Metastasis*, **25** (4), 325–334.

183. Zhang, Q., Yang, M., Shen, J., Gerhold, L.M., Hoffman, R.M., and Xing, H.R. (2010) The role of the intravascular microenvironment in spontaneous metastasis development. *Int. J. Cancer*, **126** (11), 2534–2541.

184. Yang, M., Baranov, E., Li, X.M., Wang, J.W., Jiang, P., Li, L., Moossa, A.R., Penman, S., and Hoffman, R.M. (2001) Whole-body and intravital optical imaging of angiogenesis in orthotopically implanted tumors. *Proc. Natl. Acad. Sci. U.S.A.*, **98** (5), 2616–2621.

185. Chambers, A.F., Groom, A.C., and MacDonald, I.C. (2002) Dissemination and growth of cancer cells in metastatic sites. *Nat. Rev. Cancer*, **2** (8), 563–572.

186. Yamauchi, K., Yang, M., Jiang, P., Yamamoto, N., Xu, M., Amoh, Y., Tsuji, K., Bouvet, M., Tsuchiya, H., Tomita, K., Moossa, A.R., and Hoffman, R.M. (2005) Real-time in vivo dual-color imaging of intracapillary cancer cell and nucleus deformation and migration. *Cancer Res.*, **65** (10), 4246–4252.

187. Tsuji, K., Yamauchi, K., Yang, M., Jiang, P., Bouvet, M., Endo, H., Kanai, Y., Yamashita, K., Moossa, A.R., and Hoffman, R.M. (2006) Dual-color imaging of nuclear-cytoplasmic dynamics, viability, and proliferation of cancer cells in the portal vein area. *Cancer Res.*, **66** (1), 303–306.

188. Deryugina, E.I., Conn, E.M., Wortmann, A., Partridge, J.J., Kupriyanova, T.A., Ardi, V.C., Hooper, J.D., and Quigley, J.P. (2009) Functional role of cell surface CUB domain-containing protein 1 in tumor cell dissemination. *Mol. Cancer Res.*, **7** (8), 1197–1211.

189. Roth, J.M., Caunt, M., Cretu, A., Akalu, A., Policarpio, D., Li, X., Gagne, P., Formenti, S., and Brooks, P.C. (2006) Inhibition of experimental metastasis by targeting the HUIV26 cryptic epitope in collagen. *Am. J. Pathol.*, **168** (5), 1576–1586.

190. Koop, S., Khokha, R., Schmidt, E.E., MacDonald, I.C., Morris, V.L., Chambers, A.F., and Groom, A.C. (1994) Overexpression of metalloproteinase inhibitor in B16F10 cells does not affect extravasation but reduces tumor growth. *Cancer Res.*, **54** (17), 4791–4797.

191. Acuff, H.B., Carter, K.J., Fingleton, B., Gorden, D.L., and Matrisian, L.M. (2006) Matrix metalloproteinase-9 from bone marrow-derived cells contributes to survival but not growth of tumor cells in the lung microenvironment. *Cancer Res.*, **66** (1), 259–266.
192. Itoh, T., Tanioka, M., Matsuda, H., Nishimoto, H., Yoshioka, T., Suzuki, R., and Uehira, M. (1999) Experimental metastasis is suppressed in MMP-9-deficient mice. *Clin. Exp. Metastasis*, **17** (2), 177–181.
193. Lorger, M. and Felding-Habermann, B. (2010) Capturing changes in the brain microenvironment during initial steps of breast cancer brain metastasis. *Am. J. Pathol.*, **176** (6), 2958–2971.
194. Wels, J., Kaplan, R.N., Rafii, S., and Lyden, D. (2008) Migratory neighbors and distant invaders: tumor-associated niche cells. *Genes Dev.*, **22** (5), 559–574.
195. Nguyen, D.X., Bos, P.D., and Massague, J. (2009) Metastasis: from dissemination to organ-specific colonization. *Nat. Rev. Cancer*, **9** (4), 274–284.
196. Nguyen, D.X. and Massague, J. (2007) Genetic determinants of cancer metastasis. *Nat. Rev. Genet.*, **8** (5), 341–352.
197. Kang, Y., Siegel, P.M., Shu, W., Drobnjak, M., Kakonen, S.M., Cordon-Cardo, C., Guise, T.A., and Massague, J. (2003) A multigenic program mediating breast cancer metastasis to bone. *Cancer Cell*, **3** (6), 537–549.
198. Minn, A.J., Kang, Y., Serganova, I., Gupta, G.P., Giri, D.D., Doubrovin, M., Ponomarev, V., Gerald, W.L., Blasberg, R., and Massague, J. (2005) Distinct organ-specific metastatic potential of individual breast cancer cells and primary tumors. *J. Clin. Invest.*, **115** (1), 44–55.
199. Minn, A.J., Gupta, G.P., Siegel, P.M., Bos, P.D., Shu, W., Giri, D.D., Viale, A., Olshen, A.B., Gerald, W.L., and Massague, J. (2005) Genes that mediate breast cancer metastasis to lung. *Nature*, **436** (7050), 518–524.
200. Lu, X., Wang, Q., Hu, G., Van Poznak, C., Fleisher, M., Reiss, M., Massague, J., and Kang, Y. (2009) ADAMTS1 and MMP1 proteolytically engage EGF-like ligands in an osteolytic signaling cascade for bone metastasis. *Genes Dev.*, **23** (16), 1882–1894.
201. Bos, P.D., Zhang, X.H., Nadal, C., Shu, W., Gomis, R.R., Nguyen, D.X., Minn, A.J., van de Vijver, M.J., Gerald, W.L., Foekens, J.A., and Massague, J. (2009) Genes that mediate breast cancer metastasis to the brain. *Nature*, **459** (7249), 1005–1009.
202. Ellsworth, R.E., Seebach, J., Field, L.A., Heckman, C., Kane, J., Hooke, J.A., Love, B., and Shriver, C.D. (2009) A gene expression signature that defines breast cancer metastases. *Clin. Exp. Metastasis*, **26** (3), 205–213.
203. Psaila, B. and Lyden, D. (2009) The metastatic niche: adapting the foreign soil. *Nat. Rev. Cancer*, **9** (4), 285–293.
204. Kaplan, R.N., Riba, R.D., Zacharoulis, S., Bramley, A.H., Vincent, L., Costa, C., MacDonald, D.D., Jin, D.K., Shido, K., Kerns, S.A., Zhu, Z., Hicklin, D., Wu, Y., Port, J.L., Altorki, N., Port, E.R., Ruggero, D., Shmelkov, S.V., Jensen, K.K., Rafii, S., and Lyden, D. (2005) VEGFR1-positive haematopoietic bone marrow progenitors initiate the pre-metastatic niche. *Nature*, **438** (7069), 820–827.
205. Yan, H.H., Pickup, M., Pang, Y., Gorska, A.E., Li, Z., Chytil, A., Geng, Y., Gray, J.W., Moses, H.L., and Yang, L. (2010) Gr-1+CD11b+ myeloid cells tip the balance of immune protection to tumor promotion in the premetastatic lung. *Cancer Res.*, **70** (15), 6139–6149.
206. Erler, J.T., Bennewith, K.L., Nicolau, M., Dornhofer, N., Kong, C., Le, Q.T., Chi, J.T., Jeffrey, S.S., and Giaccia, A.J. (2006) Lysyl oxidase is essential for hypoxia-induced metastasis. *Nature*, **440** (7088), 1222–1226.
207. Erler, J.T., Bennewith, K.L., Cox, T.R., Lang, G., Bird, D., Koong, A., Le, Q.T., and Giaccia, A.J. (2009) Hypoxia-induced lysyl oxidase is a critical mediator of bone marrow cell recruitment to form the premetastatic niche. *Cancer Cell*, **15** (1), 35–44.

208. Hiratsuka, S., Nakamura, K., Iwai, S., Murakami, M., Itoh, T., Kijima, H., Shipley, J.M., Senior, R.M., and Shibuya, M. (2002) MMP9 induction by vascular endothelial growth factor receptor-1 is involved in lung-specific metastasis. *Cancer Cell*, **2** (4), 289–300.

209. Gilbert, L.A. and Hemann, M.T. (2010) DNA damage-mediated induction of a chemoresistant niche. *Cell*, **143** (3), 355–366.

210. Chaffer, C.L., Thompson, E.W., and Williams, E.D. (2007) Mesenchymal to epithelial transition in development and disease. *Cells Tissues Organs*, **185** (1–3), 7–19.

211. Polyak, K. and Weinberg, R.A. (2009) Transitions between epithelial and mesenchymal states: acquisition of malignant and stem cell traits. *Nat. Rev. Cancer*, **9** (4), 265–273.

212. Mendes, O., Kim, H.T., and Stoica, G. (2005) Expression of MMP2, MMP9 and MMP3 in breast cancer brain metastasis in a rat model. *Clin. Exp. Metastasis*, **22** (3), 237–246.

213. Saito, N., Hatori, T., Murata, N., Zhang, Z.A., Ishikawa, F., Nonaka, H., Iwabuchi, S., and Samejima, H. (2007) A double three-step theory of brain metastasis in mice: the role of the pia mater and matrix metalloproteinases. *Neuropathol. Appl. Neurobiol.*, **33** (3), 288–298.

214. Corey, E., Quinn, J.E., Bladou, F., Brown, L.G., Roudier, M.P., Brown, J.M., Buhler, K.R., and Vessella, R.L. (2002) Establishment and characterization of osseous prostate cancer models: intra-tibial injection of human prostate cancer cells. *Prostate*, **52** (1), 20–33.

215. Bonfil, R.D., Dong, Z., Trindade Filho, J.C., Sabbota, A., Osenkowski, P., Nabha, S., Yamamoto, H., Chinni, S.R., Zhao, H., Mobashery, S., Vessella, R.L., Fridman, R., and Cher, M.L. (2007) Prostate cancer-associated membrane type 1-matrix metalloproteinase: a pivotal role in bone response and intraosseous tumor growth. *Am. J. Pathol.*, **170** (6), 2100–2111.

216. Sabbota, A.L., Kim, H.R., Zhe, X., Fridman, R., Bonfil, R.D., and Cher, M.L. (2010) Shedding of RANKL by tumor-associated MT1-MMP activates Src-dependent prostate cancer cell migration. *Cancer Res.*, **70** (13), 5558–5566.

217. Lynch, C.C., Hikosaka, A., Acuff, H.B., Martin, M.D., Kawai, N., Singh, R.K., Vargo-Gogola, T.C., Begtrup, J.L., Peterson, T.E., Fingleton, B., Shirai, T., Matrisian, L.M., and Futakuchi, M. (2005) MMP-7 promotes prostate cancer-induced osteolysis via the solubilization of RANKL. *Cancer Cell*, **7** (5), 485–496.

218. Holzel, D., Eckel, R., Emeny, R.T., and Engel, J. (2010) Distant metastases do not metastasize. *Cancer Metastasis Rev.*, **29** (4), 737–750.

219. Talmadge, J.E. and Fidler, I.J. (2010) AACR centennial series: the biology of cancer metastasis: historical perspective. *Cancer Res.*, **70** (14), 5649–5669.

220. Zhang, X.H., Wang, Q., Gerald, W., Hudis, C.A., Norton, L., Smid, M., Foekens, J.A., and Massague, J. (2009) Latent bone metastasis in breast cancer tied to Src-dependent survival signals. *Cancer Cell*, **16** (1), 67–78.

9
Plasminogen Activators and Their Inhibitors in Cancer

Joerg Hendrik Leupold and Heike Allgayer

9.1
Introduction

Metastasis, the process by which cancer cells leave the primary tumor mass and disseminate to distant anatomical sites, is a serious clinical problem and is responsible for 90% of morbidity and mortality in patients with malignant tumors. Metastasis itself is a complex molecular cascade that involves local invasion, intravasation into blood vessels, transport through the blood stream, extravasation, and growth of secondary tumors [1]. The complex process of tumor invasion and metastasis requires a network of different proteases, which are the basis for several steps of the metastatic cascade through their proteolytic activity. Among them, the urokinase plasminogen activator (uPA) system plays a critical role in enabling cells to overcome anatomic barriers, intravasate, and form metastases, and in the past two decades our understanding of the role of the uPA system has improved considerably [2]. Furthermore, apart from its proteolytic activity, the uPA system plays a pivotal role in several other processes important to tumor progression including direct or indirect effects on various signaling pathways, angiogenesis, and programmed cell death [3, 4]. It comprises the serine protease uPA, its substrate plasmin, the serpin inhibitors (SERPINs) plasminogen activator inhibitor (PAI)-1 and PAI-2, and the urokinase plasminogen activator receptor (uPAR). uPA leads to the activation of plasminogen, further proteases and to the degradation of main components of the extracellular matrix such as fibrin, fibronectin, and laminin. This effect of uPA is potentiated by binding to its specific cell receptor uPAR, which has been shown to be overexpressed in diverse human tumors, such as gastric, colorectal, breast, and other cancers [5, 6]. In addition to uPA, the second activator of this system, called the tissue-type plasminogen activator (tPA), can also mediate the proteolytic conversion of plasminogen to its active form plasmin. However, although both tPA and uPA are expressed in tumor cells, uPA is primarily associated with tumor biology, while tPA is associated with fibrinolysis in blood vessels acting mainly as a thrombolytic agent [7]. The endogenous inhibitors of this system, PAI-1 and PAI-2, regulate activity of uPA and uPAR by either direct inhibition or affecting cell surface expression and internalization. Among the inhibitors, PAI-1

Matrix Proteases in Health and Disease, First Edition. Edited by Niels Behrendt.

is a major player in the pathogenesis of many vascular diseases as well as in cancer. Interestingly, the role of PAI-1 in cancers is rather different; on one hand, high tumor levels of the type 1 inhibitor actually promote tumor progression, whereas high levels of the type 2 inhibitor decrease tumor growth and metastasis [8]. In this background, this chapter summarizes the evidence for the role of the different components of the uPA system in mediating the malignant phenotype, thereby trying to explain some essential aspects of plasminogen activators and their inhibitors on a molecular level in cancer.

9.2
The Plasminogen Activator System

9.2.1
Molecular Characteristics and Physiological Functions of the u-PA System

Under normal physiological conditions, the uPA system is involved in various tissue remodeling processes such as wound healing, fibrinolysis, embryogenesis, angiogenesis, and cell migration. Alongside, it is strongly induced during leukocyte activation and differentiation, suggesting a role in inflammatory and immune responses [9–14]. The serine protease uPA is secreted as a 411 amino acid single-chain zymogen form (pro-uPA or single-chain urokinase plasminogen activator; scuPA) with a molecular weight of 55 kDa. This form of human uPA is the major form of the enzyme produced by cells, in tissues, and extracellular fluids and has minimal intrinsic activity [15]. The scuPA is proteolytically activated, resulting in a two-chain form of urokinase plasminogen activator (tcuPA) by the proximity of plasmin or kallikrein or when it binds as a single-chain molecule to its cellular receptor (uPAR) [16]. This two-chain derivative is also called high-molecular-weight urokinase plasminogen activator (HMW-uPA). HMW-uPA can be processed further, releasing an amino-terminal fragment (ATF) and a small region linked to a large C-terminal domain called low-molecular-weight urokinase plasminogen activator (LMW-uPA). LMW-uPA has a molecular weight of 32 kDa and implies the same enzymatic properties as the full length form but does not bind to the uPA receptor [17]. tPA is also released from cells as a single-chain form, and, compared to uPA, even as the single-chain form it has a measurable proteolytic activity, although it is about 50-fold lower than that of its two-chain counterpart [18]. Binding of uPA to its receptor initiates a proteolytic cascade that results in the conversion of the inactive zymogen plasminogen into the active serine protease plasmin [10]. The emerging activity of plasmin promotes the degradation of extracelluar matrix components including fibrin, fibronectin, proteoglycans and, as the main molecules in basement membranes, laminin, and collagen IV [19, 20], whereby proteolysis of the latter is indirect since it is mediated via further downstream proteases. Pro-uPAR, uPA, and ATF exhibit similar dissociation constants for the binding to uPAR and, even if proteolytically activated in the extracellular space, the receptor-mediated membrane localization of uPA is of great relevance to the proteolytic efficiency of the whole system. It has been shown by kinetic studies that receptor-bound uPA exhibits a slower dissociation of the uPA/plasminogen complex, caused by surface binding of plasminogen, thus resulting in an overall increase of catalytic efficiency [21]. Functionally, uPA consists of three distinct domains: the serine protease domain, the kringle domain, and the epidermal growth factor (EGF)-like domain, which is homologous to the EGF-like protein family [22]. The receptor-binding activity resides in the EGF-like domain, and recent studies report an intramolecular interaction between the kringle and the EGF-like domains, which stabilize the ligand–receptor complex [23, 24]. In contrast to uPA, tPA contains a serine protease domain, a fibrinectin type II finger domain, an EGF-like domain, and two kringle domains [18]. The uPAR is a member of the lymphocyte antigen 6 (Ly-6) superfamily of proteins and is characterized by three domains (DI, DII, and DIII) of three similar repeats approximately 90 residues each and is anchored to the cell membrane via a glycosyl-phosphatidylinositol (GPI) anchor. These domains are extensively modified by N-linked glycosylation [25, 26]. Even though the function of these glycosylations is poorly understood, studies have shown that they can enhance the affinity of uPAR for uPA [26]. Within the cell membrane, which is organized into various subdomains of clustered macromolecules, uPAR is concentrated in special intrusions. These domains include adhesive structures (cellular synapses, substrate adhesions, and cell–cell junctions) and membrane invaginations (clathrin-coated pits and caveolae), as well as less well-defined domains such as lipid rafts. Furthermore, such domains are organized by specialized scaffold proteins including the intramembranous caveolins, which stabilize lipid raft domains and are the principal components of caveolae membranes [27]. Similar to other GPI-anchored proteins, uPAR has a high mobility within the plasma membrane and associates with these cholesterol- and sphingolipid-rich lipid rafts, at which its location on the cell surface depends on the functional state of the cell [28]. In resting cells, uPAR is uniformly distributed at the surface, whereas in the migrating cell, clusters of uPAR form at the leading edge [29].

The proteolytic activity of the uPA system is mainly modulated by the serpin (serine protease inhibitor) family, notably the PAI-1 and PAI-2 for uPA and α_2-antiplasmin for plasmin [29, 30]. PAI-1 is the primary inhibitor and is secreted as a 46 kDa single-chain glycoprotein containing 402 amino acid residues. It forms a covalent 1 : 1 complex with either uPA or tPA, distorting the active site of the protease, preventing its release from the serpin, and thereby inhibiting the further conversion of plasminogen to plasmin [31]. Besides this direct interaction, the activity of the plasminogen activators is terminated by their clearance from the extracellular space by endocytosis and degradation. If the inhibitory complex of PAI-1 and uPA is formed while the protease is attached to the uPAR, it can be endocytosed after binding to the multiligand receptors α_2-macroglobulin receptor/low-density-lipoprotein-receptor-related protein (α_2-MR/LRP) to the urokinase receptor. In succession, uPAR recycles to the cellular surface while PAI-1/uPA complex is degraded [18]. Individual uPA/PAI-1 complexes can be alternatively endocytosed after binding to the very-low-density lipoprotein-receptor-related protein (VLDLr) and LRP [32]. These receptors belong to the low-density lipoprotein receptor (LDLR) family and mediate endocytosis of a variety of functionally and structurally unrelated ligands, including lipoprotein particles and various protease–inhibitor complexes [33]. In the absence of a protease, the active form of PAI-1 is not stable and is converted into its latent form, which no longer exhibits protease inhibitor activity. This process can be delayed by the binding of free PAI-1 to vitronectin to maintain PAI-1 in its active conformation [34]. As in PAI-1, PAI-2 forms an inhibitory complex with receptor-bound uPA, and although uPA/PAI-2 is cleared from the cell surface through interactions with both LRP and VLDLr, these are of lower affinity than that of uPA/PAI-I. The reason for this difference is the lack of a complete LDLR-binding motif in PAI-II [35]. In addition, PAI-2 does not bind directly to LRP; therefore, it is unable to induce cell migration through binding of this receptor. High PAI-2 levels also potentially compete with vitronectin-bound PAI-1 for uPAR binding, preventing the removal of PAI-1 from vitronectin, and therefore decreasing vitronectin-dependent cell migration [36].

9.2.2
Expression in Cancer

For at least the past 30 years, our understanding of the role of the uPA system has improved considerably, and there is abundant experimental and clinical evidence in the literature to support its role in cancer progression. The uPA system is expressed in many human solid tumors and in some leukemias and lymphomas [4]. Unfortunately, expression frequently indicates poor prognosis and quantity rather than quality of the individual components of the system contributes to the invasive phenotype of malignant cells. High uPA, uPAR, and PAI-1 levels have been shown to be adverse prognostic markers in breast, colorectal, esophageal, gastric, ovarian, prostate, renal, and endometrial cancers [7]. Importantly, expression of components of this system can occur in tumor cells themselves as well as from tumor-associated stromal cells, such as fibroblasts and macrophages. Initially, in the 1970s, the role of the uPA system was studied in cultured cells transformed by oncogenic viruses or established cancer cell lines [37, 38]. A few years later, the expression of uPA and uPAR was shown in stromal cells of colon adenocarcinomas. Since that time, stromal cell expression of components of the uPA system has been demonstrated by various groups in many types of human cancer and is now well accepted [39, 40]. Interestingly, the different types of human cancers differ in the expression of components of the uPA system by stromal and/or tumor cells. uPA, for example, is primarily expressed by fibroblasts in ductal breast cancer and colon cancer tissue [41, 42], whereas in skin squamous carcinomas uPA is largely expressed by the cancer cells themselves [43]. However, in prostate cancer, uPA is expressed in macrophages [44]. The expression of the receptor is found on the cancer cells and macrophages in colon cancer, primarily on macrophages in cells of ductal breast cancer, on cancer cells in the case of squamous cell carcinoma, and on macrophages and neutrophils in prostate cancer [44–47].

As mentioned initially, the quantity of expression plays a pivotal role in the implication of the uPA system in cancer development, migration, and metastasis. For the uPAR, numerous studies have shown that an overexpression of this protein in diverse human malignant tumors is a characteristic of the invasive phenotype, or even the malignant phenotype. Evidence implicating the uPA system in tumor invasion and metastasis arose from various *in vitro* and in *vivo models*, in particular, invasion assays (matrigel or laminin degradation assay), the chorioallantoic membrane (CAM) assay, and different mouse or rat models. Among other studies, it has been shown that overexpression of uPAR was able to increase the invasive ability of osteosarcoma cells on matrigel, and a more comprehensive study proved that overexpression of urokinase receptor in breast cancer cells results in increased tumor invasion, growth, and metastasis, using matrigel assays and additional injection into the mammary fat pad of syngeneic female Fischer rats [48, 49]. Vice versa, using RNA interference (RNAi), it was demonstrated that downregulation of uPAR inhibits proliferation, adhesion, migration, and invasion in oral cancer cells, and siRNA-mediated simultaneous downregulation of uPA and its receptor was able to inhibit angiogenesis and invasiveness leading to increased apoptosis in breast cancer cells [50–52]. To directly examine the role of the uPA system in the process of metastasis *in vivo*, Ossowski [53] developed an experimental model in which cells from a human squamous carcinoma were placed on the CAM of chicken embryos. Using this model, they were able to see that the inhibition of uPA by a specific antibody, inhibiting the catalytic activity of this protease, was able to inhibit invasion, intravasation, and distant metastasis demonstrating that the invasive potential of tumor cells is correlated with uPA-associated proteolytic activity. Until today various additional studies have demonstrated similar results by investigating the importance of the uPA system in the invasive potential of cultured human lung or colon cancer cells [54–56].

9.2.3
Regulation of Expression of the u-PA System in Cancer

As in other proteases, the mechanism and regulations leading to the expression of the uPA system in malignant cells are still the objectives of intensive investigations. In general, expression of this system in diverse malignancies is controlled by epigenetic modifications, the presence and activation of diverse transcription factors, and regulation through oncogenic microRNAs. So far, among the other genes in the plasminogen activation system, only DNA methylation of the uPA gene was previously studied extensively in carcinomas. Toward this, the transcriptional regulation of the uPA gene (PLAU) through DNA methylation was verified in breast cancer, prostate cancer, and meningiomas [57–63]. With regard to transcriptional regulation, diverse studies in the past have defined the uPA promoter, important transcription factors, and several inducers of pro-uPA expression in different cancer types. Human pro-uPA is encoded by a 6.4 kb gene, which is located on chromosome 10 [64, 65]. The minimal promoter contains a TATA box and ensures a low-level basal expression. Furthermore, it contains a GC-rich region and a CAAT sequence, which is bound by the transcription factors Sp1 and CCAAT-box-binding transcription factor (CTF). In addition, an upstream enhancer region is found in the promoter, which requires the cooperation between a PEA3/AP1 site and an AP1 site to upregulate uPA gene expression [66]. This is important since transcription factors binding to PEA3/AP1 sites are largely activated by Ras/MAPK-dependent pathways, which are triggered by various stimuli and hyperactivated in most tumors [67]. In this context, numerous studies have shown that HGF (hepatocyte growth factor), EGF, herregulin, insulinlike growth factor-1 receptor (IGF-IR), and transforming growth factor-β(TGF-β) upregulate uPA by enhancing the promoter activity in tumor cells [68–72]. For example, it was shown in breast cancer that IGF-IR stimulation leads to an upregulation of uPA via phophatidyinositol 3- (PI3) kinase and the mitogen-activated protein kinase kinase (MEKK1) [70, 73]. This data was further supported by other studies showing that PI3 kinase and protein kinase C control cell motility by the regulation of uPA expression through the activation of NF-κB and AP-1 [74–76]. Additional studies on the breast cancer metastasis suppressor 1 (BRMS1) and histone deacetylase 1 (HDAC1) confirmed the importance of the NF-κB binding site within the uPA promoter, showing that BRMS1 recruits HDAC1 to the NF-κB binding site and modulates histone acetylation, which finally results in a reduced NF-κB binding activity and lower uPA expression [77]. In esophageal adenocarcinoma, TGF-β was found to stimulate expression of uPA and PAI-1 through MAPK- or Smad-dependent signaling, leading to a failure of growth cycle arrest and enhanced invasion [72]. Interestingly, another study found that in transformed cells the inhibition of reactive oxygen species (ROS) and NF-κB abrogates TGF-β1-induced uPA and matrix metalloproteinase (MMP-9) expression [78]. For mammalian LIM kinase 1 (LIMK1), a kinase playing an important role in cell motility, it was shown that blocking antibodies against uPA and uPAR suppress LIMK1-induced cell invasiveness. Furthermore, LIMK1 overexpression increased tumor growth in female athymic nude mice, promoted angiogenesis, and induced metastasis to livers and lung, most probably by increasing uPA expression [79]. Aberrant activation of the Notch receptor signaling pathway and overexpression of the Notch ligand was found to be the reason for Notch-dependent cerebral blood flow (CBF-1) binding to the uPA promoter and causing enhanced expression of uPA [80]. Similar results were obtained by a study characterizing the signal transduction pathways by which EGF regulates uPA gene expression and promotes glioblastoma invasion *in vitro* and *in vivo* using a xenograft mouse model. Here, it was found that signaling through c-Src, MAPK, and AP-1 is responsible for EGF-induced expression of uPA and glioblastoma invasion [68]. Other studies have shown that the stress-activated protein kinase (SAPK)/c-Jun N-terminal kinase (JNK) pathway or the p38 MAPK pathway led to phosphorylation and enhanced binding of Sp1 to the minimal promoter element of uPA in gastric and prostate cancer [81–83]. Furthermore, an association of GATA6 with Sp1 and binding of Ets1/2 after EGF stimulation were found to be important for the expression of uPA in breast cancer. Interestingly, overexpression of RelA transcription factor, which binds to NF-κB, contributes directly to elevated uPA gene expression in cell lines of pancreatic adenocarcinoma as well as in ovarian cancer and seems to be dependent on integrin α5/β3 vitronectin interaction [84–86].

Besides these mechanisms to regulate the transcription of the uPA gene, recent studies have also shown a regulation at the translational level through microRNAs. This was demonstrated in hepatocellular carcinoma cells, where it was shown that miR-23b can recognize target sites within the 3′untranslated region (UTR) of uPA and c-Met, leading to inhibition of the target genes, accompanied by a decrease in cell migration and proliferation [87]. The same picture is seen in cervical cancer, in which miR-23b is often downregulated and accompanied by an augmented uPA protein amount [88]. Furthermore, based on microRNA array data that compared low and high metastatic cell lines, miR-93b was identified in breast cancer regulating the translation of uPA mRNA. Using an animal model, it was shown that miR-93 significantly inhibited the growth and dissimination of xenograft tumors. In addition, immunohistohemical staining and real-time PCR proved that miR-93 is a negative regulator of the uPA gene in primary breast tumors [89].

For uPAR, diverse studies in the past have identified transcriptional regulation as the major reason for the induction of uPAR expression in cancer, although other mechanisms such as altered mRNA stability, receptor recycling, or, probably, micoRNAs are also involved [90–94]. The human gene (PLAUR) is located on the long arm of chromosome 19 (19q13) and contains seven exons and six introns [95]. The transcription from the gene yields a 1.4 kb mRNA or an alternatively spliced variant lacking the membrane attachment peptide sequence [96, 97]. Intensive investigations concerning the regulation of the uPAR promoter by transcription factor binding were undertaken and various studies have shown that members of the AP-1 family, Sp1, Sp3, an AP-2-like protein, NF-κB, and PEA3 are the most prominent interaction partners [98–103]. These transcription factors are mediators of a panel of different stimulating molecules, such as EGF, basic fibroblast growth factor (FGF), vascular endothelial growth factor (VEGF), TGF-β1, phorbol 12-myristate 13-acetate (PMA), interferons IFN-α or IFN-γ, protein kinase C (PKC), and protein kinase A (PKA)/cAMP, acting through signaling cascades involving c-Src, K-Ras, MAPK, and JNK [90–92, 104–107]. Although many publications showed a transcriptional activation of uPAR gene expression by diverse promoter motifs, there are not much data supporting an influence of potential silencer elements. One example of transcriptional suppression has been given, showing a PEA3-element as a mediator of integrin-induced downregulation of uPAR gene expression. Since integrins mediate signaling into the cell and previous studies have shown a link between integrins and uPAR at the cell surface, this study supported the notion that, besides the physical interaction of β(3)-integrin and uPAR, β(3) signaling is implicated in the regulation of uPAR gene transcription [84]. More recently, programmed cell death protein 4 (Pdcd4) was shown to inhibit invasion and intravasation in colorectal cancer cells by suppression of Sp1/Sp3-mediated uPAR gene expression [108]. So far, only one study implicates a microRNA, miR-10b, in the translational regulation of the uPAR. miR-10b was identified as a microRNA highly expressed in metastatic breast cancer, promoting cell migration and invasion [109]. On the basis of that knowledge, the expression of miRNA-10b and uPAR was compared in glioblastoma cell lines and a strong correlation was found, leading to the assumption that miR-10b can regulate uPAR through inhibition of the translation of homeobox D10 (HOXD10) [93]. Apart from these results, several other studies have shown that uPAR mRNA stability can be affected by the binding of a uPAR mRNA-binding protein (uPAR mRNABp) in human pleural mesothelioma cells and non-small-cell lung carcinoma (NSCLC) [110, 111]. This notion was further supported by the discovery that p53 acts as an mRNA-binding protein that regulates increased uPAR and PAI-1 expression in human lung epithelial cells and carcinoma cells by interacting with the 3′UTR of both mRNAs [112–114].

Similar to its interaction partners within the uPA system, PAI-1 is expressed in many cell types under the control of a variety of signaling molecules and external stimuli. Interestingly, when patients with tumors are analyzed for long-term survival, those with high levels of the inhibitor of the uPA system mostly present with a worse prognosis compared to those with lower PAI-1 levels [3]. This correlation between a high PAI-1 amount and a poor prognosis is an apparent contradiction to the knowledge that invasion and metastasis are promoted by uPA and its receptor, and indicates that increased proteolysis alone cannot be made responsible for the adverse effects of the uPA system and their inhibitors in tumors [3]. Therefore, it is important to consider the role of the uPA system in promoting invasion and metastasis in strong relation to other processes indispensable for cancer spread, such as formation of vascularized tumor stroma, angiogenesis, desmoplasia, the ability of tumor cells to prevent programed cell death, and dormancy. In this regard, different studies have shown hypoxia, steroid and peptide hormones, cytokines, and growth factors as regulators of PAI-1. These include, for example, angiotensin II (Ang II), EGF, placenta growth factor (PIGF), insulin, IGF-1, interleukin (IL)-6, IL-1α, and TGF-β [115–123]. PAI-1 is one of the most prominent and studied targets of TGF-β. This cytokine is produced by most cell types and plays a critical and dual role in the progression of human cancer, since TGF-β stimulation inhibits cancer cell proliferation in some cellular contexts and promotes it in others. During the early phase of tumor progression, TGF-β often acts as a tumor suppressor, exemplified by deletions or mutations in the core components of the TGF-β signaling pathway. On the contrary, disruption or mutation of regulators of TGF-β signaling can lead to a loss of balanced TGF-β signaling, resulting in the generation and progression of tumors. Consequently, the functional outcome of the TGF-β response is strongly context-dependent including cells, tissues, and cancer types [124, 125]. Regarding transcriptional regulation, initial functional analysis of the PAI-1 promoter in fibrosarcoma cells revealed a regulatory region containing a TATA box and a glucocorticoid responsive element (GRE) [126]. Later, multiple TGF-β-inducible elements (CAGA boxes) were found interacting with Smad3/Smad4 complexes [127, 128] and since that time different studies identified several other cis-elements that mediate TGF-β-related induction of the PAI-1 expression, among them AP1-, Sp1-, NF-1-motifs, and E-boxes [129–131]. Finally, recent studies suggest microRNAs in the translational control of PAI-1.

Using microRNA profiling in hypoxic cells, miR-449a/b was first identified as a key player in regulating PAI-1 [132]. Furthermore, functional analysis after stimulation with PIGF confirmed that the PAI-1 3′UTR is a direct target of miR-30c and 30a [133]. Moreover, miR-145 was found to downregulate PAI-1 in human breast cancer cells and endometrial carcinoma [134].

The human *PAI-II* gene is located on chromosome 19q13.2 [135]. It was first cloned from a lymphoma cell line and is expressed as a 1.9 kb transcript [136]. In 1988, another study isolated the promoter region and characterized, among others, a TATA box, several sequences that are homologous to the cAMP responsive element, and binding sites for AP-1 and AP-2 [137]. These observations were further supported by investigations showing that AP-1 activity correlates with the transactivation of the *PAI-II* gene promoter using a human sarcoma cell line [138]. Furthermore, deletion analysis of the PAI-II promoter region revealed the existence of a negative regulatory region, designated as plasminogen activator inhibitor-II-upstream silencer element-1 (PAUSE-1), which is bound by an as-yet unidentified specific binding factor [139]. On the basis of data accumulated over the past years, it is evident that PAI-II can also be induced by a wide range of stimulating molecules, for example, TGF-β, EGF, monocyte-colony-stimulating factor (M-CSF), tumor necrosis factor (TNFα), IL-1, IL-2, angiotensin II, and tumor promoters [140–144]. Furthermore, posttranscriptional mechanisms are involved in the regulation of PAI-II expression, most notably at the level of mRNA stability. In this regard, studies examining the 3′UTR and a region within exon 4 of PAI-II showed that PAI-II mRNA stability is influenced by elements located within both the coding and the untranslated region [145, 146]. A number of *in vivo* studies have assessed the prognostic relevance of tumor- and stromal-derived PAI-II in the metastatic spread of cancer. Using melanoma cells, it was shown that overexpression of PAI-II is able to prevent spontaneous metastasis in a mouse model [147]. A similar effect was seen in sarcoma cells, where overexpression of PAI-II led to a reduction in uPA-dependent cell movement *in vitro* and development of metastasis *in vivo* [148]. Even though breast cancer is the most frequently studied cancer type in which the prognostic value of PAI-II expression has been assessed, it was also proved to play a role in many other types of solid cancer, such as lung, bladder, head and neck, colorectal, gastric, oesophageal squamous carcinomas, and pancreatic carcinomas [36]. However, the most significant difference to PAI-I is the fact that tumor expression of PAI-II is clearly associated with increased survival and reduced metastatic potential [149].

9.2.4
Regulation of Cell Signaling by the u-PA System

Apart from its participation in proteolysis, an increasing number of studies in the recent years have shown that specifically uPAR can also act as a signaling molecule, mediating cell adhesion and downstream signaling via several molecular interactions [150, 151]. uPAR is involved in cell motility, invasion, proliferation, and survival of cells and modulates cell–matrix contacts, for example, via physical interactions with vitronectin. Since the vitronectin and uPA binding sites are distinct, uPAR can simultaneously bind both ligands. Interestingly, lipid raft partitioning of uPAR increases the ability to bind vitronectin, and lipid raft partitioning of uPAR is promoted by uPA binding [28]. Vitronectin circulates in the blood as a monomer but is converted into a multimeric form when incorporated into the extracellular matrix (ECM). Increased vitronectin deposition is found in reactive fibrotic tissue as well as in several tumors [152]. Interaction of uPAR with vitronectin contributes to cell adhesion, and it is important to know that uPA stabilizes the conformation of its receptors to facilitate vitronectin binding, thereby stimulating the vitronectin-dependent adhesion of cells [153, 154]. On the other hand, vitronectin by itself also binds and activates members of the integrin family, which is mediated through its arginine-glycine-aspartic acid adhesion sequence (RGD motif) [155, 156]. uPAR can regulate the activity of integrins, thus promoting or inhibiting integrin signaling and integrin-mediated cell adhesion to other ECM components, such as fibronectin and collagen [11]. Integrins are heterodimeric transmembrane receptors containing two distinct chains, α and β subunits. In humans, 18α and 8β subunits form 24 noncovalently associated α/β heterodimers. Each subunit has a large extracellular domain, a single transmembrane domain, and a short cytoplasmic domain [4]. Signaling initiated by uPAR via integrins can activate the Ras-mitogen-activated protein kinase (MAPK) pathway, the focal adhesion kinase (FAK), Src kinase, and the small GTPase Rac, among others [157–160].

Despite the controversy whether uPAR and integrins interact directly, many studies in various cancers have shown that uPAR signaling requires integrins as coreceptors. Early studies have shown that uPAR associates with β1 and β3 integrins of fibrosarcoma cells, depending on the extracelluar matrix components [161]. These results were supported by a recent study with lung cancer cells, where it was found that uPAR/β1 integrin interactions are essential to signals induced by matrix ligands or uPA to support cancer cell invasion and progression [162]. Using human fibrosarcoma and breast cancer cell lines, another study concluded that the association of uPAR with α5/β5 integrins leads to a functional interaction of these receptors, resulting in cytoskeletal rearrangements and cell migration [163]. Moreover, an approach using a synthetic peptide (p25), which interferes with the formation of uPAR/integrin complexes, revealed that uPAR is able to regulate the adhesive function of integrins in breast cancer cell lines *in vivo* [164]. Signaling through uPAR and integrins can also involve the epidermal growth factor receptor (EGFR), leading to a complex interplay between uPAR, integrins, and receptor tyrosine kinases. For example, in prostate cancer, the treatment with a specific inhibitor of EGFR significantly inhibited the invasive potential of cultured cells *in vitro* by reducing the amount of uPAR transcript, uPA production, and activity [165]. In a further study on breast cancer cells, it was found that uPAR stimulation with ATF transactivates the EGFR in mammary cancer cells through a mechanism involving Src and a metalloproteinase. In these cells, expressing low levels of uPAR, both ATF- and EGF-stimuli induced an interaction of the EGFR with uPAR, resulting in extracelluar signal-regulated kinase (ERK) activation. Interestingly,

EGFR activation by uPAR mediated cellular invasion rather than proliferation, while EGFR activation by EGF led to a proliferative response [166].

Another example of the complex interplay between the uPA system and EGFR is the mechanism of persistent growth suppression in cancer cells [167]. A known phenomenon of malignant cancers is the recurrence of certain tumors, even long after the removal of the primary tumor. These very long intervals until tumor recurrence, and the fact that disseminated tumor cells in the bone marrow of solid cancers can be found even several years after tumor surgery, led to the assumption that a certain fraction of tumor cells undergo a developmental delay designated as dormancy [168–170]. First evidence that the uPAR can play a role in tumor cell dormancy came from studies on human squamous carcinoma cells. Using a uPAR-antisense strategy, it was shown that a significant reduction in uPAR leads to the induction of tumor cell dormancy [171]. Furthermore, in nude mouse experiments it was verified that the prolonged latency period seen in malignant cells depended on a reduced uPAR expression on the cell surface [172]. The hypothesis that dormancy is inducible by a suppression of uPAR gene expression was further supported by a series of studies in head and neck cancer cells. It was found that the rapid growth of these cells is regulated by a high uPAR expression, leading to an activation of $\alpha 5/\beta 1$ integrins [159, 167]. This activation, which is mediated through ligand-independent activation of EGFR, is additionally enhanced by uPA binding to the uPAR and cell binding to fibronectin. Overexpression of uPAR initiates an intracellular signal through FAK and Src, leading to a strong ERK activation and tumorigenicity [157]. Interestingly, after downregulation of the uPAR, ERK activity is lost and p38 MAPK becomes active, which leads to a balance that favors p38 MAPK activation over ERK, inducing dormancy in this cell line and other human tumor entities [173].

Apart from EGFR signaling, studies have shown that the uPA system also interacts with the HGF and its transmembrane tyrosine kinase receptor c-Met. In human hepatocelluar carcinoma cells, stimulation with HGF enhanced the protein expression of uPA and uPAR and the proteolytic activity of uPA in a dose-dependent manner, leading to enhanced invasiveness *in vitro* [174]. Consistent with these data, in human pancreatic and colorectal carcinoma cells it was found that inhibition of the uPAR with a specific antibody inhibits c-Met- and IGF-1-receptor-mediated migration and invasion *in vitro* and *in vivo* [175–177].

So far, only a few studies addressed the question of a direct role of PAI-1 or PAI-II in signaling. One example is the signaling events initiated by the binding of uPA to uPAR, thus promoting proliferation of breast cancer cells. It was shown that the subsequent inhibition of uPA by PAI-1 reveals a cryptic high-affinity site within the PAI-1 moiety for the VLDL receptor, which sustains cell signaling events initiated by binding of uPA to its receptor [35]. As mentioned earlier, unlike PAI-1, the PAI-2 moiety of uPA-PAI-2 does not contain a high-affinity binding site for the VLDL receptor, although uPA-PAI-2 is still efficiently endocytosed via this receptor in breast cancer cells. Therefore, unlike uPA/PAI-1, endocytosis of uPA/PAI-2 does not induce mitogenic signaling events leading to cell proliferation. Finally, a more comprehensive study used primary breast cancer tissues and examined the expression of uPA, PAI-1, and a panel of signaling molecules using protein microarrays. They found that expression of PAI-1 was correlated with the uPA receptor and Akt activation, presumably via integrin and human epidermal growth factor (HER)-receptor signaling. Further network monitoring for PAI-1 in breast cancer revealed interactions with main signaling cascades and extended the findings from cell culture experiments [178].

9.2.5
Conclusion

Taken together, it is well established that the uPA system is closely related to the malignant process of invasion and metastasis. However, accumulating evidence shows that apart from participation in proteolysis, the uPAR can also act as a signaling molecule mediating cell adhesion, proliferation, dormancy, and downstream signaling via several molecular interactions. Furthermore, it is important to understand that the uPA system, by its interplay with multiple associated factors, is also responsible for other processes linked to tumor development, such as angiogenesis and apoptosis. Therefore, it is of utmost importance to understand the molecular regulation, the epigenetic modifications, and transcriptional and translational regulators influencing the individual components of this system. Knowledge of which parameters are important for the regulation of the uPA system and how it supports the development of cancer will help define new therapeutic strategies to target the individual components of the system and potential interaction partners or to inhibit regulators of gene expression to improve prognosis of cancer patients. This is illustrated in Chapter 13 by Thurison *et al.* on biomarkers in cancer.

References

1. Hanahan, D. and Weinberg, R.A. (2000) The hallmarks of cancer. *Cell*, **100**, 57–70.
2. Dano, K., Behrendt, N., Hoyer-Hansen, G., Johnsen, M., Lund, L.R., Ploug, M., and Romer, J. (2005) Plasminogen activation and cancer. *Thromb. Haemost.*, **93**, 676–681.
3. Binder, B.R. and Mihaly, J. (2008) The plasminogen activator inhibitor "paradox" in cancer. *Immunol. Lett.*, **118**, 116–124.
4. Smith, H.W. and Marshall, C.J. (2010) Regulation of cell signalling by uPAR. *Nat. Rev. Mol. Cell Biol.*, **11**, 23–36.
5. Mazar, A.P. (2001) The urokinase plasminogen activator receptor (uPAR) as a target for the diagnosis and therapy of cancer. *Anticancer Drugs*, **12**, 387–400.
6. Sidenius, N. and Blasi, F. (2003) The urokinase plasminogen activator system in cancer: recent advances and implication for prognosis and therapy. *Cancer Metastasis Rev.*, **22**, 205–222.
7. Kwaan, H.C. and McMahon, B. (2009) The role of plasminogen-plasmin system in cancer. *Cancer Treat. Res.*, **148**, 43–66.
8. Dass, K., Ahmad, A., Azmi, A.S., Sarkar, S.H., and Sarkar, F.H. (2008) Evolving role of uPA/uPAR system in human cancers. *Cancer Treat. Rev.*, **34**, 122–136.
9. Blasi, F. (1996) The urokinase receptor and cell migration. *Semin Thromb Hemost*, **22**, 513–516.
10. Dano, K., Andreasen, P.A., Grondahl-Hansen, J., Kristensen, P., Nielsen, L.S., and Skriver, L. (1985)

Plasminogen activators, tissue degradation, and cancer. *Adv. Cancer Res.*, **44**, 139–266.
11. Madsen, C.D. and Sidenius, N. (2008) The interaction between urokinase receptor and vitronectin in cell adhesion and signalling. *Eur. J. Cell Biol.*, **87**, 617–629.
12. Mondino, A., Resnati, M., and Blasi, F. (1999) Structure and function of the urokinase receptor. *Thromb. Haemost.*, **82** (Suppl. 1), 19–22.
13. Plesner, T., Behrendt, N., and Ploug, M. (1997) Structure, function and expression on blood and bone marrow cells of the urokinase-type plasminogen activator receptor, uPAR. *Stem Cells*, **15**, 398–408.
14. Romer, J., Lund, L.R., Eriksen, J., Pyke, C., Kristensen, P., and Dano, K. (1994) The receptor for urokinase-type plasminogen activator is expressed by keratinocytes at the leading edge during re-epithelialization of mouse skin wounds. *J. Invest. Dermatol.*, **102**, 519–522.
15. Cubellis, M.V., Nolli, M.L., Cassani, G., and Blasi, F. (1986) Binding of single-chain prourokinase to the urokinase receptor of human U937 cells. *J. Biol. Chem.*, **261**, 15819–15822.
16. Higazi, A.A., Mazar, A., Wang, J., Reilly, R., Henkin, J., Kniss, D., and Cines, D. (1996) Single-chain urokinase-type plasminogen activator bound to its receptor is relatively resistant to plasminogen activator inhibitor type 1. *Blood*, **87**, 3545–3549.
17. Lijnen, H.R., Stump, D.C. and Collen, D.C. (1987) Single-chain urokinase-type plasminogen activator: mechanism of action and thrombolytic properties. *Semin. Thromb. Hemost.*, **13**, 152–159.
18. Andreasen, P.A., Sottrup-Jensen, L., Kjoller, L., Nykjaer, A., Moestrup, S.K., Petersen, C.M., and Gliemann, J. (1994) Receptor-mediated endocytosis of plasminogen activators and activator/inhibitor complexes. *FEBS Lett.*, **338**, 239–245.
19. Andreasen, P.A., Egelund, R., and Petersen, H.H. (2000) The plasminogen activation system in tumor growth, invasion, and metastasis. *Cell. Mol. Life Sci.*, **57**, 25–40.
20. Duffy, M.J. (1992) The role of proteolytic enzymes in cancer invasion and metastasis. *Clin. Exp. Metastasis*, **10**, 145–155.
21. Ellis, V., Behrendt, N., and Dano, K. (1991) Plasminogen activation by receptor-bound urokinase. A kinetic study with both cell-associated and isolated receptor. *J. Biol. Chem.*, **266**, 12752–12758.
22. Appella, E., Robinson, E.A., Ullrich, S.J., Stoppelli, M.P., Corti, A., Cassani, G., and Blasi, F. (1987) The receptor-binding sequence of urokinase. A biological function for the growth-factor module of proteases. *J. Biol. Chem.*, **262**, 4437–4440.
23. Bdeir, K., Kuo, A., Mazar, A., Sachais, B.S., Xiao, W., Gawlak, S., Harris, S., Higazi, A.A., and Cines, D.B. (2000) A region in domain II of the urokinase receptor required for urokinase binding. *J. Biol. Chem.*, **275**, 28532–28538.
24. Bdeir, K., Kuo, A., Sachais, B.S., Rux, A.H., Bdeir, Y., Mazar, A., Higazi, A.A., and Cines, D.B. (2003) The kringle stabilizes urokinase binding to the urokinase receptor. *Blood*, **102**, 3600–3608.
25. Behrendt, N., Ronne, E., Ploug, M., Petri, T., Lober, D., Nielsen, L.S., Schleuning, W.D., Blasi, F., Appella, E., and Dano, K. (1990) The human receptor for urokinase plasminogen activator. NH2-terminal amino acid sequence and glycosylation variants. *J. Biol. Chem.*, **265**, 6453–6460.
26. Ploug, M., Rahbek-Nielsen, H., Nielsen, P.F., Roepstorff, P., and Dano, K. (1998) Glycosylation profile of a recombinant urokinase-type plasminogen activator receptor expressed in Chinese hamster ovary cells. *J. Biol. Chem.*, **273**, 13933–13943.
27. Lajoie, P., Goetz, J.G., Dennis, J.W., and Nabi, I.R. (2009) Lattices, rafts, and scaffolds: domain regulation of receptor signaling at the plasma membrane. *J. Cell Biol.*, **185**, 381–385.
28. Cunningham, O., Andolfo, A., Santovito, M.L., Iuzzolino, L., Blasi, F., and Sidenius, N. (2003) Dimerization

controls the lipid raft partitioning of uPAR/CD87 and regulates its biological functions. *EMBO J.*, **22**, 5994–6003.

29. Andreasen, P.A., Kjoller, L., Christensen, L. and Duffy, M.J. (1997) The urokinase-type plasminogen activator system in cancer metastasis: a review. *Int. J. Cancer*, **72**, 1–22.

30. Loskutoff, D.J., Curriden, S.A., Hu, G., and Deng, G. (1999) Regulation of cell adhesion by PAI-1. *APMIS*, **107**, 54–61.

31. Huntington, J.A., Read, R.J., and Carrell, R.W. (2000) Structure of a serpin-protease complex shows inhibition by deformation. *Nature*, **407**, 923–926.

32. Heegaard, C.W., Simonsen, A.C., Oka, K., Kjoller, L., Christensen, A., Madsen, B., Ellgaard, L., Chan, L., and Andreasen, P.A. (1995) Very low density lipoprotein receptor binds and mediates endocytosis of urokinase-type plasminogen activator-type-1 plasminogen activator inhibitor complex. *J. Biol. Chem.*, **270**, 20855–20861.

33. Gliemann, J. (1998) Receptors of the low density lipoprotein (LDL) receptor family in man. Multiple functions of the large family members via interaction with complex ligands. *Biol. Chem.*, **379**, 951–964.

34. Lademann, U.A. and Romer, M.U. (2008) Regulation of programmed cell death by plasminogen activator inhibitor type 1 (PAI-1). *Thromb. Haemost.*, **100**, 1041–1046.

35. Croucher, D.R., Saunders, D.N., Stillfried, G.E., and Ranson, M. (2007) A structural basis for differential cell signalling by PAI-1 and PAI-2 in breast cancer cells. *Biochem. J.*, **408**, 203–210.

36. Croucher, D.R., Saunders, D.N., Lobov, S., and Ranson, M. (2008) Revisiting the biological roles of PAI2 (SERPINB2) in cancer. *Nat. Rev. Cancer*, **8**, 535–545.

37. Dano, K. and Reich, E. (1978) Serine enzymes released by cultured neoplastic cells. *J. Exp. Med.*, **147**, 745–757.

38. Unkeless, J., Dano, K., Kellerman, G.M., and Reich, E. (1974) Fibrinolysis associated with oncogenic transformation. Partial purification and characterization of the cell factor, a plasminogen activator. *J. Biol. Chem.*, **249**, 4295–4305.

39. Egeblad, M. and Werb, Z. (2002) New functions for the matrix metalloproteinases in cancer progression. *Nat. Rev. Cancer*, **2**, 161–174.

40. Johnsen, M., Lund, L.R., Romer, J., Almholt, K., and Dano, K. (1998) Cancer invasion and tissue remodeling: common themes in proteolytic matrix degradation. *Curr. Opin. Cell. Biol.*, **10**, 667–671.

41. Harvey, S.R., Sait, S.N., Xu, Y., Bailey, J.L., Penetrante, R.M., and Markus, G. (1999) Demonstration of urokinase expression in cancer cells of colon adenocarcinomas by immunohistochemistry and in situ hybridization. *Am. J. Pathol.*, **155**, 1115–1120.

42. Nielsen, B.S., Sehested, M., Timshel, S., Pyke, C., and Dano, K. (1996) Messenger RNA for urokinase plasminogen activator is expressed in myofibroblasts adjacent to cancer cells in human breast cancer. *Lab. Invest.*, **74**, 168–177.

43. Romer, J., Lund, L.R., Eriksen, J., Ralfkiaer, E., Zeheb, R., Gelehrter, T.D., Dano, K., and Kristensen, P. (1991) Differential expression of urokinase-type plasminogen activator and its type-1 inhibitor during healing of mouse skin wounds. *J. Invest. Dermatol.*, **97**, 803–811.

44. Usher, P.A., Thomsen, O.F., Iversen, P., Johnsen, M., Brunner, N., Hoyer-Hansen, G., Andreasen, P., Dano, K., and Nielsen, B.S. (2005) Expression of urokinase plasminogen activator, its receptor and type-1 inhibitor in malignant and benign prostate tissue. *Int. J. Cancer*, **113**, 870–880.

45. Pyke, C., Graem, N., Ralfkiaer, E., Ronne, E., Hoyer-Hansen, G., Brunner, N., and Dano, K. (1993b) Receptor for urokinase is present in tumor-associated macrophages in ductal breast carcinoma. *Cancer Res.*, **53**, 1911–1915.

46. Pyke, C., Kristensen, P., Ralfkiaer, E., Grondahl-Hansen, J., Eriksen, J., Blasi, F., and Dano, K. (1991)

Urokinase-type plasminogen activator is expressed in stromal cells and its receptor in cancer cells at invasive foci in human colon adenocarcinomas. *Am.J. Pathol.*, **138**, 1059–1067.

47. Romer, J., Pyke, C., Lund, L.R., Ralfkiaer, E., and Dano, K. (2001) Cancer cell expression of urokinase-type plasminogen activator receptor mRNA in squamous cell carcinomas of the skin. *J. Invest. Dermatol.*, **116**, 353–358.
48. Kariko, K., Kuo, A., Boyd, D., Okada, S.S., Cines, D.B., and Barnathan, E.S. (1993) Overexpression of urokinase receptor increases matrix invasion without altering cell migration in a human osteosarcoma cell line. *Cancer Res.*, **53**, 3109–3117.
49. Xing, R.H. and Rabbani, S.A. (1996) Overexpression of urokinase receptor in breast cancer cells results in increased tumor invasion, growth and metastasis. *Int.J. Cancer*, **67**, 423–429.
50. Kunigal, S., Lakka, S.S., Gondi, C.S., Estes, N., and Rao, J.S. (2007) RNAi-mediated downregulation of urokinase plasminogen activator receptor and matrix metalloprotease-9 in human breast cancer cells results in decreased tumor invasion, angiogenesis and growth. *Int.J. Cancer*, **121**, 2307–2316.
51. Liang, X., Yang, X., Tang, Y., Zhou, H., Liu, X., Xiao, L., Gao, J., and Mao, Z. (2008) RNAi-mediated downregulation of urokinase plasminogen activator receptor inhibits proliferation, adhesion, migration and invasion in oral cancer cells. *Oral Oncol.*, **44**, 1172–1180.
52. Subramanian, R., Gondi, C.S., Lakka, S.S., Jutla, A., and Rao, J.S. (2006) siRNA-mediated simultaneous downregulation of uPA and its receptor inhibits angiogenesis and invasiveness triggering apoptosis in breast cancer cells. *Int.J. Oncol.*, **28**, 831–839.
53. Ossowski, L. (1988) In vivo invasion of modified chorioallantoic membrane by tumor cells: the role of cell surface-bound urokinase. *J. Cell Biol.*, **107**, 2437–2445.
54. Hollas, W., Blasi, F., and Boyd, D. (1991) Role of the urokinase receptor in facilitating extracellular matrix invasion by cultured colon cancer. *Cancer Res.*, **51**, 3690–3695.
55. Liu, G., Shuman, M.A., and Cohen, R.L. (1995) Co-expression of urokinase, urokinase receptor and PAI-1 is necessary for optimum invasiveness of cultured lung cancer cells. *Int.J. Cancer*, **60**, 501–506.
56. Schlechte, W., Murano, G., and Boyd, D. (1989) Examination of the role of the urokinase receptor in human colon cancer mediated laminin degradation. *Cancer Res.*, **49**, 6064–6069.
57. Kandenwein, J.A., Park-Simon, T.W., Schramm, J., and Simon, M. uPA/PAI-1 expression and uPA promoter methylation in meningiomas. (2010) *J. Neurooncol.*, **103**, 533–9.
58. Pakneshan, P., Szyf, M., and Rabbani, S.A. (2005a) Hypomethylation of urokinase (uPA) promoter in breast and prostate cancer: prognostic and therapeutic implications. *Curr. Cancer Drug Targets*, **5**, 471–488.
59. Pakneshan, P., Szyf, M., and Rabbani, S.A. (2005b) Methylation and inhibition of expression of uPA by the RAS oncogene: divergence of growth control and invasion in breast cancer cells. *Carcinogenesis*, **26**, 557–564.
60. Pakneshan, P., Xing, R.H., and Rabbani, S.A. (2003) Methylation status of uPA promoter as a molecular mechanism regulating prostate cancer invasion and growth in vitro and in vivo. *FASEB J.*, **17**, 1081–1088.
61. Pulukuri, S.M., Estes, N., Patel, J., and Rao, J.S. (2007) Demethylation-linked activation of urokinase plasminogen activator is involved in progression of prostate cancer. *Cancer Res.*, **67**, 930–939.
62. Shukeir, N., Pakneshan, P., Chen, G., Szyf, M., and Rabbani, S.A. (2006) Alteration of the methylation status of tumor-promoting genes decreases prostate cancer cell invasiveness and tumorigenesis in vitro and in vivo. *Cancer Res.*, **66**, 9202–9210.
63. Xing, R.H. and Rabbani, S.A. (1999) Transcriptional regulation of urokinase (uPA) gene expression in breast cancer

cells: role of DNA methylation. *Int. J. Cancer*, **81**, 443–450.

64. Riccio, A., Grimaldi, G., Verde, P., Sebastio, G., Boast, S., and Blasi, F. (1985) The human urokinase-plasminogen activator gene and its promoter. *Nucleic Acids Res.*, **13**, 2759–2771.
65. Tripputi, P., Blasi, F., Verde, P., Cannizzaro, L.A., Emanuel, B.S., and Croce, C.M. (1985) Human urokinase gene is located on the long arm of chromosome 10. *Proc. Natl. Acad. Sci. U.S.A.*, **82**, 4448–4452.
66. Verde, P., Boast, S., Franze, A., Robbiati, F., and Blasi, F. (1988) An upstream enhancer and a negative element in the 5′ flanking region of the human urokinase plasminogen activator gene. *Nucleic Acids Res.*, **16**, 10699–10716.
67. Nagamine, Y., Medcalf, R.L., and Munoz-Canoves, P. (2005) Transcriptional and posttranscriptional regulation of the plasminogen activator system. *Thromb. Haemost.*, **93**, 661–675.
68. Amos, S., Redpath, G.T., Dipierro, C.G., Carpenter, J.E., and Hussaini, I.M. (2010) Epidermal growth factor receptor-mediated regulation of urokinase plasminogen activator expression and glioblastoma invasion via C-SRC/MAPK/AP-1 signaling pathways. *J. Neuropathol. Exp. Neurol.*, **69**, 582–592.
69. Besser, D., Bardelli, A., Didichenko, S., Thelen, M., Comoglio, P.M., Ponzetto, C., and Nagamine, Y. (1997) Regulation of the urokinase-type plasminogen activator gene by the oncogene Tpr-Met involves GRB2. *Oncogene*, **14**, 705–711.
70. Dunn, S.E., Torres, J.V., Oh, J.S., Cykert, D.M., and Barrett, J.C. (2001) Up-regulation of urokinase-type plasminogen activator by insulin-like growth factor-I depends upon phosphatidylinositol-3 kinase and mitogen-activated protein kinase kinase. *Cancer Res.*, **61**, 1367–1374.
71. Mazumdar, A., Adam, L., Boyd, D., and Kumar, R. (2001) Heregulin regulation of urokinase plasminogen activator and its receptor: human breast epithelial cell invasion. *Cancer Res.*, **61**, 400–405.
72. Onwuegbusi, B.A., Rees, J.R., Lao-Sirieix, P., and Fitzgerald, R.C. (2007) Selective loss of TGFbeta Smad-dependent signalling prevents cell cycle arrest and promotes invasion in oesophageal adenocarcinoma cell lines. *PLoS ONE*, **2**, e177.
73. Cuevas, B.D., Uhlik, M.T., Garrington, T.P., and Johnson, G.L. (2005) MEKK1 regulates the AP-1 dimer repertoire via control of JunB transcription and Fra-2 protein stability. *Oncogene*, **24**, 801–809.
74. Milde-Langosch, K., Roder, H., Andritzky, B., Aslan, B., Hemminger, G., Brinkmann, A., Bamberger, C.M., Loning, T., and Bamberger, A.M. (2004) The role of the AP-1 transcription factors c-Fos, FosB, Fra-1 and Fra-2 in the invasion process of mammary carcinomas. *Breast Cancer Res. Treat.*, **86**, 139–152.
75. Sliva, D., English, D., Lyons, D., and Lloyd, F.P. Jr. (2002a) Protein kinase C induces motility of breast cancers by upregulating secretion of urokinase-type plasminogen activator through activation of AP-1 and NF-kappaB. *Biochem. Biophys. Res. Commun.*, **290**, 552–557.
76. Sliva, D., Rizzo, M.T., and English, D. (2002b) Phosphatidylinositol 3-kinase and NF-kappaB regulate motility of invasive MDA-MB-231 human breast cancer cells by the secretion of urokinase-type plasminogen activator. *J. Biol. Chem.*, **277**, 3150–3157.
77. Cicek, M., Fukuyama, R., Cicek, M.S., Sizemore, S., Welch, D.R., Sizemore, N., and Casey, G. (2009) BRMS1 contributes to the negative regulation of uPA gene expression through recruitment of HDAC1 to the NF-kappaB binding site of the uPA promoter. *Clin. Exp. Metastasis*, **26**, 229–237.
78. Tobar, N., Villar, V., and Santibanez, J.F. (2010) ROS-NFkappaB mediates TGF-beta1-induced expression of urokinase-type plasminogen activator, matrix metalloproteinase-9 and

cell invasion. *Mol. Cell Biochem.*, **340**, 195–202.
79. Bagheri-Yarmand, R., Mazumdar, A., Sahin, A.A., and Kumar, R. (2006) LIM kinase 1 increases tumor metastasis of human breast cancer cells via regulation of the urokinase-type plasminogen activator system. *Int. J. Cancer.*, **118**, 2703–2710.
80. Shimizu, M., Cohen, B., Goldvasser, P., Berman, H., Virtanen, C., and Reedijk, M. (2011) Plasminogen activator uPA is a direct transcriptional target of the JAG1-Notch receptor signaling pathway in breast cancer. *Cancer Res.*, **71**, 277–286.
81. Benasciutti, E., Pages, G., Kenzior, O., Folk, W., Blasi, F., and Crippa, M.P. (2004) MAPK and JNK transduction pathways can phosphorylate Sp1 to activate the uPA minimal promoter element and endogenous gene transcription. *Blood*, **104**, 256–262.
82. Ibanez-Tallon, I., Ferrai, C., Longobardi, E., Facetti, I., Blasi, F., and Crippa, M.P. (2002) Binding of Sp1 to the proximal promoter links constitutive expression of the human uPA gene and invasive potential of PC3 cells. *Blood*, **100**, 3325–3332.
83. Shin, B.A., Yoo, H.G., Kim, H.S., Kim, M.H., Hwang, Y.S., Chay, K.O., Lee, K.Y., Ahn, B.W., and Jung, Y.D. (2003) P38 MAPK pathway is involved in the urokinase plasminogen activator expression in human gastric SNU-638 cells. *Oncol. Rep.*, **10**, 1467–1471.
84. Hapke, S., Kessler, H., Arroyo de Prada, N., Benge, A., Schmitt, M., Lengyel, E., and Reuning, U. (2001) Integrin alpha(v)beta(3)/vitronectin interaction affects expression of the urokinase system in human ovarian cancer cells. *J. Biol. Chem.*, **276**, 26340–26348.
85. Reuning, U., Guerrini, L., Nishiguchi, T., Page, S., Seibold, H., Magdolen, V., Graeff, H., and Schmitt, M. (1999) Rel transcription factors contribute to elevated urokinase expression in human ovarian carcinoma cells. *Eur. J. Biochem.*, **259**, 143–148.
86. Wang, W., Abbruzzese, J.L., Evans, D.B., and Chiao, P.J. (1999) Overexpression of urokinase-type plasminogen activator in pancreatic adenocarcinoma is regulated by constitutively activated RelA. *Oncogene*, **18**, 4554–4563.
87. Salvi, A., Sabelli, C., Moncini, S., Venturin, M., Arici, B., Riva, P., Portolani, N., Giulini, S.M., De Petro, G., and Barlati, S. (2009) MicroRNA-23b mediates urokinase and c-met downmodulation and a decreased migration of human hepatocellular carcinoma cells. *FEBS J.*, **276**, 2966–2982.
88. Au Yeung, C.L., Tsang, T.Y., Yau, P.L., and Kwok, T.T. (2011) Human papillomavirus type 16 E6 induces cervical cancer cell migration through the p53/microRNA-23b/urokinase-type plasminogen activator pathway. *Oncogene.*, **30**, 2401–10.
89. Li, X.F., Yan, P.J., and Shao, Z.M. (2009) Downregulation of miR-193b contributes to enhance urokinase-type plasminogen activator (uPA) expression and tumor progression and invasion in human breast cancer. *Oncogene*, **28**, 3937–3948.
90. Gum, R., Juarez, J., Allgayer, H., Mazar, A., Wang, Y., and Boyd, D. (1998) Stimulation of urokinase-type plasminogen activator receptor expression by PMA requires JNK1-dependent and -independent signaling modules. *Oncogene*, **17**, 213–225.
91. Lengyel, E., Wang, H., Stepp, E., Juarez, J., Wang, Y., Doe, W., Pfarr, C.M., and Boyd, D. (1996) Requirement of an upstream AP-1 motif for the constitutive and phorbol ester-inducible expression of the urokinase-type plasminogen activator receptor gene. *J. Biol. Chem.*, **271**, 23176–23184.
92. Lund, L.R., Ellis, V., Ronne, E., Pyke, C., and Dano, K. (1995) Transcriptional and post-transcriptional regulation of the receptor for urokinase-type plasminogen activator by cytokines and tumour promoters in the human lung carcinoma cell line A549. *Biochem. J.*, **310** (Pt 1), 345–352.
93. Sasayama, T., Nishihara, M., Kondoh, T., Hosoda, K., and

Kohmura, E. (2009) MicroRNA-10b is overexpressed in malignant glioma and associated with tumor invasive factors, uPAR and RhoC. *Int. J. Cancer*, **125**, 1407–1413.

94. Shetty, S., Kumar, A., and Idell, S. (1997) Posttranscriptional regulation of urokinase receptor mRNA: identification of a novel urokinase receptor mRNA binding protein in human mesothelioma cells. *Mol. Cell Biol.*, **17**, 1075–1083.

95. Borglum, A.D., Byskov, A., Ragno, P., Roldan, A.L., Tripputi, P., Cassani, G., Dano, K., Blasi, F., Bolund, L., and Kruse, T.A. (1992) Assignment of the urokinase-type plasminogen activator receptor gene (PLAUR) to chromosome 19q13.1-q13.2. *Am. J. Hum. Genet.*, **50**, 492–497.

96. Pyke, C., Eriksen, J., Solberg, H., Nielsen, B.S., Kristensen, P., Lund, L.R., and Dano, K. (1993a) An alternatively spliced variant of mRNA for the human receptor for urokinase plasminogen activator. *FEBS Lett.*, **326**, 69–74.

97. Roldan, A.L., Cubellis, M.V., Masucci, M.T., Behrendt, N., Lund, L.R., Dano, K., Appella, E., and Blasi, F. (1990) Cloning and expression of the receptor for human urokinase plasminogen activator, a central molecule in cell surface, plasmin dependent proteolysis. *EMBO J.*, **9**, 467–474.

98. Allgayer, H., Wang, H., Gallick, G.E., Crabtree, A., Mazar, A., Jones, T., Kraker, A.J., and Boyd, D.D. (1999a) Transcriptional induction of the urokinase receptor gene by a constitutively active Src. Requirement of an upstream motif (-152/-135) bound with Sp1. *J. Biol. Chem.*, **274**, 18428–18437.

99. Allgayer, H., Wang, H., Shirasawa, S., Sasazuki, T., and Boyd, D. (1999b) Targeted disruption of the K-ras oncogene in an invasive colon cancer cell line down-regulates urokinase receptor expression and plasminogen-dependent proteolysis. *Br. J. Cancer*, **80**, 1884–1891.

100. Allgayer, H., Wang, H., Wang, Y., Heiss, M.M., Bauer, R., Nyormoi, O., and Boyd, D. (1999c) Transactivation of the urokinase-type plasminogen activator receptor gene through a novel promoter motif bound with an activator protein-2alpha-related factor. *J. Biol. Chem.*, **274**, 4702–4714.

101. Dang, J., Boyd, D., Wang, H., Allgayer, H., Doe, W.F., and Wang, Y. (1999) A region between -141 and -61 bp containing a proximal AP-1 is essential for constitutive expression of urokinase-type plasminogen activator receptor. *Eur. J. Biochem.*, **264**, 92–99.

102. Gum, R., Lengyel, E., Juarez, J., Chen, J.H., Sato, H., Seiki, M., and Boyd, D. (1996) Stimulation of 92-kDa gelatinase B promoter activity by ras is mitogen-activated protein kinase kinase 1-independent and requires multiple transcription factor binding sites including closely spaced PEA3/ets and AP-1 sequences. *J. Biol. Chem.*, **271**, 10672–10680.

103. Wang, Y. (2001) The role and regulation of urokinase-type plasminogen activator receptor gene expression in cancer invasion and metastasis. *Med. Res. Rev.*, **21**, 146–170.

104. Boyd, D. (1989) Examination of the effects of epidermal growth factor on the production of urokinase and the expression of the plasminogen activator receptor in a human colon cancer cell line. *Cancer Res.*, **49**, 2427–2432.

105. Lengyel, E., Wang, H., Gum, R., Simon, C., Wang, Y., and Boyd, D. (1997) Elevated urokinase-type plasminogen activator receptor expression in a colon cancer cell line is due to a constitutively activated extracellular signal-regulated kinase-1-dependent signaling cascade. *Oncogene*, **14**, 2563–2573.

106. Li, C., Liu, J.N., and Gurewich, V. (1995) Urokinase-type plasminogen activator-induced monocyte adhesion requires a carboxyl-terminal lysine and cAMP-dependent signal transduction. *J. Biol. Chem.*, **270**, 30282–30285.

107. Mandriota, S.J., Seghezzi, G., Vassalli, J.D., Ferrara, N., Wasi, S., Mazzieri, R., Mignatti, P., and Pepper, M.S. (1995) Vascular endothelial growth factor increases urokinase receptor expression

in vascular endothelial cells. *J. Biol. Chem.*, **270**, 9709–9716.

108. Leupold, J.H., Yang, H.S., Colburn, N.H., Asangani, I., Post, S., and Allgayer, H. (2007) Tumor suppressor Pdcd4 inhibits invasion/intravasation and regulates urokinase receptor (u-PAR) gene expression via Sp-transcription factors. *Oncogene*, **26**, 4550–4562.

109. Ma, L., Teruya-Feldstein, J., and Weinberg, R.A. (2007) Tumour invasion and metastasis initiated by microRNA-10b in breast cancer. *Nature*, **449**, 682–688.

110. Montuori, N., Mattiello, A., Mancini, A., Taglialatela, P., Caputi, M., Rossi, G., and Ragno, P. (2003) Urokinase-mediated posttranscriptional regulation of urokinase-receptor expression in non small cell lung carcinoma. *Int. J. Cancer*, **105**, 353–360.

111. Shetty, S. and Idell, S. (1999) Post-transcriptional regulation of urokinase receptor gene expression in human lung carcinoma and mesothelioma cells in vitro. *Mol. Cell Biochem.*, **199**, 189–200.

112. Shetty, P., Velusamy, T., Bhandary, Y.P., Shetty, R.S., Liu, M.C., and Shetty, S. (2008a) Urokinase expression by tumor suppressor protein p53: a novel role in mRNA turnover. *Am. J. Respir. Cell Mol. Biol.*, **39**, 364–372.

113. Shetty, S., Shetty, P., Idell, S., Velusamy, T., Bhandary, Y.P., and Shetty, R.S. (2008b) Regulation of plasminogen activator inhibitor-1 expression by tumor suppressor protein p53. *J. Biol. Chem.*, **283**, 19570–19580.

114. Shetty, S., Velusamy, T., Idell, S., Shetty, P., Mazar, A.P., Bhandary, Y.P., and Shetty, R.S. (2007) Regulation of urokinase receptor expression by p53: novel role in stabilization of uPAR mRNA. *Mol. Cell Biol.*, **27**, 5607–5618.

115. Arts, J., Grimbergen, J., Toet, K., and Kooistra, T. (1999) On the role of c-Jun in the induction of PAI-1 gene expression by phorbol ester, serum, and IL-1alpha in HepG2 cells. *Arterioscler. Thromb. Vasc. Biol.*, **19**, 39–46.

116. Dimova, E.Y., Moller, U., Herzig, S., Fink, T., Zachar, V., Ebbesen, P., and Kietzmann, T. (2005) Transcriptional regulation of plasminogen activator inhibitor-1 expression by insulin-like growth factor-1 via MAP kinases and hypoxia-inducible factor-1 in HepG2 cells. *Thromb. Haemost.*, **93**, 1176–1184.

117. Dong, J., Fujii, S., Li, H., Nakabayashi, H., Sakai, M., Nishi, S., Goto, D., Furumoto, T., Imagawa, S., Zaman, T.A. *et al.* (2005) Interleukin-6 and mevastatin regulate plasminogen activator inhibitor-1 through CCAAT/enhancer-binding protein-delta. *Arterioscler. Thromb. Vasc. Biol.*, **25**, 1078–1084.

118. Hopkins, W.E., Westerhausen, D.R., Sobel, B.E., and Billadello, J.J. Jr. (1991) Transcriptional regulation of plasminogen activator inhibitor type-1 mRNA in Hep G2 cells by epidermal growth factor. *Nucleic Acids Res.*, **19**, 163–168.

119. Jag, U.R., Zavadil, J., and Stanley, F.M. (2009) Insulin acts through FOXO3a to activate transcription of plasminogen activator inhibitor type 1. *Mol. Endocrinol.*, **23**, 1587–1602.

120. Miyagawa, R., Asakura, T., Nakamura, T., Okada, H., Iwaki, S., Sobel, B.E., and Fujii, S. (2010) Increased expression of plasminogen activator inhibitor type-1 (PAI-1) in HEPG2 cells induced by insulin mediated by the 3′-untranslated region of the PAI-1 gene and its pharmacologic implications. *Coron. Artery Dis.*, **21**, 144–150.

121. Motojima, M., Ando, T., and Yoshioka, T. (2000) Sp1-like activity mediates angiotensin-II-induced plasminogen-activator inhibitor type-1 (PAI-1) gene expression in mesangial cells. *Biochem. J.*, **349**, 435–441.

122. Patel, N., Sundaram, N., Yang, M., Madigan, C., Kalra, V.K., and Malik, P. (2010a) Placenta growth factor (PlGF), a novel inducer of plasminogen activator inhibitor-1 (PAI-1) in sickle cell disease (SCD). *J. Biol. Chem.*, **285**, 16713–16722.

123. Wyrzykowska, P., Stalinska, K., Wawro, M., Kochan, J., and Kasza,

A. (2010) Epidermal growth factor regulates PAI-1 expression via activation of the transcription factor Elk-1. *Biochim. Biophys. Acta*, **1799**, 616–621.

124. Ikushima, H. and Miyazono, K. (2010) TGFbeta signalling: a complex web in cancer progression. *Nat. Rev. Cancer*, **10**, 415–424.

125. Meulmeester, E. and Ten Dijke, P. (2011) The dynamic roles of TGF-beta in cancer. *J. Pathol.*, **223**, 205–218.

126. Riccio, A., Lund, L.R., Sartorio, R., Lania, A., Andreasen, P.A., Dano, K., and Blasi, F. (1988) The regulatory region of the human plasminogen activator inhibitor type-1 (PAI-1) gene. *Nucleic Acids Res.*, **16**, 2805–2824.

127. Dennler, S., Itoh, S., Vivien, D., ten Dijke, P., Huet, S., and Gauthier, J.M. (1998) Direct binding of Smad3 and Smad4 to critical TGF beta-inducible elements in the promoter of human plasminogen activator inhibitor-type 1 gene. *EMBO J.*, **17**, 3091–3100.

128. Westerhausen, D.R., Hopkins, W.E., and Billadello, J.J. Jr. (1991) Multiple transforming growth factor-beta-inducible elements regulate expression of the plasminogen activator inhibitor type-1 gene in Hep G2 cells. *J. Biol. Chem.*, **266**, 1092–1100.

129. Kim, B.C., Song, C.Y., Hong, H.K., and Lee, H.S. (2007) Role of CAGA boxes in the plasminogen activator inhibitor-1 promoter in mediating oxidized low-density lipoprotein-induced transcriptional activation in mesangial cells. *Transl. Res.*, **150**, 180–188.

130. Kutz, S.M., Higgins, C.E., Samarakoon, R., Higgins, S.P., Allen, R.R., Qi, L., and Higgins, P.J. (2006) TGF-beta 1-induced PAI-1 expression is E box/USF-dependent and requires EGFR signaling. *Exp. Cell Res.*, **312**, 1093–1105.

131. Vayalil, P.K., Iles, K.E., Choi, J., Yi, A.K., Postlethwait, E.M., and Liu, R.M. (2007) Glutathione suppresses TGF-beta-induced PAI-1 expression by inhibiting p38 and JNK MAPK and the binding of AP-1, SP-1, and Smad to the PAI-1 promoter. *Am. J. Physiol. Lung Cell Mol. Physiol.*, **293**, L1281–L1292.

132. Muth, M., Theophile, K., Hussein, K., Jacobi, C., Kreipe, H., and Bock, O. (2010) "Hypoxia-induced down-regulation of microRNA-449a/b impairs control over targeted SERPINE1 (PAI-1) mRNA – a mechanism involved in SERPINE1 (PAI-1) overexpression". *J. Transl. Med.*, **8**, 33.

133. Patel, N., Tahara, S.M., Malik, P., and Kalra, V.K. (2010b)Involvement of miR-30c and miR-301a in immediate induction of plasminogen activator inhibitor-1 by placenta growth factor in human pulmonary endothelial cells. *Biochem. J.*, **434**, 473–82.

134. Gotte, M., Mohr, C., Koo, C.Y., Stock, C., Vaske, A.K., Viola, M., Ibrahim, S.A., Peddibhotla, S., Teng, Y.H., Low, J.Y. *et al.* (2010) miR-145-dependent targeting of junctional adhesion molecule A and modulation of fascin expression are associated with reduced breast cancer cell motility and invasiveness. *Oncogene*, **29**, 6569–6580.

135. Webb, G., Baker, M.S., Nicholl, J., Wang, Y., Woodrow, G., Kruithof, E., and Doe, W.F. (1994) Chromosomal localization of the human urokinase plasminogen activator receptor and plasminogen activator inhibitor type-2 genes: implications in colorectal cancer. *J. Gastroenterol. Hepatol.*, **9**, 340–343.

136. Antalis, T.M., Clark, M.A., Barnes, T., Lehrbach, P.R., Devine, P.L., Schevzov, G., Goss, N.H., Stephens, R.W., and Tolstoshev, P. (1988) Cloning and expression of a cDNA coding for a human monocyte-derived plasminogen activator inhibitor. *Proc. Natl. Acad. Sci. U.S.A.*, **85**, 985–989.

137. Kruithof, E.K. and Cousin, E. (1988) Plasminogen activator inhibitor 2. Isolation and characterization of the promoter region of the gene. *Biochem. Biophys. Res. Commun.*, **156**, 383–388.

138. Dear, A.E., Costa, M., and Medcalf, R.L. (1997) Urokinase-mediated transactivation of the plasminogen activator inhibitor type 2 (PAI-2) gene promoter in HT-1080 cells utilises AP-1 binding sites and potentiates phorbol ester-mediated induction of endogenous PAI-2 mRNA. *FEBS Lett.*, **402**, 265–272.

139. Antalis, T.M., Costelloe, E., Muddiman, J., Ogbourne, S., and Donnan, K. (1996) Regulation of the plasminogen activator inhibitor type-2 gene in monocytes: localization of an upstream transcriptional silencer. *Blood*, **88**, 3686–3697.

140. Chambers, S.K., Wang, Y., Gertz, R.E., and Kacinski, B.M. (1995) Macrophage colony-stimulating factor mediates invasion of ovarian cancer cells through urokinase. *Cancer Res.*, **55**, 1578–1585.

141. Feener, E.P., Northrup, J.M., Aiello, L.P., and King, G.L. (1995) Angiotensin II induces plasminogen activator inhibitor-1 and -2 expression in vascular endothelial and smooth muscle cells. *J. Clin. Invest.*, **95**, 1353–1362.

142. George, F., Pourreau-Schneider, N., Arnoux, D., Boutiere, B., Dussault, N., Roux-Dosseto, M., Alessi, M.C., Martin, P.M., and Sampol, J. (1990) Modulation of tPA, PAI-1 and PAI-2 antigen and mRNA levels by EGF in the A431 cell line. *Blood Coagul. Fibrinolysis*, **1**, 689–693.

143. Hannocks, M.J., Oliver, L., Gabrilove, J.L., and Wilson, E.L. (1992) Regulation of proteolytic activity in human bone marrow stromal cells by basic fibroblast growth factor, interleukin-1, and transforming growth factor beta. *Blood*, **79**, 1178–1184.

144. Pytel, B.A., Peppel, K., and Baglioni, C. (1990) Plasminogen activator inhibitor type-2 is a major protein induced in human fibroblasts and SK-MEL-109 melanoma cells by tumor necrosis factor. *J. Cell. Physiol.*, **144**, 416–422.

145. Maurer, F., Tierney, M., and Medcalf, R.L. (1999) An AU-rich sequence in the 3′-UTR of plasminogen activator inhibitor type 2 (PAI-2) mRNA promotes PAI-2 mRNA decay and provides a binding site for nuclear HuR. *Nucleic Acids Res.*, **27**, 1664–1673.

146. Tierney, M.J. and Medcalf, R.L. (2001) Plasminogen activator inhibitor type 2 contains mRNA instability elements within exon 4 of the coding region. Sequence homology to coding region instability determinants in other mRNAs. *J. Biol. Chem.*, **276**, 13675–13684.

147. Mueller, B.M., Yu, Y.B., and Laug, W.E. (1995) Overexpression of plasminogen activator inhibitor 2 in human melanoma cells inhibits spontaneous metastasis in scid/scid mice. *Proc. Natl. Acad. Sci. U.S.A.*, **92**, 205–209.

148. Laug, W.E., Cao, X.R., Yu, Y.B., Shimada, H., and Kruithof, E.K. (1993) Inhibition of invasion of HT1080 sarcoma cells expressing recombinant plasminogen activator inhibitor 2. *Cancer Res.*, **53**, 6051–6057.

149. Foekens, J.A., Peters, H.A., Look, M.P., Portengen, H., Schmitt, M., Kramer, M.D., Brunner, N., Janicke, F., Meijer-van Gelder, M.E., and Henzen-Logmans, S.C. *et al.* (2000) The urokinase system of plasminogen activation and prognosis in 2780 breast cancer patients. *Cancer Res.*, **60**, 636–643.

150. Chapman, H.A., Wei, Y., Simon, D.I., and Waltz, D.A. (1999) Role of urokinase receptor and caveolin in regulation of integrin signaling. *Thromb. Haemost.*, **82**, 291–297.

151. Jo, M., Thomas, K.S., O'Donnell, D.M., and Gonias, S.L. (2003) Epidermal growth factor receptor-dependent and -independent cell-signaling pathways originating from the urokinase receptor. *J. Biol. Chem.*, **278**, 1642–1646.

152. Wei, Y., Waltz, D.A., Rao, N., Drummond, R.J., Rosenberg, S., and Chapman, H.A. (1994) Identification of the urokinase receptor as an adhesion receptor for vitronectin. *J. Biol. Chem.*, **269**, 32380–32388.

153. Hoyer-Hansen, G., Behrendt, N., Ploug, M., Dano, K., and Preissner, K.T. (1997) The intact urokinase receptor is required for efficient vitronectin binding: receptor cleavage prevents ligand interaction. *FEBS Lett.*, **420**, 79–85.

154. Kanse, S.M., Kost, C., Wilhelm, O.G., Andreasen, P.A., and Preissner, K.T. (1996) The urokinase receptor is a major vitronectin-binding protein on endothelial cells. *Exp. Cell Res.*, **224**, 344–353.

155. Hynes, R.O. (1992) Integrins: versatility, modulation, and signaling in cell adhesion. *Cell*, **69**, 11–25.

156. Preissner, K.T. (1991) Structure and biological role of vitronectin. *Annu. Rev. Cell Biol.*, **7**, 275–310.
157. Aguirre Ghiso, J.A. (2002) Inhibition of FAK signaling activated by urokinase receptor induces dormancy in human carcinoma cells in vivo. *Oncogene*, **21**, 2513–2524.
158. Kjoller, L. and Hall, A. (2001) Rac mediates cytoskeletal rearrangements and increased cell motility induced by urokinase-type plasminogen activator receptor binding to vitronectin. *J. Cell Biol.*, **152**, 1145–1157.
159. Liu, D., Aguirre Ghiso, J., Estrada, Y., and Ossowski, L. (2002) EGFR is a transducer of the urokinase receptor initiated signal that is required for in vivo growth of a human carcinoma. *Cancer Cell.*, **1**, 445–457.
160. Vial, E., Sahai, E., and Marshall, C.J. (2003) ERK-MAPK signaling coordinately regulates activity of Rac1 and RhoA for tumor cell motility. *Cancer Cell*, **4**, 67–79.
161. Xue, W., Mizukami, I., Todd, R.F., and Petty, H.R. III (1997) Urokinase-type plasminogen activator receptors associate with beta1 and beta3 integrins of fibrosarcoma cells: dependence on extracellular matrix components. *Cancer Res.*, **57**, 1682–1689.
162. Tang, C.H., Hill, M.L., Brumwell, A.N., Chapman, H.A., and Wei, Y. (2008) Signaling through urokinase and urokinase receptor in lung cancer cells requires interactions with beta1 integrins. *J. Cell Sci.*, **121**, 3747–3756.
163. Carriero, M.V., Del Vecchio, S., Capozzoli, M., Franco, P., Fontana, L., Zannetti, A., Botti, G., D'Aiuto, G., Salvatore, M., and Stoppelli, M.P. (1999) Urokinase receptor interacts with alpha(v)beta5 vitronectin receptor, promoting urokinase-dependent cell migration in breast cancer. *Cancer Res.*, **59**, 5307–5314.
164. van der Pluijm, G., Sijmons, B., Vloedgraven, H., van der Bent, C., Drijfhout, J.W., Verheijen, J., Quax, P., Karperien, M., Papapoulos, S., and Lowik, C. (2001) Urokinase-receptor/integrin complexes are functionally involved in adhesion and progression of human breast cancer in vivo. *Am. J. Pathol.*, **159**, 971–982.
165. Unlu, A. and Leake, R.E. (2003) The effect of EGFR-related tyrosine kinase activity inhibition on the growth and invasion mechanisms of prostate carcinoma cell lines. *Int. J. Biol. Markers*, **18**, 139–146.
166. Guerrero, J., Santibanez, J.F., Gonzalez, A., and Martinez, J. (2004) EGF receptor transactivation by urokinase receptor stimulus through a mechanism involving Src and matrix metalloproteinases. *Exp. Cell Res.*, **292**, 201–208.
167. Aguirre Ghiso, J.A., Kovalski, K., and Ossowski, L. (1999) Tumor dormancy induced by downregulation of urokinase receptor in human carcinoma involves integrin and MAPK signaling. *J. Cell. Biol.*, **147**, 89–104.
168. Holmgren, L., O'Reilly, M.S., and Folkman, J. (1995) Dormancy of micrometastases: balanced proliferation and apoptosis in the presence of angiogenesis suppression. *Nat. Med.*, **1**, 149–153.
169. Murray, C. (1995) Tumour dormancy: not so sleepy after all. *Nat. Med.*, **1**, 117–118.
170. Pantel, K., Brakenhoff, R.H., and Brandt, B. (2008) Detection, clinical relevance and specific biological properties of disseminating tumour cells. *Nat. Rev. Cancer*, **8**, 329–340.
171. Kook, Y.H., Adamski, J., Zelent, A., and Ossowski, L. (1994) The effect of antisense inhibition of urokinase receptor in human squamous cell carcinoma on malignancy. *EMBO J.*, **13**, 3983–3991.
172. Yu, W., Kim, J., and Ossowski, L. (1997) Reduction in surface urokinase receptor forces malignant cells into a protracted state of dormancy. *J. Cell Biol.*, **137**, 767–777.
173. Aguirre-Ghiso, J.A., Estrada, Y., Liu, D., and Ossowski, L. (2003) ERK(MAPK) activity as a determinant of tumor growth and dormancy; regulation by p38(SAPK). *Cancer Res.*, **63**, 1684–1695.

174. Lee, K.H., Choi, E.Y., Hyun, M.S., Jang, B.I., Kim, T.N., Lee, H.J., Eun, J.Y., Kim, H.G., Yoon, S.S., Lee, D.S. *et al.* (2008) Role of hepatocyte growth factor/c-Met signaling in regulating urokinase plasminogen activator on invasiveness in human hepatocellular carcinoma: a potential therapeutic target. *Clin. Exp. Metastasis*, **25**, 89–96.
175. Bauer, T.W., Fan, F., Liu, W., Johnson, M., Parikh, N.U., Parry, G.C., Callahan, J., Mazar, A.P., Gallick, G.E., and Ellis, L.M. (2005a) Insulinlike growth factor-I-mediated migration and invasion of human colon carcinoma cells requires activation of c-Met and urokinase plasminogen activator receptor. *Ann. Surg.*, **241**, 748–756; discussion 756–748.
176. Bauer, T.W., Liu, W., Fan, F., Camp, E.R., Yang, A., Somcio, R.J., Bucana, C.D., Callahan, J., Parry, G.C., Evans, D.B. *et al.* (2005b) Targeting of urokinase plasminogen activator receptor in human pancreatic carcinoma cells inhibits c-Met- and insulin-like growth factor-I receptor-mediated migration and invasion and orthotopic tumor growth in mice. *Cancer Res.*, **65**, 7775–7781.
177. Bauer, T.W., Somcio, R.J., Fan, F., Liu, W., Johnson, M., Lesslie, D.P., Evans, D.B., Gallick, G.E., and Ellis, L.M. (2006) Regulatory role of c-Met in insulin-like growth factor-I receptor-mediated migration and invasion of human pancreatic carcinoma cells. *Mol. Cancer. Ther*, **5**, 1676–1682.
178. Wolff, C., Malinowsky, K., Berg, D., Schragner, K., Schuster, T., Walch, A., Bronger, H., Hofler, H., and Becker, K.F. (2011) Signalling networks associated with urokinase-type plasminogen activator (uPA) and its inhibitor PAI-1 in breast cancer tissues: new insights from protein microarray analysis. *J. Pathol.*, **223**, 54–63.

10
Protease Nexin-1 – a Serpin with a Possible Proinvasive Role in Cancer

Tina M. Kousted, Jan K. Jensen, Shan Gao, and Peter A. Andreasen

10.1
Introduction – Serpins and Cancer

The serpins comprise a family of proteins defined by a high degree of sequence identity and a common fold, but not a common function, with members found throughout the animal kingdom, plants, prokaryotes, and viruses with the exception of fungi. The name serpin is derived from the fact that many members of the family are inhibitors of serine proteases, such as antithrombin (AT), α_1-proteinase inhibitor or α_1-antitrypsin (an elastase inhibitor), C1 inhibitor, protein C inhibitor, plasminogen activator inhibitor-1 (PAI-1), and protease nexin-1 (PN-1) [1, 2]. Other serpins have completely different functions, being hormone carriers, intracellular chaperones, and chromatin archit ectural proteins [2–6].

Not surprisingly, in view of the multitude of serpin functions, serpins have also been implicated in a variety of functions in malignant tumors. This is true for the serpins maspin [7], myoepithelium-derived serine protease inhibitor (MEPI) [8], and PAI-2 [9], which all tend to inhibit tumor growth and/or spread, and squamous cell carcinoma antigen-1 (SERPINB3) and squamous cell carcinoma antigen-2 (SERPINB4) [1], which promote tumor growth and/or spread. The mechanism of action of these serpins in cancer is at best only partially characterized.

The serpin most extensively studied in relation to extracellular proteolysis in cancer is PAI-1. PAI-1 has two extracellular and well-established target proteases, that is, urokinase-type plasminogen activator (uPA) and tissue-type plasminogen activator (tPA). In the present context, uPA is the most relevant target, as its main function is pericellular plasmin generation, whereas the main function of tPA is generation of plasmin for vascular fibrinolysis. The hypothesis of PAI-1 promoting tumor invasion and spread is mainly based on the observation that high levels of uPA as well as of PAI-1 are associated with a poor prognosis in several cancer types [9–14]. The mechanism behind the tumor-promoting function of PAI-1, although studied in many publications, is not clear, but an important characteristic of PAI-1 expression in tumors is that in most cancers, it is stromal cells, mostly myofibroblasts, which are the predominant PAI-1 producers. At present, the most

Matrix Proteases in Health and Disease, First Edition. Edited by Niels Behrendt.

likely theory is that PAI-1 participates in shaping a tumor stroma optimal for cancer cell invasion [15–17].

The serpin phylogenetically closest to PAI-1 is PN-1. PN-1 is also known as glia-derived neurite promoting factor (GdNPF), glia-derived nexin (GDN), or serpinE2. PN-1 has a target protease specificity overlapping with that of PAI-1. It has well-established functions in the reproductive organs, the central nervous system, and the vascular system. However, its possible role in cancer has been almost totally neglected. In this chapter, we review the literature on the possible tumor biological role of PN-1 and how it relates to its basic biochemical properties.

10.2
History of PN-1

The first reports on PN-1 date back to the 1970s, when cultured rat glioma cells [18] and primary cultures of rat brain [19] were found to secrete a factor promoting neurite extension and morphological differentiation of neuroblastoma cells. Hence, the secreted protein was termed *glia-derived neurite promoting factor* or *glia-derived nexin*. PN-1 was soon recognized as a heparin binding inhibitor of serine proteases [20–23]. In fact, the motive for coining the name "nexin," which is derived from the Latin *nexus*, meaning tying or binding together, was the ability of PN-1 to form covalent complexes with target proteases and to mediate their binding to fibroblasts [20].

PN-1 protein was first purified in 1983 from media conditioned by cultured human foreskin fibroblasts [23] and later, also cultured rat glioma cells [24]. The purified human protein was subjected to Edman degradation, and the sequence of the first 28 amino acids from the N-terminus was published in 1985 [25]. Later, as the full cDNA sequence encoding human PN-1 was isolated [26, 27], the similarities between the glia-derived and fibroblast-derived "nexin" were recognized and the two were later accepted to be identical proteins. Availability of purification procedures rendered possible the production of antibodies. In particular, monoclonal antibodies directed against human PN-1 were produced in 1988 and shown to inhibit complex formation between PN-1 and target proteases [28]. The human PN-1 gene and its 5′-flanking region were cloned in 1996 [29]. Mice with targeted deletion of the PN-1 gene were constructed in 2001. The mice developed to term and grew to adulthood, but displayed characteristic phenotypic changes compared to wild-type mice [30] (see below). The X-ray crystal structure analysis of PN-1 protein remains to be reported.

The protein databases now contain entries for PN-1 from *Homo sapiens, Pan troglodytes, Macaca mulatta, Callithrix jacchus, Sus scrofa, Bos taurus, Mus musculus, Rattus norvegicus, Mesocricetus aureatus, Ornithorhynchus anatinus, Gallus gallus, Xenopus laevis, Xenopus tropicalis*, and *Danio rerio*. Since mammals, birds, amphibians, as well as fish are represented in the list, it seems likely that PN-1 exists in all vertebrates. The sequence identity between PN-1 from fish and man amounts to ~85%, indicating a highly conserved protein.

10.3 General Biochemistry of PN-1

Phylogenetically, PN-1 (encoded by the *SERPINE2* gene) is most closely related to PAI-1 (encoded by the *SERPINE1* gene) and *myxoma virus* SERP-1 (encoded by the *SERPIN SPI-1* gene) [1]. The human *SERPINE1* and *SERPINE2* genes have the same exon–intron structure, but are located on different chromosomes: the PN-1 gene on chromosome 2, region q33–q35 [31] and the PAI-1 gene on chromosome 7, region q21.3–q22 [1, 32]. The PN-1 gene consists of nine exons. The 5′ and 3′ untranslated regions of PN-1 mRNA are 137 and 819 bp, respectively (Figure 10.1).

Early studies identified two forms of human PN-1 mRNA, which differ by the insertion of a CAG triplet at the position encoding amino acid 310, resulting in either Arg or Thr-Gly at this position. The two forms of PN-1 (designated α and βPN-1) are the likely splice variants resulting from a shifted exon–intron boundary [27, 33]. Both PN-1 variants were shown to be active serpins [33]; however, the physiological relevance of this variation remains to be thoroughly investigated. The sequences and numbering presented in this review are those of βPN-1.

PN-1 mRNA is translated into a M_r 43 000–50 000 globular glycoprotein of 379 amino acids, with three potential N-linked glycosylation sites (Asn^{99}, Asn^{140}, and Asn^{365}) (Figure 10.1). According to a tertiary structure model of PN-1 based on

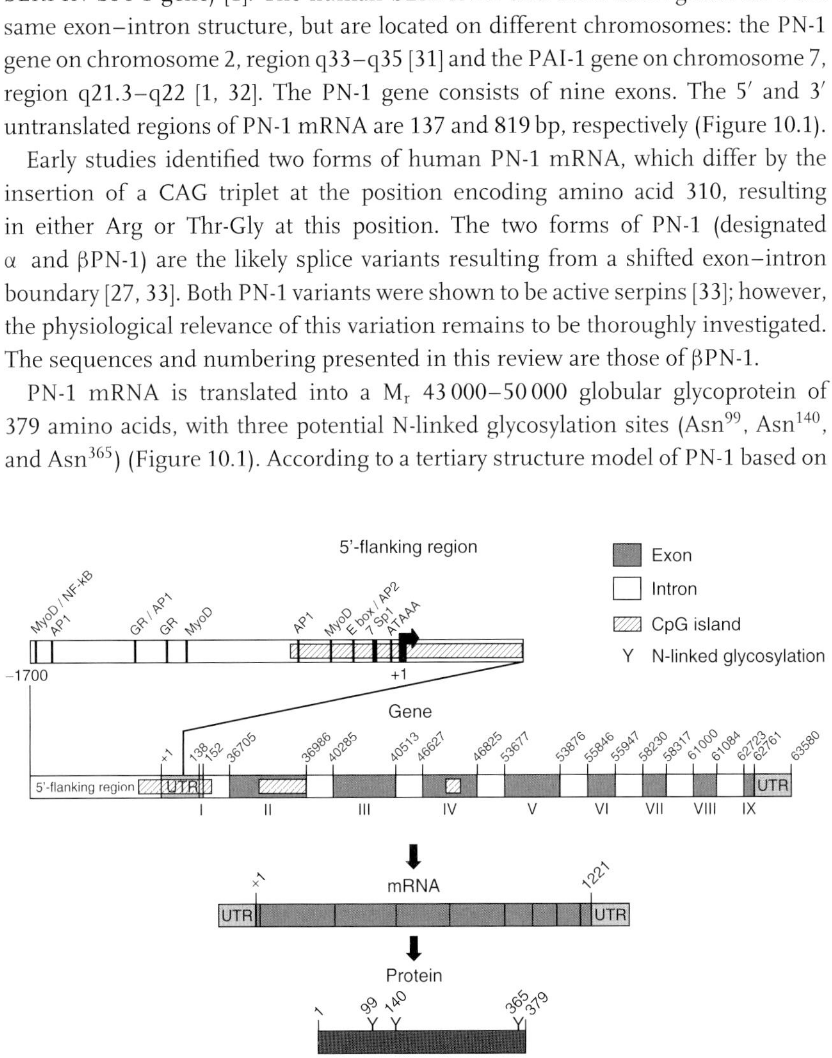

Figure 10.1 Schematic overview of the PN-1 gene, mRNA, and protein. In the 5′-flanking region of the gene, the localization of binding sites for transcription factors are indicated (GR = glucocorticoid receptor). In the coding region, the exon/intron organization and the 5′ and 3′ untranslated regions (UTRs) are shown. CpG islands are indicated by the inserted boxes with tilted-line pattern. In the final translated protein, the localization of glycosylation sites is indicated.

the available structure of PAI-1 (PDB file 3Q03, [34]) (Figure 10.2), the three potential glycosylation sites are all located in surface-exposed areas at the s2A-hE-loop, hF, and the s4B-s5B-loop. It is still not known which glycosylation sites are actually utilized. Also, according to this structural model, the three cysteines of PN-1 (Cys^{117}, Cys^{131}, and Cys^{209}) are potentially solvent accessible and located distant from each other, namely, in the hE-s1A loop, hF, and s1B, excluding the existence of intramolecular disulfide bonds in PN-1 (Figure 10.2). Experimental data supports the presence of three free thiol groups per molecule and indicates that at least one of these is in an immediately solvent-accessible conformation [35].

PN-1 has strong affinity for glycosaminoglycans, binding commercially available heparin with a K_D below 100 nM and heparan sulfate and condroitin sulfate proteoglycans obtained from neonatal rat brain with a slightly lower affinity [36, 43–45]. Among rat brain heparan sulfate proteoglycans, mainly glypican-1 and cerebroglycan bind a PN-1 affinity column, while syndecan-3 mostly flows through the column. Since PN-1 exhibited only minor selectivity among heparan sulfate chains, the differential binding of proteoglycans is believed to be due to differences in proteoglycan core proteins [44]. By means of site-directed mutagenesis, the heparin-binding site of PN-1 has been narrowed down to residues Lys71–Lys86. The PAI-1 counterparts of these positions comprise helix D and the loop connecting helix D and β-strand 2A [36, 37]. In two other serpins, AT and heparin cofactor II, the heparin binding sites are localized in the same region [46–48].

PN-1 and PN-1-protease complexes have also been reported to bind vitronectin in solid-phase enzyme assays [49]. However, the affinity of PN-1 for vitronectin and further characterization of the binding have to our knowledge not been reported.

10.4
Inhibitory Properties of PN-1

The main features of the serpin inhibitory mechanism were established by work in several laboratories around 2000 [1, 2, 50, 51]. Besides a variety of biochemical work, the now generally accepted serpin inhibitory mechanism builds, in particular, on X-ray crystal structure analysis of a number of serpin-protease Michaëlis complexes [40, 46, 47, 52–55] and two stable, covalently linked complexes [41, 56]. The serpin inhibitory mechanism, as depicted in Figure 10.2, is the following: The serpin-protease reaction is initiated by formation of a reversible docking or Michaëlis complex, in which the reactive center loop (RCL) of the serpin inserts into the active site of the protease, with the P_1-P_1' bond of the serpin arranged in a substratelike manner. In the following locking step, the P_1-P_1' bond is cleaved and the P_1 residue is coupled to the active site serine of the protease by an ester bond. The P_1 residue is now released from the P_1' residue, the N-terminal part of the RCL is inserted as strand 4 in β-sheet A (s4A), and the protease is translocated to the opposite pole of the molecule. Here, the P_1 residue of the serpin is pulled out of the active site of the protease and the active site of the protease is distorted so that the serpin remains bound to the protease at the acyl-enzyme intermediate step of the catalytic cycle. The energy needed for the distortion of the protease stems from

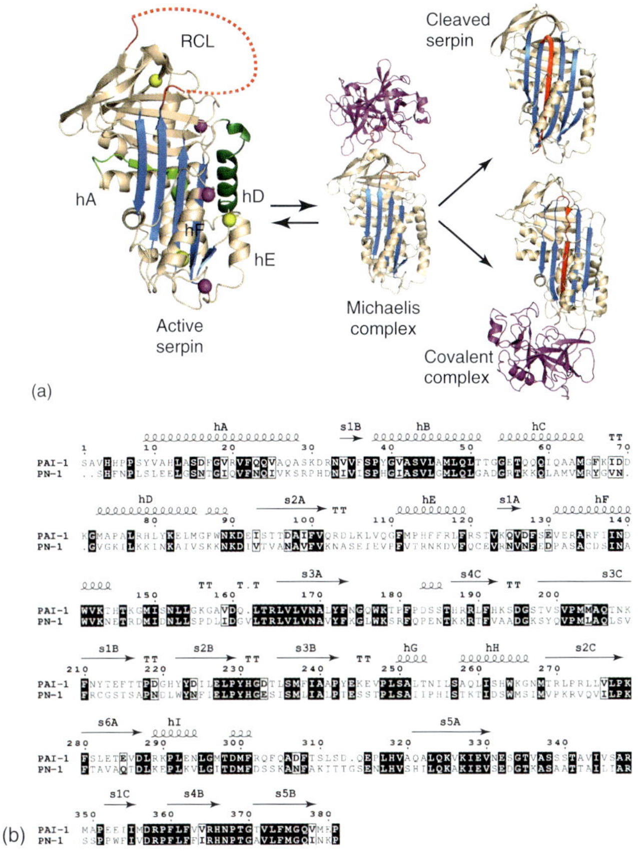

Figure 10.2 Three-dimensional visualization of PN-1 and the serpin inhibitory mechanism. (a) The left outmost figure shows a three-dimensional structure model of PN-1, constructed on the basis of the crystal structure of the PAI-1 mutant W175F (PDB file 3Q03; [34]). The localization of cysteines (yellow) and potential N-linked glycosylation sites (purple) are represented by spheres at the corresponding Cα positions. The proposed heparin binding region comprising helix D and overlapping with the primary LRP-1A binding region is shown in dark green [36, 37]. An alternatively suggested LRP-1A-binding region [38, 39] is shown in light green. The presumably highly flexible RCL, which is not visible in X-ray crystal structure analysis, is sketched by a broken red line. For orientation, β-sheet A is highlighted in blue. From left to right, the serpin inhibitory mechanism is shown: a Michaelis complex between a serine protease and a serpin (the uPA mutant S195A and the PAI-1 mutant 14-1B; PDB file 3PB1; [40]), a covalent complex formed between a serine protease and a serpin (α_1-PI and porcine pancreatic elastase; PDB file 2D26; [41]), and a reactive center cleaved serpin (PAI-1 14-1B; PDB file 3CVM, [42]). Highlighted are the reactive center loop (red), β-sheet A (blue), and the serine proteases (purple). (b) Alignment of the amino acid sequences of human PN-1 and human PAI-1. Secondary structural elements, as identified by X-ray crystal structure analysis of PAI-1 (PDB file 3Q03), are indicated above the sequences. S = β-strand; h = α-helix. Conserved and partially conserved positions are highlighted in the sequence by filled and empty squares, respectively. The alignment was created using the tools, *http://multalin.toulouse.inra.fr/multalin/* and *http://espript.ibcp.fr/ESPript/*.

Table 10.1 Second-order rate constants (M^{-1} s^{-1}) for inhibition of various proteases by human PN-1.

	PN-1	PN-1 with pentasaccharide[a]	PN-1 with heparin[b]	References
Human thrombin	$0.6–2 \times 10^6$	8×10^5	$10^8–10^9$	[25, 35, 61]
Human uPA	2×10^5	–	–	[62]
Human plasmin	10^5	9×10^3	10^5	[25, 35]
Human one chain tPA	2×10^3	–	–	[63]
Human two chain tPA	3×10^4	–	–	[63]
Human FXa	7×10^3	–	–	[25]
Human FXIa	8×10^4	–	2×10^6	[64]
Human FXIIa	2×10^3	–	–	[65]
Human plasma kallikrein	3×10^5	–	–	[65]
Human C1s	3×10^3	–	–	[65]
Human activated protein C	5×10^3	–	–	[61]
Human leukocyte elastase	N.D.	–	–	[25]
Human cathepsin G	N.D.	–	–	[25]
Bovine FXa	$5–7 \times 10^3$	5×10^3	4×10^5	[35, 61]
Bovine trypsin	5×10^6	–	–	[25, 61]
Bovine chymotrypsin	N.D.	–	–	[25]
Porcine trypsin	2×10^6	2×10^6	10^7	[35]

abbreviation: N.D., Not detected.
[a]Synthetic heparin-derived pentasaccharide corresponding to the antithrombin binding sequence of herparin [66, 67].
[b]Heterogeneous nonfractionated heparin purified from natural sources.

stabilization of the serpin in the "relaxed" conformation by insertion of the RCL into β-sheet A, as opposed to the "stressed," relatively unstable active conformation with a surface-exposed RCL (Figure 10.2).

None of the studies establishing the basic serpin inhibitory mechanism have involved PN-1, but there is no reason to believe that the same basic mechanism is not being followed. PN-1 is less specific than PAI-1, inhibiting, besides uPA, plasmin and thrombin at physiologically relevant rates, trypsin at a similar rate [1, 20, 25], and a number of other serine proteases at a slower rate (Table 10.1). PN-1 has also been shown to form inhibitory complexes with prostasin [57, 58], matripase [59], and factor-VII-activating protease (FSAP) [60]; however, the kinetics of inhibition have not yet been reported.

Binding of heparin and heparinlike glycosaminoglycans most prominently increase the rate of reaction of PN-1 with thrombin, which becomes the preferred target protease in the presence of heparin [25, 35]. The rate of reaction of PN-1 with other proteases is affected much less or not at all by the presence of heparin (Table 10.1). Glycosaminoglycans accelerate the reaction of many serpins with their target proteases. Basically, there are two different, but not mutually exclusive, mechanisms that may account for the acceleration. In one scenario, the glycosaminoglycan induces an allosteric change in the serpin, accelerating its reaction with proteases. In another scenario, the glycosaminoglycan forms a bridge between the serpin and the protease, facilitating their interaction by bringing them closer together in the appropriate orientation [68]. Besides numerous biochemical investigations [68], these mechanisms are supported by recent determinations of the three-dimensional structures of Michaëlis complexes between thrombin and AT [53] and between thrombin and protein C inhibitor [54], in which cases both the allosteric and the bridging mechanisms are active, and between factor Xa and AT [55] and between factor IXa and AT [46], in which cases the allosteric mechanism is active. Several studies indicate that glycosaminoglycans increase PN-1's rate of inhibition of certain target proteases by means of a bridging mechanism, in which the linear glycosaminoglycan simultaneously binds both serpin and protease. This conclusion is deduced from the observation of a bell-shaped dependence of the acceleration of reaction rate on the heparin concentration, which is one of the consequences of a bridging mechanism [43]. The reaction between PN-1 and various target proteases was not accelerated by prior incubation with the specific pentasaccharide of heparin [35], which is known to bind AT with high affinity and induce conformational activation of the serpin [69, 70], but is too short to allow the bridging mechanism [4, 71]. In general, PN-1 has appeared less selective for certain heparin or heparan sulfate subspecies than AT [36, 43, 44], and it is therefore likely that PN-1 is able to bind the pentasaccharide, but that conformational changes (if any) induced on binding do not significantly affect the rate of inhibition by PN-1. In this respect, it is noteworthy that a change in intrinsic tryptophan fluorescence similar to that shown for AT on heparin binding, which reflects the heparin-induced conformational changes, is not observed for PN-1 [45], indicating that heparin binding does not induce conformational changes in PN-1 similar to those induced in AT. Another characteristic of the bridging mechanism is that a minimum length of the glycosaminoglycan is required in order to accommodate both serpin and protease [4]. The minimum length of the activating glycosaminoglycan required to bridge PN-1-thrombin is not yet determined, but for AT, a 16 saccharide unit glycosaminoglycan was shown to bridge the respective complex with thrombin [71]. PN-1 appears to retain its high affinity for heparin in both the substrate-cleaved form and the complexed form [35].

10.5 Binding of PN-1 and PN-1-Protease Complexes to Endocytosis Receptors of the Low-Density Lipoprotein Receptor Family

The low-density lipoprotein receptor (LDLR) family of endocytosis receptors includes a number of receptors composed of the same type of domains, but with variable sequence and a highly variable number of each type of domain. The known mammalian members of the family are, besides LDLR itself, very low density lipoprotein receptor (VLDLR), apolipoprotein E receptor 2 (apoER2) or low density lipoprotein receptor-related protein 8 (LRP8) multiple epidermal growth

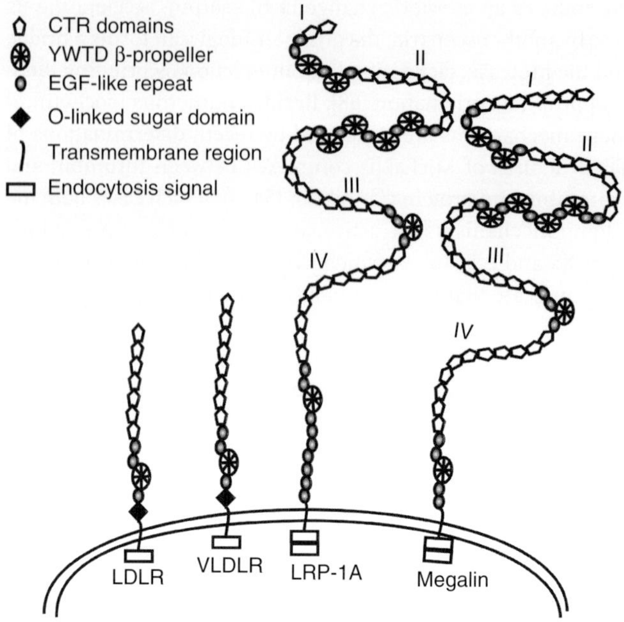

Figure 10.3 Endocytosis receptors of the LDLR family. Selected members of the LDLR family of endocytosis receptors are depicted in a schematic presentation. The extracellular N-terminal ligand binding regions of the receptors contain a varying array of consensus domains followed by a short transmembrane segment and a cytoplasmic part containing motifs mediating internalization through clathrin-coated pits. Clusters of ligand binding CTR domains are numbered by roman numerals from the N-terminus in receptors containing multiple clusters.

factor (EGF) repeat-containing protein (MEGF7), LRP-1A, LRP-1B, and megalin or LRP-2 (Figure 10.3). The domains include complement-type repeats (CTRs), YWTD repeats, and EGF precursor domains. All the receptors have a transmembrane α-helix and a cytoplasmic C-terminal domain with one or more sequences mediating endocytosis via clathrin-coated pits. Generally, these receptors are not only endocytosis receptors but also have signaling functions [72–75].

The classic ligands for receptors of the LDLR family are apolipoproteins B100 and E [72]. In 1990, however, it was discovered that receptors of the family also have non-apolipoprotein ligands, as the receptor for protease-complexed α_2-macroglobulin was found to be identical to LRP-1A [76, 77], the amino acid sequence of which had been reported two years earlier [78]. In 1992, it was reported that tPA-PAI-1 complex [79] and uPA-PAI-1 complex [80] can bind to and be endocytosed by LRP-1A. Later, a variety of protease-serpin complexes were found to bind to and be endocytosed by members of this receptor family [81].

Ligand binding by receptors of the LDLR family is mainly mediated through the CTR domains. The fact that each of these receptors binds many different ligands with nanomolar K_D values is believed to be due to each ligand interacting with a specific combination of two or more CTR domains: The more CTRs interacting simultaneously, the higher the affinity [82]. It has been suggested that in order for a certain ligand to have affinity to a CTR domain, it must contain common binding motifs centered on a basic residue, primarily a lysine [83].

The first report of PN-1 binding to a cellular receptor resulting in its endocytosis and degradation was presented in 1987 [84]. The receptor responsible for this observation was later shown to be LRP-1A [85]. The uPA-PN-1 complex appears to represent a special case as far as the binding to LRP-1A and VLDLR is concerned. The uPA-PN-1 complex also binds to LRP-1A as the uPA-PAI-1 complex and appears to have a particularly high affinity to VLDLR, 5- to 15-fold higher than that of uPA-PAI-1 complex [86]. It is the only protease-serpin complex known to be able to compete the binding of the high-affinity chaperone, $M_r \sim 40\,000$ receptor-associated protein (RAP) [87].

The early observation that a binding site for receptors of the LDLR family was present in PN-1 was based on cell culture experiments in which internalization and degradation of PN-1 complexes with thrombin or uPA was shown to be dependent on LRP-1A. Several trials were later done to more precisely localize the receptor binding site in PN-1 [37, 38, 88–90]. Subsequently, PN-1 residues localized in its heparin binding site in α-helix D were strongly implicated in receptor binding primarily by the use of a heparin binding site mutant K71–K86 (a mutant with all seven lysines between K71 and K86, situated in hD and the following loop, mutated to glutamic acid) [37]. The P47–I58 region of PN-1 has been suggested to be implicated in binding of thrombin-PN-1 complexes to LRP-1A, although the evidence was, in this case, limited to peptide competition [38, 90]. It was later proposed that PN-1's receptor binding site is cryptic, uncovered only on formation of a complex with a target protease [39].

In order to study the involvement of the individual CTR domains within the ligand binding CTR cluster II (CTR3–10) of LRP-1A in the binding of serpin-protease complexes, two- and three-domain fragments, covering seven of the eight CTRs, were expressed, refolded, and analyzed. A general high affinity for a reactive center loop-cleaved form of PAI-1 and PN-1 was observed to combinations of CTRs from throughout the cluster. It became evident that, in both cases, two CTR domains constitute the majority of the binding interaction, as addition of a third CTR domain resulted in only a marginal enhancement of binding affinity [91]. In PAI-1, the binding site for LRP-1A is located in hD and hE, areas easily spanned by two CTR domains [92]. Furthermore, it was observed that both the native and cleaved forms of both PN-1 and PAI-1 bind cluster II with similar high affinity (K_D 3–10 nM). Complexation of PN-1 or PAI-1 with a serine protease further improves the overall affinity through a few additional favorable interactions, enough to secure a more efficient receptor binding and endocytosis of the complexes as compared to the free serpins at physiologically relevant concentrations [91]. The observed similarity in binding behavior between the two serpins, consistent with the high abundance of lysine residues in and around hD, strongly suggest that the LRP-1A binding site in PN-1 is localized in the same secondary structural elements as in PAI-1.

Our knowledge about CTR domain-ligand interactions would be greatly increased by determinations of three-dimensional structures of physiologically relevant receptor-ligand complexes, but to date, only two structures have been determined, namely, that of a five-CTR domain from VLDLR in complex with a minor group human rhinovirus [93] and a two-CTR domain from LDLR in complex with domain 3 from M_r ~40 000 RAP [94].

10.6
Pericellular Functions of PN-1 in Cell Cultures

PN-1 is generally considered to be a cell-associated serpin, which serves to regulate protease activity at or near the cell surface. This idea was founded already in the mid-1980s, when studies on PN-1 in cultures of human foreskin fibroblasts established the importance of heparinlike glycosaminoglycans for target specificity and pericellular localization of PN-1 [20, 95, 96]. It has been hypothesized that PN-1 associates with cell surface heparinlike glycosaminoglycans shortly after secretion, most likely on the PN-1-secreting cells themselves. The binding of PN-1 to cell surface glycosaminoglycans would subsequently direct the inhibitory potency of PN-1 toward certain target proteases, either because their binding to PN-1 is stimulated by glycosaminoglycans or because they themselves are localized on the cell surface. Unlike most other heparin binding serpins, PN-1 retains high affinity for heparin after complex formation with target proteases [35, 37]. The continued association with cell surface may favor rapid internalization and degradation of inhibitor-protease complexes by facilitating their interaction with endocytosis receptors of the LDLR family. This scenario has been proposed for PN-1-thrombin complexes based on *in vitro* studies of fibroblasts [37, 88] and astrocytes [89]. The potential overlap between the LRP-1A and glycosaminoglycan binding sites on PN-1 suggests a sequence of events in which the protease-PN-1 complexes are shifted from their binding to the glycosaminoglycans to the endocytosis receptor. An additional role of heparinlike glycosaminoglycans, at a later step in the pathway than cell surface binding, was suggested by Crisp *et al.* [90], who hypothesized that the interaction with endosomal heparins may be crucial for proper sorting of PN-1-protease complexes to the lysosomes, explaining the reduced rate of catabolism of complexes formed between a non-heparin-binding PN-1 variant and uPA, despite unaltered cell binding and internalization rates. Complexes between PN-1 and uPA may also associate with cell surfaces through the urokinase-type plasminogen activator receptor (uPAR), which then facilitates the interaction between the inhibitor-protease complex and proper endocytosis receptors. Data on both uPAR-dependent [85, 97] and uPAR-independent internalization [86] of PN-1-uPA complex have been reported. One idea is that PN-1-uPA complex preferentially binds uPAR, but if uPAR is inaccessible, PN-1-uPA complex may associate with the cell surface by alternative mechanisms, most likely through heparinlike glycosaminoglycans [90].

There is evidence suggesting that PN-1 may induce different signaling pathways depending on the interacting molecules on the cell surface. Thus, the interaction of PN-1 with LRP-1A has been reported to induce protein kinase A activation in murine fibroblasts [98] and MAPK/ERK signaling in murine mammary cells [99], whereas PN-1-induced MAPK/ERK signaling in LRP-1A-deficient murine fibroblasts by an alternative mechanism involving the proteoglycan syndecan-1 [98]. PN-1 may also affect cell signaling indirectly by competing with other ligands for binding and signaling through LRP-1A. This is speculated to be the mechanism by which PN-1 antagonizes sonic hedgehog-induced proliferation of cerebellar granular neuron precursors, providing an explanation for the cerebellar abnormalities observed in PN-1-deficient mice [100].

The ability of PN-1 to protect extracellular matrix against pericellular proteolysis [101] provides a possible explanation for the observed proadhesive, antiapoptotic, and antimigratory effects of PN-1, which have been reported based on *in vitro* cell-based assays [102, 103] and both *ex vivo* [104] and *in vivo* [105] neuronal migration studies. However, inhibition of pericellular proteolysis by PN-1 does not explain the effects of PN-1 on uPAR-dependent cell adhesion to vitronectin [45, 106, 107], in which case other molecular mechanisms such as integrin-matrix interactions, cell signaling, or intracellular trafficking of cell surface receptors may contribute to the overall picture.

In addition to *in vitro* and *in vivo* studies, suggesting a role of PN-1 in cancer cell migration and/or invasion (Section 10.9), these results support a role of PN-1 in the regulation of cell adhesion and motility, although the data thus far does not give a consistent or conclusive picture of the underlying mechanisms.

10.7 PN-1 Expression Patterns

10.7.1 Expression of PN-1 in Cultured Cells

Early studies indicated that a factor, later shown to be PN-1, is secreted from cultured human, mouse, and rat glial cells [18]. PN-1 is now known to be synthesized and secreted by a wide range of cultured cells derived from normal tissues, including human fibroblasts [20], human endothelial cells [108], human skeletal muscle myotubes [109], rat vascular smooth muscle cells [110], and mouse renal epithelial cells [111]. Also, expression of PN-1 was found in a number of human cell lines of neoplastic origin, the fibrosarcoma cell line HT1080 showing high levels of expression, while weaker expression was found in the breast carcinoma cell lines MCF-7 and T47D, the monocyte-derived cell line U937, and T-cell leukemia cell line Jurkat [112].

In cultured cells, PN-1 expression has been shown to be regulated by a variety of hormones, growth factors, and cytokines. The following factors have been shown to upregulate PN-1 expression in the indicated cell lines or primary cell cultures: TGF-β in SK-N-SH neural cells [113], human astrocytes [113], and murine renal epithelial cells [111]; platelet-derived growth factor in SK-N-SH neural cells and human astrocytes [113]; interleukin-1 in skin fibroblast [114], SK-N-SH cells [113], human astrocytes [115], and human U373-MG astrocytoma cells [115]; interleukin-6 and interleukin-10 in human astrocytes [115]; vasoactive intestinal peptide in Schwann cells [116] and astroglial cultures [117]; chromogranin A in a mouse anterior pituitary tumor cell line 6T3 [118]; and human chorionic gonadotropin (hCG) in cells in the bovine preovulatory follicular wall [119]. Furthermore, inhibition of hepatocyte growth factor (HGF)-c-MET signaling, by means of specific blocking antibodies targeting c-MET, was shown to suppress PN-1 expression in human colon xenograft tumors [120]. Although the signaling pathways involved in the upregulation of PN-1 expression have not yet been thoroughly mapped, the involvement of the MAPK/ERK signaling pathway was supported by the use of a specific MAPKK/MEK inhibitor in human colorectal cancer cell lines [121].

While angiotensin II and calcitonin gene-related peptide downregulate the PN-1 expression in cultured Schwann cells [122], angiotensin II had no effect on PN-1 expression in rat aortic smooth muscle cells [110]. Aldosterone caused a twofold reduction of PN-1 expression in murine renal epithelial cells [111]. In addition to PN-1, TGF-β1 and aldosterone also regulate the expression of sodium channels and prostasin in renal epithelial cells, and it has been hypothesized that PN-1 may participate in the coordinated regulation of sodium reabsorption in the kidney through inhibition of prostasin, an activator of renal sodium channels [123].

10.7.2
Mechanisms of Transcriptional Regulation of PN-1 Expression

The following regulatory elements have been identified within a 1.8 kb segment upstream of the transcriptional start site (including 179 bp of exon 1) of the human *SERPINE2* gene: seven Sp1 binding sites, three MyoD binding sites, three AP-1 sites, two AP-2 sites, one NF-κB binding site, and two glucocorticoid response elements (GREs) [29] (Figure 10.1). Furthermore, a perfect consensus sequence (GCGGGGGCG), representing the binding site of the Krox family of transcription factors, has been identified in the promoter of the rat and human PN-1 gene [124, 125]. A TNF-α-responsive NF-κB p65 binding element has been identified in the distal promoter of the mouse PN-1 gene, which plays an important role in TNF-α-induced expression of PN-1 in mouse embryonic fibroblasts [126]. The high degree of homology within this region in both human and rat PN-1 genes suggests that similar cis-acting sites regulate PN-1 transcriptional activity in those species [124, 125].

The *SERPINE2* gene promoter region is highly GC rich. A CpG island, containing 155 CpGs, spanning from −539 to +776 bp relative to the transcription initiation site, has been identified (Figure 10.1) [112]. Within recent years, evidence, suggesting that PN-1 expression could be regulated by epigenetic mechanisms has been obtained. Thus, methylation at specific sites in the promoter region of the *SERPINE2* gene appears to downregulate PN-1 expression in cultured rat hepatoma cells [124], whereas inhibition of CpG methylation and/or histone deacetylase increased PN-1 expression in three human cell lines [112]. Also, when comparing human fetal and adult lungs, an inverse correlation between methylation and expression of the *SERPINE2* gene was found [127].

10.7.3
Expression of PN-1 in the Intact Organism

In rodents, PN-1 protein, as evaluated by immunohistochemistry or immunoblotting, is most abundantly expressed in the male reproductive organs, namely, in the seminal vesicle and epididymis, but is also widely expressed in the nervous system as well as in other organs and tissues, including lung, liver, spleen, kidney, heart, prostate, ovary, muscle, and cartilage [58, 128–133]. The expression of PN-1 has been confirmed at the mRNA level by *in situ* or Northern blot hybridization in various rodent organs and tissues [128, 129, 132–134].

PN-1 is present in human blood platelets but is not detectable in human plasma [135–138]. In the human brain, PN-1 protein and mRNA were detected in the proximity of vessels, more specifically in astroglial processes and smooth muscle cells of the vessel wall [128]. Likewise, immunoreactivity of PN-1 was found in the arterial media smooth muscle cells of rats [110]. PN-1 mRNA and protein in the vascular wall were found increased (threefold and twofold, respectively) in a chronic hypertension rat model during long-term inhibition of nitric oxide synthase compared to control rats [110].

PN-1 protein levels were greatly increased in lung, prostate, and pancreas and slightly increased in seminal vesicles of *MMP-9* knockout mice as compared to wild-type mice, which supports *in vitro* evidence that PN-1 is a potential substrate for MMP-9 and suggests MMP-9 mediated cleavage to be an alternative mechanism of regulation, which is posttranscriptional [131].

10.8
Functions of PN-1 in Normal Physiology

10.8.1
Reproductive Organs

Highly informative indications on the physiological role of PN-1 come from studies performed on mice deficient in the *SERPINE2* gene. These mice develop to term, grow to adulthood, but suffer from male infertility [30], suggesting the essential role of PN-1 in the male reproductive organs. In accordance, among a wide variety of mouse tissues expressing PN-1 [129], the level was highest in male reproductive organs, most prominently in the seminal vesicle as mentioned above; its expression there is under androgen control [132, 139]. On the basis of abnormal protein composition of seminal fluid and the inability to form functional copulatory plugs, both of which characterize PN-1-deficient male mice, it was speculated that

PN-1 could serve to regulate the proteolytic activity in the seminal fluid [30]. A recent study indicates that PN-1 in murine seminal fluid could contribute to the regulation of sperm cell capacitation, an important step in the maturation of sperm cells following ejaculation, which involves destabilization of the acrosomal sperm head membrane and is a prerequisite for oocyte fertilization [139]. However, in spite of the fact that PN-1 is abundant in the male reproductive tract, its target enzyme there is not known [140]. In humans, immunoblot analyses indicate that PN-1 is present only in low amounts in semen from fertile males, whereas abnormally high level of PN-1 is found in semen from males with fertility problems and altered seminal vesicle secretory activity [30]. Thus, it seems unlikely that PN-1 plays the same role in human as in murine male fertility.

Interestingly, disruption of the gene for another serpin with a protease specificity overlapping with that of PN-1, that is, protein C inhibitor, also results in murine male infertility, due to impaired spermatogenesis. Protein C inhibitor is present in human semen in relatively high concentrations ($\sim$160 $\mu g\,ml^{-1}$) and could serve a role similar to the one that PN-1 plays in rodent seminal fluid [140].

A number of studies have suggested a role of PN-1 in female reproductive tissues at times of extensive tissue remodeling, including the ovarian follicle at ovulation, the placenta during implantation, and the mammary gland during pregnancy and lactation [141]. The fact that female PN-1-deficient mice remain fertile suggests that a level of functional redundancy between PN-1 and other serpins may exist.

10.8.2 Neurobiological Functions

Among possible PN-1-expressing cells in the central and peripheral nervous system are fibroblasts, glial cells, including astrocytes and Schwann cells, and a subset of neuronal cells, which are also known to express PN-1 in culture [18, 20, 24, 115, 116, 130, 142]. By *in situ* hybridization and immunohistochemistry, constitutive expression of PN-1 was found in a number of structures of the rat brain, including the olfactory system, which is characterized by a constant neuronal degeneration and regeneration [130, 143]. Furthermore, distinct expression patterns of PN-1 during mouse embryogenesis and in the developing postnatal brain have been reported [129, 144], suggesting a role of PN-1 in the maturation and maintenance of the central nervous system.

On the basis of studies on PN-1-deficient mice, PN-1 has been implicated in the regulation of neuronal progenitor cell proliferation and survival in the developing cerebellum and adult hippocampus [100, 145]. Both overexpression and lack of PN-1 in mouse brain have been shown to alter hippocampal long-term potentiation [146], and overexpression of PN-1 leads to progressive neuronal and motor dysfunction [147]. Further studies of PN-1-deficient mice have indicated that dysregulation of the NMDA receptor and increased brain tPA activity are likely to be involved in the impaired functioning of sensory pathways observed in PN-1-deficient mice, supporting the importance of balanced serine protease/serpin levels for maintaining normal brain function [148].

In rodents, PN-1 is highly upregulated at sites of neuronal cell damage both within the central and peripheral nervous system [149, 150]. Interestingly, PN-1 deficiency significantly slowed the kinetics of reinnervation and functional recovery following injury of the sciatic nerve [145].

On the basis of the observation that PN-1 is predominantly localized to blood vessels and capillaries within the human brain, PN-1 is thought to serve a neuroprotective role against the deleterious effects posed by extravasated serine proteases [128]. Thrombin has been shown to exert a number of negative effects on cells of the nervous system both *in vitro* and *in vivo*, and elevation of thrombin levels in brain tissue has been associated with neuropathologies such as Alzheimer's disease, multiple schlerosis, and ischemia [151, 152]. In accordance with the assumption of a neuroprotective role of PN-1, increased expression of PN-1 accompanied the appearance of thrombin in brain tissue following cerebrovascular injury [153]. Also, PN-1 was more frequently found in complex with thrombin than in free form near vessels within certain areas of the brain of patients with Alzheimer's disease [154–157], suggesting that PN-1 is sequestrated by extravasated thrombin following disruption of the blood–brain barrier.

Thus, the bulk of experimental work concerning the neurobiological role of PN-1 has until now focused on the consequences of PN-1/thrombin imbalance within the nervous system. There is no doubt that PN-1 plays important roles during development and maintenance of the nervous system, but details regarding the underlying mechanisms are still largely unknown.

10.8.3 Vascular Functions

Within the vascular system, potential sources of PN-1 are vascular smooth muscle cells [110], endothelial cells [108], and platelets, which have been shown to store PN-1 and carry PN-1 on their surfaces [135–137]. Given the expected low to subnanomolar concentrations of PN-1 [138] and active thrombin in human plasma and the micromolar concentrations of AT, the relevance of PN-1 as an inhibitor of circulating thrombin seems unlikely. However, in a recent study, Boulaftali and coworkers [158] found that platelets derived from PN-1-deficient mice exhibited increased responsiveness to thrombin-induced activation and aggregation and, despite no apparent vascular disorders, exhibited significantly faster thrombus formation rates on vascular injury. This supports the notion that locally accumulated PN-1 on the surface of platelets may exert an antithrombotic effect at thrombin doses or exposure times below a certain threshold [138].

A potential antithrombotic role of PN-1 is supported by the finding that PN-1 in the presence of heparin is a potent inhibitor of Factor XIa, which is involved in thrombin generation through the intrinsic coagulation pathway [64]. Furthermore, PN-1 has been reported to associate with thrombomodulin on the surface of endothelial cells, potentially through interaction with the thrombomodulin chondroitin sulfate moiety. The association of both thrombin and PN-1 with thrombomodulin accelerated inhibition of thrombin and delayed thrombin-induced fibrin clotting, protein C activation, and activation of thrombin activable fibrinolysis inhibitor (TAFI) *in vitro* [159].

Interestingly, Boulaftali and coworkers [160] recently reported that PN-1-deficient mice, in addition to increased rates of thrombosis, also exhibited increased responsiveness to tPA-induced recanalization of occluded vessels following vascular injury. They hypothesize that platelet PN-1 together with platelet PAI-1, may contribute to the stabilization of the thrombus by inhibiting local plasmin generation and activity.

PN-1 has been found in high concentrations within atherosclerotic lesions, in which platelets and monocytes/macrophages were identified to be the main source [106, 137]. Also, immunostaining revealed that PN-1 colocalizes with vitronectin, uPAR, and uPA in atherosclerotic vessels, where the role of PN-1 has been speculated to involve regulation of serine proteases present in vascular lesions [45, 106, 137].

Together, these results suggest that PN-1 may exert local antithrombotic and/or antifibrinolytic effects at or near surfaces of PN-1-secreting cells of the vascular system, in addition to controlling proteolytic activity in the vessel wall during pathological conditions such as atherosclerosis.

10.9 Functions of PN-1 in Cancer

Being previously almost neglected, the possibility of a role of PN-1 in cancer has within recent years received increasing attention. The studies reported thus far all suggest that PN-1 may indeed exert important tumor biological functions.

10.9.1 PN-1 Expression is Upregulated in Human Cancers, and a High Expression Is a Marker for a Poor Prognosis

Among human cancers, overexpression of PN-1 in the tumor, as compared to the corresponding normal tissue, was found in the majority of a small number of investigated oral squamous cell carcinomas as well as pancreatic, breast, colorectal, and adipose tissue cancer samples [121, 161–165]. In human pancreatic tumors, PN-1 was found to be exclusively expressed by cancer cells [162]. However, the cellular sources of PN-1 in human tumors, in general, still remain to be thoroughly investigated.

Not only is PN-1 overexpressed in human tumors as compared to the corresponding normal tissue, but its levels are, in many cases, correlated with a poor prognosis. The increased PN-1 mRNA levels in breast tumor tissue appeared to correlate with increased expression of both PAI-1 and uPA, both markers for a poor prognosis [163]. Analysis of three unrelated data sets revealed that PN-1 is among the genes predicting poor prognosis for breast cancer patients with estrogen receptor-α-negative tumors [166]. Overexpression of PN-1 was also found in the estrogen receptor-α-negative group compared to the estrogen receptor-α-positive group in seven data sets, representing a number of breast cancer cell lines and tumor samples [167]. A separate analysis of breast cancer gene expression data sets revealed that PN-1 mRNA levels increased significantly with breast tumor grade and that PN-1 mRNA levels were significantly higher in estrogen receptor-α-negative tumors than in estrogen receptor-α-positive tumors [99]. Moreover, analysis of 126 breast cancer patients after five-years follow-up showed that those with elevated tumor PN-1 mRNA were more likely to develop lung metastasis, but not metastasis at other sites. Taken together, these observations indicate that the breast tumor PN-1-level may serve as a marker of poor prognosis and organ-specific metastatic potential [99].

The analysis of small series of colorectal cancers showed that the PN-1 mRNA levels increase as the tumors progress from adenomas to colorectal cancer [164]. In a similar study, PN-1 mRNA levels were elevated in both adenomas and colorectal carcinomas compared to adjacent healthy tissue, although seemingly independent of tumor grade and stage [121].

The potential value of PN-1 as a prognostic marker in certain cancers is supported by the comparison of PN-1 and PAI-1 mRNA levels, measured by RT-PCR, in a small number of oral squamous cell carcinomas and matching healthy tissue. Accordingly, although the overall levels of both PN-1 [161] and PAI-1 mRNA [168] was elevated in tumor samples compared to healthy tissue (approximately threefold and approximately sixfold, respectively), only tumor PN-1 levels correlated significantly with lymph node metastasis. Furthermore, there appeared to be no correlation between PN-1 and PAI-1 mRNA levels in the individual tumor samples, indicating that tumor levels of PN-1 may contain prognostic information independent of that provided by the PAI-1 level.

Thus, taken together, these studies suggest that a high PN-1 expression in tumors is often correlated with a poor prognosis, but whether the PN-1 expression level is an independent prognostic factor remains to be established.

10.9.2
Studies with Cell Cultures and Animal Tumor Models Indicate a Proinvasive Role of PN-1

When comparing cell line variants, which differ strongly in their tumorigenic, invasive, and/or metastatic potential, overexpression of PN-1 has been found almost consistently in the most aggressive variant. Such a correlation was demonstrated for variants of human pancreatic adenocarcinoma cell lines [162], human lung adenocarcinoma cell lines [169], human mammary epithelial cell lines [163], human testicular seminoma cell lines [170], mouse mammary carcinoma cell lines [99], and murine cell lines transformed with a strain of adenovirus 12 with tumorigenic potential [171]. So far, only one exception to this overall picture has been reported: Proteomic analysis revealed that PN-1 levels were decreased in a human malignant breast cancer cell line compared to a control breast cell line isolated from the same individual, but from tissue peripheral to the breast carcinoma [172]. A causal link between PN-1 levels and aggressiveness of the individual cancer cell lines has not been fully established.

Human colorectal cancer cell lines with relatively common oncogenic mutations in two members of the MAPK/ERK pathway, *KRAS* or *BRAF*, showed increased expression of PN-1, the silencing of which by RNAi markedly decreased adhesion and migration *in vitro* and formation of tumors in nude mice. Treatment with the MAPKK/MEK inhibitor U0126 revealed that overexpression of PN-1 by the colorectal cancer cells is likely a direct effect of elevated MAPK/ERK signaling [121].

In a separate study, a microarray gene expression analysis revealed that PN-1 was among the target genes affected by a deficient HGF-c-Met signaling in human colon xenograft tumors on nude mice. Blockage of the HGF-c-Met signaling significantly suppressed PN-1 expression (approximately fourfold) while inhibiting growth of xenograft tumors. However, a causal link between PN-1 expression level and the observed effect was not established [120].

A number of independent studies have pointed to the pivotal role of PN-1 in invasion and metastasis of human xenograft tumors on nude mice. In one study, the metastatic potential of an otherwise rarely metastatic xenografted seminoma tumor was enhanced by intraperitoneal administration of media conditioned by cultured transfected seminoma cells overexpressing PN-1 or a highly metastatic seminoma sub cell line with endogenous PN-1 expression. Conversely, administration of media from the metastatic sub cell line, in which PN-1 expression had been silenced by RNAi, reduced lymph node metastasis in the xenograft model. The mechanism(s) by which PN-1 affected the metastatic potential of the xenograft tumor still remains to be investigated [170]. Also, recombinant overexpression of PN-1 in human pancreatic xenograft tumors enhanced the local invasiveness, but not tumor size, compared to the nontransfected and rarely metastatic pancreatic cancer cell line. PN-1 overexpression was accompanied by a massive increase in extracellular matrix deposition, containing collagen I, fibronectin, and laminin, thus resembling the desmoplastic reaction commonly observed in chronic pancreatitis and pancreatic cancer [162]. The presence of human pancreatic stellate cells, which have been identified as the primary source of fibrosis in pancreatic cancer, was shown to enhance the invasive potential and also favor growth of PN-1-expressing pancreatic xenograft tumors [173]. One hypothesis is that PN-1 could play a role in the formation of a microenvironment, which favors tumor growth and invasion [174].

That PN-1 may play a central proinvasive role in tumor progression was further supported by findings reported by Fayard *et al.* [99], who found that PN-1 expression by a murine mammary carcinoma cell line was essential for metastasis, but not tumor growth *in vivo*. Interestingly, exogenously added PN-1 appeared to induce MMP-9 expression in these cells, likely through an LRP-1A-dependent activation of the MAPK/ERK signaling pathway. Thus, Fayard *et al.* [99] suggested that the effect of elevated PN-1 expression on the metastatic potential of the carcinoma cells *in vivo* was somehow linked to the functions of MMP-9.

An additional dimension to the PN-1-MMP-9 relationship was added recently, as Xu *et al.* [131, 175] reported that PN-1 may be a novel substrate of MMP-9. Based on *in vitro* cell migration assays, MMP-9 was suggested to regulate uPA activity and hence migration of prostate cancer cells through cleavage and degradation of PN-1. Overexpression of PN-1 or silencing of MMP-9 appeared to reduce cell invasion, which in contrast to the animal xenograft tumor models points to a protective role of PN-1 against tumor invasion. Xu *et al.* [131] suggested that a regulatory feedback loop exists in which PN-1 and MMP-9 exert opposing regulatory mechanisms on each other, the final outcome thus depending on their relative abundance.

Prostasin, a GPI-anchored serine protease with tumor suppressor functions, which is known to be downregulated in prostate cancer, has been demonstrated to be a target for PN-1 on surfaces of prostate epithelial cells [58]. On the basis of RT-PCR measurements of prostasin and its inhibitors, PN-1 and hepatocyte growth factor activator inhibitor-1 (HAI-1), in a small series of colorectal adenomas, carcinomas, and corresponding normal tissue, the inhibition of prostasin activity by PN-1 has recently been suggested to be involved in the progression of human colorectal cancers [164]. Reportedly, the expression of PN-1 was progressively upregulated with increasing tumor grade, whereas mRNA levels of prostasin and isoforms of its major inhibitor, HAI-1, remained relatively stable or decreased during carcinogenesis [164]. The physiological and pathological importance of the PN-1-prostasin interaction still remains to be determined.

Conclusively, the level of PN-1 in human tumors is correlated with a poor prognosis. In cell culture and animal models, there was often a correlation between the PN-1 expression and tumor cell growth, invasion and/or metastasis, and a causal link between PN-1 expression and tumor aggressiveness was established

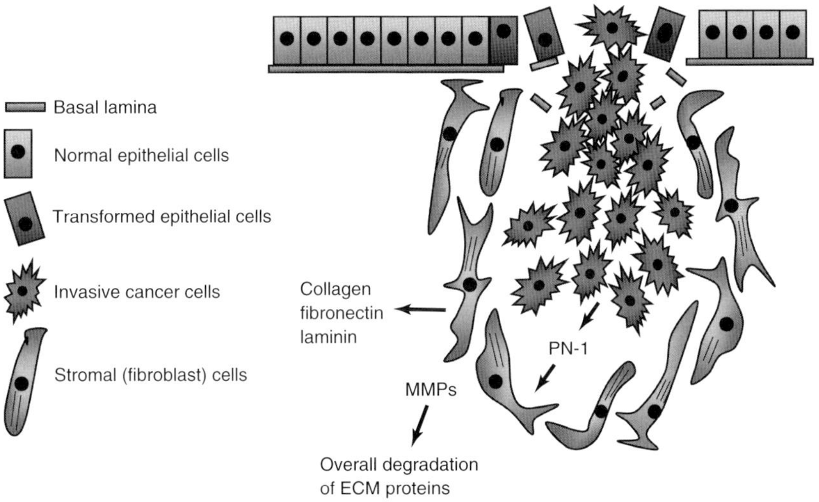

Figure 10.4 A proposal for a proinvasive role of PN-1 in malignant tumors. Overexpression of PN-1 in tumors leads to increased levels of some extracellular matrix (ECM) proteins and concomitantly, through increased levels of matrix metalloproteases (MMPs), to decreased levels of other extracellular matrix proteins. The result is a pathological stroma composition favoring cancer cell proliferation, migration, and invasion. The proposal is mainly based on the work of Buchholz and coworkers [162], Neesse and coworkers [173], and Fayard and coworkers [99].

in some cases. However, the mechanism behind the putative proinvasive role of PN-1 remains to be established and may indeed be variable. In an attempt to put all the different types of observations together, one may suggest that PN-1 participates in replacement of a normal stroma with a pathological stroma favoring cancer cell proliferation and invasiveness. This hypothesis may unite the, at first sight, apparently conflicting results between induction of MMP9 and increased deposition of the extracellular matrix (Figure 10.4).

10.10 Conclusions

High levels of PN-1 expression in human tumors were consistently found to be correlated with a poor prognosis, although it remains to be established whether PN-1 levels are of independent prognostic value. The development of hypotheses about the function of PN-1 in cancers is hampered by an almost complete lack of knowledge about which cell types express PN-1 in the tumors. Nevertheless, most observations with tumor model systems point to the possibility that PN-1 plays a proinvasive role in cancer. The knowledge accumulated about the role of PN-1 in cancer, in fact, has a strong similarity with the knowledge about the role of PAI-1 in cancer, although the latter is much more extensive. In both cases, a high tumor level of the serpin is associated with a poor prognosis, and mechanistic studies point to stromal expression and stromal function of the serpin being important for its proinvasive effect. The proinvasive property is, at first glance, in apparent conflict with the fact that PN-1 and PAI-1 are protease inhibitors and the idea that proteases contribute to cancer cell invasion by catalyzing degradation of basement membranes. But both PN-1 and PAI-1 may promote invasion by leading to accumulation of substrates for certain proteases while allowing or even favoring the action of other proteases and thus changing the composition of extracellular matrix of the stroma and making it optimal for cancer cell invasion. Alternatively, the proinvasive effects of PN-1 and PAI-1 may depend on non-antiproteolytic effects. When complexed with target proteases, both PN-1 and PAI-1 bind with high affinity to receptors of the LDLR family, which are not only endocytosis receptors but also signaling receptors. Some cell biological effects of the two serpins may also be related to their ability to bind to vitronectin. However, many aspects of the exact mechanism of PN-1 (and PAI-1) in tumors remain unknown. Further development of ideas about the tumor biological functions of PN-1 and of its pathophysiological functions in general depends on a further progression of the knowledge about the general cell biological functions and the biochemistry of PN-1. From studies with cell cultures, PN-1 expression is known to be regulated by a large number of hormones, cytokines, and growth factors, although the signal transduction pathways are known in only a few cases. Which are the regulatory mechanisms leading to overexpression in cancers?

References

1. Irving, J.A., Pike, R.N., Lesk, A.M., and Whisstock, J.C. (2000) Phylogeny of the serpin superfamily: implications of patterns of amino acid conservation for structure and function. *Genome Res.*, **10**, 1845–1864.
2. Huntington, J.A. (2006) Shape-shifting serpins–advantages of a mobile mechanism. *Trends Biochem. Sci.*, **31**, 427–435.
3. Silverman, G.A., Bird, P.I., Carrell, R.W., Church, F.C., Coughlin, P.B., Gettins, P.G., Irving, J.A., Lomas, D.A., Luke, C.J., Moyer, R.W., Pemberton, P.A., Remold-O'Donnell, E., Salvesen, G.S., Travis, J., and Whisstock, J.C. (2001) The serpins are an expanding superfamily of structurally similar but functionally diverse proteins. Evolution, mechanism of inhibition, novel functions, and a revised nomenclature. *J. Biol. Chem.*, **276**, 33293–33296.
4. Gettins, P.G. (2002) Serpin structure, mechanism, and function. *Chem. Rev.*, **102**, 4751–4804.
5. Silverman, G.A., Whisstock, J.C., Bottomley, S.P., Huntington, J.A., Kaiserman, D., Luke, C.J., Pak, S.C., Reichhart, J.M., and Bird, P.I. (2010) Serpins flex their muscle: I. Putting the clamps on proteolysis in diverse biological systems. *J. Biol. Chem.*, **285**, 24299–24305.
6. Ragg, H. (2007) The role of serpins in the surveillance of the secretory pathway. *Cell. Mol. Life Sci.*, **64**, 2763–2770.
7. Zhang, M., Martin, K.J., Sheng, S., and Sager, R. (1998) Expression genetics: a different approach to cancer diagnosis and prognosis. *Trends Biotechnol.*, **16**, 66–71.
8. Xiao, G., Liu, Y.E., Gentz, R., Sang, Q.A., Ni, J., Goldberg, I.D., and Shi, Y.E. (1999) Suppression of breast cancer growth and metastasis by a serpin myoepithelium-derived serine proteinase inhibitor expressed in the mammary myoepithelial cells. *Proc. Natl. Acad. Sci. U.S.A.*, **96**, 3700–3705.
9. Andreasen, P.A., Kjoller, L., Christensen, L., and Duffy, M.J. (1997) The urokinase-type plasminogen activator system in cancer metastasis: a review. *Int. J. Cancer*, **72**, 1–22.
10. Andreasen, P.A., Egelund, R., and Petersen, H.H. (2000) The plasminogen activation system in tumor growth, invasion, and metastasis. *Cell. Mol. Life Sci.*, **57**, 25–40.
11. Janicke, F., Prechtl, A., Thomssen, C., Harbeck, N., Meisner, C., Untch, M., Sweep, C.G., Selbmann, H.K., Graeff, H., and Schmitt, M. (2001) Randomized adjuvant chemotherapy trial in high-risk, lymph node-negative breast cancer patients identified by urokinase-type plasminogen activator and plasminogen activator inhibitor type 1. *J. Natl. Cancer Inst.*, **93**, 913–920.
12. Harbeck, N., Kates, R.E., and Schmitt, M. (2002) Clinical relevance of invasion factors urokinase-type plasminogen activator and plasminogen activator inhibitor type 1 for individualized therapy decisions in primary breast cancer is greatest when used in combination. *J. Clin. Oncol.*, **20**, 1000–1007.
13. Look, M.P., van Putten, W.L., Duffy, M.J., Harbeck, N., Christensen, I.J., Thomssen, C., Kates, R., Spyratos, F., Ferno, M., Eppenberger-Castori, S., Sweep, C.G., Ulm, K., Peyrat, J.P., Martin, P.M., Magdelenat, H., Brunner, N., Duggan, C., Lisboa, B.W., Bendahl, P.O., Quillien, V., Daver, A., Ricolleau, G., Meijer-van Gelder, M.E., Manders, P., Fiets, W.E., Blankenstein, M.A., Broet, P., Romain, S., Daxenbichler, G., Windbichler, G., Cufer, T., Borstnar, S., Kueng, W., Beex, L.V., Klijn, J.G., O'Higgins, N., Eppenberger, U., Janicke, F., Schmitt, M., and Foekens, J.A. (2002) Pooled analysis of prognostic impact of urokinase-type plasminogen activator and its inhibitor PAI-1 in 8377 breast cancer patients. *J. Natl. Cancer Inst.*, **94**, 116–128.
14. Harris, L., Fritsche, H., Mennel, R., Norton, L., Ravdin, P., Taube, S., Somerfield, M.R., Hayes, D.F., and Bast, R.C. Jr., American Society of Clinical Oncology (2007) American

Society of Clinical Oncology 2007 update of recommendations for the use of tumor markers in breast cancer. *J. Clin. Oncol.*, **25**, 5287–5312.

15. Durand, M.K., Bodker, J.S., Christensen, A., Dupont, D.M., Hansen, M., Jensen, J.K., Kjelgaard, S., Mathiasen, L., Pedersen, K.E., Skeldal, S., Wind, T., and Andreasen, P.A. (2004) Plasminogen activator inhibitor-I and tumour growth, invasion, and metastasis. *Thromb. Haemostasis*, **91**, 438–449.
16. Andreasen, P.A. (2007) PAI-1 - a potential therapeutic target in cancer. *Curr. Drug Targets*, **8**, 1030–1041.
17. Dupont, D.M., Madsen, J.B., Kristensen, T., Bodker, J.S., Blouse, G.E., Wind, T., and Andreasen, P.A. (2009) Biochemical properties of plasminogen activator inhibitor-1. *Front. Biosci.*, **14**, 1337–1361.
18. Monard, D., Solomon, F., Rentsch, M., and Gysin, R. (1973) Glia-induced morphological differentiation in neuroblastoma cells. *Proc. Natl. Acad. Sci. U.S.A.*, **70**, 1894–1897.
19. Schurch-Rathgeb, Y.M.D. (1978) Brain development influences the appearance of glial factor-like activity in rat brain primary cultures. *Nature*, **273**, 308–309.
20. Baker, J.B., Low, D.A., Simmer, R.L., and Cunningham, D.D. (1980) Protease-nexin: a cellular component that links thrombin and plasminogen activator and mediates their binding to cells. *Cell*, **21**, 37–45.
21. Low, D.A., Baker, J.B., Koonce, W.C., and Cunningham, D.D. (1981) Released protease-nexin regulates cellular binding, internalization, and degradation of serine proteases. *Proc. Natl. Acad. Sci. U.S.A.*, **78**, 2340–2344.
22. Monard, D., Niday, E., Limat, A., and Solomon, F. (1983) Inhibition of protease activity can lead to neurite extension in neuroblastoma cells. *Prog. Brain Res.*, **58**, 359–364.
23. Scott, R.W. and Baker, J.B. (1983) Purification of human protease nexin. *J. Biol. Chem.*, **258**, 10439–10444.
24. Guenther, J., Nick, H., and Monard, D. (1985) A glia-derived neurite-promoting factor with protease inhibitory activity. *EMBO J.*, **4**, 1963–1966.
25. Scott, R.W., Bergman, B.L., Bajpai, A., Hersh, R.T., Rodriguez, H., Jones, B.N., Barreda, C., Watts, S., and Baker, J.B. (1985) Protease nexin. Properties and a modified purification procedure. *J. Biol. Chem.*, **260**, 7029–7034.
26. Gloor, S., Odink, K., Guenther, J., Nick, H., and Monard, D. (1986) A glia-derived neurite promoting factor with protease inhibitory activity belongs to the protease nexins. *Cell*, **47**, 687–693.
27. Sommer, J., Gloor, S.M., Rovelli, G.F., Hofsteenge, J., Nick, H., Meier, R., and Monard, D. (1987) cDNA sequence coding for a rat glia-derived nexin and its homology to members of the serpin superfamily. *Biochemistry*, **26**, 6407–6410.
28. Wagner, S.L., Van Nostrand, W.E., Lau, A.L., and Cunningham, D.D. (1988) Monoclonal antibodies to protease nexin 1 that differentially block its inhibition of target proteases. *Biochemistry*, **27**, 2173–2176.
29. Guttridge, D.C. and Cunningham, D.D. (1996) Characterization of the human protease nexin-1 promoter and its regulation by Sp1 through a G/C-rich activation domain. *J. Neurochem.*, **67**, 498–507.
30. Murer, V., Spetz, J.F., Hengst, U., Altrogge, L.M., de Agostini, A., and Monard, D. (2001) Male fertility defects in mice lacking the serine protease inhibitor protease nexin-1. *Proc. Natl. Acad. Sci. U.S.A.*, **98**, 3029–3033.
31. Carter, R.E., Cerosaletti, K.M., Burkin, D.J., Fournier, R.E., Jones, C., Greenberg, B.D., Citron, B.A., and Festoff, B.W. (1995) The gene for the serpin thrombin inhibitor (PI7), protease nexin I, is located on human chromosome 2q33-q35 and on syntenic regions in the mouse and sheep genomes. *Genomics*, **27**, 196–199.
32. Klinger, K.W., Winqvist, R., Riccio, A., Andreasen, P.A., Sartorio, R., Nielsen, L.S., Stuart, N., Stanislovitis, P., Watkins, P., Douglas, R. *et al.* (1987) Plasminogen activator inhibitor type 1 gene is located at region q21.3-q22 of

chromosome 7 and genetically linked with cystic fibrosis. *Proc. Natl. Acad. Sci. U.S.A.*, **84**, 8548–8552.

33. McGrogan, M., Kennedy, J., Li, M.P., Hsu, C., Scott, R.W., Simonsen, C.C., and Baker, J.B. (1988) Molecular cloning and expression of two forms of human protease nexin I. *Biotechnology*, **6**, 172–177.
34. Jensen, J.K., Thompson, L.C., Bucci, J.C., Nissen, P., Gettins, P.G., Peterson, C.B., Andreasen, P.A., and Morth, J.P. (2011) Crystal structure of plasminogen activator inhibitor-1 in an active conformation with normal thermodynamic stability. *J. Biol. Chem.*, **286**, 29709–29717.
35. Evans, D.L., McGrogan, M., Scott, R.W., and Carrell, R.W. (1991) Protease specificity and heparin binding and activation of recombinant protease nexin I. *J. Biol. Chem.*, **266**, 22307–22312.
36. Rovelli, G., Stone, S.R., Guidolin, A., Sommer, J., and Monard, D. (1992) Characterization of the heparin-binding site of glia-derived nexin/protease nexin-1. *Biochemistry*, **31**, 3542–3549.
37. Stone, S.R., Brown-Luedi, M.L., Rovelli, G., Guidolin, A., McGlynn, E., and Monard, D. (1994) Localization of the heparin-binding site of glia-derived nexin/protease nexin-1 by site-directed mutagenesis. *Biochemistry*, **33**, 7731–7735.
38. Knauer, M.F., Hawley, S.B., and Knauer, D.J. (1997) Identification of a binding site in protease nexin I (PN1) required for the receptor mediated internalization of PN1-thrombin complexes. *J. Biol. Chem.*, **272**, 12261–12264.
39. Knauer, M.F., Crisp, R.J., Kridel, S.J., and Knauer, D.J. (1999) Analysis of a structural determinant in thrombin-protease nexin 1 complexes that mediates clearance by the low density lipoprotein receptor-related protein. *J. Biol. Chem.*, **274**, 275–281.
40. Lin, Z., Jiang, L., Yuan, C., Jensen, J.K., Zhang, X., Luo, Z., Furie, B.C., Furie, B., Andreasen, P.A., and Huang, M. (2011) Structural basis for recognition of urokinase-type plasminogen activator by plasminogen activator inhibitor-1. *J. Biol. Chem.*, **286**, 7027–32.
41. Dementiev, A., Dobo, J., and Gettins, P.G. (2006) Active site distortion is sufficient for proteinase inhibition by serpins: structure of the covalent complex of alpha1-proteinase inhibitor with porcine pancreatic elastase. *J. Biol. Chem.*, **281**, 3452–3457.
42. Jensen, J.K. and Gettins, P.G. (2008) High-resolution structure of the stable plasminogen activator inhibitor type-1 variant 14-1B in its proteinase-cleaved form: a new tool for detailed interaction studies and modeling. *Protein Sci.*, **17**, 1844–1849.
43. Wallace, A., Rovelli, G., Hofsteenge, J., and Stone, S.R. (1989) Effect of heparin on the glia-derived-nexin-thrombin interaction. *Biochem. J.*, **257**, 191–196.
44. Herndon, M.E., Stipp, C.S., and Lander, A.D. (1999) Interactions of neural glycosaminoglycans and proteoglycans with protein ligands: assessment of selectivity, heterogeneity and the participation of core proteins in binding. *Glycobiology*, **9**, 143–155.
45. Richard, B., Pichon, S., Arocas, V., Venisse, L., Berrou, E., Bryckaert, M., Jandrot-Perrus, M., and Bouton, M.C. (2006) The serpin protease nexin-1 regulates vascular smooth muscle cell adhesion, spreading, migration and response to thrombin. *J. Thromb. Haemostasis*, **4**, 322–328.
46. Johnson, D.J., Langdown, J., and Huntington, J.A. (2010) Molecular basis of factor IXa recognition by heparin-activated antithrombin revealed by a 1.7-A structure of the ternary complex. *Proc. Natl. Acad. Sci. U.S.A.*, **107**, 645–650.
47. Baglin, T.P., Carrell, R.W., Church, F.C., Esmon, C.T., and Huntington, J.A. (2002) Crystal structures of native and thrombin-complexed heparin cofactor II reveal a multistep allosteric mechanism. *Proc. Natl. Acad. Sci. U.S.A.*, **99**, 11079–11084.
48. Ehrlich, H.J., Gebbink, R.K., Keijer, J., and Pannekoek, H. (1992) Elucidation of structural requirements on plasminogen activator inhibitor 1 for

binding to heparin. *J. Biol. Chem.*, **267**, 11606–11611.

49. Rovelli, G., Stone, S.R., Preissner, K.T., and Monard, D. (1990) Specific interaction of vitronectin with the cell-secreted protease inhibitor glia-derived nexin and its thrombin complex. *Eur. J. Biochem.*, **192**, 797–803.
50. Ye, S. and Goldsmith, E.J. (2001) Serpins and other covalent protease inhibitors. *Curr. Opin. Struct. Biol.*, **11**, 740–745.
51. Wind, T., Hansen, M., Jensen, J.K., and Andreasen, P.A. (2002) The molecular basis for anti-proteolytic and non-proteolytic functions of plasminogen activator inhibitor type-1: roles of the reactive centre loop, the shutter region, the flexible joint region and the small serpin fragment. *Biol. Chem.*, **383**, 21–36.
52. Ye, S., Cech, A.L., Belmares, R., Bergstrom, R.C., Tong, Y., Corey, D.R., Kanost, M.R., and Goldsmith, E.J. (2001) The structure of a Michaelis serpin-protease complex. *Nat. Struct. Biol.*, **8**, 979–983.
53. Li, W., Johnson, D.J., Esmon, C.T., and Huntington, J.A. (2004) Structure of the antithrombin-thrombin-heparin ternary complex reveals the antithrombotic mechanism of heparin. *Nat. Struct. Mol. Biol.*, **11**, 857–862.
54. Li, W., Adams, T.E., Nangalia, J., Esmon, C.T., and Huntington, J.A. (2008) Molecular basis of thrombin recognition by protein C inhibitor revealed by the 1.6-A structure of the heparin-bridged complex. *Proc. Natl. Acad. Sci. U.S.A.*, **105**, 4661–4666.
55. Johnson, D.J., Li, W., Adams, T.E., and Huntington, J.A. (2006) Antithrombin-S195A factor Xa-heparin structure reveals the allosteric mechanism of antithrombin activation. *EMBO J.*, **25**, 2029–2037.
56. Huntington, J.A., Read, R.J., and Carrell, R.W. (2000) Structure of a serpin-protease complex shows inhibition by deformation. *Nature*, **407**, 923–926.
57. Chen, L.M., Skinner, M.L., Kauffman, S.W., Chao, J., Chao, L., Thaler, C.D., and Chai, K.X. (2001) Prostasin is a glycosylphosphatidylinositol-anchored active serine protease. *J. Biol. Chem.*, **276**, 21434–21442.
58. Chen, L.M., Zhang, X., and Chai, K.X. (2004) Regulation of prostasin expression and function in the prostate. *Prostate*, **59**, 1–12.
59. Myerburg, M.M., McKenna, E.E., Luke, C.J., Frizzell, R.A., Kleyman, T.R., and Pilewski, J.M. (2008) Prostasin expression is regulated by airway surface liquid volume and is increased in cystic fibrosis. *Am. J. Physiol. Lung Cell. Mol. Physiol.*, **294**, L932–L941.
60. Muhl, L., Nykjaer, A., Wygrecka, M., Monard, D., Preissner, K.T., and Kanse, S.M. (2007) Inhibition of PDGF-BB by factor VII-activating protease (FSAP) is neutralized by protease nexin-1, and the FSAP-inhibitor complexes are internalized via LRP. *Biochem. J.*, **404**, 191–196.
61. Djie, M.Z., Stone, S.R., and Le Bonniec, B.F. (1997) Intrinsic specificity of the reactive site loop of alpha1-antitrypsin, alpha1-antichymotrypsin, antithrombin III, and protease nexin I. *J. Biol. Chem.*, **272**, 16268–16273.
62. Crisp, R.J., Knauer, M.F., and Knauer, D.J. (2002) Protease nexin 1 is a potent urinary plasminogen activator inhibitor in the presence of collagen type IV. *J. Biol. Chem.*, **277**, 47285–47291.
63. Eaton, D.L., Scott, R.W., and Baker, J.B. (1984) Purification of human fibroblast urokinase proenzyme and analysis of its regulation by proteases and protease nexin. *J. Biol. Chem.*, **259**, 6241–6247.
64. Knauer, D.J., Majumdar, D., Fong, P.C., and Knauer, M.F. (2000) SERPIN regulation of factor XIa. The novel observation that protease nexin 1 in the presence of heparin is a more potent inhibitor of factor XIa than C1 inhibitor. *J. Biol. Chem.*, **275**, 37340–37346.
65. Van Nostrand, W.E., Wagner, S.L., and Cunningham, D.D. (1988) Purification of a form of protease nexin 1 that binds heparin with a low affinity. *Biochemistry*, **27**, 2176–2181.

66. Torri, G., Casu, B., Gatti, G., Petitou, M., Choay, J., Jacquinet, J.C., and Sinay, P. (1985) Mono- and bidimensional 500 MHz 1H-NMR spectra of a synthetic pentasaccharide corresponding to the binding sequence of heparin to antithrombin-III: evidence for conformational peculiarity of the sulfated iduronate residue. *Biochem. Biophys. Res. Commun.*, **128**, 134–140.
67. Choay, J. (1989) Chemically synthesized heparin-derived oligosaccharides. *Ann. N.Y. Acad. Sci.*, **556**, 61–74.
68. Huntington, J.A. (2003) Mechanisms of glycosaminoglycan activation of the serpins in hemostasis. *J. Thromb. Haemostasis*, **1**, 1535–1549.
69. Lindahl, U., Backstrom, G., Thunberg, L., and Leder, I.G. (1980) Evidence for a 3-O-sulfated D-glucosamine residue in the antithrombin-binding sequence of heparin. *Proc. Natl. Acad. Sci. U.S.A.*, **77**, 6551–6555.
70. Olson, S.T., Bjork, I., Sheffer, R., Craig, P.A., Shore, J.D., and Choay, J. (1992) Role of the antithrombin-binding pentasaccharide in heparin acceleration of antithrombin-proteinase reactions. Resolution of the antithrombin conformational change contribution to heparin rate enhancement. *J. Biol. Chem.*, **267**, 12528–12538.
71. Dementiev, A., Petitou, M., Herbert, J.M., and Gettins, P.G. (2004) The ternary complex of antithrombin-anhydrothrombin-heparin reveals the basis of inhibitor specificity. *Nat. Struct. Mol. Biol.*, **11**, 863–867.
72. Gliemann, J., Nykjaer, A., Petersen, C.M., Jorgensen, K.E., Nielsen, M., Andreasen, P.A., Christensen, E.I., Lookene, A., Olivecrona, G., and Moestrup, S.K. (1994) The multiligand alpha 2-macroglobulin receptor/low density lipoprotein receptor-related protein (alpha 2MR/LRP). Binding and endocytosis of fluid phase and membrane-associated ligands. *Ann. N.Y. Acad. Sci.*, **737**, 20–38.
73. Nykjaer, A. and Willnow, T.E. (2002) The low-density lipoprotein receptor gene family: a cellular Swiss army knife? *Trends Cell Biol.*, **12**, 273–280.
74. Lillis, A.P., Mikhailenko, I., and Strickland, D.K. (2005) Beyond endocytosis: LRP function in cell migration, proliferation and vascular permeability. *J. Thromb. Haemostasis*, **3**, 1884–1893.
75. Strickland, D.K. and Medved, L. (2006) Low-density lipoprotein receptor-related protein (LRP)-mediated clearance of activated blood coagulation co-factors and proteases: clearance mechanism or regulation? *J. Thromb. Haemostasis*, **4**, 1484–1486.
76. Kristensen, T., Moestrup, S.K., Gliemann, J., Bendtsen, L., Sand, O., and Sottrup-Jensen, L. (1990) Evidence that the newly cloned low-density-lipoprotein receptor related protein (LRP) is the alpha 2-macroglobulin receptor. *FEBS Lett.*, **276**, 151–155.
77. Strickland, D.K., Ashcom, J.D., Williams, S., Burgess, W.H., Migliorini, M., and Argraves, W.S. (1990) Sequence identity between the alpha 2-macroglobulin receptor and low density lipoprotein receptor-related protein suggests that this molecule is a multifunctional receptor. *J. Biol. Chem.*, **265**, 17401–17404.
78. Herz, J., Hamann, U., Rogne, S., Myklebost, O., Gausepohl, H., and Stanley, K.K. (1988) Surface location and high affinity for calcium of a 500-kd liver membrane protein closely related to the LDL-receptor suggest a physiological role as lipoprotein receptor. *EMBO J.*, **7**, 4119–4127.
79. Orth, K., Madison, E.L., Gething, M.J., Sambrook, J.F., and Herz, J. (1992) Complexes of tissue-type plasminogen activator and its serpin inhibitor plasminogen-activator inhibitor type 1 are internalized by means of the low density lipoprotein receptor-related protein/alpha 2-macroglobulin receptor. *Proc. Natl. Acad. Sci. U.S.A.*, **89**, 7422–7426.
80. Nykjaer, A., Petersen, C.M., Moller, B., Jensen, P.H., Moestrup, S.K., Holtet, T.L., Etzerodt, M., Thogersen, H.C., Munch, M., Andreasen, P.A. *et al.* (1992) Purified alpha 2-macroglobulin receptor/LDL receptor-related protein binds urokinase.plasminogen activator

inhibitor type-1 complex. Evidence that the alpha 2-macroglobulin receptor mediates cellular degradation of urokinase receptor-bound complexes. *J. Biol. Chem.*, **267**, 14543–14546.

81. Strickland, D.K. and Ranganathan, S. (2003) Diverse role of LDL receptor-related protein in the clearance of proteases and in signaling. *J. Thromb. Haemostasis*, **1**, 1663–1670.
82. Jensen, J.K., Dolmer, K., Schar, C., and Gettins, P.G. (2009) Receptor-associated protein (RAP) has two high-affinity binding sites for the low-density lipoprotein receptor-related protein (LRP): consequences for the chaperone functions of RAP. *Biochem. J.*, **421**, 273–282.
83. Jensen, G.A., Andersen, O.M., Bonvin, A.M., Bjerrum-Bohr, I., Etzerodt, M., Thogersen, H.C., O'Shea, C., Poulsen, F.M., and Kragelund, B.B. (2006) Binding site structure of one LRP-RAP complex: implications for a common ligand-receptor binding motif. *J. Mol. Biol.*, **362**, 700–716.
84. Howard, E.W. and Knauer, D.J. (1987) Characterization of the receptor for protease nexin-I:protease complexes on human fibroblasts. *J. Cell. Physiol.*, **131**, 276–283.
85. Conese, M., Olson, D., and Blasi, F. (1994) Protease nexin-1-urokinase complexes are internalized and degraded through a mechanism that requires both urokinase receptor and alpha 2-macroglobulin receptor. *J. Biol. Chem.*, **269**, 17886–17892.
86. Kasza, A., Petersen, H.H., Heegaard, C.W., Oka, K., Christensen, A., Dubin, A., Chan, L., and Andreasen, P.A. (1997) Specificity of serine proteinase/serpin complex binding to very-low-density lipoprotein receptor and alpha2-macroglobulin receptor/low-density-lipoprotein-receptor-related protein. *Eur. J. Biochem.*, **248**, 270–281.
87. Rettenberger, P.M., Oka, K., Ellgaard, L., Petersen, H.H., Christensen, A., Martensen, P.M., Monard, D., Etzerodt, M., Chan, L., and Andreasen, P.A. (1999) Ligand binding properties of the very low density lipoprotein receptor. Absence of the third complement-type repeat encoded by exon 4 is associated with reduced binding of Mr 40,000 receptor-associated protein. *J. Biol. Chem.*, **274**, 8973–8980.
88. Knauer, M.F., Kridel, S.J., Hawley, S.B., and Knauer, D.J. (1997) The efficient catabolism of thrombin-protease nexin 1 complexes is a synergistic mechanism that requires both the LDL receptor-related protein and cell surface heparins. *J. Biol. Chem.*, **272**, 29039–29045.
89. Mentz, S., de Lacalle, S., Baerga-Ortiz, A., Knauer, M.F., Knauer, D.J., and Komives, E.A. (1999) Mechanism of thrombin clearance by human astrocytoma cells. *J. Neurochem.*, **72**, 980–987.
90. Crisp, R.J., Knauer, D.J., and Knauer, M.F. (2000) Roles of the heparin and low density lipid receptor-related protein-binding sites of protease nexin 1 (PN1) in urokinase-PN1 complex catabolism. The PN1 heparin-binding site mediates complex retention and degradation but not cell surface binding or internalization. *J. Biol. Chem.*, **275**, 19628–19637.
91. Jensen, J.K., Dolmer, K., and Gettins, P.G. (2009) Specificity of binding of the low density lipoprotein receptor-related protein to different conformational states of the clade E serpins plasminogen activator inhibitor-1 and proteinase nexin-1. *J. Biol. Chem.*, **284**, 17989–17997.
92. Skeldal, S., Larsen, J.V., Pedersen, K.E., Petersen, H.H., Egelund, R., Christensen, A., Jensen, J.K., Gliemann, J., and Andreasen, P.A. (2006) Binding areas of urokinase-type plasminogen activator-plasminogen activator inhibitor-1 complex for endocytosis receptors of the low-density lipoprotein receptor family, determined by site-directed mutagenesis. *FEBS J.*, **273**, 5143–5159.
93. Verdaguer, N., Fita, I., Reithmayer, M., Moser, R., and Blaas, D. (2004) X-ray structure of a minor group human rhinovirus bound to a fragment of its cellular receptor protein. *Nat. Struct. Mol. Biol.*, **11**, 429–434.

94. Fisher, C., Beglova, N., and Blacklow, S.C. (2006) Structure of an LDLR-RAP complex reveals a general mode for ligand recognition by lipoprotein receptors. *Mol. Cell*, **22**, 277–283.
95. Farrell, D.H. and Cunningham, D.D. (1986) Human fibroblasts accelerate the inhibition of thrombin by protease nexin. *Proc. Natl. Acad. Sci. U.S.A.*, **83**, 6858–6862.
96. Farrell, D.H. and Cunningham, D.D. (1987) Glycosaminoglycans on fibroblasts accelerate thrombin inhibition by protease nexin-1. *Biochem. J.*, **245**, 543–550.
97. Conese, M., Nykjaer, A., Petersen, C.M., Cremona, O., Pardi, R., Andreasen, P.A., Gliemann, J., Christensen, E.I., and Blasi, F. (1995) Alpha-2 Macroglobulin receptor/Ldl receptor-related protein(Lrp)-dependent internalization of the urokinase receptor. *J. Cell Biol.*, **131**, 1609–1622.
98. Li, X., Herz, J., and Monard, D. (2006) Activation of ERK signaling upon alternative protease nexin-1 internalization mediated by syndecan-1. *J. Cell. Biochem.*, **99**, 936–951.
99. Fayard, B., Bianchi, F., Dey, J., Moreno, E., Djaffer, S., Hynes, N.E., and Monard, D. (2009) The serine protease inhibitor protease nexin-1 controls mammary cancer metastasis through LRP-1-mediated MMP-9 expression. *Cancer Res.*, **69**, 5690–5698.
100. Vaillant, C., Michos, O., Orolicki, S., Brellier, F., Taieb, S., Moreno, E., Te, H., Zeller, R., and Monard, D. (2007) Protease nexin 1 and its receptor LRP modulate SHH signalling during cerebellar development. *Development*, **134**, 1745–1754.
101. Bergman, B.L., Scott, R.W., Bajpai, A., Watts, S., and Baker, J.B. (1986) Inhibition of tumor-cell-mediated extracellular matrix destruction by a fibroblast proteinase inhibitor, protease nexin I. *Proc. Natl. Acad. Sci. U.S.A.*, **83**, 996–1000.
102. Rossignol, P., Bouton, M.C., Jandrot-Perrus, M., Bryckaert, M., Jacob, M.P., Bezeaud, A., Guillin, M.C., Michel, J.B., and Meilhac, O. (2004) A paradoxical pro-apoptotic effect of thrombin on smooth muscle cells. *Exp. Cell Res.*, **299**, 279–285.
103. Rossignol, P., Ho-Tin-Noe, B., Vranckx, R., Bouton, M.C., Meilhac, O., Lijnen, H.R., Guillin, M.C., Michel, J.B., and Angles-Cano, E. (2004) Protease nexin-1 inhibits plasminogen activation-induced apoptosis of adherent cells. *J. Biol. Chem.*, **279**, 10346–10356.
104. Lindner, J., Guenther, J., Nick, H., Zinser, G., Antonicek, H., Schachner, M., and Monard, D. (1986) Modulation of granule cell migration by a glia-derived protein. *Proc. Natl. Acad. Sci. U.S.A.*, **83**, 4568–4571.
105. Drapkin, P.T., Monard, D., and Silverman, A.J. (2002) The role of serine proteases and serine protease inhibitors in the migration of gonadotropin-releasing hormone neurons. *BMC Dev. Biol.*, **2**, 1.
106. Kanse, S.M., Chavakis, T., Al-Fakhri, N., Hersemeyer, K., Monard, D., and Preissner, K.T. (2004) Reciprocal regulation of urokinase receptor (CD87)-mediated cell adhesion by plasminogen activator inhibitor-1 and protease nexin-1. *J. Cell Sci.*, **117**, 477–485.
107. Czekay, R.P. and Loskutoff, D.J. (2009) Plasminogen activator inhibitors regulate cell adhesion through a uPAR-dependent mechanism. *J. Cell. Physiol.*, **220**, 655–663.
108. Leroy-Viard, K., Jandrot-Perrus, M., Tobelem, G., and Guillin, M.C. (1989) Covalent binding of human thrombin to a human endothelial cell-associated protein. *Exp. Cell Res.*, **181**, 1–10.
109. Mbebi, C., Hantai, D., Jandrot-Perrus, M., Doyennette, M.A., and Verdiere-Sahuque, M. (1999) Protease nexin I expression is up-regulated in human skeletal muscle by injury-related factors. *J. Cell. Physiol.*, **179**, 305–314.
110. Bouton, M.C., Richard, B., Rossignol, P., Philippe, M., Guillin, M.C., Michel, J.B., and Jandrot-Perrus, M. (2003) The serpin protease-nexin 1 is present in rat aortic smooth muscle cells and is upregulated in L-NAME hypertensive

rats. *Arterioscler. Thromb. Vasc. Biol.*, **23**, 142–147.

111. Wakida, N., Kitamura, K., Tuyen, D.G., Maekawa, A., Miyoshi, T., Adachi, M., Shiraishi, N., Ko, T., Ha, V., Nonoguchi, H., and Tomita, K. (2006) Inhibition of prostasin-induced ENaC activities by PN-1 and regulation of PN-1 expression by TGF-beta1 and aldosterone. *Kidney Int.*, **70**, 1432–1438.
112. Gao, S. and Andreasen, P.A. (2011) DNA methylation profiles of protease nexin 1 (SERPINE2) gene in human cell lines. *Chin. J. Cancer Res.*, **23**, 92–98.
113. Vaughan, P.J. and Cunningham, D.D. (1993) Regulation of protease nexin-1 synthesis and secretion in cultured brain cells by injury-related factors. *J. Biol. Chem.*, **268**, 3720–3727.
114. Guttridge, D.C., Lau, A.L., and Cunningham, D.D. (1993) Protease nexin-1, a thrombin inhibitor, is regulated by interleukin-1 and dexamethasone in normal human fibroblasts. *J. Biol. Chem.*, **268**, 18966–18974.
115. Hultman, K., Blomstrand, F., Nilsson, M., Wilhelmsson, U., Malmgren, K., Pekny, M., Kousted, T., Jern, C., and Tjarnlund-Wolf, A. (2010) Expression of plasminogen activator inhibitor-1 and protease nexin-1 in human astrocytes: Response to injury-related factors. *J. Neurosci. Res.*, **88**, 2441–2449.
116. Kasza, A., Kowanetz, M., Poslednik, K., Witek, B., Kordula, T., and Koj, A. (2001) Epidermal growth factor and pro-inflammatory cytokines regulate the expression of components of plasminogen activation system in U373-MG astrocytoma cells. *Cytokine*, **16**, 187–190.
117. Festoff, B.W., Nelson, P.G., and Brenneman, D.E. (1996) Prevention of activity-dependent neuronal death: vasoactive intestinal polypeptide stimulates astrocytes to secrete the thrombin-inhibiting neurotrophic serpin, protease nexin I. *J. Neurobiol.*, **30**, 255–266.
118. Koshimizu, H., Kim, T., Cawley, N.X., and Loh, Y.P. (2010) Reprint of: Chromogranin A: a new proposal for trafficking, processing and induction of granule biogenesis. *Regul. Pept.*, **165**, 95–101.
119. Cao, M., Buratini, J. Jr., Lussier, J.G., Carriere, P.D., and Price, C.A. (2006) Expression of protease nexin-1 and plasminogen activators during follicular growth and the periovulatory period in cattle. *Reproduction*, **131**, 125–137.
120. van der Horst, E.H., Chinn, L., Wang, M., Velilla, T., Tran, H., Madrona, Y., Lam, A., Ji, M., Hoey, T.C., and Sato, A.K. (2009) Discovery of fully human anti-MET monoclonal antibodies with antitumor activity against colon cancer tumor models in vivo. *Neoplasia*, **11**, 355–364.
121. Bergeron, S., Lemieux, E., Durand, V., Cagnol, S., Carrier, J.C., Lussier, J.G., Boucher, M.J., and Rivard, N. (2010) The serine protease inhibitor serpinE2 is a novel target of ERK signaling involved in human colorectal tumorigenesis. *Mol. Cancer*, **9**, 271.
122. Bleuel, A., Degasparo, M., Whitebread, S., Puttner, I., and Monard, D. (1995) Regulation of protease Nexin-1 expression in cultured Schwann-cells is mediated by Angiotensin-Ii receptors. *J. Neurosci.*, **15**, 750–761.
123. Kitamura, K. and Tomita, K. (2010) Regulation of renal sodium handling through the interaction between serine proteases and serine protease inhibitors. *Clin. Exp. Nephrol.*, **14**, 405–410.
124. Erno, H. and Monard, D. (1993) Molecular organization of the rat glia-derived nexin/protease nexin-1 promoter. *Gene Expr.*, **3**, 163–174.
125. Erno, H., Kury, P., Botteri, F.M., and Monard, D. (1996) A Krox binding site regulates protease nexin-1 promoter activity in embryonic heart, cartilage and parts of the nervous system. *Mech. Dev.*, **60**, 139–150.
126. Suzuki, S., Singhirunnusorn, P., Nakano, H., Doi, T., Saiki, I., and Sakurai, H. (2006) Identification of TNF-alpha-responsive NF-kappaB p65-binding element in the distal promoter of the mouse serine protease inhibitor SerpinE2. *FEBS Lett.*, **580**, 3257–3262.

127. Cortese, R., Hartmann, O., Berlin, K., and Eckhardt, F. (2008) Correlative gene expression and DNA methylation profiling in lung development nominate new biomarkers in lung cancer. *Int. J. Biochem. Cell Biol.*, **40**, 1494–1508.
128. Choi, B.H., Suzuki, M., Kim, T., Wagner, S.L., and Cunningham, D.D. (1990) Protease nexin-1. Localization in the human brain suggests a protective role against extravasated serine proteases. *Am. J. Pathol.*, **137**, 741–747.
129. Mansuy, I.M., van der Putten, H., Schmid, P., Meins, M., Botteri, F.M., and Monard, D. (1993) Variable and multiple expression of Protease Nexin-1 during mouse organogenesis and nervous system development. *Development*, **119**, 1119–1134.
130. Reinhard, E., Suidan, H.S., Pavlik, A., and Monard, D. (1994) Glia-derived nexin/protease nexin-1 is expressed by a subset of neurons in the rat brain. *J. Neurosci. Res.*, **37**, 256–270.
131. Xu, D., McKee, C.M., Cao, Y., Ding, Y., Kessler, B.M., and Muschel, R.J. (2010) Matrix metalloproteinase-9 regulates tumor cell invasion through cleavage of protease nexin-1. *Cancer Res.*, **70**, 6988–6998.
132. Vassalli, J.D., Huarte, J., Bosco, D., Sappino, A.P., Sappino, N., Velardi, A., Wohlwend, A., Erno, H., Monard, D., and Belin, D. (1993) Protease-nexin I as an androgen-dependent secretory product of the murine seminal vesicle. *EMBO J.*, **12**, 1871–1878.
133. Hasan, S., Hosseini, G., Princivalle, M., Dong, J.C., Birsan, D., Cagide, C., and de Agostini, A.I. (2002) Coordinate expression of anticoagulant heparan sulfate proteoglycans and serine protease inhibitors in the rat ovary: a potent system of proteolysis control. *Biol. Reprod.*, **66**, 144–158.
134. Simpson, C.S., Johnston, H.M., and Morris, B.J. (1994) Neuronal expression of protease-nexin 1 mRNA in rat brain. *Neurosci. Lett.*, **170**, 286–290.
135. Gronke, R.S., Bergman, B.L., and Baker, J.B. (1987) Thrombin interaction with platelets. Influence of a platelet protease nexin. *J. Biol. Chem.*, **262**, 3030–3036.
136. Gronke, R.S., Knauer, D.J., Veeraraghavan, S., and Baker, J.B. (1989) A form of protease nexin I is expressed on the platelet surface during platelet activation. *Blood*, **73**, 472–478.
137. Mansilla, S., Boulaftali, Y., Venisse, L., Arocas, V., Meilhac, O., Michel, J.B., Jandrot-Perrus, M., and Bouton, M.C. (2008) Macrophages and platelets are the major source of protease nexin-1 in human atherosclerotic plaque. *Arterioscler. Thromb. Vasc. Biol.*, **28**, 1844–1850.
138. Baker, J.B. and Gronke, R.S. (1986) Protease nexins and cellular regulation. *Semin. Thromb. Hemost.*, **12**, 216–220.
139. Lu, C.H., Lee, R.K., Hwu, Y.M., Chu, S.L., Chen, Y.J., Chang, W.C., Lin, S.P., and Li, S.H. (2010) SERPINE2, a serine protease inhibitor extensively expressed in adult male mouse reproductive tissues, may serve as a murine sperm decapacitation factor. *Biol. Reprod.*, **84**, 514–25.
140. Lwaleed, B.A., Greenfield, R., Stewart, A., Birch, B., and Cooper, A.J. (2004) Seminal clotting and fibrinolytic balance: a possible physiological role in the male reproductive system. *Thromb. Haemost.*, **92**, 752–766.
141. Price, C.A. (2008) Protease nexin-1 in reproductive tissues: a review. *ARBS Annu. Rev. Biomed. Sci.*, **10**, 75–83.
142. Rosenblatt, D.E., Cotman, C.W., Nieto-Sampedro, M., Rowe, J.W., and Knauer, D.J. (1987) Identification of a protease inhibitor produced by astrocytes that is structurally and functionally homologous to human protease nexin-I. *Brain Res.*, **415**, 40–48.
143. Reinhard, E., Meier, R., Halfter, W., Rovelli, G., and Monard, D. (1988) Detection of glia-derived nexin in the olfactory system of the rat. *Neuron*, **1**, 387–394.
144. Kury, P., Schaeren-Wiemers, N., and Monard, D. (1997) Protease nexin-1 is expressed at the mouse met-/mesencephalic junction and FGF signaling regulates its promoter activity in primary met-/mesencephalic cells. *Development*, **124**, 1251–1262.

145. Lino, M.M., Vaillant, C., Orolicki, S., Sticker, M., Kvajo, M., and Monard, D. (2010) Newly generated cells are increased in hippocampus of adult mice lacking a serine protease inhibitor. *BMC Neurosci.*, **11**, 70.

146. Luthi, A., van der Putten, H., Botteri, F.M., Mansuy, I.M., Meins, M., Frey, U., Sansig, G., Portet, C., Schmutz, M., Schroder, M., Nitsch, C., Laurent, J.P., and Monard, D. (1997) Endogenous serine protease inhibitor modulates epileptic activity and hippocampal long-term potentiation. *J. Neurosci.*, **17**, 4688–4699.

147. Meins, M., Piosik, P., Schaeren-Wiemers, N., Franzoni, S., Troncoso, E., Kiss, J.Z., Brosamle, C., Schwab, M.E., Molnar, Z., and Monard, D. (2001) Progressive neuronal and motor dysfunction in mice overexpressing the serine protease inhibitor protease nexin-1 in postmitotic neurons. *J. Neurosci.*, **21**, 8830–8841.

148. Kvajo, M., Albrecht, H., Meins, M., Hengst, U., Troncoso, E., Lefort, S., Kiss, J.Z., Petersen, C.C., and Monard, D. (2004) Regulation of brain proteolytic activity is necessary for the in vivo function of NMDA receptors. *J. Neurosci.*, **24**, 9734–9743.

149. Monard, D., Suidan, H.S., and Nitsch, C. (1992) Relevance of the balance between glia-derived nexin and thrombin following lesion in the nervous system. *Ann. N.Y. Acad. Sci.*, **674**, 237–242.

150. Monard, D. (1993) Tinkering with certain blood components can engender distinct functions in the nervous system. *Perspect Dev. Neurobiol.*, **1**, 165–168.

151. Festoff, B.W., Smirnova, I.V., Ma, J., and Citron, B.A. (1996) Thrombin, its receptor and protease nexin I, its potent serpin, in the nervous system. *Semin. Thromb. Hemost.*, **22**, 267–271.

152. Turgeon, V.L. and Houenou, L.J. (1997) The role of thrombin-like (serine) proteases in the development, plasticity and pathology of the nervous system. *Brain Res. Brain Res. Rev.*, **25**, 85–95.

153. Wu, H., Zhao, R., Qi, J., Cong, Y., Wang, D., Liu, T., Gu, Y., Ban, X., and Huang, Q. (2008) The expression and the role of protease nexin-1 on brain edema after intracerebral hemorrhage. *J. Neurol. Sci.*, **270**, 172–183.

154. Wagner, S.L., Geddes, J.W., Cotman, C.W., Lau, A.L., Gurwitz, D., Isackson, P.J., and Cunningham, D.D. (1989) Protease nexin-1, an antithrombin with neurite outgrowth activity, is reduced in Alzheimer disease. *Proc. Natl. Acad. Sci. U.S.A.*, **86**, 8284–8288.

155. Rosenblatt, D.E., Geula, C., and Mesulam, M.M. (1989) Protease nexin I immunostaining in Alzheimer's disease. *Ann. Neurol.*, **26**, 628–634.

156. Vaughan, P.J., Su, J., Cotman, C.W., and Cunningham, D.D. (1994) Protease nexin-1, a potent thrombin inhibitor, is reduced around cerebral blood vessels in Alzheimer's disease. *Brain Res.*, **668**, 160–170.

157. Choi, B.H., Kim, R.C., Vaughan, P.J., Lau, A., Van Nostrand, W.E., Cotman, C.W., and Cunningham, D.D. (1995) Decreases in protease nexins in Alzheimer's disease brain. *Neurobiol. Aging*, **16**, 557–562.

158. Boulaftali, Y., Adam, F., Venisse, L., Ollivier, V., Richard, B., Taieb, S., Monard, D., Favier, R., Alessi, M.C., Bryckaert, M., Arocas, V., Jandrot-Perrus, M., and Bouton, M.C. (2009) Anticoagulant and antithrombotic properties of platelet protease nexin-1. *Blood.*, **115**, 97–106.

159. Bouton, M.C., Venisse, L., Richard, B., Pouzet, C., Arocas, V., and Jandrot-Perrus, M. (2007) Protease nexin-1 interacts with thrombomodulin and modulates its anticoagulant effect. *Circ. Res.*, **100**, 1174–1181.

160. Boulaftali, Y., Ho-Tin-Noe, B., Pena, A., Loyau, S., Venisse, L., Francois, D., Richard, B., Arocas, V., Collet, J.P., Jandrot-Perrus, M., and Bouton, M.C. (2011) Platelet protease nexin-1, a serpin that strongly influences fibrinolysis and thrombolysis. *Circulation*, **123**, 1326–1334.

161. Gao, S., Krogdahl, A., Sorensen, J.A., Kousted, T.M., Dabelsteen, E., and Andreasen, P.A. (2008) Overexpression of protease nexin-1 mRNA and protein

in oral squamous cell carcinomas. *Oral oncol.*, **44**, 309–313.
162. Buchholz, M., Biebl, A., Neesse, A., Wagner, M., Iwamura, T., Leder, G., Adler, G., and Gress, T.M. (2003) SERPINE2 (protease nexin I) promotes extracellular matrix production and local invasion of pancreatic tumors in vivo. *Cancer Res.*, **63**, 4945–4951.
163. Candia, B.J., Hines, W.C., Heaphy, C.M., Griffith, J.K., and Orlando, R.A. (2006) Protease nexin-1 expression is altered in human breast cancer. *Cancer Cell Int.*, **6**, 16.
164. Selzer-Plon, J., Bornholdt, J., Friis, S., Bisgaard, H.C., Lothe, I.M., Tveit, K.M., Kure, E.H., Vogel, U., and Vogel, L.K. (2009) Expression of prostasin and its inhibitors during colorectal cancer carcinogenesis. *BMC Cancer*, **9**, 201.
165. Thelin-Jarnum, S., Lassen, C., Panagopoulos, I., Mandahl, N., and Aman, P. (1999) Identification of genes differentially expressed in TLS-CHOP carrying myxoid liposarcomas. *Int. J. Cancer*, **83**, 30–33.
166. Martin, K.J., Patrick, D.R., Bissell, M.J., and Fournier, M.V. (2008) Prognostic breast cancer signature identified from 3D culture model accurately predicts clinical outcome across independent datasets. *PLoS ONE*, **3**, e2994.
167. Bianchini, G., Qi, Y., Alvarez, R.H., Iwamoto, T., Coutant, C., Ibrahim, N.K., Valero, V., Cristofanilli, M., Green, M.C., Radvanyi, L., Hatzis, C., Hortobagyi, G.N., Andre, F., Gianni, L., Symmans, W.F., and Pusztai, L. (2010) Molecular anatomy of breast cancer stroma and its prognostic value in estrogen receptor-positive and -negative cancers. *J. Clin. Oncol.*, **28**, 4316–4323.
168. Gao, S., Nielsen, B.S., Krogdahl, A., Sorensen, J.A., Tagesen, J., Dabelsteen, S., Dabelsteen, E., and Andreasen, P.A. (2010) Epigenetic alterations of the SERPINE1 gene in oral squamous cell carcinomas and normal oral mucosa. *Genes Chromosomes Cancer*, **49**, 526–538.
169. Yang, S., Dong, Q., Yao, M., Shi, M., Ye, J., Zhao, L., Su, J., Gu, W., Xie, W., Wang, K., Du, Y., Li, Y., and Huang, Y. (2009) Establishment of an experimental human lung adenocarcinoma cell line SPC-A-1BM with high bone metastases potency by (99m)Tc-MDP bone scintigraphy. *Nucl. Med. Biol.*, **36**, 313–321.
170. Nagahara, A., Nakayama, M., Oka, D., Tsuchiya, M., Kawashima, A., Mukai, M., Nakai, Y., Takayama, H., Nishimura, K., Jo, Y., Nagai, A., Okuyama, A., and Nonomura, N. (2010) SERPINE2 is a possible candidate promotor for lymph node metastasis in testicular cancer. *Biochem. Biophys. Res. Commun.*, **391**, 1641–1646.
171. Guan, H., Smirnov, D.A., and Ricciardi, R.P. (2003) Identification of genes associated with adenovirus 12 tumorigenesis by microarray. *Virology*, **309**, 114–124.
172. Liang, X., Huuskonen, J., Hajivandi, M., Manzanedo, R., Predki, P., Amshey, J.R., and Pope, R.M. (2009) Identification and quantification of proteins differentially secreted by a pair of normal and malignant breast-cancer cell lines. *Proteomics*, **9**, 182–193.
173. Neesse, A., Wagner, M., Ellenrieder, V., Bachem, M., Gress, T.M., and Buchholz, M. (2007) Pancreatic stellate cells potentiate proinvasive effects of SERPINE2 expression in pancreatic cancer xenograft tumors. *Pancreatology*, **7**, 380–385.
174. Duner, S., Lopatko Lindman, J., Ansari, D., Gundewar, C., and Andersson, R. (2011) Pancreatic cancer: the role of pancreatic stellate cells in tumor progression. *Pancreatology*, **10**, 673–681.
175. Xu, D., Suenaga, N., Edelmann, M.J., Fridman, R., Muschel, R.J., and Kessler, B.M. (2008) Novel MMP-9 substrates in cancer cells revealed by a label-free quantitative proteomics approach. *Mol. Cell Proteomics*, **7**, 2215–2228.

11
Secreted Cysteine Cathepsins – Versatile Players in Extracellular Proteolysis

Fee Werner, Kathrin Sachse, and Thomas Reinheckel

11.1
Introduction

The endolysosomal cell compartment is a major place for processing and degradation of cellular and extracellular macromolecules. To accomplish this task, the endolysosomal membranes encompass more than 50 highly potent hydrolytic enzymes with lysosomal proteases, the cysteine cathepsins, among them. The biogenesis and function of the lysosomal degradation compartment is tightly regulated by a multitude of cellular mechanisms in order to keep the destructive potential of the hydrolases in check [1]. However, more recent investigations provided clear evidence that lysosomal proteases can act at so-called ectopic intra- and extracellular locations. Most often, these "noncanonical" functions of lysosomal proteases are associated with pathogenic processes. Here, we summarize the modes and conditions for extracellular activity and functions of lysosomal cysteine cathepsins with special focus on the role of extracellular cathepsins in cancer.

11.2
Structure and Function of Cysteine Cathepsins

Cathepsins are papainlike peptidases (clan CA, family C1) [2] of the endolysosomal compartment. In the lysosome cathepsins exhibit the housekeeper function of "bulk proteolysis" that is terminal degradation of endocytosed material and intracellular debris. At present, 11 human cysteine cathepsins (cathepsin B, C, F, H, L, K, O, S, V, W, and X/Z) have been identified together with 18 homologous proteases in the mouse genome. Seven of the human cathepsins (B, C, F, H, L, O, and Z) are ubiquitously expressed, whereas others are tissue specific and have highly specific functions, for example, cathepsin K in osteoclasts and cathepsin S in immune cells. Most cysteine cathepsins are endopeptidases, cathepsin Z and cathepsin C, however, are exopeptidases, and cathepsin B and cathepsin H are capable of both enzymatic functions [3]. Cathepsins are globular enzymes with a two-domain structure: the L-domain is built by three α-helices, while the R-domain contains

Matrix Proteases in Health and Disease, First Edition. Edited by Niels Behrendt.

an antiparallel β-sheet. The two domains interact via polar residues and open up to the active site cleft, where substrates bind in an extended conformation [4]. All cysteine cathepsins share a typical catalytic triad in their active center containing a histidine, a cysteine, and an asparagin residue. All cathepsins are synthesized as zymogens with a proregion blocking the active site cleft. Proteolytic removal of the propeptide is required for activation of the enzymes. *In vitro* activation can be achieved by the aspartic proteases pepsin or cathepsin D, by elastase, and by the cysteine cathepsins themselves [5]. Activation, stability, and catalytic activity of the cathepsins largely depend on the pH of the surrounding milieu. In general, acidic conditions at pH 4–6 are optimal for catheptic activities [4]. Because of this preference the cysteine cathepsins are often called *acidic* proteases.

11.3
Synthesis, Processing, and Sorting of Cysteine Cathepsins

Cathepsins are synthesized as preprocathepsins at the rough endoplasmic reticulum (ER). The signaling peptide cotranslationally targets them to the ER. There the ER-import signal peptide is cleaved off and the remaining inactive procathepsins are N-glycosylated. The proenzyme is transported to the Golgi for glycosylation trimming and mannose-6-phosphorylation. The M6P-receptor in the trans-Golgi network mediates sorting into clathrin-coated vesicles [6]. On delivery into the lysosomal pathway, activation is triggered by the acidic pH in this cellular compartment. The selective proteolytic cleavage required for removal of the propeptide of cysteine cathepsins is performed by already active lysosomal proteases [4]. However, the network of reciprocal activations of lysosomal proteases in cells or living organisms is complex and to date poorly understood.

Proteolytic removal of the propeptide yields a mature single-chain cathepsin with a molecular weight of approximately 30 kDa, for example, ~32 kDa for cathepsin B. In the lysosome, the N-glycoside modifications of cathepsins are gradually removed by lysosomal glycosidases, resulting in a carbohydrate-free enzyme, for example, 27.8 kDa for cathepsin B in its single-chain form. Further processing in mature lysosomes leads to a two-chain form, which is linked by a disulfide-bond, for example, cathepsin B with 22 + 5 kDa [7].

As described, the majority of cysteine cathepsins is targeted to endolysosomal compartments (Figure 11.1, route A), where the enzymes encounter their substrates in the form of endocytosed or autophagocytosed material. However, cathepsins can take alternative sorting pathways toward the extracellular space. Procathepsins that have not been sorted to the endolysosome can simply follow the secretory route from ER to Golgi to the plasma membrane in clathrin-coated vesicles (Figure 11.1, route B). This results in constitutive secretion of procathepsins that can be proteolytically activated in the extracellular space. In addition, in exocrine or endocrine glands, regulated secretion in response to a certain stimulus is conceivable (Figure 11.1, route C). In the thyroid gland, redistribution and enhanced secretion of cysteine cathepsins has been found on TSH stimulation [8].

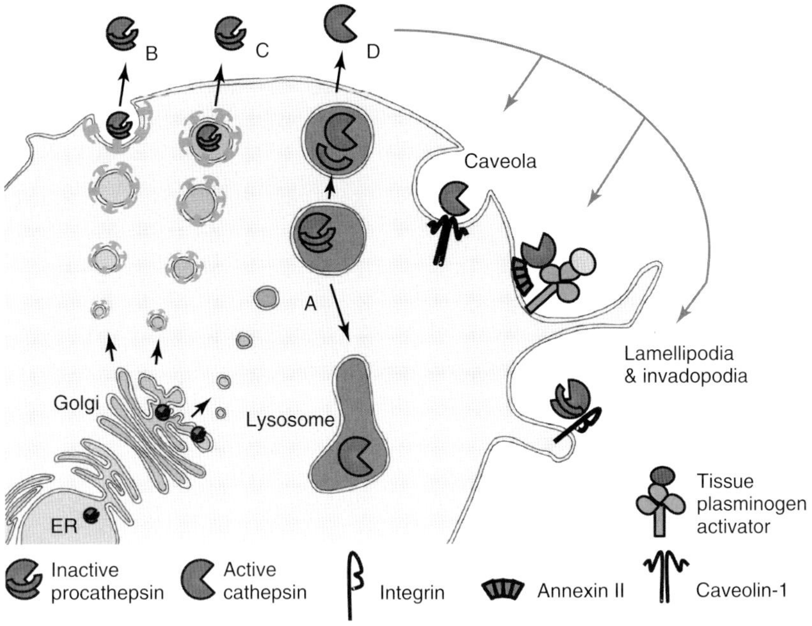

Figure 11.1 *Cathepsin pathways towards the extracellular space.* Procathepsins in the trans-Golgi network are for the most part directed to the lysosome (route A), where proteolytic activation takes place. Active cathepsins encounter their intracellular substrates in the endolysosome. Alternatively, inactive procathepsins can leave the Golgi via constitutively secreted clathrin-coated vesicles (route B) and get activated in the extracellular space. Regulated secretion (route C) occurs in specific secretory cell types of endocrine and exocrine tissues. Mature active cathepsins can be exocytosed from the lysosome (Route D). The association of secreted cathepsins to the cell membrane occurs at specialized microdomains. In caveolae, cathepsins are bound to the structural protein caveolin-1. At lamellipodia, procathepsin Z binds to integrins. Also at invadopodia of tumor cells, cathepsin B is found in a complex with annexin II that mediates binding to the tissue plasminogen activator and plasminogen.

In addition, activated cathepsins may be released from cells by secretory lysosomes (Figure 11.1, route D). For instance, the redistribution of acidic vesicles containing lysosomal enzymes toward the cell periphery and their enhanced secretion have been reported in response to calcium ions [9].

Some extracellular cysteine cathepsins are membrane associated and restricted to defined membrane microdomains [10]. Procathepsin Z is attached to the cell membrane and accumulates in vesicles at lamellipodia. The proenzyme is able to bind to integrins via an arginine-glycine-aspartic acid (RGD) motif in its propeptide [11] and may therefore act as a modulator of cell adhesion in endothelial tube formation. Cathepsin B is often located at focal adhesions and invadopodia of tumor cells. It is bound in a cell surface complex with the annexin II tetramer, which could serve as a linker to several extracellular cysteine cathepsin substrates, for example, tissue-type plasminogen activator (tPA), plasminogen, plasmin, collagen I, and tenascin C [12]. The localization of cathepsin B in caveolae is due to binding to caveolin-1, in a trimolecular complex with urokinase-type plasminogen activator receptor (uPAR) and β1-integrin [13]. In this study, the knockdown of caveolin-1 abolished formation of such proteolytic pits and decreased extracellular collagen IV degradation and cell invasion. Cell surface labeling by activity-based probes for cysteine cathepsins detected cathepsins B and Z at the outer side of the plasma membrane of primary breast cancer cells from the MMTV-PyMT (polyoma middle T) cancer mouse model [14, 15], and cathepsin B was detected at the surface of migrating HaCaT keratinocytes [16].

11.4 Extracellular Enzymatic Activity of Lysosomal Cathepsins

Cathepsin activity in the endolysosomal compartment requires a reducing, acidic milieu with pH values reaching as low as 3.8 in mature lysosomes [4]. However, secreted cathepsins show proteolytic activity in the extracellular space, too. This has been described for secreted cathepsin B in breast cancer, where hypoxic conditions lead to lactate acidosis and consecutive acidification of the tumor cell surroundings [17]. Apart from that, recent findings prove that cathepsins are also active at neutral pH and under oxidizing conditions. For example, processing of thyroglobulin in the thyroid gland by cysteine cathepsins comprises a sequence of cleavage events in compartments of distinct pH values and redox potentials. Thyroglobulin is liberated from its intraluminal macromolecular storage by cathepsins at the neutral pH and the low redox potential of the extracellular space. Further processing of the internalized thyroglobulin occurs in the slightly acidic endocytotic compartment under reducing conditions [18]. Experiments with cathepsin-deficient mice proved the functional relevance of cysteine cathepsins for thyroid hormone production [19]. Thus, cathepsins show enzymatic activity in both the acidic and neutral pH range. Nevertheless, extracellular cathepsin functions remain restricted by enzyme stability. Some cathepsins are relatively unstable at neutral pH, for example, cathepsin L and cathepsin B with half-lives of 1.3 and 15 min, respectively, whereas cathepsin S was found to be extremely stable [4]. However, substrates, the composition of the extracellular environment as well as inhibitor binding to the active site increase the stability of the cathepsin enzymes [20].

11.5 Endogenous Cathepsin Inhibitors as Regulators of Extracellular Cathepsins

Considering the high proteolytic activity of cysteine cathepsins in the extracellular compartment and their broad substrate spectra, tight control of their activity is required. Endogenous inhibitors of cysteine cathepsins are the stefins/cystatines [21]. They can be subdivided, according to their location, into three subfamilies.

Type I cystatins (= stefins) are cytosolic proteins, while type II cystatins are secreted into the extracellular environment [22]. Type III cystatins (= kininogens) are large plasma proteins that act as regulators of coagulation and acute phase response besides their function as cathepsin inhibitors. The members of all three subfamilies of cysteine cathepsin inhibitors are active under neutral and alkaline conditions. In several pathologies the cathepsin/cystatin balance is disturbed in favor of proteolysis. For instance, cathepsins B, L, and S were found upregulated in atherosclerosis while their inhibitor cystatin C was reduced [23].

11.6
Extracellular Substrates of Cysteine Cathepsins

Extracellular matrix (ECM) degradation by cysteine cathepsins is implicated in wound healing, angiogenesis, bone remodeling, and in some pathologic conditions. For example, cathepsin L secreted by endothelial progenitor cells was identified as a crucial factor for neovascularization of ischemic tissue *in vivo* [24]. In order to execute such important tasks, cysteine cathepsins are capable of degrading triple-helical collagens, that is, collagen I and collagen II, in connective tissues [25] as well as the main components of the basal membrane, that is, collagen IV, laminin, and fibronectin [26]. Cathepsin K from osteoclasts plays a pivotal role in bone remodeling as discussed in Chapter 4. Some cathepsins may also process ECM molecules more specifically and mediate the generation of bioactive peptides. It has been demonstrated that cathepsin L liberates endostatin, an inhibitor of angiogenesis, and tumor growth from collagen XVIII [27]. Besides directly degrading ECM components, cysteine cathepsins can achieve ECM breakdown by triggering a proteolytic cascade. This involves activation of the urokinase-type plasminogen activator (uPA) [28] and plasmin [29]. Apart from ECM components, cell surface proteins such as E-cadherin may be substrates of cysteine cathepsins. Proteolytic ectodomain shedding of E-cadherin was suggested to enhance invasiveness in a mouse model of pancreatic islet cancer, because cathepsins B, L, and S were capable of cleaving E-cadherin in its extracellular domain *in vitro* [30]. Cathepsin-mediated ectodomain shedding on the cell surface was also described for the epidermal growth factor (EGF) receptor [31].

11.7
Cysteine Cathepsins in Cancer: Clinical Associations

Besides their physiological roles, cysteine cathepsins are also implicated in various pathological processes such as cancer [10, 32, 33]. Several human tumor entities show increased levels of specific cathepsins. For instance, elevated expression of cathepsin B could be detected in breast, prostate, colorectal, lung, oral, and ovarian cancers as well as in meningioma, melanoma, and glioma [34–39], whereas an increased level of cathepsin S was found in prostate and gastric cancers [40, 41]. Cathepsin V levels were found to be upregulated in breast, colorectal, ovarian, renal, and squamous cell carcinomas [42], while cathepsins Z and H are implicated in inflammatory breast cancer [43]. Elevated levels of secreted cathepsins in malignant tissues [44–46] have frequently been associated with cancer progression and poor prognosis for patient outcome, indicating cathepsins as prognostic markers and therapeutic targets [47–51].

11.8
Cysteine Cathepsins in Cancer: Evidence from Animal Models

Much of the functional evidence for the role of cathepsins in tumor progression and metastasis has been obtained from several genetic mouse models of human cancer. In these models the downregulation of cathepsins revealed specific roles for individual proteases during tumorigenesis [14, 15, 30, 52].

The *RIP1-Tag2 mouse model of pancreatic islet cancer* makes use of a recombinant insulin/simian virus 40 large T antigen (Tag) [53]. Owing to the rat insulin promoter, the expression of the oncogene is restricted to the insulin-producing β-cells in the islets of Langerhans. The Tag oncoprotein can bind and inactivate the tumor suppressors p53 [54]. In this model, a cell-permeable broad spectrum inhibitor for cysteine cathepsins caused delayed and reduced progression to invasive cancer, delayed angiogenic switch, tumor growth, and vascularity [55]. Null mutations of cathepsin B, L, or S in this model resulted in reduced tumor growth due to decreased proliferation, increased apoptosis, or impaired formation of tumor vasculature. In contrast, the absence of cathepsin C had no impact on any of these tumor-relevant characteristics [30], whereas a deletion of cathepsin H again led to decreased tumor burden [56], indicating a tumor promoting role of cathepsins in this model.

Specific functions of cathepsins during breast cancer have been studied in the transgenic *MMTV-PyMT* mouse model for metastasizing breast cancer [57]. The viral PyMT oncogene is expressed under the control of a mouse mammary duct promoter and induces spontaneous tumorigenesis in all mammary glands of the female mouse. In this model, Vasiljeva *et al.* [15, 58] have shown that mice deficient in cathepsin B exhibit reduced tumor volume, delayed cancer progression, as well as reduced number and volume of lung metastases, due to diminished migratory and invasive potential of the tumor cells. They also identified tumor infiltrating macrophages as additional source of cathepsin B. Knockout experiments for combined loss of cathepsin B and Z in this model revealed synergistic effects of both cathepsins, leading to significant reduction in tumor and metastatic burden [14], whereas the overexpression of human cathepsin B resulted in increased tumor weight and accelerated number of infiltrating immune cells [59].

Squamous epidermal carcinogenesis was studied in the *K14-HPV16 mouse model*. In this model, the human papilloma virus type 16 (HPV16) early region genes, including the well-known oncogenes E6 and E7, are expressed under the control of human keratin-14 promoter (K14). Expression of the oncogenes leads to epidermal hyperproliferation and dysplastic lesions. These premalignant tumor stages progress to squamous cell carcinomas that metastasize to the lymph nodes [60]. Elevated expression of cathepsin S, as well as reduced levels of the endogenous inhibitor for cathepsins, cystatin C, was detected in skin tissues of this mouse model during tumor development. Both supported cancer progression by influencing proliferation, apoptosis, and neovascularization within premalignant lesions [61]. However, the consequences of cathepsin L ablation indicate antitumor properties of this protease because of its role in termination of growth factor signaling within the endosomal/lysosomal compartment of keratinocytes [52].

11.9
Molecular Dysregulation of Cathepsins in Cancer Progression

The expression of proteases in cancer is regulated at different molecular levels, affecting intrinsic and extrinsic cellular programs [62–64]. For the cysteine cathepsins, these mechanisms include gene amplification [65], increased gene expression regulated by transcription factors such as NF-κb [66], transcript variants due to alternative promoters, and alternative splicing [67–70] along with posttranscriptional regulation affecting stability of mRNA [71, 72] and protein [73].

In addition to genomic, transcriptional, and translational mechanisms, altered protease trafficking results in upregulation of extracellular cathepsins during neoplastic transformation [10, 12, 50, 74]. For cathepsins B, L, and Z, a translocation to the tumor cell surface and increased secretion of the mature enzyme or its zymogen into the extracellular space was shown particularly by cells at the invasive front [34, 75, 76]. This redistribution of cathepsins could also be detected in samples of human breast, colon, and esophageal carcinomas as well as gliomas [77–80].

The extracellular activity of cathepsins is influenced by the ratio of cathepsins to their endogenous inhibitors such as cystatins and stefins [81, 82]. Overexpression of these inhibitors leads to a reduction in tumor-promoting characteristics such as invasion, migration, and vascularity [83, 84]. In contrast, silencing of inhibitors such as cystatin M/E or cystatin C contributes to cancer progression by increased migration, invasion, or proliferation and reduced apoptosis of tumor cells [37, 61, 85].

11.10
Extracellular Cathepsins in Cancer

Although cathepsins function most efficiently at acidic conditions within the endolysosomes, they also show proteolytic activity in neutral or slightly acidic extracellular milieu [18, 86]. Cathepsins B, C, L, and Z can be released by tumor cells [87–89]. However, the primary source of secreted cathepsins seems to be the tumor stroma [90], especially tumor infiltrating immune cells such as macrophages and mast cells [15, 30, 91, 92] as well as fibroblasts [93]. Secreted cathepsins are capable of degrading ECM proteins directly [26, 30, 94–96] or indirectly by activating proteolytic cascades of downstream proteases [10, 97]. The cleavage of ECM proteins by cathepsins enables tumor cells to disseminate from the primary tumor, invade into blood vessels and lymphatics, extravasate, and form micrometastases in distant organs [98, 99]. Invasion, migration, and metastasis belong to the key hallmarks of cancer [100], revealing a crucial role for cathepsins in tumor-promoting processes.

The implication of extracellular cathepsins in malignant progression was confirmed by different *in vitro* and *in vivo* studies. Overexpression as well as abrogation of certain cathepsins revealed specific tumor-promoting roles of cathepsin B [58, 101–104], cathepsin S [105, 106], cathepsin K [107, 108], and cathepsin H [56, 85] in cancer cell invasion, migration, proliferation, metastatic spreading, apoptosis, tumor burden, and angiogenesis.

As cathepsins are mainly secreted by tumor-infiltrating immune cells, tumor progression and tumor cell growth also depend on a disordered relationship between tumor and stromal cells [109, 110]. First studies also revealed an impact of cytokines, such as interleukin-4 (IL-4) and IL-6, released from cells of the tumor microenvironment, on cathepsin secretion and activation [111, 112]. In this regard, further studies are required to elucidate mechanisms of the tumor microenvironment to impede tumor development and metastasis by influencing expression and activity of cathepsins.

11.11 Conclusions and Further Directions

At present, the secretion of cysteine cathepsins from cells and their extracellular proteolytic activity, especially in the context of cancer, are well-established facts. However, puzzling questions remain. For instance, most of cellular routes of cathepsin secretion make it likely that the proteolytically inactive proforms of cathepsins are secreted and, in fact, only the cathepsin zymogens are readily detectable in cell-conditioned media. As described, the proteolytic conversion of the procathepsin to the active protease is required for efficient proteolytic activity. However, there is no clear answer regarding the conditions and the proteases that activate the cathepsins outside of cells. In addition, there is considerable biochemical "test tube" evidence that cathepsins are able to process or degrade many ECM proteins, cytokines, and chemokines. However, less of these data are available from cell culture experiments, and there is only circumstantial evidence for substrates from animal models or patients. The challenge will be to identify extracellular cathepsin substrates in the natural micromilieu of cells and to link these cleavage events to physiological and pathological processes.

Acknowledgments

Supported by the European Union Framework Program (FP7 "MICROENVIMET" No 201279), by the Deutsche Forschungsgemeinschaft SFB 850 Project B7, the Centre of Chronic Immunodeficiency (CCI) Freiburg grant TP8, and the Excellence Initiative of the German Federal and State Governments (EXC 294).

References

1. Saftig, P. and Klumperman, J. (2009) Lysosome biogenesis and lysosomal membrane proteins: trafficking meets function. *Nat. Rev. Mol. Cell Biol.*, **10** (9), 623–635.
2. Rawlings, N.D., Morton, F.R., Kok, C.Y., Kong, J., and Barrett, A.J. (2008) MEROPS: the peptidase database. *Nucleic Acids Res.*, **36** (Database issue), D320–D325.
3. Barrett, A.J. and Kirschke, H. (1981) Cathepsin B, Cathepsin H, and cathepsin L. *Methods Enzymol.*, **80** (Pt C), 535–561.
4. Turk, B., Turk, D., and Turk, V. (2000) Lysosomal cysteine proteases: more than scavengers. *Biochim. Biophys. Acta*, **1477** (1–2), 98–111.
5. Mach, L., Mort, J.S., and Glossl, J. (1994) Maturation of human procathepsin B. Proenzyme activation and proteolytic processing of the precursor to the mature proteinase, in vitro, are primarily unimolecular processes. *J. Biol. Chem.*, **269** (17), 13030–13035.
6. Mach, L., Stuwe, K., Hagen, A., Ballaun, C., and Glossl, J. (1992) Proteolytic processing and glycosylation of cathepsin B. The role of the primary structure of the latent precursor and of the carbohydrate moiety for cell-type-specific molecular forms of the enzyme. *Biochem. J.*, **282** (Pt 2), 577–582.
7. Mach, L. (2002) Biosynthesis of lysosomal proteinases in health and disease. *Biol. Chem.*, **383** (5), 751–756.
8. Linke, M., Jordans, S., Mach, L., Herzog, V., and Brix, K. (2002) Thyroid stimulating hormone upregulates secretion of cathepsin B from thyroid epithelial cells. *Biol. Chem.*, **383** (5), 773–784.
9. Andrews, N.W. (2000) Regulated secretion of conventional lysosomes. *Trends Cell Biol.*, **10** (8), 316–321.
10. Mohamed, M.M. and Sloane, B.F. (2006) Cysteine cathepsins: multifunctional enzymes in cancer. *Nat. Rev. Cancer*, **6** (10), 764–775.
11. Lechner, A.M., Assfalg-Machleidt, I., Zahler, S., Stoeckelhuber, M., Machleidt, W., Jochum, M., and Nagler, D.K. (2006) RGD-dependent binding of procathepsin X to integrin alphavbeta3 mediates cell-adhesive properties. *J. Biol. Chem.*, **281** (51), 39588–39597.
12. Mai, J., Waisman, D.M., and Sloane, B.F. (2000) Cell surface complex of cathepsin B/annexin II tetramer in malignant progression. *Biochim. Biophys. Acta*, **1477** (1–2), 215–230.
13. Cavallo-Medved, D., Mai, J., Dosescu, J., Sameni, M., and Sloane, B.F. (2005) Caveolin-1 mediates the expression and localization of cathepsin B, pro-urokinase plasminogen activator and their cell-surnameface receptors in human colorectal carcinoma cells. *J. Cell Sci.*, **118** (Pt 7), 1493–1503.
14. Sevenich, L., Schurigt, U., Sachse, K., Gajda, M., Werner, F., Muller, S., Vasiljeva, O., Schwinde, A., Klemm, N., Deussing, J., Peters, C., and Reinheckel, T. (2010) Synergistic antitumor effects of combined cathepsin B and cathepsin Z deficiencies on breast cancer progression and metastasis in mice. *Proc. Natl. Acad. Sci. U.S.A.*, **107** (6), 2497–2502.
15. Vasiljeva, O., Papazoglou, A., Kruger, A., Brodoefel, H., Korovin, M., Deussing, J., Augustin, N., Nielsen, B.S., Almholt, K., Bogyo, M., Peters, C., and Reinheckel, T. (2006) Tumor

cell-derived and macrophage-derived cathepsin B promotes progression and lung metastasis of mammary cancer. *Cancer Res.*, **66** (10), 5242–5250.

16. Buth, H., Wolters, B., Hartwig, B., Meier-Bornheim, R., Veith, H., Hansen, M., Sommerhoff, C.P., Schaschke, N., Machleidt, W., Fusenig, N.E., Boukamp, P., and Brix, K. (2004) HaCaT keratinocytes secrete lysosomal cysteine proteinases during migration. *Eur. J. Cell Biol.*, **83** (11–12), 781–795.
17. Gocheva, V. and Joyce, J.A. (2007) Cysteine cathepsins and the cutting edge of cancer invasion. *Cell Cycle*, **6** (1), 60–64.
18. Jordans, S., Jenko-Kokalj, S., Kuhl, N.M., Tedelind, S., Sendt, W., Bromme, D., Turk, D., and Brix, K. (2009) Monitoring compartment-specific substrate cleavage by cathepsins B, K, L, and S at physiological pH and redox conditions. *BMC Biochem.*, **10**, 23.
19. Friedrichs, B., Tepel, C., Reinheckel, T., Deussing, J., von Figura, K., Herzog, V., Peters, C., Saftig, P., and Brix, K. (2003) Thyroid functions of mouse cathepsins B, K, and L. *J. Clin. Invest.*, **111** (11), 1733–1745.
20. Turk, B., Dolenc, I., Lenarcic, B., Krizaj, I., Turk, V., Bieth, J.G., and Bjork, I. (1999) Acidic pH as a physiological regulator of human cathepsin L activity. *Eur. J. Biochem.*, **259** (3), 926–932.
21. Turk, V. (1992) Cystatins as regulators of intracellular proteolysis. *J. Nutr. Sci. Vitaminol. (Tokyo)*, 292–297 (Spec No).
22. Abrahamson, M., Alvarez-Fernandez, M., and Nathanson, C.M. (2003) Cystatins. *Biochem. Soc. Symp.*, **70**, 179–199.
23. Liu, J., Sukhova, G.K., Sun, J.S., Xu, W.H., Libby, P., and Shi, G.P. (2004) Lysosomal cysteine proteases in atherosclerosis. *Arterioscler. Thromb. Vasc. Biol.*, **24** (8), 1359–1366.
24. Urbich, C., Heeschen, C., Aicher, A., Sasaki, K., Bruhl, T., Farhadi, M.R., Vajkoczy, P., Hofmann, W.K., Peters, C., Pennacchio, L.A., Abolmaali, N.D., Chavakis, E., Reinheckel, T., Zeiher, A.M., and Dimmeler, S. (2005) Cathepsin L is required for endothelial progenitor cell-induced neovascularization. *Nat. Med.*, **11** (2), 206–213.
25. Maciewicz, R.A. and Etherington, D.J. (1988) A comparison of four cathepsins (B, L, N and S) with collagenolytic activity from rabbit spleen. *Biochem. J.*, **256** (2), 433–440.
26. Buck, M.R., Karustis, D.G., Day, N.A., Honn, K.V., and Sloane, B.F. (1992) Degradation of extracellular-matrix proteins by human cathepsin B from normal and tumour tissues. *Biochem. J.*, **282** (Pt 1), 273–278.
27. Felbor, U., Dreier, L., Bryant, R.A., Ploegh, H.L., Olsen, B.R., and Mothes, W. (2000) Secreted cathepsin L generates endostatin from collagen XVIII. *Embo J.*, **19** (6), 1187–1194.
28. Kobayashi, H., Schmitt, M., Goretzki, L., Chucholowski, N., Calvete, J., Kramer, M., Gunzler, W.A., Janicke, F., and Graeff, H. (1991) Cathepsin B efficiently activates the soluble and the tumor cell receptor-bound form of the proenzyme urokinase-type plasminogen activator (Pro-uPA). *J. Biol. Chem.*, **266** (8), 5147–5152.
29. Goretzki, L., Schmitt, M., Mann, K., Calvete, J., Chucholowski, N., Kramer, M., Gunzler, W.A., Janicke, F., and Graeff, H. (1992) Effective activation of the proenzyme form of the urokinase-type plasminogen activator (pro-uPA) by the cysteine protease cathepsin L. *FEBS Lett.*, **297** (1–2), 112–118.
30. Gocheva, V., Zeng, W., Ke, D., Klimstra, D., Reinheckel, T., Peters, C., Hanahan, D., and Joyce, J.A. (2006) Distinct roles for cysteine cathepsin genes in multistage tumorigenesis. *Genes Dev.*, **20** (5), 543–556.
31. Schraufstatter, I.U., Trieu, K., Zhao, M., Rose, D.M., Terkeltaub, R.A., and Burger, M. (2003) IL-8-mediated cell migration in endothelial cells depends on cathepsin B activity and transactivation of the epidermal growth factor receptor. *J. Immunol.*, **171** (12), 6714–6722.

32. Reiser, J., Adair, B., and Reinheckel, T. (2010) Specialized roles for cysteine cathepsins in health and disease. *J. Clin. Invest.*, **120** (10), 3421–3431.
33. Vasiljeva, O., Reinheckel, T., Peters, C., Turk, D., Turk, V., and Turk, B. (2007) Emerging roles of cysteine cathepsins in disease and their potential as drug targets. *Curr. Pharm. Des.*, **13** (4), 387–403.
34. Jedeszko, C. and Sloane, B.F. (2004) Cysteine cathepsins in human cancer. *Biol. Chem.*, **385** (11), 1017–1027.
35. Kolwijck, E., Massuger, L.F., Thomas, C.M., Span, P.N., Krasovec, M., Kos, J., and Sweep, F.C. (2010b) Cathepsins B, L and cystatin C in cyst fluid of ovarian tumors. *J. Cancer Res. Clin. Oncol.*, **136** (5), 771–778.
36. Lah, T.T., Kalman, E., Najjar, D., Gorodetsky, E., Brennan, P., Somers, R., and Daskal, I. (2000) Cells producing cathepsins D, B, and L in human breast carcinoma and their association with prognosis. *Hum. Pathol*, **31** (2), 149–160.
37. Lah, T.T., Nanni, I., Trinkaus, M., Metellus, P., Dussert, C., De Ridder, L., Rajcevic, U., Blejec, A., and Martin, P.M. (2010) Toward understanding recurrent meningioma: the potential role of lysosomal cysteine proteases and their inhibitors. *J. Neurosurg.*, **112** (5), 940–950.
38. Scorilas, A., Fotiou, S., Tsiambas, E., Yotis, J., Kotsiandri, F., Sameni, M., Sloane, B.F., and Talieri, M. (2002) Determination of cathepsin B expression may offer additional prognostic information for ovarian cancer patients. *Biol. Chem.*, **383** (7–8), 1297–1303.
39. Talieri, M., Papadopoulou, S., Scorilas, A., Xynopoulos, D., Arnogianaki, N., Plataniotis, G., Yotis, J., and Agnanti, N. (2004) Cathepsin B and cathepsin D expression in the progression of colorectal adenoma to carcinoma. *Cancer Lett.*, **205** (1), 97–106.
40. Fernandez, P.L., Farre, X., Nadal, A., Fernandez, E., Peiro, N., Sloane, B.F., Shi, G.P., Chapman, H.A., Campo, E., and Cardesa, A. (2001) Expression of cathepsins B and S in the progression of prostate carcinoma. *Int. J. Cancer*, **95** (1), 51–55.
41. Nagler, D.K., Kruger, S., Kellner, A., Ziomek, E., Menard, R., Buhtz, P., Krams, M., Roessner, A., and Kellner, U. (2004) Up-regulation of cathepsin X in prostate cancer and prostatic intraepithelial neoplasia. *Prostate*, **60** (2), 109–119.
42. Haider, A.S., Peters, S.B., Kaporis, H., Cardinale, I., Fei, J., Ott, J., Blumenberg, M., Bowcock, A.M., Krueger, J.G., and Carucci, J.A. (2006) Genomic analysis defines a cancer-specific gene expression signature for human squamous cell carcinoma and distinguishes malignant hyperproliferation from benign hyperplasia. *J. Invest. Dermatol.*, **126** (4), 869–881.
43. Decock, J., Obermajer, N., Vozelj, S., Hendrickx, W., Paridaens, R., and Kos, J. (2008) Cathepsin B, cathepsin H, cathepsin X and cystatin C in sera of patients with early-stage and inflammatory breast cancer. *Int. J. Biol. Markers*, **23** (3), 161–168.
44. Brix, K., Linke, M., Tepel, C., and Herzog, V. (2001) Cysteine proteinases mediate extracellular prohormone processing in the thyroid. *Biol. Chem.*, **382** (5), 717–725.
45. Koblinski, J.E., Dosescu, J., Sameni, M., Moin, K., Clark, K., and Sloane, B.F. (2002) Interaction of human breast fibroblasts with collagen I increases secretion of procathepsin B. *J. Biol. Chem.*, **277** (35), 32220–32227.
46. Mort, J.S., Recklies, A.D., and Poole, A.R. (1985) Release of cathepsin B precursors from human and murine tumours. *Prog. Clin. Biol. Res.*, **180**, 243–245.
47. Berdowska, I. (2004) Cysteine proteases as disease markers. *Clin. Chim. Acta*, **342** (1–2), 41–69.
48. Devetzi, M., Scorilas, A., Tsiambas, E., Sameni, M., Fotiou, S., Sloane, B.F., and Talieri, M. (2009) Cathepsin B protein levels in endometrial cancer: potential value as a tumour biomarker. *Gynecol. Oncol.*, **112** (3), 531–536.

49. Nouh, M.A., Mohamed, M.M., El-Shinawi, M., Shaalan, M.A., Cavallo-Medved, D., Khaled, H.M., and Sloane, B.F. (2011) Cathepsin B: a potential prognostic marker for inflammatory breast cancer. *J. Transl. Med.*, **9**, 1.
50. Palermo, C. and Joyce, J.A. (2008) Cysteine cathepsin proteases as pharmacological targets in cancer. *Trends Pharmacol. Sci.*, **29** (1), 22–28.
51. Thomssen, C., Schmitt, M., Goretzki, L., Oppelt, P., Pache, L., Dettmar, P., Janicke, F., and Graeff, H. (1995) Prognostic value of the cysteine proteases cathepsins B and cathepsin L in human breast cancer. *Clin. Cancer Res.*, **1** (7), 741–746.
52. Dennemarker, J., Lohmuller, T., Mayerle, J., Tacke, M., Lerch, M.M., Coussens, L.M., Peters, C., and Reinheckel, T. (2010) Deficiency for the cysteine protease cathepsin L promotes tumor progression in mouse epidermis. *Oncogene*, **29** (11), 1611–1621.
53. Hanahan, D. (1985) Heritable formation of pancreatic beta-cell tumours in transgenic mice expressing recombinant insulin/simian virus 40 oncogenes. *Nature*, **315** (6015), 115–122.
54. Ludlow, J.W. (1993) Interactions between SV40 large-tumor antigen and the growth suppressor proteins pRB and p53. *FASEB J.*, **7** (10), 866–871.
55. Joyce, J.A., Baruch, A., Chehade, K., Meyer-Morse, N., Giraudo, E., Tsai, F.Y., Greenbaum, D.C., Hager, J.H., Bogyo, M., and Hanahan, D. (2004) Cathepsin cysteine proteases are effectors of invasive growth and angiogenesis during multistage tumorigenesis. *Cancer Cell*, **5** (5), 443–453.
56. Gocheva, V., Chen, X., Peters, C., Reinheckel, T., and Joyce, J.A. (2010a) Deletion of cathepsin H perturbs angiogenic switching, vascularization and growth of tumors in a mouse model of pancreatic islet cell cancer. *Biol. Chem.*, **391** (8), 937–945.
57. Guy, C.T., Cardiff, R.D., and Muller, W.J. (1992) Induction of mammary tumors by expression of polyomavirus middle T oncogene: a transgenic mouse model for metastatic disease. *Mol. Cell Biol.*, **12** (3), 954–961.
58. Vasiljeva, O., Korovin, M., Gajda, M., Brodoefel, H., Bojic, L., Kruger, A., Schurigt, U., Sevenich, L., Turk, B., Peters, C., and Reinheckel, T. (2008) Reduced tumour cell proliferation and delayed development of high-grade mammary carcinomas in cathepsin B-deficient mice. *Oncogene*, **27** (30), 4191–4199.
59. Sevenich, L., Werner, F., Gajda, M., Schurigt, U., Sieber, C., Muller, S., Follo, M., Peters, C., and Reinheckel, T. (2011) Transgenic expression of human cathepsin B promotes progression and metastasis of polyoma-middle-T-induced breast cancer in mice. *Oncogene*, **30** (1), 54–64.
60. Coussens, L.M., Hanahan, D., and Arbeit, J.M. (1996) Genetic predisposition and parameters of malignant progression in K14-HPV16 transgenic mice. *Am.J. Pathol.*, **149** (6), 1899–1917.
61. Yu, W., Liu, J., Shi, M.A., Wang, J., Xiang, M., Kitamoto, S., Wang, B., Sukhova, G.K., Murphy, G.F., Orasanu, G., Grubb, A., and Shi, G.P. (2010) Cystatin C deficiency promotes epidermal dysplasia in K14-HPV16 transgenic mice. *PLoS ONE*, **5** (11), e13973.
62. Balkwill, F., Charles, K.A., and Mantovani, A. (2005) Smoldering and polarized inflammation in the initiation and promotion of malignant disease. *Cancer Cell*, **7** (3), 211–217.
63. Coussens, L.M. and Werb, Z. (2002) Inflammation and cancer. *Nature*, **420** (6917), 860–867.
64. Yan, S. and Sloane, B.F. (2003) Molecular regulation of human cathepsin B: implication in pathologies. *Biol. Chem.*, **384** (6), 845–854.
65. Lin, L., Aggarwal, S., Glover, T.W., Orringer, M.B., Hanash, S., and Beer, D.G. (2000) A minimal critical region of the 8p22-23 amplicon in esophageal adenocarcinomas defined using sequence tagged site-amplification mapping and quantitative polymerase

chain reaction includes the GATA-4 gene. *Cancer Res.*, **60** (5), 1341–1347.

66. Hamer, I., Delaive, E., Dieu, M., Abdel-Sater, F., Mercy, L., Jadot, M., and Arnould, T. (2009) Up-regulation of cathepsin B expression and enhanced secretion in mitochondrial DNA-depleted osteosarcoma cells. *Biol. Cell*, **101** (1), 31–41.
67. Arora, S. and Chauhan, S.S. (2002) Identification and characterization of a novel human cathepsin L splice variant. *Gene*, **293** (1–2), 123–131.
68. Caserman, S., Kenig, S., Sloane, B.F., and Lah, T.T. (2006) Cathepsin L splice variants in human breast cell lines. *Biol. Chem.*, **387** (5), 629–634.
69. Gong, Q., Chan, S.J., Bajkowski, A.S., Steiner, D.F., and Frankfater, A. (1993) Characterization of the cathepsin B gene and multiple mRNAs in human tissues: evidence for alternative splicing of cathepsin B pre-mRNA. *DNA Cell Biol.*, **12** (4), 299–309.
70. Tedelind, S., Poliakova, K., Valeta, A., Hunegnaw, R., Yemanaberhan, E.L., Heldin, N.E., Kurebayashi, J., Weber, E., Kopitar-Jerala, N., Turk, B., Bogyo, M., and Brix, K. (2010) Nuclear cysteine cathepsin variants in thyroid carcinoma cells. *Biol. Chem.*, **391** (8), 923–935.
71. Frohlich, E., Schlagenhauff, B., Mohrle, M., Weber, E., Klessen, C., and Rassner, G. (2001) Activity, expression, and transcription rate of the cathepsins B, D, H, and L in cutaneous malignant melanoma. *Cancer*, **91** (5), 972–982.
72. Jean, D., Rousselet, N., and Frade, R. (2006) Expression of cathepsin L in human tumor cells is under the control of distinct regulatory mechanisms. *Oncogene*, **25** (10), 1474–1484.
73. Hazen, L.G., Bleeker, F.E., Lauritzen, B., Bahns, S., Song, J., Jonker, A., Van Driel, B.E., Lyon, H., Hansen, U., Kohler, A., and Van Noorden, C.J. (2000) Comparative localization of cathepsin B protein and activity in colorectal cancer. *J. Histochem. Cytochem.*, **48** (10), 1421–1430.
74. Nomura, T. and Katunuma, N. (2005) Involvement of cathepsins in the invasion, metastasis and proliferation of cancer cells. *J. Med. Invest.*, **52** (1–2), 1–9.
75. Emmert-Buck, M.R., Roth, M.J., Zhuang, Z., Campo, E., Rozhin, J., Sloane, B.F., Liotta, L.A., and Stetler-Stevenson, W.G. (1994) Increased gelatinase A (MMP-2) and cathepsin B activity in invasive tumor regions of human colon cancer samples. *Am. J. Pathol.*, **145** (6), 1285–1290.
76. Frosch, B.A., Berquin, I., Emmert-Buck, M.R., Moin, K., and Sloane, B.F. (1999) Molecular regulation, membrane association and secretion of tumor cathepsin B. *Acta Pathol. Microbiol. Immunol. Scand.*, **107** (1), 28–37.
77. Hughes, S.J., Glover, T.W., Zhu, X.X., Kuick, R., Thoraval, D., Orringer, M.B., Beer, D.G., and Hanash, S. (1998) A novel amplicon at 8p22-23 results in overexpression of cathepsin B in esophageal adenocarcinoma. *Proc. Natl. Acad. Sci. U. S. A.*, **95** (21), 12410–12415.
78. Mikkelsen, T., Yan, P.S., Ho, K.L., Sameni, M., Sloane, B.F., and Rosenblum, M.L. (1995) Immunolocalization of cathepsin B in human glioma: implications for tumor invasion and angiogenesis. *J. Neurosurg.*, **83** (2), 285–290.
79. Sameni, M., Elliott, E., Ziegler, G., Fortgens, P.H., Dennison, C., and Sloane, B.F. (1995) Cathepsin B and D are localized at the surface of human breast cancer cells. *Pathol. Oncol. Res.*, **1** (1), 43–53.
80. Sloane, B.F., Dunn, J.R., and Honn, K.V. (1981) Lysosomal cathepsin B: correlation with metastatic potential. *Science*, **212** (4499), 1151–1153.
81. Chang, S.H., Kanasaki, K., Gocheva, V., Blum, G., Harper, J., Moses, M.A., Shih, S.C., Nagy, J.A., Joyce, J., Bogyo, M., Kalluri, R., and Dvorak, H.F. (2009) VEGF-A induces angiogenesis by perturbing the cathepsin-cysteine protease inhibitor balance in venules, causing basement membrane degradation and

mother vessel formation. *Cancer Res.*, **69** (10), 4537–4544.
82. Kolwijck, E., Kos, J., Obermajer, N., Span, P.N., Thomas, C.M., Massuger, L.F., and Sweep, F.C. (2010a) The balance between extracellular cathepsins and cystatin C is of importance for ovarian cancer. *Eur. J. Clin. Invest.*, **40** (7), 591–599.
83. Kopitz, C., Anton, M., Gansbacher, B., and Kruger, A. (2005) Reduction of experimental human fibrosarcoma lung metastasis in mice by adenovirus-mediated cystatin C overexpression in the host. *Cancer Res.*, **65** (19), 8608–8612.
84. Wang, B., Sun, J., Kitamoto, S., Yang, M., Grubb, A., Chapman, H.A., Kalluri, R., and Shi, G.P. (2006) Cathepsin S controls angiogenesis and tumor growth via matrix-derived angiogenic factors. *J. Biol. Chem.*, **281** (9), 6020–6029.
85. Vigneswaran, N., Wu, J., Nagaraj, N., James, R., Zeeuwen, P., and Zacharias, W. (2006) Silencing of cystatin M in metastatic oral cancer cell line MDA-686Ln by siRNA increases cysteine proteinases and legumain activities, cell proliferation and in vitro invasion. *Life Sci.*, **78** (8), 898–907.
86. Linebaugh, B.E., Sameni, M., Day, N.A., Sloane, B.F., and Keppler, D. (1999) Exocytosis of active cathepsin B enzyme activity at pH 7.0, inhibition and molecular mass. *Eur. J. Biochem.*, **264** (1), 100–109.
87. Cavallo-Medved, D., Dosescu, J., Linebaugh, B.E., Sameni, M., Rudy, D., and Sloane, B.F. (2003) Mutant K-ras regulates cathepsin B localization on the surface of human colorectal carcinoma cells. *Neoplasia*, **5** (6), 507–519.
88. Collette, J., Ulku, A.S., Der, C.J., Jones, A., and Erickson, A.H. (2004) Enhanced cathepsin L expression is mediated by different Ras effector pathways in fibroblasts and epithelial cells. *Int. J. Cancer*, **112** (2), 190–199.
89. Roshy, S., Sloane, B.F., and Moin, K. (2003) Pericellular cathepsin B and malignant progression. *Cancer Metastasis Rev.*, **22** (2–3), 271–286.
90. Mueller, M.M. and Fusenig, N.E. (2004) Friends or foes – bipolar effects of the tumour stroma in cancer. *Nat. Rev. Cancer*, **4** (11), 839–849.
91. Allavena, P., Garlanda, C., Borrello, M.G., Sica, A., and Mantovani, A. (2008) Pathways connecting inflammation and cancer. *Curr. Opin. Genet. Dev.*, **18** (1), 3–10.
92. Lindahl, C., Simonsson, M., Bergh, A., Thysell, E., Antti, H., Sund, M., and Wikstrom, P. (2009) Increased levels of macrophage-secreted cathepsin S during prostate cancer progression in TRAMP mice and patients. *Cancer Genomics Proteomics*, **6** (3), 149–159.
93. Orimo, A., Gupta, P.B., Sgroi, D.C., Arenzana-Seisdedos, F., Delaunay, T., Naeem, R., Carey, V.J., Richardson, A.L., and Weinberg, R.A. (2005) Stromal fibroblasts present in invasive human breast carcinomas promote tumor growth and angiogenesis through elevated SDF-1/CXCL12 secretion. *Cell*, **121** (3), 335–348.
94. Ishidoh, K. and Kominami, E. (1995) Procathepsin L degrades extracellular matrix proteins in the presence of glycosaminoglycans in vitro. *Biochem. Biophys. Res. Commun.*, **217** (2), 624–631.
95. Mai, J., Sameni, M., Mikkelsen, T., and Sloane, B.F. (2002) Degradation of extracellular matrix protein tenascin-C by cathepsin B: an interaction involved in the progression of gliomas. *Biol. Chem.*, **383** (9), 1407–1413.
96. Wilson, S.R., Peters, C., Saftig, P., and Bromme, D. (2009) Cathepsin K activity-dependent regulation of osteoclast actin ring formation and bone resorption. *J. Biol. Chem.*, **284** (4), 2584–2592.
97. Guo, M., Mathieu, P.A., Linebaugh, B., Sloane, B.F., and Reiners, J.J. Jr. (2002) Phorbol ester activation of a proteolytic cascade capable of activating latent transforming growth factor-betaL a process initiated by the exocytosis of cathepsin B. *J. Biol. Chem.*, **277** (17), 14829–14837.
98. Chambers, A.F., Groom, A.C., and MacDonald, I.C. (2002) Dissemination

and growth of cancer cells in metastatic sites. *Nat. Rev. Cancer*, **2** (8), 563–572.

99. Polyak, K., Haviv, I., and Campbell, I.G. (2009) Co-evolution of tumor cells and their microenvironment. *Trends Genet.*, **25** (1), 30–38.

100. Hanahan, D. and Weinberg, R.A. (2000) The hallmarks of cancer. *Cell*, **100** (1), 57–70.

101. Andl, C.D., McCowan, K.M., Allison, G.L., and Rustgi, A.K. (2010) Cathepsin B is the driving force of esophageal cell invasion in a fibroblast-dependent manner. *Neoplasia*, **12** (6), 485–498.

102. Matarrese, P., Ascione, B., Ciarlo, L., Vona, R., Leonetti, C., Scarsella, M., Mileo, A.M., Catricala, C., Paggi, M.G., and Malorni, W. (2010) Cathepsin B inhibition interferes with metastatic potential of human melanoma: an in vitro and in vivo study. *Mol. Cancer*, **9**, 207.

103. Nalla, A.K., Gorantla, B., Gondi, C.S., Lakka, S.S., and Rao, J.S. (2010) Targeting MMP-9, uPAR, and cathepsin B inhibits invasion, migration and activates apoptosis in prostate cancer cells. *Cancer Gene Ther.* **17** (9), 599–613.

104. Tummalapalli, P., Spomar, D., Gondi, C.S., Olivero, W.C., Gujrati, M., Dinh, D.H., and Rao, J.S. (2007) RNAi-mediated abrogation of cathepsin B and MMP-9 gene expression in a malignant meningioma cell line leads to decreased tumor growth, invasion and angiogenesis. *Int. J. Oncol.*, **31** (5), 1039–1050.

105. Burden, R.E., Gormley, J.A., Jaquin, T.J., Small, D.M., Quinn, D.J., Hegarty, S.M., Ward, C., Walker, B., Johnston, J.A., Olwill, S.A., and Scott, C.J. (2009) Antibody-mediated inhibition of cathepsin S blocks colorectal tumor invasion and angiogenesis. *Clin. Cancer Res.*, **15** (19), 6042–6051.

106. Yang, Y., Lim, S.K., Choong, L.Y., Lee, H., Chen, Y., Chong, P.K., Ashktorab, H., Wang, T.T., Salto-Tellez, M., Yeoh, K.G., and Lim, Y.P. (2010) Cathepsin S mediates gastric cancer cell migration and invasion via a putative network of metastasis-associated proteins. *J. Proteome. Res.*, **9** (9), 4767–4778.

107. Kleer, C.G., Bloushtain-Qimron, N., Chen, Y.H., Carrasco, D., Hu, M., Yao, J., Kraeft, S.K., Collins, L.C., Sabel, M.S., Argani, P., Gelman, R., Schnitt, S.J., Krop, I.E., and Polyak, K. (2008) Epithelial and stromal cathepsin K and CXCL14 expression in breast tumor progression. *Clin. Cancer Res.*, **14** (17), 5357–5367.

108. Xie, L., Moroi, Y., Hayashida, S., Tsuji, G., Takeuchi, S., Shan, B., Nakahara, T., Uchi, H., Takahara, M., and Furue, M. (2011) Cathepsin K-upregulation in fibroblasts promotes matrigel invasive ability of squamous cell carcinoma cells via tumor-derived IL-1alpha. *J. Dermatol. Sci.*, **61** (1), 45–50.

109. Grivennikov, S.I., Greten, F.R., and Karin, M. (2010) Immunity, inflammation, and cancer. *Cell*, **140** (6), 883–899.

110. Mantovani, A., Allavena, P., Sica, A., and Balkwill, F. (2008) Cancer-related inflammation. *Nature*, **454** (7203), 436–444.

111. Gocheva, V., Wang, H.W., Gadea, B.B., Shree, T., Hunter, K.E., Garfall, A.L., Berman, T., and Joyce, J.A. (2010b) IL-4 induces cathepsin protease activity in tumor-associated macrophages to promote cancer growth and invasion. *Genes Dev.*, **24** (3), 241–255.

112. Mohamed, M.M., Cavallo-Medved, D., Rudy, D., Anbalagan, A., Moin, K., and Sloane, B.F. (2010) Interleukin-6 increases expression and secretion of cathepsin B by breast tumor-associated monocytes. *Cell Physiol. Biochem.*, **25** (2–3), 315–324.

12
ADAMs in Cancer

Dorte Stautz, Sarah Louise Dombernowsky, and Marie Kveiborg

12.1
ADAMs–Multifunctional Proteins

The ADAMs (a disintegrin and metalloproteases) belong to the adamalysin subfamily of metzincins, which are zinc-dependent endopeptidases characterized by a common zinc-binding consensus motif and a Met-turn. ADAMs comprise a family of nearly 40 cell-surface-associated proteins from various species (denoted ADAM1 to ADAM40[1]), where 24 have been identified in humans (4 are presumed pseudogenes) [1–3].

12.1.1
Structure and Biochemistry

Structurally, ADAMs are multidomain proteins averaging 800 amino acids in length. Most ADAMs share a well-conserved domain structure consisting from the N-terminus of a signal sequence, pro-, metalloprotease, disintegrin, cysteine-rich, and epidermal growth factor (EGF)-like domains, a single transmembrane region, and a C-terminal cytoplasmic domain (Figure 12.1). Almost all ADAMs exist in two or more isoforms derived from gene duplication, alternative splicing, or proteolytic processing, and are not necessarily conserved between species [1]. ADAM isoforms vary greatly, but a secreted form that lacks the transmembrane and cytoplasmic domains and instead has a unique C-terminal amino acid stretch is often found. So far, ADAM9, -10, -11, -12, and -28 have been found in this form.

ADAMs are synthesized and exported through the endoplasmic reticulum and matured via cleavage of the prodomain by furin or other proprotein convertases during Golgi transit, which is a prerequisite for enzymatic activity [1, 5].

1) An updated list of ADAMs can be found on *http://people.virginia.edu/~jw7g/Table_of_the_ADAMs.html* (8 September 2011).

Matrix Proteases in Health and Disease, First Edition. Edited by Niels Behrendt.

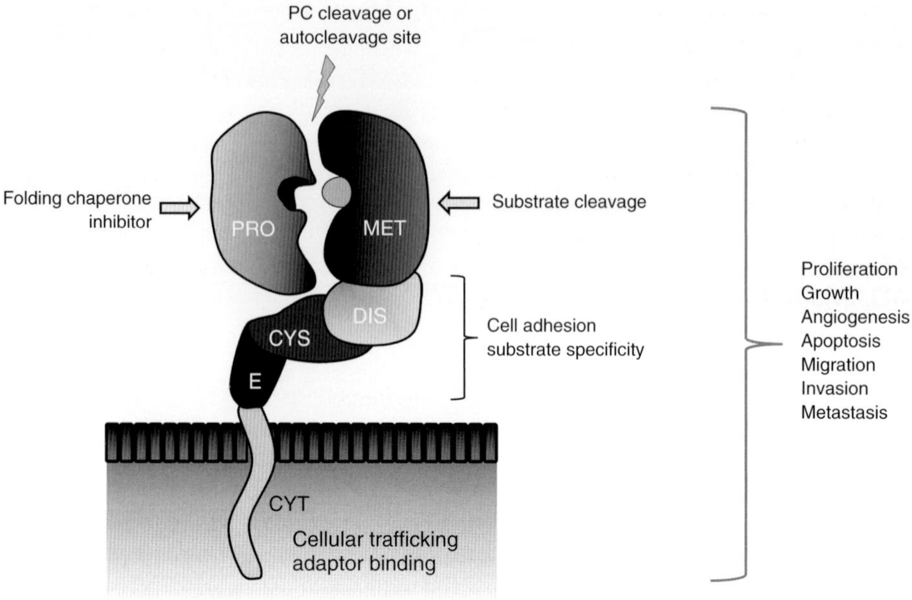

Figure 12.1 General domain structure of the ADAMs family of metalloproteases and their roles in cancer progression. PRO: prodomain; MET: metalloprotease domain; DIS: disintegrin domain; CYS: cysteine-rich domain; E: EGF-like domain; CYT: cytoplasmic tail; PC: proprotein convertase. (Figure modified from Kveiborg *et al.* [4]).

For ADAM8, -19, and -28, evidence also exists for autocatalytic activation. The prodomain maintains enzyme latency through a cysteine-switch mechanism and is proposed to aid in correct protein folding. The extracellular domains are extensively glycosylated and are stabilized by several dynamic disulfide bonds [6].

12.1.2
Biological Functions

One of the key functions of ADAMs is their protease activity, mediated by the metalloprotease domain. About half of the human ADAMs are predicted to be active proteases (ADAM8, -9, -10, -12, -15, -17, -19, -20, -21, -28, -30, and -33) containing the common zinc-binding consensus motif (HEXGHXXGXXHD) [7, 8]. The substrates of ADAMs are mostly membrane bound and include cytokines, chemokines, adhesion molecules, growth factors, and receptors. Substrate cleavage releases the ectodomain to the extracellular environment by a mechanism termed *ectodomain shedding* [1, 2, 9, 10]. When studied *in vitro*, many ADAMs appear to have multiple substrates, with extensive redundancy, but less is known of their *in vivo* specificities.

The biological functions of ADAMs also include cell adhesion, as the disintegrin domain of several ADAMs has been proposed to mediate cell–cell and cell–matrix interactions via integrins [1, 2, 10, 11]. Although ADAM–integrin interactions have been thoroughly studied and seem to support cell adhesion and migration, the biological relevance of this interaction is still unclear [3].

Finally, the cytoplasmic tails of ADAMs appear to be involved in the regulation of their protease activity, cell signaling, protein maturation, and subcellular localization [1]. These domains vary greatly in sequence and length from 14 (ADAM11) to 235 (ADAM19) amino acids, and often contain potential phosphorylation sites and/or specialized motifs for protein–protein interactions, possibly conferring unique functions to individual ADAMs. Many interaction partners have been identified, including adaptor proteins and protein kinases [1].

The expression patterns of ADAMs vary considerably, but many are testis specific (ADAM2, -4, -5, -6, -7, -18, -20, -21, -29, -30, and -32), some are brain specific (ADAM11, -22, and -23), and some display a more broad expression profile (ADAM9, -10, -12, -15, -17, -19, and -28) [1, 3, 12]. Functionally, ADAMs have been linked to sperm–egg interactions, cell fate determination, cell migration, axon guidance, muscle development, and diverse aspects of immunity [1, 13]. The generation of knockout mice has demonstrated that some ADAMs (ADAM8, -9, -11, -12, -15, and -33) are not necessary for normal development, as these knockouts display only mild phenotypes, whereas other ADAMs (ADAM10, -17, -19, -22, and -23) are critical for proper development [1, 14, 15]. However, it is still unknown whether this is due to elimination of critical proteolytic activities or of other functions, such as cell adhesion or cell signaling. Whether some degree of compensatory or redundant functions among different ADAMs may explain the mild defects of some ADAM-deficient mice has been a matter of debate [16, 17].

12.1.3
Pathological Functions

On the basis of the diversity of ADAM functions, dysregulated expression and/or activity of ADAMs may impart a variety of pathological consequences. Indeed, accumulating data implicate ADAMs in several diseases, such as multiple sclerosis, Alzheimer's disease, rheumatoid arthritis, atherosclerosis, inflammation, asthma, Down syndrome, diabetes, and cancer (for a recent review, see Edwards *et al.* [1] or the review series in *Curr. Pharm. Des.* 2009, vol. 15, issue 20). The most striking role of ADAMs is, however, in cancer, and, to date, 12 of the ADAMs have been linked to human cancer (Table 12.1).

12.2
ADAMs in Tumors and Cancer Progression

In 2000, Hanahan and Weinberg [85] introduced the hallmarks of cancer. This important review was revised in 2011 [86] and describes common capabilities

Table 12.1 Human cancers with altered ADAMs expression.

ADAM	Cancer	Level	Prognostic marker value	Catalytic activity	References
ADAM8	Brain, head and neck, kidney, lung, pancreas, and prostate	↑	Yes	Yes	[18–23]
ADAM9	Breast, cervix, gastric, kidney, liver, head and neck, pancreas, prostate, and skin	↑	Yes	Yes	[19, 24–33]
ADAM10	Colon, gastric, endometrial, liver, mantle lymphoma, oral cavity, pancreatic, prostate, salivary gland, and skin	↑	Yes	Yes	[34–42]
ADAM12	Bladder, bone, brain, breast, colon, gastric, liver, lung, oral cavity, laryngeal, and skin	↑	Yes	Yes	[27, 43–52]
ADAM15	Brain, breast, colon, gastric, lung, pancreas, prostate, and skin	↑↓	Yes	Yes	[18, 27, 53–58]
ADAM17	Brain, breast, colon, gall bladder, gastric, head and neck, kidney, liver, lung, ovarian, pancreas, prostate, and skin	↑	Yes	Yes	[18, 20, 52, 59–68]
ADAM19	Brain and kidney	↑	Possible	Yes	[18, 20]
ADAM22	Brain	↑↓	Possible	No	[69, 70]
ADAM23	Brain, breast, colon, gastric, head and neck, lung ovarian, and pancreas	↑↓	Yes	No	[71–78]
ADAM28	Bone, breast, head and neck, kidney, and lung	↑	Yes	Yes	[19, 20, 79–81]
ADAM29	Colon and leukemia	↑	Yes	No	[82, 83]
ADAM33	Breast and gastric	↓	Yes	Yes	[84]

Note: For additional information, see the oncomine database <*www.oncomine.com*>.

shared by most malignant tumors. Four of the hallmarks are of relevance to ADAM biology and offer a good way to evaluate the involvement of ADAMs in tumors and cancer progression. Findings for the ADAMs for each of these four hallmarks – self-sufficiency in growth signals, evasion of apoptosis, sustained angiogenesis, and tissue invasion and metastasis – are discussed below. In addition, the roles of ADAMs in cancer-related inflammation and tumor–stroma interactions are discussed.

12.2.1 Self-Sufficiency in Growth Signals

A feature of tumor cells is reduced dependency on exogenous growth signals. Many cancer cells acquire the ability to produce growth factors to which they are themselves responsive, creating a positive feedback loop. In addition, many growth factor receptors are themselves deregulated during tumor pathogenesis.

The ErbB receptor tyrosine kinases are important regulators of cell growth, survival, migration, angiogenesis, and invasion. Excessive activity of ErbB receptors has been implicated in the formation and progression of multiple cancer types, and epidermal growth factor receptor (EGFR) and ErbB2 are currently among the best-validated targets for cancer treatment [87]. Most mature soluble ErbB ligands are generated by cleavage of their membrane-bound precursor forms by ADAMs, enabling receptor binding and activation [88]. ADAM10 and ADAM17 seem to be the major sheddases of ErbB ligands [16]; however, gain-of-function experiments have shown that other ADAMs, including ADAM8, -9, -10, -12, and -19 may also release ErbB ligands [89]. ADAM-mediated release of ErbB ligands can be activated by a number of physiological and pharmacological stimuli (e.g., downstream of activated G-protein coupled receptors (GPCRs)) [1, 2].

The importance of ADAM-mediated shedding of ErbB ligands in tumor growth has been shown in several model systems. For instance, ADAM17-mediated shedding of transforming growth factor (TGF)-α was shown to be important for cancer cell malignancy and *in vivo* tumor formation [90, 91], and ADAM17 inhibition decreased proliferation and reversed the malignant phenotype of human breast cancer cells in a three-dimensional culture model [92]. Shedding of ErbB2 by ADAM10 has been shown to leave a truncated transmembrane protein with constitutive kinase activity, providing ligand-independent growth and survival signals to the cell. Moreover, serum levels of the ErbB2 extracellular domain in metastatic breast cancer patients have been found to correlate with poor prognosis, and selective ADAM10 inhibitors (in combination with low doses of Herceptin®) decreased proliferation in ErbB2-overexpressing breast cancer cell lines [93]. Furthermore, ADAM9-deficient mice have been shown to exhibit reduced tumor growth, possibly due to decreased EGF and fibroblast growth factor receptor (FGFR)2iiib shedding [94], whereas ADAM28 was shown to stimulate insulin-like growth factor (IGF)-I-induced cell proliferation *in vitro* and *in vivo* [80].

12.2.2 Evasion of Apoptosis

The ability of a tumor to grow depends not only on the rate of proliferation but also on the degree of cell death. A general feature of cancers is acquired resistance to programmed cell death. Several ADAMs have been shown to modulate apoptosis in a variety of ways. ADAM12 was found to have dual effects on cell survival in a mouse model of breast cancer, decreasing tumor cell apoptosis and increasing stromal cell apoptosis *in vitro* and *in vivo* [46]. ADAM10 has been shown to cleave the Fas-ligand

(FasL), which may either cause apoptosis or alternatively decrease the apoptotic response by blocking Fas receptors [95, 96]. Another well-known apoptotic inducer is tumor necrosis factor (TNF)-α, which has been found to be released primarily by ADAM17 [97] but also by ADAM19 [98]. Moreover, ADAM17 and ADAM8 have been found to be able to shed TNF receptors [99–101], inhibiting TNF-α-mediated apoptosis of cancer cells either directly or by scavenging extracellular TNF-α [101].

The shedding of ErbB ligands, as described above, has been shown to lead to antiapoptotic signals through activation of Akt [102], and it has been suggested that ADAMs may, in this way, be involved in resistance to chemotherapy. For example, ADAM12-overexpressing cells have been found to be resistant to etoposide-induced apoptosis and the use of a heparin-binding EGF-like growth factor (HB-EGF) neutralizing antibody was found to restore apoptosis [103]. In addition, ADAM9 knockdown increased apoptosis and sensitized prostate cancer cells to therapeutic treatments [31, 104]. In colon cancer cells, ADAM17-mediated growth factor shedding and growth factor receptor activation have been shown to protect against chemotherapy-induced apoptosis [105]. Furthermore, the dual ADAM10 and ADAM17 inhibitor INCB3619 has been shown to induce apoptosis in lung cancer cells, and heregulin reversed the effect, indicating that evasion from apoptosis resulted from ErbB activation [65].

The survival of most cells also depends on cell–cell or cell–matrix adherence, and abrogation of these adherence-based survival signals can cause apoptosis. In this context, it is interesting to note that several of the ADAMs are able to shed molecules involved in cell adhesion. Whether this feature affects apoptosis remains to be determined.

12.2.3 Sustained Angiogenesis

Early neoplasias must acquire angiogenic potential in order to grow, and several ADAMs have been shown to promote angiogenesis. Vascular-endothelial (VE)-cadherin is the major adhesion molecule of endothelial adherence junctions. It plays an essential role in controlling endothelial permeability, vascular integrity, leukocyte transmigration, and angiogenesis. As a possible mechanism of ADAM regulation of angiogenesis, ADAM9, 10, and 17 have been found to shed VE-cadherin. ADAM10 has been found to be upregulated by vascular endothelial growth factor (VEGF) [106], and shedding of VE-cadherin by ADAM10 has been shown to lead to an increase in vascular permeability and migration of endothelial cells [106, 107].

Pathological neovascularization and tumor growth have been found to be significantly reduced in ADAM9 knockout mice, and, because ADAM9 can shed EphB4, Tie-2, Flk-1, CD40, vascular cell adhesion molecule (VCAM), and VE-cadherin, it might affect neovascularization in this way [108]. Several studies have shown that ADAM17 promotes angiogenesis in *in vitro* assays [109–112], and ADAM17 knockout mice have been shown to have hypovascularization of the lungs [113]. Inactivation of ADAM17 in endothelial cells had no evident effect on developmental angiogenesis, whereas it has been found to significantly reduce pathological neovascularization in a mouse model for retinopathy, and to reduce the growth of injected tumor cells [112]. Proposed mechanisms for the effect of ADAM17 on angiogenesis include shedding of neuregulin [111] or M6P/IGF2R, which may interfere with plasminogen activation [114], as well as several other proteins known to affect angiogenesis (VE-cadherin, VCAM, EphB4, extracellular matrix metalloprotease inducer (EMMPRIN), IGF receptor 1, and platelet endothelial cell adhesion molecule (PECAM)) [112]. GPNMB/Osteoactivin, a protein linked to breast cancer, can be shed by ADAM10, and the soluble extracellular fragment promoted migration of endothelial cells *in vitro* [115]. ADAM28 digested connective tissue growth factor (CTGF) complexed with VEGF165, releasing biologically active VEGF165 from the complex. Furthermore, ADAM28, CTGF, and VEGF have been shown to be coexpressed in breast carcinoma tissues [116].

Inhibitory effects of ADAMs on angiogenesis have also been described. ADAM8-deficient mice had increased retinal revascularization and growth of heterotopically injected tumor cells, and ADAM8 overexpression has been shown to increase the ectodomain shedding of several known angiogenic regulators [117]. Likewise, ADAM17 can shed Flk-1, while ADAM10 can shed its coreceptor neuropilin-1, potentially inhibiting angiogenesis by downregulating cell-surface levels of these receptors and generating soluble decoy receptors [118]. Expression of the ADAM15 disintegrin domain in mice has been shown to reduce tumor growth, angiogenesis, and metastasis [119], implicating adhesive rather than proteolytic effects. In addition, ADAM15 deficiency in mice has been shown to reduce retinal neovascularization and growth of ectopic tumors [120, 121].

12.2.4
Tissue Invasion and Metastasis

The ability of tumor cells to invade tissues and metastasize distinguishes malignant cells from those that are benign; these two properties of malignant tumor cells are the main cause of cancer deaths. ADAMs play an important role in cancer cell invasion and metastasis through the cleavage of growth factors and adhesion molecules, as well as by modulating integrin-mediated adhesion. Knockout or knockdown of individual ADAMs has been shown to reduce tumor growth, migration, invasion, and/or metastasis in a variety of cancer types [22, 40, 42, 80, 91, 122–128], whereas the overexpression of ADAMs has been shown to increase tumor aggressiveness [18, 31, 46, 127–129]. However, opposing effects have been observed, as overexpression of ADAM15 has been found to reduce anchorage-dependent and -independent cell growth, migration, and invasion in melanoma cells [58], whereas suppression of ADAM15 in prostate cancer cells has been shown to decrease tumor growth and metastasis to lung and bone in severe combined immune deficiency (SCID) mice [130]. However, increased ADAM15 expression in breast and prostate cancer tissue was found to be associated with angioinvasion [57].

ADAM10, -15, and -17 have been shown to shed a number of adhesion molecules [1], which may disrupt cell–cell and cell–matrix interactions, thereby promoting cancer cell invasion and metastasis. Overexpression of ADAM10 in colon cancer cells has been found to enhance cleavage of L1 cell adhesion molecule (L1-CAM) and to induce liver metastasis [131], while in glioblastoma cells, ADAM10-mediated cleavage of N-cadherin has been shown to be involved in cell migration [132]. N-cadherin can also be shed by ADAM15, and ADAM15-deficient prostate cancer cells have been shown to display reduced cleavage of N-cadherin and decreased migration [130]. ADAM15 can also cleave E-cadherin, and the soluble E-cadherin fragment has been suggested to increase migration and proliferation by stabilizing and transactivating the ErbB2/ErbB3 dimer [133]. ADAM17-mediated shedding of activated leukocyte cell adhesion molecule (ALCAM) has been demonstrated to be involved in the motility of ovarian carcinoma cells [134].

ADAMs may also modulate tumor cell adhesion by interactions with integrins and cell-surface proteoglycans. The disintegrin domain of ADAM12 has been shown to support integrin-mediated attachment and migration of various tumor cells [135–137], whereas ADAM23, which is frequently silenced in tumors, has been found to repress $\alpha v\beta 3$ integrin activation and to reduce metastasis [138].

An important step in a malignant cell's development of the ability to metastasize is the process of epithelial-to-mesenchymal transition (EMT), where differentiated epithelial cells change into a depolarized and migratory phenotype. High ADAM expression has often been found to be associated with poorly differentiated tumors, suggesting involvement of ADAMs in this cellular change [29, 30, 35, 46, 64, 94]. ADAM17-mediated cleavage of vasorin has been reported to modulate TGF-β-mediated EMT in epithelial cells [139]. In addition, knockdown of ADAM9 in prostate cancer cells has been shown to induce an epithelial phenotype characterized by increased expression of E-cadherin and several integrins and decreased invasive ability [104]. On the contrary, knockdown of ADAM10 in renal carcinoma cells induced an EMT phenotype [140].

12.2.5
Cancer-Related Inflammation

In 2009, Colotta *et al.* introduced cancer-related inflammation as a seventh hallmark of cancer [141]. ADAMs are important regulators of inflammatory responses [2, 142], and ADAM10 and ADAM17 have been found to be involved in cancer-related inflammation [143, 144].

ADAM17 has been found to control TNF-α-mediated inflammation by shedding of both TNF-α and TNF receptors [97, 100, 145], and viable hypomorphic ADAM17-deficient mice have been found to exhibit increased susceptibility to inflammation due to impaired shedding of EGFR ligands [146]. Both of these shedding events may well affect tumor formation and/or progression [147]. ADAM10 and ADAM17 both shed the interleukin (IL)-6 receptor [148], which is known to regulate immune responses and contribute to various pathological states when dysregulated [149]. In addition, high expression of ADAM17 in intestinal tumors has been shown to correlate with low levels of membrane-bound IL-6 receptor, and blocking soluble IL-6 receptor has been found to suppress colitis-associated cancer growth *in vivo* [150, 151]. Finally, the fact that many cytokines and inflammatory stimuli associated with immune responses can stimulate ADAM expression and proteolytic activity (reviewed in Murphy [5]) suggests that inflammatory signals may promote ADAM-mediated signaling events associated with an increased risk of carcinogenesis.

12.2.6
Tumor–Stroma Interactions

Tumors are composed not only of neoplastic cells but also of stromal cells, which together with angiogenic and inflammatory cells reside in the extracellular matrix [152]. The major source of ADAM expression in tumors appears to be the neoplastic cells, yet contributions from the stromal compartment have also been reported. For example, both ADAM12 and ADAM28 have been reported to be almost exclusively produced by tumor cells in human carcinomas [46, 80, 81]. However, in mouse models of breast, prostate, and colon cancer, ADAM12 has also been found to be localized to a subpopulation of stromal cells adjacent to epithelial tumor cells [124]. Interestingly, it was recently shown that ADAM12 endogenously expressed in tumor-associated stroma in a mouse model of breast cancer did not influence tumor progression, but that ADAM12 expression by tumor cells enhanced tumor progression in these mice [125]. In basal cell carcinomas of the skin, ADAM10, -12, and -17 have been localized to the peripheral invading tumor margin, but have also been observed in a proportion of peritumoral inflammatory cells, fibroblasts, and capillary endothelial cells [52]. ADAM17 has also been found to be overexpressed in both neoplastic and endothelial cells in human primary colon carcinoma [153].

The importance of tumor–stroma interactions in tumor progression is becoming increasingly clear, and ADAMs may well play an important role in this context. Expression of ADAM12 in mammary carcinoma cells in mice has been shown to modulate both tumor and stromal cell survival [46]. A soluble ADAM9 isoform expressed by stromal liver myofibroblasts has been suggested to enable invasion and colonization of colon cancer cells in the liver by binding integrins at the tumor cell surface [154]. Similarly, ADAM9 has been found to be expressed at the tumor–stroma border in human melanoma cells, where it appears to mediate the cell–cell interaction between fibroblasts and melanoma cells that is required for invasive growth [33].

12.3
ADAMs in Cancer–Key Questions Yet to Be Answered

Although the molecular mechanisms whereby ADAMs contribute to the malignant behavior of human cancer cells are becoming increasingly clear, several open questions remain, some of which we address here.

12.3.1
ADAM Upregulation

Little attention has been given to the cause of ADAM mRNA and protein upregulation in cancers. However, gene amplification, epigenetic regulation, and transcriptional activation appear to be involved. For example, the chromosomal regions of the *adam9* (8p11–12) and *adam15* (1q21.3) genes have been found to be amplified in several carcinomas [155–159], and *adam29* has been identified as one of the most frequently amplified genes in colorectal cancer [82].

At the transcriptional level, several cancer-related growth factors and cytokines, such as TGF-β1 and TNF-α, induce ADAM expression [5]. In addition, cellular stress, such as hypoxia and reactive oxygen species generated by cancer cells, can stimulate the invasiveness of cancer cells and has been shown to induce expression and activity of ADAMs [126, 160].

Recently, microRNAs (miR) and epigenetic modifications have been added to the list of ADAM regulators. ADAM17 has been found to be a target for miR-122, and downregulation of ADAM17 or restoration of miR-122 in hepatocellular carcinoma cells has been shown to reduce *in vivo* tumorigenesis, angiogenesis, and local invasion in the liver of nude mice [161]. Likewise, miR-29a was shown to selectively regulate the expression of the membrane-bound isoform of ADAM12 in response to Notch signaling in breast cancer cells [162] and upregulation of miR-1274a by sorafenib reduced ADAM9 expression in hepatocellular carcinoma cells [163]. As for epigenetic silencing, aberrant TGF-β1 signaling has been reported to result in SMAD-dependent epigenetic changes that repressed the *adam19* gene [164]. In addition, *adam33* gene silencing by hypermethylation has been observed in a majority of breast cancer cell lines and tissue samples [84]. Interestingly, *adam23* expression has been found to be epigenetically silenced in several cancers, and *adam23* has therefore been suggested to be a tumor suppressor gene [74, 165]. However, a recent study has found ADAM23 to be upregulated in myeloma cells, with high expression correlating with poor overall survival [166], arguing against ADAM23 being a tumor suppressor.

12.3.2
Isoforms

Most ADAMs exist in two or more isoforms, and while the biological significance of these are largely unknown, these isoforms should be considered when evaluating the role of ADAMs in cancer. Interestingly, several ADAM isoforms have been specifically linked to cancer. For example, two novel splice variants of ADAM8 have been found to be significantly upregulated in non-small-cell lung cancer cell lines, but one form was particularly associated with an aggressive phenotype [21]. Moreover, soluble ADAM9 (ADAM9-S) has been observed to promote cell migration in a metalloprotease-dependent manner, whereas membrane-bound ADAM9 (ADAM9-L) has been found to suppress cell migration independent of catalytic activity [25]. In addition, splice variants of ADAM15, which differ only in the cytoplasmic tails, have been found to be differentially expressed in breast cancer and melanoma cells, and some correlated with poor survival while others did not [53, 58, 167]. ADAM22, which is a nonproteolytic ADAM, has been found to exist in at least 13 alternatively spliced variant forms, all of which affect the cytoplasmic tail. Interestingly, some ADAM22 isoforms have been found to be more abundant in normal brain, whereas others were upregulated in gliomas [70, 168]. Furthermore, an ADAM22 variant found in normal brain, when overexpressed in high-grade glioma cell lines, has been demonstrated to inhibit cell proliferation [69], suggesting that specific sites in the cytoplasmic tail of ADAM22 possess a tumor-suppressive function.

12.3.3 Proteolytic versus Nonproteolytic Effect

A key question in ADAMs-mediated cancer progression is whether their proteolytic activity and/or nonproteolytic properties are implicated in that process. Several studies have indicated that both effects may play a role, yet the main functionality seems to be the catalytic activity, with ADAM10- and ADAM17-mediated ectodomain shedding in particular being linked to cancer progression and aggressiveness [92, 169]. On the other hand, some data support nonproteolytic protumorigenic effects of ADAM9 and ADAM12. For example, *in vitro* data have indicated that ADAM12 contributes to tumor progression partly through protease-independent regulation of cell survival [46], and more recently, it was reported that the catalytic activity of ADAM12 is dispensable for its growth-promoting effects on mouse mammary tumor cells both *in vitro* and *in vivo* [125]. For ADAM9-S both the protease activity and integrin binding have been implicated in colon cancer cell invasion *in vitro* [154].

12.4 The Clinical Potential of ADAMs

12.4.1 Diagnostic or Prognostic Biomarkers

As stated earlier, several ADAMs are overexpressed in a variety of human cancers. In addition, a number of studies have found that the expression levels of ADAMs often correlate with histological grade and disease stage, making them potential cancer biomarkers (for example, [18, 35, 43, 170]).

Interestingly, some of the soluble ADAM isoforms have been found to be upregulated in serum and/or urine from cancer patients. For example, ADAM12-S in urine from breast and bladder cancer patients has been found to correlate with progression of disease [43, 171, 172]. ADAM28 was also found upregulated in urine from bladder cancer patients [173], and ADAM28 levels in serum may predict survival in non-small-cell lung cancer patients [174]. In addition, soluble ADAM8 in serum from lung cancer patients has been suggested as a diagnostic marker to be used in conjunction with conventional markers, improving the overall sensitivity for detection of non-small-cell lung cancer [175].

Furthermore, in a single study, ADAM9 and ADAM11 were identified as having a predictive value especially in stroma-enriched primary breast tumors, where they were significantly associated with increased benefit of tamoxifen therapy [176].

12.4.2
ADAMs as Therapeutic Targets

ADAMs are considered promising therapeutic targets in cancer, and a number of synthetic inhibitors have been developed. The focus has largely been on strategies targeting both ADAM10 and ADAM17 or on specific ADAM17 inhibitors (for a list, refer to Arribas and Esselens [177]), such as the ADAM10- and ADAM17-selective compound, INCB3619. In xenograft mouse models, INCB3619 has been shown to synergize with cisplatin, reducing tumor growth of head and neck carcinomas, and to improve the therapeutic responses to other cytotoxic agents in models of pancreatic, non-small-cell lung, and colon cancer [178]. As ADAM10 and ADAM17 are the main sheddases of ErbB ligands, it has been suggested that dual inhibition of ADAM 10/17 and ErbB receptor activities would be synergistic, since elevated levels of ErbB ligands may circumvent the effectiveness of ErbB-targeted therapeutics. Indeed, combining ADAM17 knockdown/inhibition with ErbB-targeting chemotherapy has been shown to reduce cell growth and to increase apoptosis in colon cancer cells *in vitro* [60, 105]. Combinatorial treatment with INCB3619 and the anti-HER-2 antibody trastuzumab or the EGFR tyrosine kinase inhibitor gefitinib has been found to reduce tumor growth in breast and non-small-cell lung cancer *in vitro* and *in vivo* [65, 179]. INCB3619 and a related inhibitor, INCB7839, have been shown to be highly synergistic with lapatinib and lapatinib-like (GW2974) inhibitors in cultured breast cancer cells and xenografts [180]. INCB7839 has been evaluated in Phase Ib trials with promising outcomes, and is currently in Phase II trials[2)] [177]. A selective ADAM17 inhibitor, WAY-022, has been shown to decrease DNA replication and cell growth in colorectal cancer cells. Combinations of WAY-022 with either cetuximab or gefitinib have been found to result in increased apoptosis and growth inhibition of colorectal cancer cells *in vitro* [60].

It has also been shown that isolated domains of ADAMs can have potent effects on tumorigenesis and might serve as highly specific inhibitors of ADAM function. The recombinant disintegrin domain (RDD) of ADAM15 has been observed to inhibit tumor growth, angiogenesis, and metastasis to the lung in mouse models using breast carcinoma or melanoma cells [119]. The RDD of ADAM9 has been shown to inhibit the invasion of breast cancer cells in a Matrigel™ assay [181].

2) Incyte Corporation, Wilmington, DE.

Combinatorial treatment with ADAM15 RDD, thrombospondin 1, and soluble VEGF receptor in melanoma-bearing mice has been shown to significantly decrease cell proliferation, reduce metastasis, and increase long-term tumor-free survival [182]. In addition, the isolated recombinant prodomains of ADAM10 and ADAM17 have been demonstrated to be efficient inhibitors of their catalytic activity [183–185] and ADAM10 prodomain has been shown to reduce cell proliferation in a bladder cancer cell line [186].

12.5 Concluding Remarks

This chapter touches on several aspects of the relationship of ADAMs to cancer; however, as this is a rapidly expanding field, we have regrettably not been able to include all studies. Clearly, ADAMs play important roles in cancer – largely as sheddases regulating the activity of a plethora of cell-surface molecules and downstream signaling pathways, but seemingly also through nonproteolytic effects. While most ADAMs promote tumorigenesis, a few examples of ADAMs promoting tumor suppression exist. Thus, great care should be taken when evaluating the effect of ADAMs in cancer, as several ADAMs are upregulated and implicated in the same types of cancer. Future studies should therefore consider the mosaic of ADAMs expressed, including splice variants and processed forms, as well as their individual functions. Despite the challenges that lie ahead, the clinical potential of ADAMs remains promising, both as diagnostic and prognostic biomarkers and as targets in cancer therapy.

References

1. Edwards, D.R., Handsley, M.M., and Pennington, C.J. (2008) The ADAM metalloproteinases. *Mol. Aspects Med.*, **29**, 258–289.
2. Murphy, G. (2008) The ADAMs: signalling scissors in the tumour microenvironment. *Nat. Rev. Cancer*, **8**, 929–941.
3. Klein, T. and Bischoff, R. (2011) Active metalloproteases of the a disintegrin and metalloprotease (ADAM) family: biological function and structure. *J. Proteome Res.*, **10**, 17–33.
4. Kveiborg, M., Albrechtsen, R., Couchman, J.R., and Wewer, U.M. (2008) Cellular roles of ADAM12 in health and disease. *Int. J. Biochem. Cell Biol.*, **40**, 1685–1702.
5. Murphy, G. (2009) Regulation of the proteolytic disintegrin metalloproteinases, the 'Sheddases'. *Semin. Cell Dev. Biol.*, **20**, 138–145.
6. Takeda, S. (2009) Three-dimensional domain architecture of the ADAM family proteinases. *Semin. Cell Dev. Biol.*, **20**, 146–152.
7. Murphy, G.J., Murphy, G., and Reynolds, J.J. (1991) The origin of matrix metalloproteinases and their familial relationships. *FEBS Lett.*, **289**, 4–7.
8. Bode, W., Gomis-Ruth, F.X., and Stockler, W. (1993) Astacins, serralysins, snake venom and matrix metalloproteinases exhibit identical zinc-binding environments

(HEXXHXXGXXH and Met-turn) and topologies and should be grouped into a common family, the 'metzincins'. *FEBS Lett.*, **331**, 134–140.
9. Blobel, C.P. (2005) ADAMs: key components in EGFR signalling and development. *Nat. Rev. Mol. Cell Biol.*, **6**, 32–43.
10. Reiss, K., Ludwig, A., and Saftig, P. (2006) Breaking up the tie: disintegrin-like metalloproteinases as regulators of cell migration in inflammation and invasion. *Pharmacol. Ther.*, **111**, 985–1006.
11. Arribas, J., Bech-Serra, J.J., and Santiago-Josefat, B. (2006) ADAMs, cell migration and cancer. *Cancer Metastasis Rev.*, **25**, 57–68.
12. Mochizuki, S. and Okada, Y. (2007) ADAMs in cancer cell proliferation and progression. *Cancer Sci.*, **98**, 621–628.
13. Alfandari, D., McCusker, C., and Cousin, H. (2009) ADAM function in embryogenesis. *Semin. Cell Dev. Biol.*, **20**, 153–163.
14. Weskamp, G., Cai, H., Brodie, T.A., Higashyama, S., Manova, K., Ludwig, T., and Blobel, C.P. (2002) Mice lacking the metalloprotease-disintegrin MDC9 (ADAM9) have no evident major abnormalities during development or adult life. *Mol. Cell Biol.*, **22**, 1537–1544.
15. Mitchell, K.J., Pinson, K.I., Kelly, O.G., Brennan, J., Zupicich, J., Scherz, P., Leighton, P.A., Goodrich, L.V., Lu, X., Avery, B.J., Tate, P., Dill, K., Pangilinan, E., Wakenight, P., Tessier-Lavigne, M., and Skarnes, W.C. (2001) Functional analysis of secreted and transmembrane proteins critical to mouse development. *Nat. Genet.*, **28**, 241–249.
16. Sahin, U., Weskamp, G., Kelly, K., Zhou, H.M., Higashiyama, S., Peschon, J., Hartmann, D., Saftig, P., and Blobel, C.P. (2004) Distinct roles for ADAM10 and ADAM17 in ectodomain shedding of six EGFR ligands. *J. Cell Biol.*, **164**, 769–779.
17. Horiuchi, K., Zhou, H.M., Kelly, K., Manova, K., and Blobel, C.P. (2005) Evaluation of the contributions of ADAMs 9, 12, 15, 17, and 19 to heart development and ectodomain shedding of neuregulins beta1 and beta2. *Dev. Biol.*, **283**, 459–471.
18. Wildeboer, D., Naus, S., Amy Sang, Q.X., Bartsch, J.W., and Pagenstecher, A. (2006) Metalloproteinase disintegrins ADAM8 and ADAM19 are highly regulated in human primary brain tumors and their expression levels and activities are associated with invasiveness. *J. Neuropathol. Exp. Neurol.*, **65**, 516–527.
19. Stokes, A., Joutsa, J., Ala-Aho, R., Pitchers, M., Pennington, C.J., Martin, C., Premachandra, D.J., Okada, Y., Peltonen, J., Grenman, R., James, H.A., Edwards, D.R., and Kahari, V.M. (2010) Expression profiles and clinical correlations of degradome components in the tumor microenvironment of head and neck squamous cell carcinoma. *Clin. Cancer Res.*, **16**, 2022–2035.
20. Roemer, A., Schwettmann, L., Jung, M., Stephan, C., Roigas, J., Kristiansen, G., Loening, S.A., Lichtinghagen, R., and Jung, K. (2004) The membrane proteases adams and hepsin are differentially expressed in renal cell carcinoma. Are they potential tumor markers? *J. Urol.*, **172**, 2162–2166.
21. Hernandez, I., Moreno, J.L., Zandueta, C., Montuenga, L., and Lecanda, F. (2010) Novel alternatively spliced ADAM8 isoforms contribute to the aggressive bone metastatic phenotype of lung cancer. *Oncogene*, **29**, 3758–3769.
22. Valkovskaya, N., Kayed, H., Felix, K., Hartmann, D., Giese, N.A., Osinsky, S.P., Friess, H., and Kleeff, J. (2007) ADAM8 expression is associated with increased invasiveness and reduced patient survival in pancreatic cancer. *J. Cell Mol. Med.*, **11**, 1162–1174.
23. Fritzsche, F.R., Jung, M., Xu, C., Rabien, A., Schicktanz, H., Stephan, C., Dietel, M., Jung, K., and Kristiansen, G. (2006) ADAM8 expression in prostate cancer is associated with parameters of unfavorable prognosis. *Virchows Arch.*, **449**, 628–636.
24. O'Shea, C., McKie, N., Buggy, Y., Duggan, C., Hill, A.D., McDermott, E., O'Higgins, N., and Duffy, M.J. (2003)

Expression of ADAM-9 mRNA and protein in human breast cancer. *Int. J. Cancer*, **105**, 754–761.

25. Fry, J.L. and Toker, A. (2010) Secreted and membrane-bound isoforms of protease ADAM9 have opposing effects on breast cancer cell migration. *Cancer Res.*, **70**, 8187–8198.
26. Zubel, A., Flechtenmacher, C., Edler, L., and Alonso, A. (2009) Expression of ADAM9 in CIN3 lesions and squamous cell carcinomas of the cervix. *Gynecol. Oncol.*, **114**, 332–336.
27. Carl-McGrath, S., Lendeckel, U., Ebert, M., Roessner, A., and Rocken, C. (2005) The disintegrin-metalloproteinases ADAM9, ADAM12, and ADAM15 are upregulated in gastric cancer. *Int. J. Oncol.*, **26**, 17–24.
28. Fritzsche, F.R., Wassermann, K., Jung, M., Tolle, A., Kristiansen, I., Lein, M., Johannsen, M., Dietel, M., Jung, K., and Kristiansen, G. (2008b) ADAM9 is highly expressed in renal cell cancer and is associated with tumour progression. *BMC Cancer*, **8**, 179.
29. Tao, K., Qian, N., Tang, Y., Ti, Z., Song, W., Cao, D., and Dou, K. (2010) Increased expression of a disintegrin and metalloprotease-9 in hepatocellular carcinoma: implications for tumor progression and prognosis. *Jpn. J. Clin. Oncol.*, **40**, 645–651.
30. Grutzmann, R., Luttges, J., Sipos, B., Ammerpohl, O., Dobrowolski, F., Alldinger, I., Kersting, S., Ockert, D., Koch, R., Kalthoff, H., Schackert, H.K., Saeger, H.D., Kloppel, G., and Pilarsky, C. (2004) ADAM9 expression in pancreatic cancer is associated with tumour type and is a prognostic factor in ductal adenocarcinoma. *Br. J. Cancer*, **90**, 1053–1058.
31. Sung, S.Y., Kubo, H., Shigemura, K., Arnold, R.S., Logani, S., Wang, R., Konaka, H., Nakagawa, M., Mousses, S., Amin, M., Anderson, C., Johnstone, P., Petros, J.A., Marshall, F.F., Zhau, H.E., and Chung, L.W. (2006) Oxidative stress induces ADAM9 protein expression in human prostate cancer cells. *Cancer Res.*, **66**, 9519–9526.
32. Fritzsche, F.R., Jung, M., Tolle, A., Wild, P., Hartmann, A., Wassermann, K., Rabien, A., Lein, M., Dietel, M., Pilarsky, C., Calvano, D., Grutzmann, R., Jung, K., and Kristiansen, G. (2008a) ADAM9 expression is a significant and independent prognostic marker of PSA relapse in prostate cancer. *Eur. Urol.*, **54**, 1097–1106.
33. Zigrino, P., Mauch, C., Fox, J.W., and Nischt, R. (2005) Adam-9 expression and regulation in human skin melanoma and melanoma cell lines. *Int. J. Cancer*, **116**, 853–859.
34. Knosel, T., Emde, A., Schluns, K., Chen, Y., Jurchott, K., Krause, M., Dietel, M., and Petersen, I. (2005) Immunoprofiles of 11 biomarkers using tissue microarrays identify prognostic subgroups in colorectal cancer. *Neoplasia*, **7**, 741–747.
35. Wang, Y.Y., Ye, Z.Y., Li, L., Zhao, Z.S., Shao, Q.S., and Tao, H.Q. (2011) ADAM 10 is associated with gastric cancer progression and prognosis of patients. *J. Surg. Oncol.*, **103**, 116–123.
36. Fogel, M., Gutwein, P., Mechtersheimer, S., Riedle, S., Stoeck, A., Smirnov, A., Edler, L., Ben-Arie, A., Huszar, M., and Altevogt, P. (2003) L1 expression as a predictor of progression and survival in patients with uterine and ovarian carcinomas. *Lancet*, **362**, 869–875.
37. Kohga, K., Takehara, T., Tatsumi, T., Miyagi, T., Ishida, H., Ohkawa, K., Kanto, T., Hiramatsu, N., and Hayashi, N. (2009) Anticancer chemotherapy inhibits MHC class I-related chain a ectodomain shedding by downregulating ADAM10 expression in hepatocellular carcinoma. *Cancer Res.*, **69**, 8050–8057.
38. Armanious, H., Gelebart, P., Anand, M., Belch, A., and Lai, R. (2011) Constitutive activation of metalloproteinase ADAM10 in mantle cell lymphoma promotes cell growth and activates the TNFalpha/NFkappaB pathway. *Blood*, **117**, 6237–6246.
39. Ko, S.Y., Lin, S.C., Wong, Y.K., Liu, C.J., Chang, K.W., and Liu, T.Y. (2007) Increase of disintergin metalloprotease

10 (ADAM10) expression in oral squamous cell carcinoma. *Cancer Lett.*, **245**, 33–43.
40. Gaida, M.M., Haag, N., Gunther, F., Tschaharganeh, D.F., Schirmacher, P., Friess, H., Giese, N.A., Schmidt, J., and Wente, M.N. (2010) Expression of A disintegrin and metalloprotease 10 in pancreatic carcinoma. *Int. J. Mol. Med.*, **26**, 281–288.
41. Arima, T., Enokida, H., Kubo, H., Kagara, I., Matsuda, R., Toki, K., Nishimura, H., Chiyomaru, T., Tatarano, S., Idesako, T., Nishiyama, K., and Nakagawa, M. (2007) Nuclear translocation of ADAM-10 contributes to the pathogenesis and progression of human prostate cancer. *Cancer Sci.*, **98**, 1720–1726.
42. Lee, S.B., Schramme, A., Doberstein, K., Dummer, R., Abdel-Bakky, M.S., Keller, S., Altevogt, P., Oh, S.T., Reichrath, J., Oxmann, D., Pfeilschifter, J., Mihic-Probst, D., and Gutwein, P. (2010) ADAM10 is upregulated in melanoma metastasis compared with primary melanoma. *J. Invest. Dermatol.*, **130**, 763–773.
43. Frohlich, C., Albrechtsen, R., Dyrskjot, L., Rudkjaer, L., Orntoft, T.F., and Wewer, U.M. (2006) Molecular profiling of ADAM12 in human bladder cancer. *Clin. Cancer Res.*, **12**, 7359–7368.
44. Tian, B.L., Wen, J.M., Zhang, M., Xie, D., Xu, R.B., and Luo, C.J. (2002) The expression of ADAM12 (meltrin alpha) in human giant cell tumours of bone. *Mol. Pathol.*, **55**, 394–397.
45. Kodama, T., Ikeda, E., Okada, A., Ohtsuka, T., Shimoda, M., Shiomi, T., Yoshida, K., Nakada, M., Ohuchi, E., and Okada, Y. (2004) ADAM12 is selectively overexpressed in human glioblastomas and is associated with glioblastoma cell proliferation and shedding of heparin-binding epidermal growth factor. *Am. J. Pathol.*, **165**, 1743–1753.
46. Kveiborg, M., Frohlich, C., Albrechtsen, R., Tischler, V., Dietrich, N., Holck, P., Kronqvist, P., Rank, F., Mercurio, A.M., and Wewer, U.M. (2005) A role for ADAM12 in breast tumor progression and stromal cell apoptosis. *Cancer Res.*, **65**, 4754–4761.
47. Iba, K., Albrechtsen, R., Gilpin, B.J., Loechel, F., and Wewer, U.M. (1999) Cysteine-rich domain of human ADAM 12 (meltrin alpha) supports tumor cell adhesion. *Am. J. Pathol.*, **154**, 1489–1501.
48. Le Pabic, H., Bonnier, D., Wewer, U.M., Coutand, A., Musso, O., Baffet, G., Clement, B., and Theret, N. (2003) ADAM12 in human liver cancers: TGF-beta-regulated expression in stellate cells is associated with matrix remodeling. *Hepatology*, **37**, 1056–1066.
49. Mino, N., Miyahara, R., Nakayama, E., Takahashi, T., Takahashi, A., Iwakiri, S., Sonobe, M., Okubo, K., Hirata, T., Sehara, A., and Date, H. (2009) A disintegrin and metalloprotease 12 (ADAM12) is a prognostic factor in resected pathological stage I lung adenocarcinoma. *J. Surg. Oncol.*, **100**, 267–272.
50. Markowski, J., Tyszkiewicz, T., Jarzab, M., Oczko-Wojciechowska, M., Gierek, T., Witkowska, M., Paluch, J., Kowalska, M., Wygoda, Z., Lange, D., and Jarzab, B. (2009) Metal-proteinase ADAM12, kinesin 14 and checkpoint suppressor 1 as new molecular markers of laryngeal carcinoma. *Eur. Arch. Otorhinolaryngol.*, **266**, 1501–1507.
51. Kornberg, L.J., Villaret, D., Popp, M., Lui, L., McLaren, R., Brown, H., Cohen, D., Yun, J., and McFadden, M. (2005) Gene expression profiling in squamous cell carcinoma of the oral cavity shows abnormalities in several signaling pathways. *Laryngoscope*, **115**, 690–698.
52. Oh, S.T., Schramme, A., Stark, A., Tilgen, W., Gutwein, P., and Reichrath, J. (2009) The disintegrin-metalloproteinases ADAM 10, 12 and 17 are upregulated in invading peripheral tumor cells of basal cell carcinomas. *J. Cutan. Pathol.*, **36**, 395–401.
53. Zhong, J.L., Poghosyan, Z., Pennington, C.J., Scott, X., Handsley, M.M., Warn, A., Gavrilovic, J., Honert, K., Kruger, A., Span, P.N., Sweep, F.C.,

and Edwards, D.R. (2008) Distinct functions of natural ADAM-15 cytoplasmic domain variants in human mammary carcinoma. *Mol. Cancer Res.*, **6**, 383–394.

54. Toquet, C., Colson, A., Jarry, A., Bezieau, S., Volteau, C., Boisseau, P., Merlin, D., Laboisse, C.L., and Mosnier, J.F. (2012) ADAM15 to alpha5beta1 integrin switch in colon carcinoma cells: A late event in cancer progression associated with tumor dedifferentiation and poor prognosis. *Int. J. Cancer.*, **130**, 278–287.
55. Schutz, A., Hartig, W., Wobus, M., Grosche, J., Wittekind, C., and Aust, G. (2005) Expression of ADAM15 in lung carcinomas. *Virchows Arch.*, **446**, 421–429.
56. Yamada, D., Ohuchida, K., Mizumoto, K., Ohhashi, S., Yu, J., Egami, T., Fujita, H., Nagai, E., and Tanaka, M. (2007) Increased expression of ADAM 9 and ADAM 15 mRNA in pancreatic cancer. *Anticancer Res.*, **27**, 793–799.
57. Kuefer, R., Day, K.C., Kleer, C.G., Sabel, M.S., Hofer, M.D., Varambally, S., Zorn, C.S., Chinnaiyan, A.M., Rubin, M.A., and Day, M.L. (2006) ADAM15 disintegrin is associated with aggressive prostate and breast cancer disease. *Neoplasia*, **8**, 319–329.
58. Ungerer, C., Doberstein, K., Burger, C., Hardt, K., Boehncke, W.H., Bohm, B., Pfeilschifter, J., Dummer, R., Mihic-Probst, D., and Gutwein, P. (2010) ADAM15 expression is downregulated in melanoma metastasis compared to primary melanoma. *Biochem. Biophys. Res. Commun.*, **401**, 363–369.
59. McGowan, P.M., McKiernan, E., Bolster, F., Ryan, B.M., Hill, A.D., McDermott, E.W., Evoy, D., O'Higgins, N., Crown, J., and Duffy, M.J. (2008) ADAM-17 predicts adverse outcome in patients with breast cancer. *Ann. Oncol.*, **19**, 1075–1081.
60. Merchant, N.B., Voskresensky, I., Rogers, C.M., Lafleur, B., Dempsey, P.J., Graves-Deal, R., Revetta, F., Foutch, A.C., Rothenberg, M.L., Washington, M.K., and Coffey, R.J. (2008) TACE/ADAM-17: a component of the epidermal growth factor receptor axis and a promising therapeutic target in colorectal cancer. *Clin. Cancer Res.*, **14**, 1182–1191.
61. Nakagawa, M., Nabeshima, K., Asano, S., Hamasaki, M., Uesugi, N., Tani, H., Yamashita, Y., and Iwasaki, H. (2009) Up-regulated expression of ADAM17 in gastrointestinal stromal tumors: coexpression with EGFR and EGFR ligands. *Cancer Sci.*, **100**, 654–662.
62. Wu, K., Liao, M., Liu, B., and Deng, Z. (2011) ADAM-17 over-expression in gallbladder carcinoma correlates with poor prognosis of patients. *Med. Oncol.*, **28**, 475–480.
63. Kornfeld, J.W., Meder, S., Wohlberg, M., Friedrich, R.E., Rau, T., Riethdorf, L., Loning, T., Pantel, K., and Riethdorf, S. (2011a) Overexpression of TACE and TIMP3 mRNA in head and neck cancer: association with tumour development and progression. *Br. J. Cancer*, **104**, 138–145.
64. Ding, X., Yang, L.Y., Huang, G.W., Wang, W., and Lu, W.Q. (2004) ADAM17 mRNA expression and pathological features of hepatocellular carcinoma. *World J. Gastroenterol.*, **10**, 2735–2739.
65. Zhou, B.B., Peyton, M., He, B., Liu, C., Girard, L., Caudler, E., Lo, Y., Baribaud, F., Mikami, I., Reguart, N., Yang, G., Li, Y., Yao, W., Vaddi, K., Gazdar, A.F., Friedman, S.M., Jablons, D.M., Newton, R.C., Fridman, J.S., Minna, J.D., and Scherle, P.A. (2006) Targeting ADAM-mediated ligand cleavage to inhibit HER3 and EGFR pathways in non-small cell lung cancer. *Cancer Cell*, **10**, 39–50.
66. Tanaka, Y., Miyamoto, S., Suzuki, S.O., Oki, E., Yagi, H., Sonoda, K., Yamazaki, A., Mizushima, H., Maehara, Y., Mekada, E., and Nakano, H. (2005) Clinical significance of heparin-binding epidermal growth factor-like growth factor and a disintegrin and metalloprotease 17 expression in human ovarian cancer. *Clin. Cancer Res.*, **11**, 4783–4792.
67. Ringel, J., Jesnowski, R., Moniaux, N., Luttges, J., Ringel, J., Choudhury, A.,

Batra, S.K., Kloppel, G., and Lohr, M. (2006) Aberrant expression of a disintegrin and metalloproteinase 17/tumor necrosis factor-alpha converting enzyme increases the malignant potential in human pancreatic ductal adenocarcinoma. *Cancer Res.*, **66**, 9045–9053.

68. Karan, D., Lin, F.C., Bryan, M., Ringel, J., Moniaux, N., Lin, M.F., and Batra, S.K. (2003) Expression of ADAMs (a disintegrin and metalloproteases) and TIMP-3 (tissue inhibitor of metalloproteinase-3) in human prostatic adenocarcinomas. *Int. J. Oncol.*, **23**, 1365–1371.
69. D'Abaco, G.M., Ng, K., Paradiso, L., Godde, N.J., Kaye, A., and Novak, U. (2006) ADAM22, expressed in normal brain but not in high-grade gliomas, inhibits cellular proliferation via the disintegrin domain. *Neurosurgery*, **58**, 179–186.
70. Godde, N.J., D'Abaco, G.M., Paradiso, L., and Novak, U. (2007) Differential coding potential of ADAM22 mRNAs. *Gene*, **403**, 80–88.
71. Costa, F.F., Colin, C., Shinjo, S.M., Zanata, S.M., Marie, S.K., Sogayar, M.C., and Camargo, A.A. (2005) ADAM23 methylation and expression analysis in brain tumors. *Neurosci. Lett.*, **380**, 260–264.
72. Verbisck, N.V., Costa, E.T., Costa, F.F., Cavalher, F.P., Costa, M.D., Muras, A., Paixao, V.A., Moura, R., Granato, M.F., Ierardi, D.F., Machado, T., Melo, F., Ribeiro, K.B., Cunha, I.W., Lima, V.C., Maciel, M.S., Carvalho, A.L., Soares, F.F., Zanata, S., Sogayar, M.C., Chammas, R., and Camargo, A.A. (2009a) ADAM23 negatively modulates alpha(v)beta(3) integrin activation during metastasis. *Cancer Res.*, **69**, 5546–5552.
73. Choi, J.S., Kim, K.H., Jeon, Y.K., Kim, S.H., Jang, S.G., Ku, J.L., and Park, J.G. (2009) Promoter hypermethylation of the ADAM23 gene in colorectal cancer cell lines and cancer tissues. *Int. J. Cancer*, **124**, 1258–1262.
74. Takada, H., Imoto, I., Tsuda, H., Nakanishi, Y., Ichikura, T., Mochizuki, H., Mitsufuji, S., Hosoda, F., Hirohashi, S., Ohki, M., and Inazawa, J. (2005) ADAM23, a possible tumor suppressor gene, is frequently silenced in gastric cancers by homozygous deletion or aberrant promoter hypermethylation. *Oncogene*, **24**, 8051–8060.
75. Calmon, M.F., Colombo, J., Carvalho, F., Souza, F.P., Filho, J.F., Fukuyama, E.E., Camargo, A.A., Caballero, O.L., Tajara, E.H., Cordeiro, J.A., and Rahal, P. (2007) Methylation profile of genes CDKN2A (p14 and p16), DAPK1, CDH1, and ADAM23 in head and neck cancer. *Cancer Genet. Cytogenet.*, **173**, 31–37.
76. Hu, C., Lv, H., Pan, G., Cao, H., Deng, Z., Hu, C., Wen, J., and Zhou, J. (2011) The expression of ADAM23 and its correlation with promoter methylation in non-small-cell lung carcinoma. *Int. J. Exp. Pathol.*, **92**, 333–339.
77. Ghilardi, C., Chiorino, G., Dossi, R., Nagy, Z., Giavazzi, R., and Bani, M. (2008) Identification of novel vascular markers through gene expression profiling of tumor-derived endothelium. *BMC Genomics*, **9**, 201.
78. Hagihara, A., Miyamoto, K., Furuta, J., Hiraoka, N., Wakazono, K., Seki, S., Fukushima, S., Tsao, M.S., Sugimura, T., and Ushijima, T. (2004) Identification of 27 5' CpG islands aberrantly methylated and 13 genes silenced in human pancreatic cancers. *Oncogene*, **23**, 8705–8710.
79. Matsuura, S., Oda, Y., Matono, H., Izumi, T., Yamamoto, H., Tamiya, S., Iwamoto, Y., and Tsuneyoshi, M. (2010) Overexpression of A disintegrin and metalloproteinase 28 is correlated with high histologic grade in conventional chondrosarcoma. *Hum. Pathol.*, **41**, 343–351.
80. Mitsui, Y., Mochizuki, S., Kodama, T., Shimoda, M., Ohtsuka, T., Shiomi, T., Chijiiwa, M., Ikeda, T., Kitajima, M., and Okada, Y. (2006) ADAM28 is overexpressed in human breast carcinomas: implications for carcinoma cell proliferation through cleavage of insulin-like growth factor binding protein-3. *Cancer Res.*, **66**, 9913–9920.
81. Ohtsuka, T., Shiomi, T., Shimoda, M., Kodama, T., Amour, A., Murphy, G.,

Ohuchi, E., Kobayashi, K., and Okada, Y. (2006) ADAM28 is overexpressed in human non-small cell lung carcinomas and correlates with cell proliferation and lymph node metastasis. *Int. J. Cancer*, **118**, 263–273.

82. Ashktorab, H., Schaffer, A.A., Daremipouran, M., Smoot, D.T., Lee, E., and Brim, H. (2010) Distinct genetic alterations in colorectal cancer. *PLoS One*, **5**, e8879.
83. Oppezzo, P., Vasconcelos, Y., Settegrana, C., Jeannel, D., Vuillier, F., Legarff-Tavernier, M., Kimura, E.Y., Bechet, S., Dumas, G., Brissard, M., Merle-Beral, H., Yamamoto, M., Dighiero, G., and Davi, F. (2005) The LPL/ADAM29 expression ratio is a novel prognosis indicator in chronic lymphocytic leukemia. *Blood*, **106**, 650–657.
84. Seniski, G.G., Camargo, A.A., Ierardi, D.F., Ramos, E.A., Grochoski, M., Ribeiro, E.S., Cavalli, I.J., Pedrosa, F.O., de Souza, E.M., Zanata, S.M., Costa, F.F., and Klassen, G. (2009) ADAM33 gene silencing by promoter hypermethylation as a molecular marker in breast invasive lobular carcinoma. *BMC Cancer*, **9**, 80.
85. Hanahan, D. and Weinberg, R.A. (2000) The hallmarks of cancer. *Cell*, **100**, 57–70.
86. Hanahan, D. and Weinberg, R.A. (2011) Hallmarks of cancer: the next generation. *Cell*, **144**, 646–674.
87. De Luca, A., Carotenuto, A., Rachiglio, A., Gallo, M., Maiello, M.R., Aldinucci, D., Pinto, A., and Normanno, N. (2008) The role of the EGFR signaling in tumor microenvironment. *J. Cell. Physiol.*, **214**, 559–567.
88. Blobel, C.P., Carpenter, G., and Freeman, M. (2009) The role of protease activity in ErbB biology. *Exp. Cell Res.*, **315**, 671–682.
89. Horiuchi, K., Le, G.S., Schulte, M., Yamaguchi, T., Reiss, K., Murphy, G., Toyama, Y., Hartmann, D., Saftig, P., and Blobel, C.P. (2007) Substrate selectivity of epidermal growth factor-receptor ligand sheddases and their regulation by phorbol esters and calcium influx. *Mol. Biol. Cell*, **18**, 176–188.
90. Borrell-Pages, M., Rojo, F., Albanell, J., Baselga, J., and Arribas, J. (2003) TACE is required for the activation of the EGFR by TGF-alpha in tumors. *EMBO J.*, **22**, 1114–1124.
91. Franovic, A., Robert, I., Smith, K., Kurban, G., Pause, A., Gunaratnam, L., and Lee, S. (2006) Multiple acquired renal carcinoma tumor capabilities abolished upon silencing of ADAM17. *Cancer Res.*, **66**, 8083–8090.
92. Kenny, P.A. and Bissell, M.J. (2007) Targeting TACE-dependent EGFR ligand shedding in breast cancer. *J. Clin. Invest.*, **117**, 337–345.
93. Liu, P.C., Liu, X., Li, Y., Covington, M., Wynn, R., Huber, R., Hillman, M., Yang, G., Ellis, D., Marando, C., Katiyar, K., Bradley, J., Abremski, K., Stow, M., Rupar, M., Zhuo, J., Li, Y.L., Lin, Q., Burns, D., Xu, M., Zhang, C., Qian, D.Q., He, C., Sharief, V., Weng, L., Agrios, C., Shi, E., Metcalf, B., Newton, R., Friedman, S., Yao, W., Scherle, P., Hollis, G., and Burn, T.C. (2006a) Identification of ADAM10 as a major source of HER2 ectodomain sheddase activity in HER2 overexpressing breast cancer cells. *Cancer Biol. Ther.*, **5**, 657–664.
94. Peduto, L., Reuter, V.E., Shaffer, D.R., Scher, H.I., and Blobel, C.P. (2005) Critical function for ADAM9 in mouse prostate cancer. *Cancer Res.*, **65**, 9312–9319.
95. Kirkin, V., Cahuzac, N., Guardiola-Serrano, F., Huault, S., Luckerath, K., Friedmann, E., Novac, N., Wels, W.S., Martoglio, B., Hueber, A.O., and Zornig, M. (2007) The fas ligand intracellular domain is released by ADAM10 and SPPL2a cleavage in T-cells. *Cell Death Differ.*, **14**, 1678–1687.
96. Schulte, M., Reiss, K., Lettau, M., Maretzky, T., Ludwig, A., Hartmann, D., de Strooper, B., Janssen, O., and Saftig, P. (2007) ADAM10 regulates FasL cell surface expression and modulates FasL-induced cytotoxicity and activation-induced cell death. *Cell Death Differ.*, **14**, 1040–1049.

97. Black, R.A., Rauch, C.T., Kozlosky, C.J., Peschon, J.J., Slack, J.L., Wolfson, M.F., Castner, B.J., Stocking, K.L., Reddy, P., Srinivasan, S., Nelson, N., Boiani, N., Schooley, K.A., Gerhart, M., Davis, R., Fitzner, J.N., Johnson, R.S., Paxton, R.J., March, C.J., and Cerretti, D.P. (1997) A metalloproteinase disintegrin that releases tumour-necrosis factor-alpha from cells. *Nature*, **385**, 729–733.
98. Zheng, Y., Saftig, P., Hartmann, D., and Blobel, C. (2004) Evaluation of the contribution of different ADAMs to tumor necrosis factor alpha (TNFalpha) shedding and of the function of the TNFalpha ectodomain in ensuring selective stimulated shedding by the TNFalpha convertase (TACE/ADAM17). *J. Biol. Chem.*, **279**, 42898–42906.
99. Peschon, J.J., Slack, J.L., Reddy, P., Stocking, K.L., Sunnarborg, S.W., Lee, D.C., Russell, W.E., Castner, B.J., Johnson, R.S., Fitzner, J.N., Boyce, R.W., Nelson, N., Kozlosky, C.J., Wolfson, M.F., Rauch, C.T., Cerretti, D.P., Paxton, R.J., March, C.J., and Black, R.A. (1998) An essential role for ectodomain shedding in mammalian development. *Science*, **282**, 1281–1284.
100. Reddy, P., Slack, J.L., Davis, R., Cerretti, D.P., Kozlosky, C.J., Blanton, R.A., Shows, D., Peschon, J.J., and Black, R.A. (2000) Functional analysis of the domain structure of tumor necrosis factor-alpha converting enzyme. *J. Biol. Chem.*, **275**, 14608–14614.
101. Koller, G., Schlomann, U., Golfi, P., Ferdous, T., Naus, S., and Bartsch, J.W. (2009) ADAM8/MS2/CD156, an emerging drug target in the treatment of inflammatory and invasive pathologies. *Curr. Pharm. Des.*, **15**, 2272–2281.
102. Yarden, Y. and Sliwkowski, M.X. (2001) Untangling the ErbB signalling network. *Nat. Rev. Mol. Cell Biol.*, **2**, 127–137.
103. Rocks, N., Estrella, C., Paulissen, G., Quesada-Calvo, F., Gilles, C., Gueders, M.M., Crahay, C., Foidart, J.M., Gosset, P., Noel, A., and Cataldo, D.D. (2008) The metalloproteinase ADAM-12 regulates bronchial epithelial cell proliferation and apoptosis. *Cell Prolif.*, **41**, 988–1001.
104. Josson, S., Anderson, C.S., Sung, S.Y., Johnstone, P.A., Kubo, H., Hsieh, C.L., Arnold, R., Gururajan, M., Yates, C., and Chung, L.W. (2011) Inhibition of ADAM9 expression induces epithelial phenotypic alterations and sensitizes human prostate cancer cells to radiation and chemotherapy. *Prostate*, **71**, 232–240.
105. Kyula, J.N., Van Schaeybroeck, S., Doherty, J., Fenning, C.S., Longley, D.B., and Johnston, P.G. (2010) Chemotherapy-induced activation of ADAM-17: a novel mechanism of drug resistance in colorectal cancer. *Clin. Cancer Res.*, **16**, 3378–3389.
106. Donners, M.M., Wolfs, I.M., Olieslagers, S., Mohammadi-Motahhari, Z., Tchaikovski, V., Heeneman, S., van Buul, J.D., Caolo, V., Molin, D.G., Post, M.J., and Waltenberger, J. (2010) A disintegrin and metalloprotease 10 is a novel mediator of vascular endothelial growth factor-induced endothelial cell function in angiogenesis and is associated with atherosclerosis. *Arterioscler. Thromb. Vasc. Biol.*, **30**, 2188–2195.
107. Schulz, B., Pruessmeyer, J., Maretzky, T., Ludwig, A., Blobel, C.P., Saftig, P., and Reiss, K. (2008) ADAM10 regulates endothelial permeability and T-Cell transmigration by proteolysis of vascular endothelial cadherin. *Circ. Res.*, **102**, 1192–1201.
108. Guaiquil, V., Swendeman, S., Yoshida, T., Chavala, S., Campochiaro, P.A., and Blobel, C.P. (2009) ADAM9 is involved in pathological retinal neovascularization. *Mol. Cell Biol.*, **29**, 2694–2703.
109. Gooz, P., Gooz, M., Baldys, A., and Hoffman, S. (2009) ADAM-17 regulates endothelial cell morphology, proliferation, and in vitro angiogenesis. *Biochem. Biophys. Res. Commun.*, **380**, 33–38.
110. Kwak, H.I., Mendoza, E.A., and Bayless, K.J. (2009) ADAM17 co-purifies with TIMP-3 and modulates endothelial invasion responses in

three-dimensional collagen matrices. *Matrix Biol.*, **28**, 470–479.

111. Kalinowski, A., Plowes, N.J., Huang, Q., Berdejo-Izquierdo, C., Russell, R.R., and Russell, K.S. (2010) Metalloproteinase-dependent cleavage of neuregulin and autocrine stimulation of vascular endothelial cells. *FASEB J.*, **24**, 2567–2575.

112. Weskamp, G., Mendelson, K., Swendeman, S., Le Gall, S., Ma, Y., Lyman, S., Hinoki, A., Eguchi, S., Guaiquil, V., Horiuchi, K., and Blobel, C.P. (2010) Pathological neovascularization is reduced by inactivation of ADAM17 in endothelial cells but not in pericytes. *Circ. Res.*, **106**, 932–940.

113. Zhao, J., Chen, H., Peschon, J.J., Shi, W., Zhang, Y., Frank, S.J., and Warburton, D. (2001) Pulmonary hypoplasia in mice lacking tumor necrosis factor-alpha converting enzyme indicates an indispensable role for cell surface protein shedding during embryonic lung branching morphogenesis. *Dev. Biol.*, **232**, 204–218.

114. Leksa, V., Loewe, R., Binder, B., Schiller, H.B., Eckerstorfer, P., Forster, F., Soler-Cardona, A., Ondrovicova, G., Kutejova, E., Steinhuber, E., Breuss, J., Drach, J., Petzelbauer, P., Binder, B.R., and Stockinger, H. (2011) Soluble M6P/IGF2R released by TACE controls angiogenesis via blocking plasminogen activation. *Circ. Res.*, **108**, 676–685.

115. Rose, A.A., Annis, M.G., Dong, Z., Pepin, F., Hallett, M., Park, M., and Siegel, P.M. (2010) ADAM10 releases a soluble form of the GPNMB/Osteoactivin extracellular domain with angiogenic properties. *PLoS ONE*, **5**, e12093.

116. Mochizuki, S., Tanaka, R., Shimoda, M., Onuma, J., Fujii, Y., Jinno, H., and Okada, Y. (2010) Connective tissue growth factor is a substrate of ADAM28. *Biochem. Biophys. Res. Commun.*, **402**, 651–657.

117. Guaiquil, V.H., Swendeman, S., Zhou, W., Guaiquil, P., Weskamp, G., Bartsch, J.W., and Blobel, C.P. (2010) ADAM8 is a negative regulator of retinal neovascularization and of the growth of heterotopically injected tumor cells in mice. *J. Mol. Med.*, **88**, 497–505.

118. Swendeman, S., Mendelson, K., Weskamp, G., Horiuchi, K., Deutsch, U., Scherle, P., Hooper, A., Rafii, S., and Blobel, C.P. (2008) VEGF-A stimulates ADAM17-dependent shedding of VEGFR2 and crosstalk between VEGFR2 and ERK signaling. *Circ. Res.*, **103**, 916–918.

119. Trochon-Joseph, V., Martel-Renoir, D., Mir, L.M., Thomaidis, A., Opolon, P., Connault, E., Li, H., Grenet, C., Fauvel-Lafeve, F., Soria, J., Legrand, C., Soria, C., Perricaudet, M., and Lu, H. (2004) Evidence of antiangiogenic and antimetastatic activities of the recombinant disintegrin domain of metargidin. *Cancer Res.*, **64**, 2062–2069.

120. Horiuchi, K., Weskamp, G., Lum, L., Hammes, H.P., Cai, H., Brodie, T.A., Ludwig, T., Chiusaroli, R., Baron, R., Preissner, K.T., Manova, K., and Blobel, C.P. (2003) Potential role for ADAM15 in pathological neovascularization in mice. *Mol. Cell Biol.*, **23**, 5614–5624.

121. Xie, B., Shen, J., Dong, A., Swaim, M., Hackett, S.F., Wyder, L., Worpenberg, S., Barbieri, S., and Campochiaro, P.A. (2008) An Adam15 amplification loop promotes vascular endothelial growth factor-induced ocular neovascularization. *FASEB J.*, **22**, 2775–2783.

122. Xu, Q., Liu, X., Cai, Y., Yu, Y., and Chen, W. (2010a) RNAi-mediated ADAM9 gene silencing inhibits metastasis of adenoid cystic carcinoma cells. *Tumour. Biol.*, **31**, 217–224.

123. Xu, Q., Liu, X., Chen, W., and Zhang, Z. (2010b) Inhibiting adenoid cystic carcinoma cells growth and metastasis by blocking the expression of ADAM 10 using RNA interference. *J. Transl. Med.*, **8**, 136.

124. Peduto, L., Reuter, V.E., Sehara-Fujisawa, A., Shaffer, D.R., Scher, H.I., and Blobel, C.P. (2006) ADAM12 is highly expressed in carcinoma-associated stroma and is required for mouse prostate tumor progression. *Oncogene*, **25**, 5462–5466.

125. Frohlich, C., Nehammer, C., Albrechtsen, R., Kronqvist, P.,

Kveiborg, M., Sehara-Fujisawa, A., Mercurio, A.M., and Wewer, U.M. (2011) ADAM12 produced by tumor cells rather than stromal cells accelerates breast tumor progression. *Mol. Cancer Res.*, **9**, 1449–1461.

126. Zheng, X., Jiang, F., Katakowski, M., Kalkanis, S.N., Hong, X., Zhang, X., Zhang, Z.G., Yang, H., and Chopp, M. (2007) Inhibition of ADAM17 reduces hypoxia-induced brain tumor cell invasiveness. *Cancer Sci.*, **98**, 674–684.

127. McGowan, P.M., Ryan, B.M., Hill, A.D., McDermott, E., O'Higgins, N., and Duffy, M.J. (2007) ADAM-17 expression in breast cancer correlates with variables of tumor progression. *Clin. Cancer Res.*, **13**, 2335–2343.

128. Mochizuki, S. and Okada, Y. (2009) ADAM28 as a target for human cancers. *Curr. Pharm. Des.*, **15**, 2349–2358.

129. Shintani, Y., Higashiyama, S., Ohta, M., Hirabayashi, H., Yamamoto, S., Yoshimasu, T., Matsuda, H., and Matsuura, N. (2004) Overexpression of ADAM9 in non-small cell lung cancer correlates with brain metastasis. *Cancer Res.*, **64**, 4190–4196.

130. Najy, A.J., Day, K.C., and Day, M.L. (2008a) ADAM15 supports prostate cancer metastasis by modulating tumor cell-endothelial cell interaction. *Cancer Res.*, **68**, 1092–1099.

131. Gavert, N., Sheffer, M., Raveh, S., Spaderna, S., Shtutman, M., Brabletz, T., Barany, F., Paty, P., Notterman, D., Domany, E., and Ben-Ze'ev, A. (2007) Expression of L1-CAM and ADAM10 in human colon cancer cells induces metastasis. *Cancer Res.*, **67**, 7703–7712.

132. Kohutek, Z.A., diPierro, C.G., Redpath, G.T., and Hussaini, I.M. (2009) ADAM-10-mediated N-cadherin cleavage is protein kinase C-alpha dependent and promotes glioblastoma cell migration. *J. Neurosci.*, **29**, 4605–4615.

133. Najy, A.J., Day, K.C., and Day, M.L. (2008b) The ectodomain shedding of E-cadherin by ADAM15 supports ErbB receptor activation. *J. Biol. Chem.*, **283**, 18393–18401.

134. Rosso, O., Piazza, T., Bongarzone, I., Rossello, A., Mezzanzanica, D., Canevari, S., Orengo, A.M., Puppo, A., Ferrini, S., and Fabbi, M. (2007) The ALCAM shedding by the metalloprotease ADAM17/TACE is involved in motility of ovarian carcinoma cells. *Mol. Cancer Res.*, **5**, 1246–1253.

135. Thodeti, C.K., Frohlich, C., Nielsen, C.K., Holck, P., Sundberg, C., Kveiborg, M., Mahalingam, Y., Albrechtsen, R., Couchman, J.R., and Wewer, U.M. (2005) Hierarchy of ADAM12 binding to integrins in tumor cells. *Exp. Cell Res.*, **309**, 438–450.

136. Huang, J., Bridges, L.C., and White, J.M. (2005) Selective modulation of integrin-mediated cell migration by distinct ADAM family members. *Mol. Biol. Cell*, **16**, 4982–4991.

137. Lydolph, M.C., Morgan-Fisher, M., Hoye, A.M., Couchman, J.R., Wewer, U.M., and Yoneda, A. (2009) Alpha9beta1 integrin in melanoma cells can signal different adhesion states for migration and anchorage. *Exp. Cell Res.*, **315**, 3312–3324.

138. Verbisck, N.V., Costa, E.T., Costa, F.F., Cavalher, F.P., Costa, M.D., Muras, A., Paixao, V.A., Moura, R., Granato, M.F., Ierardi, D.F., Machado, T., Melo, F., Ribeiro, K.B., Cunha, I.W., Lima, V.C., Maciel, M.S., Carvalho, A.L., Soares, F.F., Zanata, S., Sogayar, M.C., Chammas, R., and Camargo, A.A. (2009b) ADAM23 negatively modulates alpha(v)beta(3) integrin activation during metastasis. *Cancer Res.*, **69**, 5546–5552.

139. Malapeira, J., Esselens, C., Bech-Serra, J.J., Canals, F., and Arribas, J. (2011) ADAM17 (TACE) regulates TGFbeta signaling through the cleavage of vasorin. *Oncogene*, **30**, 1912–1922.

140. Doberstein, K., Pfeilschifter, J., and Gutwein, P. (2011) The transcription factor PAX2 regulates ADAM10 expression in renal cell carcinoma. *Carcinogenesis.*, **32**, 1713–1723.

141. Colotta, F., Allavena, P., Sica, A., Garlanda, C., and Mantovani, A. (2009) Cancer-related inflammation, the seventh hallmark of cancer: links to

genetic instability. *Carcinogenesis*, **30**, 1073–1081.

142. Garton, K.J., Gough, P.J., and Raines, E.W. (2006) Emerging roles for ectodomain shedding in the regulation of inflammatory responses. *J. Leukoc. Biol.*, **79**, 1105–1116.

143. Pruessmeyer, J. and Ludwig, A. (2009) The good, the bad and the ugly substrates for ADAM10 and ADAM17 in brain pathology, inflammation and cancer. *Semin. Cell Dev. Biol.*, **20**, 164–174.

144. Chalaris, A., Garbers, C., Rabe, B., Rose-John, S., and Scheller, J. (2011) The soluble interleukin 6 receptor: generation and role in inflammation and cancer. *Eur. J. Cell Biol.*, **90**, 484–494.

145. Solomon, K.A., Pesti, N., Wu, G., and Newton, R.C. (1999) Cutting edge: a dominant negative form of TNF-alpha converting enzyme inhibits proTNF and TNFRII secretion. *J. Immunol.*, **163**, 4105–4108.

146. Chalaris, A., Adam, N., Sina, C., Rosenstiel, P., Lehmann-Koch, J., Schirmacher, P., Hartmann, D., Cichy, J., Gavrilova, O., Schreiber, S., Jostock, T., Matthews, V., Hasler, R., Becker, C., Neurath, M.F., Reiss, K., Saftig, P., Scheller, J., and Rose-John, S. (2010) Critical role of the disintegrin metalloprotease ADAM17 for intestinal inflammation and regeneration in mice. *J. Exp. Med.*, **207**, 1617–1624.

147. Balkwill, F. (2009) Tumour necrosis factor and cancer. *Nat. Rev. Cancer*, **9**, 361–371.

148. Matthews, V., Schuster, B., Schutze, S., Bussmeyer, I., Ludwig, A., Hundhausen, C., Sadowski, T., Saftig, P., Hartmann, D., Kallen, K.J., and Rose-John, S. (2003) Cellular cholesterol depletion triggers shedding of the human interleukin-6 receptor by ADAM10 and ADAM17 (TACE). *J. Biol. Chem.*, **278**, 38829–38839.

149. Rose-John, S., Waetzig, G.H., Scheller, J., Grotzinger, J., and Seegert, D. (2007) The IL-6/sIL-6R complex as a novel target for therapeutic approaches. *Expert Opin. Ther. Targets*, **11**, 613–624.

150. Becker, C., Fantini, M.C., Schramm, C., Lehr, H.A., Wirtz, S., Nikolaev, A., Burg, J., Strand, S., Kiesslich, R., Huber, S., Ito, H., Nishimoto, N., Yoshizaki, K., Kishimoto, T.,\ Galle, P.R., Blessing, M., Rose-John, S., and Neurath, M.F. (2004) TGF-beta suppresses tumor progression in colon cancer by inhibition of IL-6 trans-signaling. *Immunity*, **21**, 491–501.

151. Matsumoto, S., Hara, T., Mitsuyama, K., Yamamoto, M., Tsuruta, O., Sata, M., Scheller, J., Rose-John, S., Kado, S., and Takada, T. (2010) Essential roles of IL-6 trans-signaling in colonic epithelial cells, induced by the IL-6/soluble-IL-6 receptor derived from lamina propria macrophages, on the development of colitis-associated premalignant cancer in a murine model. *J. Immunol.*, **184**, 1543–1551.

152. Egeblad, M., Nakasone, E.S., and Werb, Z. (2010) Tumors as organs: complex tissues that interface with the entire organism. *Dev. Cell*, **18**, 884–901.

153. Blanchot-Jossic, F., Jarry, A., Masson, D., Bach-Ngohou, K., Paineau, J., Denis, M.G., Laboisse, C.L., and Mosnier, J.F. (2005) Up-regulated expression of ADAM17 in human colon carcinoma: co-expression with EGFR in neoplastic and endothelial cells. *J. Pathol.*, **207**, 156–163.

154. Mazzocca, A., Coppari, R., De Franco, R., Cho, J.Y., Libermann, T.A., Pinzani, M., and Toker, A. (2005) A secreted form of ADAM9 promotes carcinoma invasion through tumor-stromal interactions. *Cancer Res.*, **65**, 4728–4738.

155. Moelans, C.B., de Weger, R.A., Monsuur, H.N., Vijzelaar, R., and van Diest, P.J. (2010) Molecular profiling of invasive breast cancer by multiplex ligation-dependent probe amplification-based copy number analysis of tumor suppressor and oncogenes. *Mod. Pathol.*, **23**, 1029–1039.

156. Alers, J.C., Rochat, J., Krijtenburg, P.J., Hop, W.C., Kranse, R., Rosenberg, C., Tanke, H.J., Schroder, F.H., and van Dekken, H. (2000) Identification of genetic markers for prostatic cancer progression. *Lab. Invest.*, **80**, 931–942.

157. Balazs, M., Adam, Z., Treszl, A., Begany, A., Hunyadi, J., and Adany, R. (2001) Chromosomal imbalances in primary and metastatic melanomas revealed by comparative genomic hybridization. *Cytometry*, **46**, 222–232.

158. Glinsky, G.V., Krones-Herzig, A., and Glinskii, A.B. (2003) Malignancy-associated regions of transcriptional activation: gene expression profiling identifies common chromosomal regions of a recurrent transcriptional activation in human prostate, breast, ovarian, and colon cancers. *Neoplasia*, **5**, 218–228.

159. Karkkainen, I., Karhu, R., and Huovila, A.P. (2000) Assignment of the ADAM15 gene to human chromosome band 1q21.3 by in situ hybridization. *Cytogenet. Cell Genet.*, **88**, 206–207.

160. Mongaret, C., Alexandre, J., Thomas-Schoemann, A., Bermudez, E., Chereau, C., Nicco, C., Goldwasser, F., Weill, B., Batteux, F., and Lemare, F. (2011) Tumor invasion induced by oxidative stress is dependent on membrane ADAM 9 protein and its secreted form. *Int. J. Cancer*, **129**, 791–798.

161. Tsai, W.C., Hsu, P.W., Lai, T.C., Chau, G.Y., Lin, C.W., Chen, C.M., Lin, C.D., Liao, Y.L., Wang, J.L., Chau, Y.P., Hsu, M.T., Hsiao, M., Huang, H.D., and Tsou, A.P. (2009) MicroRNA-122, a tumor suppressor microRNA that regulates intrahepatic metastasis of hepatocellular carcinoma. *Hepatology*, **49**, 1571–1582.

162. Li, H., Solomon, E., Duhachek, M.S., Sun, D., and Zolkiewska, A. (2011) Metalloprotease-disintegrin ADAM12 expression is regulated by Notch signaling via microRNA-29. *J. Biol. Chem.*, **286**, 21500–21510.

163. Zhou, C., Liu, J., Li, Y., Liu, L., Zhang, X., Ma, C.Y., Hua, S.C., Yang, M., and Yuan, Q. (2011) microRNA-1274a, a modulator of sorafenib induced a disintegrin and metalloproteinase 9 (ADAM9) down-regulation in hepatocellular carcinoma. *FEBS Lett.*, **585**, 1828–1834.

164. Chan, M.W., Huang, Y.W., Hartman-Frey, C., Kuo, C.T., Deatherage, D., Qin, H., Cheng, A.S., Yan, P.S., Davuluri, R.V., Huang, T.H., Nephew, K.P., and Lin, H.J. (2008) Aberrant transforming growth factor beta1 signaling and SMAD4 nuclear translocation confer epigenetic repression of ADAM19 in ovarian cancer. *Neoplasia*, **10**, 908–919.

165. Kim, K.E., Song, H., Hahm, C., Yoon, S.Y., Park, S., Lee, H.R., Hur, D.Y., Kim, T., Kim, C.H., Bang, S.I., Bang, J.W., Park, H., and Cho, D.H. (2009) Expression of ADAM33 is a novel regulatory mechanism in IL-18-secreted process in gastric cancer. *J. Immunol.*, **182**, 3548–3555.

166. Bret, C., Hose, D., Reme, T., Kassambara, A., Seckinger, A., Meissner, T., Schved, J.F., Kanouni, T., Goldschmidt, H., and Klein, B. (2011) Gene expression profile of ADAMs and ADAMTSs metalloproteinases in normal and malignant plasma cells and in the bone marrow environment. *Exp. Hematol.*, **39**, 546–557.

167. Ortiz, R.M., Karkkainen, I., and Huovila, A.P. (2004) Aberrant alternative exon use and increased copy number of human metalloprotease-disintegrin ADAM15 gene in breast cancer cells. *Genes Chromosomes Cancer*, **41**, 366–378.

168. Harada, T., Nishie, A., Torigoe, K., Ikezaki, K., Shono, T., Maehara, Y., Kuwano, M., and Wada, M. (2000) The specific expression of three novel splice variant forms of human metalloprotease-like disintegrin-like cysteine-rich protein 2 gene inBrain tissues and gliomas. *Jpn. J. Cancer Res.*, **91**, 1001–1006.

169. Moss, M.L., Stoeck, A., Yan, W., and Dempsey, P.J. (2008) ADAM10 as a target for anti-cancer therapy. *Curr. Pharm. Biotechnol.*, **9**, 2–8.

170. Kornfeld, J.W., Meder, S., Wohlberg, M., Friedrich, R.E., Rau, T., Riethdorf, L., Loning, T., Pantel, K., and Riethdorf, S. (2011b) Overexpression of TACE and TIMP3 mRNA in head and neck cancer: association with tumour development and progression. *Br. J. Cancer*, **104**, 138–145.

171. Roy, R., Wewer, U.M., Zurakowski, D., Pories, S.E., and Moses, M.A. (2004)

ADAM 12 cleaves extracellular matrix proteins and correlates with cancer status and stage. *J. Biol. Chem.*, **279**, 51323–51330.

172. Pories, S.E., Zurakowski, D., Roy, R., Lamb, C.C., Raza, S., Exarhopoulos, A., Scheib, R.G., Schumer, S., Lenahan, C., Borges, V., Louis, G.W., Anand, A., Isakovich, N., Hirshfield-Bartek, J., Wewer, U., Lotz, M.M., and Moses, M.A. (2008) Urinary metalloproteinases: noninvasive biomarkers for breast cancer risk assessment. *Cancer Epidemiol. Biomarkers Prev.*, **17**, 1034–1042.
173. Yang, M.H., Chu, P.Y., Chen, S.C., Chung, T.W., Chen, W.C., Tan, L.B., Kan, W.C., Wang, H.Y., Su, S.B., and Tyan, Y.C. (2011) Characterization of ADAM28 as a biomarker of bladder transitional cell carcinomas by urinary proteome analysis. *Biochem. Biophys. Res. Commun.*, **411**, 714–720.
174. Kuroda, H., Mochizuki, S., Shimoda, M., Chijiiwa, M., Kamiya, K., Izumi, Y., Watanabe, M., Horinouchi, H., Kawamura, M., Kobayashi, K., and Okada, Y. (2010) ADAM28 is a serological and histochemical marker for non-small-cell lung cancers. *Int. J. Cancer*, **127**, 1844–1856.
175. Ishikawa, N., Daigo, Y., Yasui, W., Inai, K., Nishimura, H., Tsuchiya, E., Kohno, N., and Nakamura, Y. (2004) ADAM8 as a novel serological and histochemical marker for lung cancer. *Clin. Cancer Res.*, **10**, 8363–8370.
176. Sieuwerts, A.M., Meijer-van Gelder, M.E., Timmermans, M., Trapman, A.M., Garcia, R.R., Arnold, M., Goedheer, A.J., Portengen, H., Klijn, J.G., and Foekens, J.A. (2005) How ADAM-9 and ADAM-11 differentially from estrogen receptor predict response to tamoxifen treatment in patients with recurrent breast cancer: a retrospective study. *Clin. Cancer Res.*, **11**, 7311–7321.
177. Arribas, J. and Esselens, C. (2009) ADAM17 as a therapeutic target in multiple diseases. *Curr. Pharm. Des.*, **15**, 2319–2335.
178. Fridman, J.S., Caulder, E., Hansbury, M., Liu, X., Yang, G., Wang, Q., Lo, Y., Zhou, B.B., Pan, M., Thomas, S.M., Grandis, J.R., Zhuo, J., Yao, W., Newton, R.C., Friedman, S.M., Scherle, P.A., and Vaddi, K. (2007) Selective inhibition of ADAM metalloproteases as a novel approach for modulating ErbB pathways in cancer. *Clin. Cancer Res.*, **13**, 1892–1902.
179. Liu, X., Fridman, J.S., Wang, Q., Caulder, E., Yang, G., Covington, M., Liu, C., Marando, C., Zhuo, J., Li, Y., Yao, W., Vaddi, K., Newton, R.C., Scherle, P.A., and Friedman, S.M. (2006b) Selective inhibition of ADAM metalloproteases blocks HER-2 extracellular domain (ECD) cleavage and potentiates the anti-tumor effects of trastuzumab. *Cancer Biol. Ther.*, **5**, 648–656.
180. Witters, L., Scherle, P., Friedman, S., Fridman, J., Caulder, E., Newton, R., and Lipton, A. (2008) Synergistic inhibition with a dual epidermal growth factor receptor/HER-2/neu tyrosine kinase inhibitor and a disintegrin and metalloprotease inhibitor. *Cancer Res.*, **68**, 7083–7089.
181. Cominetti, M.R., Martin, A.C., Ribeiro, J.U., Djaafri, I., Fauvel-Lafeve, F., Crepin, M., and Selistre-de-Araujo, H.S. (2009) Inhibition of platelets and tumor cell adhesion by the disintegrin domain of human ADAM9 to collagen I under dynamic flow conditions. *Biochimie*, **91**, 1045–1052.
182. Daugimont, L., Vandermeulen, G., Desfresne, F., Bouzin, C., Mir, L., Bouquet, C., Feron, O., and Preat, V. (2011) Antitumoral and antimetastatic effect of antiangiogenic plasmids in B16 melanoma: higher efficiency of the recombinant disintegrin domain of ADAM 15. *Eur. J. Pharm. Biopharm.*, **78**, 314–319.
183. Moss, M.L., Bomar, M., Liu, Q., Sage, H., Dempsey, P., Lenhart, P.M., Gillispie, P.A., Stoeck, A., Wildeboer, D., Bartsch, J.W., Palmisano, R., and Zhou, P. (2007) The ADAM10 prodomain is a specific inhibitor of

ADAM10 proteolytic activity and inhibits cellular shedding events. *J. Biol. Chem.*, **282**, 35712–35721.

184. Gonzales, P.E., Solomon, A., Miller, A.B., Leesnitzer, M.A., Sagi, I., and Milla, M.E. (2004) Inhibition of the tumor necrosis factor-alpha-converting enzyme by its pro domain. *J. Biol. Chem.*, **279**, 31638–31645.

185. Li, X., Yan, Y., Huang, W., Yang, Y., Wang, H., and Chang, L. (2009) The regulation of TACE catalytic function by its prodomain. *Mol. Biol. Rep.*, **36**, 641–651.

186. Crawford, H.C., Dempsey, P.J., Brown, G., Adam, L., and Moss, M.L. (2009) ADAM10 as a therapeutic target for cancer and inflammation. *Curr. Pharm. Des.*, **15**, 2288–2299.

13
Urokinase-Type Plasminogen Activator, Its Receptor and Inhibitor as Biomarkers in Cancer

Tine Thurison, Ida K. Lund, Martin Illemann, Ib J. Christensen, and Gunilla Høyer-Hansen

13.1
Introduction

Identification of biomarkers is needed in order to determine cancer patient prognosis as well as to predict response to specific treatment modalities. Early detection of cancer will likely increase curability of the disease, and therefore a search for biomarkers that can discriminate benign from malignant tumors is warranted. The definition of a biomarker is "a characteristic that is objectively measured and evaluated as an indicator of normal biological processes, pathogenic processes, or pharmacologic responses to a therapeutic intervention" [1].

The plasminogen activation system is involved in the degradation of the extracellular matrix and is therefore a prerequisite for cancer invasion and metastasis. Three key components of this system, the urokinase-type plasminogen activator (uPA), its physiological inhibitor plasminogen activator inhibitor 1 (PAI-1), and its cellular receptor urokinase-type plasminogen activator receptor (uPAR), are all upregulated in cancer and are often located at the invasive front of the tumor [2, 3]. They are primarily expressed by stromal cells and also by some cancer cells and have proved to be promising prognostic, diagnostic, and predictive biomarkers in different cancers [4]. The zymogen form of uPA (pro-uPA) is attached to the cell surface by binding to uPAR, where pro-uPA is converted to active uPA by cleavage mediated by the broad-spectrum serine protease plasmin. uPA in turn cleaves plasminogen resulting in plasmin, which degrades the extracellular matrix and activates matrix metalloproteases (MMPs). The proteolytic activity of uPA is inhibited primarily by PAI-1, while plasmin in solution is inhibited by α_2-antiplasmin. uPAR, consisting of three homologous domains (I, II, and III), is a cell surface protein anchored to the plasma membrane by a glycosylphosphatidylinositol (GPI) moiety located on domain III. Each domain in uPAR is composed of 80–90 residues, which are connected by two linker regions of 15–20 residues each [5]. The binding site of uPA is located in domain I and II of uPAR, and the high-affinity binding of uPA critically depends on 9 residues in domain I and 21 residues in domain II [6]. The inhibitor, PAI-1, is a glycoprotein of the serpin family. It can irreversibly bind and inhibit free, as well as receptor-bound uPA, thereby serving as a regulator

Matrix Proteases in Health and Disease, First Edition. Edited by Niels Behrendt.

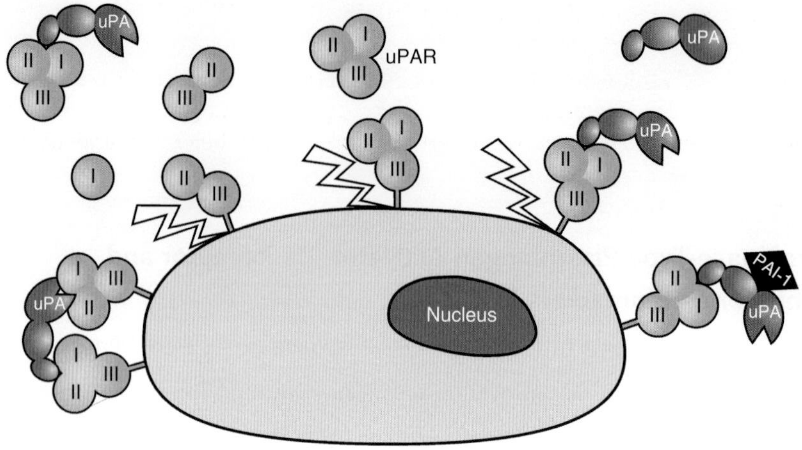

Figure 13.1 uPAR, consisting of three structurally homologous domains (I, II, and III), is a cell surface protein anchored to the cell membrane by a glycolipid anchor on domain III. The domains are connected by linker regions. uPA bound to uPAR can cleave neighboring uPAR molecules, liberating uPAR(I), while leaving uPAR(II-III) on the cell surface. Specific phospholipases (arrows) can cleave the glycolipid anchor, liberating soluble forms of uPAR(I-III) and uPAR(II-III). uPAR(I-III) (both with and without uPA and PAI-1 bound), uPAR(II-III), and uPAR(I) can be identified in tumor tissue and blood samples from cancer patients.

of focal plasminogen activation [7, 8]. In addition to binding to uPAR, uPA can cleave uPAR in the linker region between domain I and II, liberating domain I, uPAR(I), and leaving the cleaved uPAR(II-III) on the cell surface. There are two cleavage sites for uPA in the linker region between domain I and II in uPAR, that is, between R^{83} and A^{84} and between R^{89} and S^{90} [9]. Soluble forms of uPAR are present in blood [10]. Only intact uPAR(I-III) has been detected in blood from healthy individuals, whereas both uPAR(I-III) and uPAR(II-III) are present in blood from patients with acute myeloid leukemia [11] and in cystic fluid from ovarian cancer patients [12, 13] as evidenced by Western blotting. These uPAR forms are shed from the cell surface, and the mechanisms and the enzymes responsible are yet to be fully elucidated. GPI-specific phospholipase D can release uPAR from the cell surface by cleavage of the GPI anchor [14]. However, protease-dependent cleavage has also been suggested [15, 16]. Thus, two GPI-anchored [uPAR(I-III) and uPAR(II-III)] and three soluble forms of uPAR [uPAR(I-III), uPAR(II-III), and uPAR(I)] have been identified in blood and tumor tissue samples from cancer patients (Figure 13.1). In addition, complexes of uPAR(I-III) and (pro-)uPA have been detected [17]. Nonbound pro-uPA and uPA, as well as uPA in complex with PAI-1, are also present in blood and tumor tissue [18].

The prognostic and predictive value of uPA, PAI-1, and uPAR in different cancers has previously been analyzed in studies, where the biomarker levels have been determined in blood or tissue extracts using immunoassays (reviewed in Christensen *et al.* [4]). In addition, prognostic, diagnostic, and predictive values of the different forms of uPAR have been demonstrated in different cancers. The prognostic values of uPA, PAI-1, and uPAR have also been assessed by immunohistochemical staining of patient tumor tissue samples. In this chapter, we focus on recent studies of the prognostic, diagnostic, and predictive values of uPA, PAI-1, and the different uPAR forms, determined at the protein level using immunoassays and/or immunohistochemistry (IHC). The studies reviewed are summarized in Table 13.1.

13.2 Breast Cancer

In 2007, the American Society of Clinical Oncology (ASCO) recommended measurements of uPA and PAI-1 for identification of high-risk patients in newly diagnosed node-negative breast cancer patients [39]. The uPA and PAI-1 measured in extracts of tumor tissue biopsies are validated at the highest level of evidence (LOE 1) regarding their clinical utility in node-negative breast cancer [19, 20, 40]. The LOE 1 validation of uPA and PAI-1 has been obtained both from a single, prospective, controlled study specifically designed to test these markers [19] and from a pooled analysis of 18 studies from the Receptor and Biomarker Group under the European Organization for Research and Treatment of Cancer. The results demonstrated a significant association between high levels of uPA and PAI-1 and shorter relapse-free survival as well as overall survival [20]. The German multicenter Chemo N_0 trial randomized node-negative patients with high levels of uPA and/or PAI-1 to combination chemotherapy or observation, whereas all of those with low levels of the two markers were only observed [19]. This study demonstrated that patients with low concentrations of uPA and PAI-1 in their primary tumor (uPA levels $\leq 3\ ng\,mg^{-1}$ of protein and PAI-1 levels $\leq 14\ ng\,mg^{-1}$ of protein) had a good prognosis without adjuvant chemotherapy after tumor resection. In the group of patients with high uPA and/or PAI-1 levels, those who received adjuvant chemotherapy had significantly longer disease-free survival than those who did not. In recurrent breast cancer, high tumor tissue levels of uPA, PAI-1, and uPAR predict poor response to endocrine treatment with tamoxifen [41].

From the studies leading to the LOE 1 validation it was recommended to use a minimum of 300 mg of fresh tumor tissue for enzyme-linked immunosorbent assay (ELISA) measurements of uPA and PAI-1. Ultrasound-guided core needle biopsies taken preoperatively with smaller tumor biopsy volumes are becoming more common. These biopsies can potentially be used for biomarker determination, and hence information from these could be provided before operation. Thomssen and coworkers [42] therefore investigated the feasibility of using these core needle biopsies for ELISA measurements of uPA and PAI-1. The results showed a significant correlation between uPA and PAI-1 protein levels measured in the biopsies and corresponding larger tumor specimens, leading to the suggestion that core needle biopsies can be used for routine determination of uPA and PAI-1 levels in primary breast cancer. To account for tumor heterogeneity, two or three

Table 13.1 Summary of studies of the prognostic, diagnostic, and predictive values of uPA, PAI-1, and the different uPAR forms, determined at protein level using immunoassays and immunohistochemical methods.

Cancer type	Method of quantification	Biomarker(s)	Outcome	Comment	References
Breast	ELISA, tumor tissue extracts	uPA, PAI-1	Prognostic	LOE 1 validation	Jänicke *et al.* [19], Look *et al.*[20]
	ELISA, tissue extracts	uPA, PAI-1, uPAR	Prognostic	Laser capture microdissection	Hildenbrandt *et al.* [21]
	IHC on TMA	uPAR	Prognostic	Tumor cell expression	Kotzsch *et al.* [22]
	IHC on whole sections	uPAR	Prognostic	Stromal cell expression	Giannopoulou *et al.* [23]
Colorectal	IHC on TMA	uPA, uPAR	uPA prognostic	Multimarker study, mismatch-repair-proficient CRC patients	Zlobec *et al.* [24]
	IHC on TMA	uPA, uPAR	uPA prognostic	Mismatch-repair-proficient CRC patients	Minoo *et al.* [25]
	TR-FIA, serum	uPAR(I-III) + uPAR(II-III), uPAR(I)	Prognostic	–	Lomholt *et al.* [26]
	TR-FIA, plasma	uPAR(I-III) + uPAR(II-III), uPAR(I)	Prognostic	–	Thurison *et al.* [27]
	TR-FIA, plasma	All uPAR forms	Diagnostic	Case control study	Lomholt *et al.* [28]
	ELISA, serum	uPA, PAI-1	Diagnostic	Case control study	Herszenyi *et al.* [29]
Lung	TR-FIA, plasma and serum	uPAR(I-III), uPAR(I)	Prognostic	Different subtypes	Almasi *et al.* [30]
	TR-FIA, serum	uPAR(I-III), uPAR(I)	Prognostic	Different subtypes	Almasi *et al.* [68]
	ELISA, tumor tissue extracts	uPA, PAI-1	No association to survival	Lower levels in nonangiogenic tumors	Offersen *et al.* [32]

(*continued overleaf*)

Table 13.1 (*Continued*)

Cancer type	Method of quantification	Biomarker(s)	Outcome	Comment	References
Gynecological	IHC on whole sections	uPA, uPAR	Prognostic	–	Wang *et al.* [33]
	TR-FIA, plasma	uPAR(I-III) + uPAR(II-III)	Diagnostic	In combination with CA125	Henic *et al.* [34]
		uPAR(I-III), uPAR(I-III) + uPAR(II-III), uPAR(I)	Prognostic		
	ELISA, cytosolic fluids	uPA, PAI-1	Prognostic	–	Steiner *et al.* [35]
Prostate	ELISA, plasma	uPA, uPAR	Prognostic	–	Shariat *et al.* [36]
	ELISA, plasma	uPA, PAI-1, uPAR	Prognostic	Multimarker study	Shariat *et al.* [37]
	TR-FIA, serum	uPAR(I)/ uPAR(I-III)	Diagnostic	In combination with free PSA	Piironen *et al.* [38]
	TR-FIA, serum	uPAR(I-III) + uPAR(II-III),	Predictive,	PEP vs. TAB treatment	Almasi *et al.* [31]
		uPAR(I-III), uPAR(I-III) + uPAR(II-III), uPAR(I)	Prognostic		

ELISA, enzyme-linked immunosorbent assay; LOE, level of evidence; IHC, immunohistochemistry; TMA, tissue microarray; CRC, colorectal cancer; TR-FIA, time-resolved fluorescence immunoassay; CA125, cancer antigen 125; PSA, prostate-specific antigen; PEP, polyestradiol phosphate; TAB, total androgen blockade.

biopsies from each patient should be analyzed. As only 42 patients were included in this study, a larger study cohort is needed to fully establish the feasibility of the method.

In order to address the problem of tumor heterogeneity, tumor cells and stromal tissue from 60 patients with invasive breast carcinomas were separated by laser capture microdissection. Subsequently, the non-dissected tumor sample and the preparations either enriched in tumor cells or stromal tissue were measured by ELISA to determine the uPA, PAI-1, and uPAR levels [21]. No statistically significant difference between the levels of uPA, PAI-1, and uPAR in the isolated tumor cells, in the isolated stromal tissue, or in the non-dissected tumor sample was found. There was, nevertheless, for all three markers a trend towards higher levels in the tumor stroma. Patients were divided into groups with high or low levels of the three factors according to previously defined cut off points: uPA 3.0 ng mg^{-1} of protein, PAI-1 14.0 ng mg^{-1} [43, 44], and uPAR 4.0 ng mg^{-1} [45]. Cox regression analysis demonstrated that patients with high tumor tissue levels of either of the biomarkers had a poorer prognosis for overall and relapse-free survival than those with low levels. The outcome of this study is important as it suggests that tumor heterogeneity does not present a problem when using the uPA and PAI-1 levels in tumor tissue extracts for assessment of patient prognosis. Considering the more than 2000 tissue section samples dissected in the study to get enough material for the ELISA measurements it is possible that complete separation of the tumor cells from the stromal tissue was not obtained in every case, and a similar study with the same outcome is thus needed to confirm these results.

Tumor tissue measurements of uPA, PAI-1, and uPAR by immunoassays rely on the use of tissue extracts or tissue lysates made from fresh or fresh-frozen tissue, which is available in very limited amounts in a clinical setting. Owing to the pathologist's priorities in selecting samples for diagnosis, attempts have been made to score immunohistochemical staining of biomarkers on fixated tissue and correlate the results to clinical data. Among the methods developed is tissue microarray (TMA). The fact that every tissue sample within the TMA arises from different tumors, where the handling and fixation of individual tumor tissue samples might not be identical, can give rise to apparent alterations in the levels of protein expression. Furthermore, the antigen-retrieval method chosen during the histological staining of the TMA cannot be optimized for all samples within the TMA. In addition, the tissue samples within TMAs only comprise a relatively small area of a tumor. This area may not include the invasive front of the tumor, where components of the plasminogen activation system are primarily upregulated. These method limitations have to be considered when evaluating TMA results.

In a study using TMAs of tumor tissue samples from 270 patients with invasive ductal breast carcinoma, uPAR expression was evaluated using an uPAR domain-II-specific monoclonal antibody [22]. The antibody used reacts with both intact uPAR [uPAR(I-III)] and cleaved uPAR [uPAR(II-III)]. The uPAR-expressing cell types within the tumor tissue were found to be both cancer cells and stromal cells. The uPAR immunoreactivity of cancer cells was found on the cell surface as expected but surprisingly cytoplasmatic staining was also observed. The expression level of uPAR in stromal cells was found not to be associated with patient outcome. In contrast, high uPAR expression in tumor cells was significantly correlated to poor prognosis.

In Cox multivariate regression analysis, uPAR expression in tumor cells (high vs. low uPAR expression) remained a significant prognostic factor when adjusted for clinicopathological parameters, including lymph node status, tumor size, grade, and age. Stromal and combined stromal/tumor cell uPAR expression was not a significant prognostic factor in multivariate analysis. In another study, stromal uPAR expression was found to be an independent prognostic factor for shorter relapse-free survival [23]. The discrepancy between these two studies cannot result from the antibodies used in the immunohistochemical analyses, as in both studies the same uPAR domain-II-specific monoclonal antibody IID7, raised against nonglycosylated, recombinant human uPAR, was used [46]. The differences could, however, be due to the histological subtypes of breast carcinomas included in the two studies. Kotzsch *et al.* [22] exclusively analyzed invasive ductal carcinomas, whereas Giannopoulou *et al.* [23] included both ductal and lobular carcinomas, with uPAR expression in cancer cells most often seen in the lobular subtype. In addition, Giannopoulou *et al.* stained whole tissue biopsies, whereas Kotzsch *et al.* used TMAs.

An attractive alternative to analyses of tumors is the use of plasma or serum, since blood can be collected in a non-invasive manner and is homogeneous as opposed to tumor tissue. In order to evaluate if soluble uPAR measured in blood has prognostic significance, preoperatively collected serum samples from 274 primary breast cancer patients were analyzed by ELISA [47]. This study provided clear evidence that uPAR levels measured in serum are statistically significant associated with overall survival, demonstrating that soluble uPAR in blood is a prognostic marker in breast cancer.

13.3
Colorectal Cancer

Colorectal cancer (CRC) is one of the leading causes of cancer-specific deaths [48]. At the time of resection, approximately 40% of the patients are diagnosed with stage II CRC (T3 or T4 tumor classification, negative lymph node status, and no metastasis) [49]. Even though the primary treatment of stage II CRC is surgery, increasing evidence has shown a possible survival benefit by administrating adjuvant chemotherapy to these patients [50–52]. As stage II CRC is a heterogeneous disease, an improved method to identify high-risk patients, who will benefit from adjuvant therapy, is required. Several studies have demonstrated that high levels of uPA, PAI-1, or uPAR, measured in plasma or tumor tissue extracts, have prognostic value in CRC patients (reviewed in Christensen *et al.* [4]).

Zlobec *et al.* [24] conducted a study with the objective of defining a multimarker prognostic model of five-year survival in lymph node-negative, mismatch-repair-proficient CRC patients, defined by having no mutations in the mismatch-repair genes MLH1, MSH2, and MSH6. Thirteen candidate proteins were included in the study, selected to represent (i) novel and promising prognostic factors, (ii) mediators of cell cycle arrest, and (iii) proteins of signaling pathways involved in tumor progression. uPA and uPAR were both included in the first category. The 13 protein markers were analyzed by IHC in 587 CRC tumors in TMAs, and the results scored semiquantitatively by evaluating the ratio of marker-positive tumor cells to the total number of tumor cells. uPA and uPAR were both significantly associated with more adverse outcome in univariate analysis, while only uPA remained significant in multivariate analysis. In an additional study by the same group of investigators, the expression of uPA and uPAR in TMAs from 811 mismatch-repair-proficient and 164 mismatch-repair-deficient

CRC patients were investigated [25]. In the mismatch-repair-proficient cases, high uPA and uPAR expression was associated with the more advanced T stages, and five-year disease-specific survival was significantly lower in these patients compared to those with low uPA and uPAR expression. uPA maintained its prognostic value in mismatch-repair-proficient CRC when adjusting for tumor stage, grade, and vascular invasion. In the mismatch-repair-deficient CRC patients, neither uPA nor uPAR expression was correlated to survival. The results from both studies indicate that uPA is a significant prognostic marker with independent value in mismatch-repair-proficient CRC patients. The latest study, however, illustrates the technical problems with TMAs addressed above (Section 13.2). The overexpression of uPA and uPAR was scored on TMAs, where nuclear staining was also observed [25]. This lack of agreement with the expected localization of uPA and uPAR was not discussed. Since no immunohistochemical stainings of the TMAs are shown in the first study, we do not know if the uPA and uPAR stainings were similar in that case.

In the above studies, uPAR was stained with an antibody recognizing both intact and cleaved uPAR. At present, there are no antibodies that differentially stain the various uPAR forms in IHC, but time-resolved fluorescence immunoassays (TR-FIAs), quantifying the individual uPAR forms in tumor tissue extracts and blood, have been designed [27, 53]. The prognostic value of the individual uPAR forms was evaluated in serum samples from a cohort of 518 CRC patients [26]. In a previous study of the same patient cohort, the preoperative plasma level of uPAR had been shown to independently predict survival of CRC patients [54]. The ELISA applied in the first study measures the combined amounts of intact uPAR(I-III) with and without uPA bound and cleaved uPAR(II-III). There was a high correlation between the ELISA-measured uPAR levels in the first study [54] and the levels of uPAR(I-III) and uPAR(II-III) measured by the TR-FIA assay (TR-FIA 2) used in the second study [26]. The conclusion that can be drawn from both of these studies is that high levels of the uPAR forms are associated with poor prognosis and low levels with good prognosis. In multivariate analysis, TR-FIA-2-measured uPAR(I-III) + uPAR(II-III) as well as uPAR(I), measured with a different assay (TR-FIA 3), were shown to be independent predictors of prognosis, with high levels associated with poor outcome. Only 24% of the patients had uPAR(I) levels above the limit of quantification of the applied TR-FIA 3 assay. Therefore, when analyzing the prognostic impact of uPAR(I), patients were dichotomized by the limit of quantification. In contrast, all patients had measurable uPAR(I-III) + uPAR(II-III) levels, thereby increasing the power of the statistical analysis to detect survival differences. Low and high levels of uPAR(I) could, however, discriminate between patients with good and poor overall survival in patients with Dukes B tumors. Dukes B is similar to the CRC stage II classification used at present. Among the Dukes B patients with uPAR(I) levels above the limit of quantification, 65% died during follow-up, as compared with 45% of those with uPAR(I) levels below the limit of quantification, indicating that uPAR(I) could be used to identify high-risk Dukes B patients.

To enable quantification of uPAR(I) in all patient samples, a new uPAR(I) assay with improved functional sensitivity was designed [27]. This new assay, TR-FIA 4, has a fourfold improved functional sensitivity compared to TR-FIA 3, leading to a limit of quantification below the uPAR(I) levels in all patient samples. In order to investigate if the prognostic potential of the different uPAR forms and to test uPAR(I) measured with TR-FIA 4 would result in a superior separation of the patients with good and bad prognosis, the prognostic impact of the different uPAR forms was evaluated in plasma from a study cohort of 298 CRC patients using the four TR-FIA assays (TR-FIA 1 measuring uPAR(I-III), TR-FIA 2 measuring uPAR(I-III) + uPAR(II-III), and TR-FIA 3 and TR-FIA 4 measuring uPAR(I)). The uPAR forms were all significantly associated with survival, with high concentrations predicting short overall survival. In multivariate analysis with model reduction, uPAR(I-III) + uPAR(II-III), measured with TR-FIA 2, and uPAR(I), measured with TR-FIA 4 (but not TR-FIA 3), were independent prognostic factors, adjusted for stage, gender, and age. These factors can thus be used to identify high-risk CRC patients independent of stage. Of the CRC stage II patients with TR-FIA-4-measured uPAR(I) levels above the 70th percentile, 54% died during follow-up, whereas only 27% of those with levels below the 70th percentile died [27]. This indicates that TR-FIA-4-measured uPAR(I) efficiently separates high-risk stage II CRC patients from those with low risk.

In a case control study, 308 patients undergoing endoscopical examination due to symptoms related to CRC were matched by localization of lesion (adenomas), age, and gender [28]. Patients were classified as those with CRC, adenomas, other non-malignant findings, and without findings and were then compared with respect to uPAR levels. Intact uPAR(I-III), uPAR(I-III) + uPAR(II-III), and uPAR(I) were measured in citrate plasma from these patients using TR-FIAs 1, 2, and 3. Plasma levels of all the different uPAR forms were significantly higher in cancer patients than those in the other groups, suggesting that the uPAR forms could aid in the detection of CRC. Similar conclusions were drawn concerning the plasma levels of uPA and PAI-1 [29]. These were found to be significantly elevated in patients with CRC, when the levels were measured in 56 CRC patients, 25 with ulcerative colitis, 26 with colorectal adenomas, and 35 tumor-free control subjects.

13.4 Lung Cancer

Lung cancer generally has a poor prognosis, and despite intensive clinical research, it has not yet been possible to radically improve survival [55]. Thus, there is a substantial need for biomarkers, which could aid in diagnosis, prognosis, and/or selection of patients for therapy. Several studies on lung cancer indicate that the prognostic value of uPA, PAI-1, and uPAR measured by ELISA in tumor tissue extracts is dependent on histological subtype. In a cohort of 106 patients with pulmonary adenocarcinoma (AC), PAI-1 was found to be an independent prognostic marker [56]. In squamous cell carcinoma (SCC) of the lung, high levels of uPAR in tumor tissue extracts were significantly associated with short overall survival, while neither uPA nor PAI-1 showed any prognostic value in this histological subtype. Interestingly, the levels of PAI-1 and uPAR differ between tumors of different histological subtypes [57].

In a more recent study, the prognostic significance of uPA and PAI-1 in tumor tissue extracts from a cohort of 118 untreated non-small cell lung cancer (NSCLC) patients comprising the three major histological subtypes of lung cancer (68 SCC, 35 AC, 15 large cell carcinoma (LCC)) was investigated [32]. The applied uPA ELISA measures pro-uPA, uPA, as well as uPA in complex with PAI-1 and/or uPAR. The total PAI-1 amount was quantified by an ELISA measuring PAI-1, PAI-1/uPA complexes, and PAI-1/uPA/uPAR complexes [58]. The association between angiogenesis and prognosis had previously been studied in the patient cohort [59]. Multivariate analysis showed no statistical association between uPA or PAI-1 levels and survival in the entire cohort. In this study, no subanalyses of the association with the different histological subtypes were performed, and therefore the influence of this parameter is not included. Significantly lower levels of both uPA and PAI-1 were found in tumors previously defined as non-angiogenic compared to the angiogenic tumors, which could indicate that uPA and PAI-1 stimulate angiogenesis in NSCLC tumors [32].

The prognostic impact of the liberated uPAR(I) measured by TR-FIA 3 was assessed in a cohort of SCC patients [60], where the ELISA-measured total uPAR amount previously was demonstrated to be an independent prognostic marker [61]. The levels of uPAR(I) in the primary tumor extracts independently predicted overall survival in the cohort of SCC patients and was a stronger prognostic marker than the ELISA-measured total amount of uPAR. The assays measuring the other uPAR forms could not be validated for use on the tumor extracts, possibly because of the presence of high amounts of uPA [60]. In a subsequent study, serum and plasma samples from 32 NSCLC patients of mixed histological subtypes (16 SCC, 13 AC, and 3 LCC) were analyzed [30]. Interestingly, for patients of all three histological subtypes, significant associations between the level of uPAR(I-III) measured in plasma and uPAR(I) measured in both plasma and serum and overall survival were demonstrated, with high levels associated with shorter overall survival. This finding was recently reproduced in a cohort of 171 NSCLC patients of mixed histological subtypes, where uPAR(I-III) and uPAR(I) in serum were found to be independent prognostic factors in patients radically operated for NSCLC [68].

13.5 Gynecological Cancers

Current treatment regimes for ovarian cancer include surgery, adjuvant chemotherapy, and hormonal therapy; even so, the majority of patients have relapses [62]. The most common ovarian cancer, epithelial ovarian cancer (EOC), accounts for 90% of the cases, and as EOC can spread directly to the adjacent organs, intra-abdominal dissemination is the primary cause of death for these patients [63].

High levels of uPA and PAI-1 measured in EOC tumor tissue extracts were demonstrated to be significantly associated with short relapse-free as well as overall survival [64]. Analysis of the expression pattern of uPA and uPAR in EOC showed uPA staining in 92% and uPAR staining in 88% of the EOC specimens, with similar results in metastatic lesions [33]. When grading the staining intensity and correlating this to progression-free survival, a significant difference was found between patients with tumors with high and low expression of uPA and uPAR. uPA and uPAR were mainly found to be membrane associated, and in high-grade tumors (tumor stage >2), both stromal and tumor cell expressions of uPA and uPAR were found. The intense specific staining suggested a high expression level of uPA and uPAR in these EOC specimens.

Owing to the substantial difference in survival between patients diagnosed with early- and late-stage ovarian tumors, there is a great need for biomarkers with diagnostic potential. To address this, preoperatively collected plasma samples from 335 patients with adnexal masses were analyzed for the levels of the different uPAR forms [34]. Plasma levels of all uPAR forms were higher in patients diagnosed with borderline and malignant ovarian tumors than in those with benign tumors. In multivariate analysis, uPAR(I) was demonstrated to be an independent predictor of prognosis in ovarian cancer. Since the level of uPAR(I) is measured in blood taken before surgery, this information could be used to guide how extensive the operation should be. Furthermore, the product of uPAR(I-III) + uPAR(II-III) and cancer antigen 125 (CA125) levels was identified as a diagnostic marker for invasive and borderline ovarian tumors. CA125 alone has too low a sensitivity to detect early-stage tumors, since approximately 20% of the ovarian cancers are CA125 negative and will accordingly not be detected [65]. Combining uPAR(I-III) + uPAR(II-III) with CA125 improves the sensitivity and thus the utility of these biomarkers for ovarian cancer detection.

In contrast to ovarian cancers, endometrial cancers are often detected at early stages, leading to better prognosis for the patients [66]. Nevertheless, 20% of the patients with stage I and II endometrial tumors will eventually progress, and there is therefore a need for markers to select patients who will benefit from adjuvant therapy. Steiner *et al.* [35] determined uPA and PAI-1 concentrations in cytosolic fluids from endometrial carcinomas of 69 patients. uPA and PAI-1 were measured using ELISAs, and the concentrations were correlated to relapse-free time and overall survival. Multivariate analysis demonstrated that high uPA and PAI-1 levels were significantly correlated to shorter relapse-free time in endometrial cancer patients. Significant correlation with overall survival was only found for uPA.

13.6 Prostate Cancer

Several studies have focused on the impact of uPA, PAI-1, and uPAR on prostate cancer progression, both by cellular localization studies and quantitative measurements of circulating levels. In prostate AC, uPAR was determined by IHC on whole sections to be localized mainly to macrophages and present in 50% of the specimens. PAI-1 staining was found in approximately 80% of the samples and localized to myofibroblasts, macrophages, and endothelial cells, while no cancer cells were demonstrated positive for either protein [2]. In benign prostatic hyperplasia, uPAR and PAI-1 were expressed by the same cells as in the prostate AC. In a different study using TMAs, the immunohistochemical localization of uPA and uPAR was investigated in 120 prostate cancer specimens from untreated cancer patients [67]. The study also included tissue from 10 matched lymph node metastases, prostate glands from 15 patients with benign prostate hyperplasia, and prostate tissue from 40 controls. All control tissues were negative for uPA, while 15% scored positive for uPAR. In the benign prostate hyperplasia group, 27 and 47% were positive for uPA and uPAR, respectively. Interestingly, these authors found uPAR located on epithelial cells, which is in contrast to the results by Usher and coworkers [2]. Among the prostate cancer tissue samples, 69 and 83% were found to be positive for uPA and uPAR, respectively. In the lymph node metastasis, uPA expression was found in 100% of the cases and uPAR expression in 90%. Only the radically operated patients ($n = 72$) were included in the correlation to clinical outcome. Of these patients, only 7% experienced biochemical recurrence as defined by the prostate-specific antigen (PSA) level. No patients died of prostate cancer during follow-up.

Circulating levels of uPA and uPAR were measured by ELISA in preoperatively collected plasma samples from 429 treatment-naïve patients with localized prostate AC [36]. In 76 of these patients, having undetectable postoperative PSA levels, uPA and uPAR were also measured in plasma collected six to eight weeks after radical prostatectomy. The results demonstrated that the levels of uPA and uPAR were significantly lower in postoperative than in preoperative plasma samples, suggesting that the elevated circulating levels of uPA and uPAR at least partly originate from the prostate tumor. The study cohort also included 44 controls without cancer, 19 with lymph node metastasis, and 10 with bone metastasis. The levels of both uPA and uPAR were higher in patients with prostate cancer compared to the healthy controls, and the levels were highest in patients with bone metastases. Furthermore, a significant, independent correlation to biochemical recurrence (i.e., elevated PSA levels) was found for uPA but not for uPAR.

In the search for independent but complementary biomarker combinations, which may provide a better prediction of patient outcome than a single biomarker, plasma samples from the same patient cohort previously analyzed for preoperative levels of uPA and uPAR [36] were used [37]. Biomarkers in the study included proteins involved in cell signaling, angiogenesis, and cancer progression; the latter category including uPA, PAI-1, and uPAR. uPA was, as in the prior study, shown to be an independent predictor of biochemical recurrence. In addition, transforming growth factor-β1, interleukin-6, soluble interleukin-6 receptor, vascular cell adhesion molecule-1, vascular endothelial growth factor, and endoglin remained significant in the multivariate analysis. The result from this study indicates that these biomarkers in combination can help identify prostate cancer patients with increased risk of biochemical recurrence.

The above studies did not demonstrate an independent association of the total amount of plasma uPAR with biochemical recurrence. As uPA, however, is an independent prognostic marker and as uPA cleaves uPAR, the biomarker potential of the cleaved uPAR forms in prostate cancer was investigated. Preoperative serum levels of the different uPAR forms were determined in 390 men referred to urology department examination, of which 224 were diagnosed with prostate cancer [38]. The aim was to elucidate if the uPAR forms could increase the specificity and sensitivity of prostate cancer detection, when combined with PSA. Only a weak correlation between PSA and the individual uPAR forms was found. Both uPAR(I) and uPAR(II-III) were found to be significantly elevated in serum from prostate cancer patients compared to men with benign prostatic disease. By combining the ratio of uPAR(I)/uPAR(I-III) and free PSA, the predictive accuracy was increased, and thus the specific cancer detection, both for the entire study cohort ($n = 390$) and importantly for the clinically more difficult group with moderately increased PSA levels, that is, between 2 and 10 ng ml^{-1}.

In metastatic prostate cancer, serum levels of the different uPAR forms were associated with overall survival in univariate analysis [31]. In a cohort of 131 treatment-naïve metastatic prostate cancer patients randomized to treatment with either total androgen blockade (TAB) or polyestradiol phosphate (PEP), the levels of the different uPAR forms were determined in serum samples taken before treatment. Overall survival was similar in the two treatment groups. The uPAR forms were all found to be statistically significant prognostic markers in the TAB-treated patients. In multivariate analysis, uPAR(I-III) + uPAR(II-III) was found to be an independent prognostic marker in TAB-treated but not in PEP-treated patients. High levels of uPAR(I-III) + uPAR(II-III) were found to be predictive of the effect of PEP versus TAB treatment. Thus, a concentration of uPAR(I-III) + uPAR(II-III) in the blood above the median predicts better response to PEP than to TAB treatment, with a significant gain in overall survival (median difference 11.3 months), for metastatic prostate cancer patients [31]. This is the first report providing evidence for the utility of the uPAR forms as predictive markers for treatment response.

13.7 Conclusion and Perspectives

Tumor biomarkers are molecules that indicate the presence or prognosis of malignancy. Thus, tumor biomarkers can be useful for early diagnosis of cancer and relapsed disease, as well as to determine prognosis and to predict response to therapy. The overall goal is to detect a cancer in its early stage, preferably by application of a simple and minimally invasive method in order to avoid unnecessary biopsies. To achieve this, sensitive biomarkers with high specificity are needed.

Fortunately, tumors are now often detected earlier than previously, leading to smaller tumor volumes at diagnosis. The pathologists have the highest priority for tumor tissue for diagnosis and consequently smaller tumor samples, if any, are available for protein extraction. Thus, it is imperative to develop alternative methods, independent of protein extraction from tumor tissue for biomarker quantification. In this respect, biomarker determinations in blood by immunoassays and on paraffin-embedded tissue using immunohistochemical staining are valuable tools. Blood sampling is a minimally invasive procedure with a reasonable sample volume. Blood is furthermore homogeneous in contrast to tumor tissue. The use of immunoassays requires thorough validation and quality control in order to ensure that the quantitative measurements of a biomarker in a particular biological matrix are reliable. Validation procedures must include determination of specificity, cross-reactivity, linearity, functional and analytic sensitivity as well as recovery, and precision. Similarly, validation and high-level quality control are needed when applying immunohistochemical methods, including TMA techniques, for semiquantitative biomarker scoring. Analyses of a biomarker in independent patient cohorts, as well as with different methods of quantification, yielding similar and significant results support its clinical utility. In two studies on preoperative plasma samples from two different CRC patient cohorts, the combined concentration of uPAR(I-III) and uPAR(II-III), measured with two different immunoassays was found to independently predict survival [27, 54]. These results confirm the significance of the uPAR forms as prognostic biomarkers in CRC.

We hypothesized that cleavage of uPAR is an indication of an active plasminogen activation system, and hence that high levels of the cleaved uPAR forms is a hallmark of aggressive tumor progression. Several studies have supported this hypothesis, demonstrating the cleaved uPAR forms to be superior to intact uPAR as prognostic biomarkers. Recent studies have furthermore provided evidence for application of the cleaved uPAR forms as diagnostic biomarkers [34, 38] as well as predictive biomarkers for the outcome of treatment [31]. However, further studies are needed to verify these prognostic, diagnostic, and predictive potentials of the cleaved uPAR forms, preferably by participation in prospective randomized controlled clinical trials.

uPA and PAI-1 are important prognostic biomarkers in breast cancer, and they are recommended by ASCO for identification of high-risk node-negative patients [39]. These biomarkers have not achieved similar validation for utility in other cancer types, even though several studies have proved their prognostic significance. In CRC patients, serum levels of uPA and PAI-1 have been demonstrated to be elevated compared to the levels in healthy controls [29]. In gynecological cancers, the immunohistochemical expression levels of uPA correlate to progression-free survival [33], and the levels of uPA and PAI-1 measured by ELISA on cytosolic extracts correlate to relapse-free time [35]. Finally, ELISA-measured plasma uPA levels independently predict biochemical recurrence in prostate cancer patients [36]. The clinical utility of uPA and PAI-1 in breast cancer has been established, and there is substantial evidence that these biomarkers as well as uPAR and its cleaved forms could be useful in many cancer diseases. There is a need for well-designed studies as well as for pooled meta-analyses in order to bring these biomarkers to LOE 1.

Acknowledgment

The graphical assistance provided by photographer John Post is gratefully acknowledged. This work was supported by the Danish Cancer Research Foundation, the Lundbeck Foundation, Capital Region of Denmark Research Foundation, the Danish Cancer Society, and the European Community's Seventh Framework Program FP7/2007–2011 under grant agreement n°201279.

Abbreviations

uPA	urokinase-type plasminogen activator
PAI-1	plasminogen activator inhibitor 1
uPAR	urokinase-type plasminogen activator receptor
Pro-uPA	zymogen form of uPA
MMP	matrix metalloprotease
GPI	glycosylphosphatidylinositol
ASCO	American Society of Clinical Oncology
LOE	level of evidence
ELISA	enzyme-linked immunosorbent assay
TMA	tissue microarray
IHC	immunohistochemistry
CRC	colorectal cancer
TR-FIA	time-resolved fluorescence immunoassay
CA125	cancer antigen 125
PSA	prostate-specific antigen
PEP	polyestradiol phosphate
TAB	total androgen blockade
AC	adenocarcinoma
SCC	squamous cell carcinoma
NSCLC	non-small-cell lung cancer
LCC	large cell carcinoma
EOC	epithelial ovarian cancer

References

1. Biomarkers Definition Working Group (2001) Biomarkers and surrogate endpoints: preferred definitions and conceptual framework. *Clin. Pharmacol. Ther.*, **69**, 89–95.
2. Usher, P.A., Thomsen, O.F., Iversen, P., Johnsen, M., Brünner, N., Høyer-Hansen, G., Andreasen, P., Danø, K., and Nielsen, B.S. (2005) Expression of urokinase plasminogen activator, its receptor and type-1 inhibitor in malignant and benign prostate tissue. *Int. J. Cancer*, **113**, 870–880.
3. Illemann, M., Bird, N., Majeed, A., Lærum, O.D., Lund, L.R., Danø, K., and Nielsen, B.S. (2009) Two distinct expression patterns of urokinase, urokinase receptor and plasminogen activator inhibitor-1 in colon cancer liver metastases. *Int. J. Cancer*, **124**, 1860–1870.
4. Christensen, I.J., Pappot, H., and Høyer-Hansen, G. (2008) in *The Cancer Degradomme: Proteases and Cancer Biology*, Chapter 28 (eds D. Edwards, G. Høyer-Hansen, F. Blasi, and B.F. Sloane) Springer, pp. 569–586.

5. Ploug, M. (2003) Structure-function relationships in the interaction between the urokinase-type plasminogen activator and its receptor. *Curr. Pharm. Des*, **9**, 1499–1528.
6. Gårdsvoll, H., Gilquin, B., Le Du, M.H., Menez, A., Jørgensen, T.J., and Ploug, M. (2006) Characterization of the functional epitope on the urokinase receptor. Complete alanine scanning mutagenesis supplemented by chemical cross-linking. *J. Biol. Chem.*, **281**, 19260–19272.
7. Estreicher, A., Muhlhauser, J., Carpentier, J.L., Orci, L., and Vassalli, J.D. (1990) The receptor for urokinase type plasminogen activator polarizes expression of the protease to the leading edge of migrating monocytes and promotes degradation of enzyme inhibitor complexes. *J. Cell Biol.*, **111**, 783–792.
8. Cubellis, M.V., Andreasen, P., Ragno, P., Mayer, M., Danø, K., and Blasi, F. (1989) Accessibility of receptor-bound urokinase to type-1 plasminogen activator inhibitor. *Proc. Natl. Acad. Sci. U.S.A.*, **86**, 4828–4832.
9. Høyer-Hansen, G., Ploug, M., Behrendt, N., Rønne, E., and Danø, K. (1997) Cell-surface acceleration of urokinase-catalyzed receptor cleavage. *Eur. J. Biochem.*, **243**, 21–26.
10. Pappot, H., Høyer-Hansen, G., Rønne, E., Hansen, H.H., Brünner, N., Danø, K., and Grøndahl-Hansen, J. (1997) Elevated plasma levels of urokinase plasminogen activator receptor in non-small cell lung cancer patients. *Eur. J. Cancer*, **33**, 867–872.
11. Mustjoki, S., Sidenius, N., Sier, C.F., Blasi, F., Elonen, E., Alitalo, R., and Vaheri, A. (2000) Soluble urokinase receptor levels correlate with number of circulating tumor cells in acute myeloid leukemia and decrease rapidly during chemotherapy. *Cancer Res.*, **60**, 7126–7132.
12. Wahlberg, K., Høyer-Hansen, G., and Casslen, B. (1998) Soluble receptor for urokinase plasminogen activator in both full-length and a cleaved form is present in high concentration in cystic fluid from ovarian cancer. *Cancer Res.*, **58**, 3294–3298.
13. Sier, C.F., Nicoletti, I., Santovito, M.L., Frandsen, T., Aletti, G., Ferrari, A., Lissoni, A., Giavazzi, R., Blasi, F., and Sidenius, N. (2004) Metabolism of tumour-derived urokinase receptor and receptor fragments in cancer patients and xenografted mice. *Thromb. Haemost.*, **91**, 403–411.
14. Wilhelm, O.G., Wilhelm, S., Escott, G.M., Lutz, V., Magdolen, V., Schmitt, M., Rifkin, D.B., Wilson, E.L., Graeff, H., and Brünner, G. (1999) Cellular glycosylphosphatidylinositol-specific phospholipase D regulates urokinase receptor shedding and cell surface expression. *J. Cell Physiol.*, **180**, 225–235.
15. Beaufort, N., Leduc, D., Rousselle, J.C., Magdolen, V., Luther, T., Namane, A., Chignard, M., and Pidard, D. (2004a) Proteolytic regulation of the urokinase receptor/CD87 on monocytic cells by neutrophil elastase and cathepsin G. *J. Immunol.*, **172**, 540–549.
16. Beaufort, N., Leduc, D., Rousselle, J.C., Namane, A., Chignard, M., and Pidard, D. (2004b) Plasmin cleaves the juxtamembrane domain and releases truncated species of the urokinase receptor (CD87) from human bronchial epithelial cells. *FEBS Lett.*, **574**, 89–94.
17. Høyer-Hansen, G. and Lund, I.K. (2007) Urokinase receptor variants in tissue and body fluids. *Adv. Clin. Chem.*, **44**, 65–102.
18. Andreasen, P.A., Kjøller, L., Christensen, L., and Duffy, M.J. (1997) The urokinase-type plasminogen activator system in cancer metastasis: a review. *Int. J. Cancer*, **72**, 1–22.
19. Jänicke, F., Prechtl, A., Thomssen, C., Harbeck, N., Meisner, C., Untch, M., Sweep, C.G., Selbmann, H.K., Graeff, H., and Schmitt, M. (2001) Randomized adjuvant chemotherapy trial in high-risk, lymph node-negative breast cancer patients identified by urokinase-type plasminogen activator and plasminogen activator inhibitor type 1. *J. Natl. Cancer Inst.*, **93**, 913–920.
20. Look, M.P., van Putten, W.L., Duffy, M.J., Harbeck, N., Christensen, I.J., Thomssen, C., Kates, R., Spyratos, F., Fernø, M., Eppenberger-Castori, S., Sweep, C.G., Ulm, K., Peyrat, J.P.,

Martin, P.M., Magdelenat, H., Brünner, N., Duggan, C., Lisboa, B.W., Bendahl, P.O., Quillien, V., Daver, A., Ricolleau, G., Meijer-Van Gelder, M.E., Manders, P., Fiets, W.E., Blankenstein, M.A., Broet, P., Romain, S., Daxenbichler, G., Windbichler, G., Cufer, T., Borstnar, S., Kueng, W., Beex, L.V., Klijn, J.G., O'Higgins, N., Eppenberger, U., Jänicke, F., Schmitt, M., and Foekens, J.A. (2002) Pooled analysis of prognostic impact of urokinase-type plasminogen activator and its inhibitor PAI-1 in 8377 breast cancer patients. *J. Natl. Cancer Inst.*, **94**, 116–128.

21. Hildenbrand, R., Schaaf, A., Dorn-Beineke, A., Allgayer, H., Sutterlin, M., Marx, A., and Stroebel, P. (2009) Tumor stroma is the predominant uPA-, uPAR-, PAI-1-expressing tissue in human breast cancer: prognostic impact. *Histol. Histopathol.*, **24**, 869–877.
22. Kotzsch, M., Bernt, K., Friedrich, K., Luther, E., Albrecht, S., Gatzweiler, A., Magdolen, V., Baretton, G., Zietz, C., and Luther, T. (2010) Prognostic relevance of tumour cell-associated uPAR expression in invasive ductal breast carcinoma. *Histopathology*, **57**, 461–471.
23. Giannopoulou, I., Mylona, E., Kapranou, A., Mavrommatis, J., Markaki, S., Zoumbouli, C., Keramopoulos, A., and Nakopoulou, L. (2007) The prognostic value of the topographic distribution of uPAR expression in invasive breast carcinomas. *Cancer Lett.*, **246**, 262–267.
24. Zlobec, I., Minoo, P., Baumhoer, D., Baker, K., Terracciano, L., Jass, J.R., and Lugli, A. (2008) Multimarker phenotype predicts adverse survival in patients with lymph node-negative colorectal cancer. *Cancer*, **112**, 495–502.
25. Minoo, P., Baker, K., Baumhoer, D., Terracciano, L., Lugli, A., and Zlobec, I. (2010) Urokinase-type plasminogen activator is a marker of aggressive phenotype and an independent prognostic factor in mismatch repair-proficient colorectal cancer. *Hum. Pathol.*, **41**, 70–78.
26. Lomholt, A.F., Christensen, I.J., Høyer-Hansen, G., and Nielsen, H.J. (2010) Prognostic value of intact and cleaved forms of the urokinase plasminogen activator receptor in a retrospective study of 518 colorectal cancer patients. *Acta Oncol.*, **49**, 805–811.
27. Thurison, T., Lomholt, A.F., Rasch, M.G., Lund, I.K., Nielsen, H.J., Christensen, I.J., and Høyer-Hansen, G. (2010) A new assay for measurement of the liberated domain I of the urokinase receptor in plasma improves the prediction of survival in colorectal cancer. *Clin. Chem.*, **56**, 1636–1640.
28. Lomholt, A.F., Høyer-Hansen, G., Nielsen, H.J., and Christensen, I.J. (2009) Intact and cleaved forms of the urokinase receptor enhance discrimination of cancer from non-malignant conditions in patients presenting with symptoms related to colorectal cancer. *Br. J. Cancer*, **101**, 992–997.
29. Herszenyi, L., Farinati, F., Cardin, R., Istvan, G., Molnar, L.D., Hritz, I., De, P.M., Plebani, M., and Tulassay, Z. (2008) Tumor marker utility and prognostic relevance of cathepsin B, cathepsin L, urokinase-type plasminogen activator, plasminogen activator inhibitor type-1, CEA and CA 19-9 in colorectal cancer. *BMC Cancer*, **8**, 194.
30. Almasi, C.E., Høyer-Hansen, G., Christensen, I.J., and Pappot, H. (2009) Prognostic significance of urokinase plasminogen activator receptor and its cleaved forms in blood from patients with non-small cell lung cancer. *APMIS.*, **117**, 755–761.
31. Almasi, C.E., Brasso, K., Iversen, P., Pappot, H., Høyer-Hansen, G., Danø, K., and Christensen, I.J. (2011) Prognostic and predictive value of intact and cleaved forms of the urokinase plasminogen activator receptor in metastatic prostate cancer. *Prostate*, **71**, 899–907.
32. Offersen, B.V., Pfeiffer, P., Andreasen, P., and Overgaard, J. (2007) Urokinase plasminogen activator and plasminogen activator inhibitor type-1 in non small-cell lung cancer: relation to prognosis and angiogenesis. *Lung Cancer*, **56**, 43–50.
33. Wang, L., Madigan, M.C., Chen, H., Liu, F., Patterson, K.I., Beretov, J., O'Brien, P.M., and Li, Y. (2009) Expression of urokinase plasminogen activator and its

receptor in advanced epithelial ovarian cancer patients. *Gynecol. Oncol.*, **114**, 265–272.

34. Henic, E., Borgfeldt, C., Christensen, I.J., Casslen, B., and Høyer-Hansen, G. (2008) Cleaved forms of the urokinase plasminogen activator receptor in plasma have diagnostic potential and predict postoperative survival in patients with ovarian cancer. *Clin. Cancer Res.*, **14**, 5785–5793.
35. Steiner, E., Pollow, K., Hasenclever, D., Schormann, W., Hermes, M., Schmidt, M., Puhl, A., Brulport, M., Bauer, A., Petry, I.B., Koelbl, H., and Hengstler, J.G. (2008) Role of urokinase-type plasminogen activator (uPA) and plasminogen activator inhibitor type 1 (PAI-1) for prognosis in endometrial cancer. *Gynecol. Oncol.*, **108**, 569–576.
36. Shariat, S.F., Roehrborn, C.G., McConnell, J.D., Park, S., Alam, N., Wheeler, T.M., and Slawin, K.M. (2007) Association of the circulating levels of the urokinase system of plasminogen activation with the presence of prostate cancer and invasion, progression, and metastasis. *J. Clin. Oncol.*, **25**, 349–355.
37. Shariat, S.F., Karam, J.A., Walz, J., Roehrborn, C.G., Montorsi, F., Margulis, V., Saad, F., Slawin, K.M., and Karakiewicz, P.I. (2008) Improved prediction of disease relapse after radical prostatectomy through a panel of preoperative blood-based biomarkers. *Clin. Cancer Res.*, **14**, 3785–3791.
38. Piironen, T., Haese, A., Huland, H., Steuber, T., Christensen, I.J., Brünner, N., Danø, K., Høyer-Hansen, G., and Lilja, H. (2006) Enhanced discrimination of benign from malignant prostatic disease by selective measurements of cleaved forms of urokinase receptor in serum. *Clin. Chem.*, **52**, 838–844.
39. Harris, L., Fritsche, H., Mennel, R., Norton, L., Ravdin, P., Taube, S., Somerfield, M.R., Hayes, D.F., and Bast, R.C. Jr. (2007) American Society of Clinical Oncology 2007 update of recommendations for the use of tumor markers in breast cancer. *J. Clin. Oncol.*, **25**, 5287–5312.
40. Hayes, D.F., Bast, R.C., Desch, C.E., Fritsche, H., Kemeny, N.E., Jessup, J.M., Locker, G.Y., Macdonald, J.S., Mennel, R.G., Norton, L., Ravdin, P., Taube, S., and Winn, R.J. Jr. (1996) Tumor marker utility grading system: a framework to evaluate clinical utility of tumor markers. *J. Natl. Cancer Inst.*, **88**, 1456–1466.
41. Meijer-van Gelder, M.E., Look, M.P., Peters, H.A., Schmitt, M., Brünner, N., Harbeck, N., Klijn, J.G., and Foekens, J.A. (2004) Urokinase-type plasminogen activator system in breast cancer: association with tamoxifen therapy in recurrent disease. *Cancer Res.*, **64**, 4563–4568.
42. Thomssen, C., Harbeck, N., Dittmer, J., Abraha-Spaeth, S.R., Papendick, N., Paradiso, A., Lisboa, B., Jänicke, F., Schmitt, M., and Vetter, M. (2009) Feasibility of measuring the prognostic factors uPA and PAI-1 in core needle biopsy breast cancer specimens. *J. Natl. Cancer Inst.*, **101**, 1028–1029.
43. Foekens, J.A., Peters, H.A., Look, M.P., Portengen, H., Schmitt, M., Kramer, M.D., Brünner, N., Jänicke, F., Meijer-Van Gelder, M.E., Henzen-Logmans, S.C., van Putten, W.L., and Klijn, J.G. (2000) The urokinase system of plasminogen activation and prognosis in 2780 breast cancer patients. *Cancer Res.*, **60**, 636–643.
44. Harbeck, N., Kates, R.E., Gauger, K., Willems, A., Kiechle, M., Magdolen, V., and Schmitt, M. (2004) Urokinase-type plasminogen activator (uPA) and its inhibitor PAI-I: novel tumor-derived factors with a high prognostic and predictive impact in breast cancer. *Thromb. Haemost.*, **91**, 450–456.
45. Schmalfeldt, B., Kuhn, W., Reuning, U., Pache, L., Dettmar, P., Schmitt, M., Jänicke, F., Hofler, H., and Graeff, H. (1995) Primary tumor and metastasis in ovarian cancer differ in their content of urokinase-type plasminogen activator, its receptor, and inhibitors types 1 and 2. *Cancer Res.*, **55**, 3958–3963.
46. Luther, T., Magdolen, V., Albrecht, S., Kasper, M., Riemer, C., Kessler, H., Graeff, H., Müller, M., and Schmitt, M. (1997) Epitope-mapped monoclonal antibodies as tools for functional and morphological analyses of the human

urokinase receptor in tumor tissue. *Am. J. Pathol.*, **150**, 1231–1244.

47. Riisbro, R., Christensen, I.J., Piironen, T., Greenall, M., Larsen, B., Stephens, R.W., Han, C., Høyer-Hansen, G., Smith, K., Brünner, N., and Harris, A.L. (2002) Prognostic significance of soluble urokinase plasminogen activator receptor in serum and cytosol of tumor tissue from patients with primary breast cancer. *Clin. Cancer Res.*, **8**, 1132–1141.
48. Ferlay, J., Autier, P., Boniol, M., Heanue, M., Colombet, M., and Boyle, P. (2007) Estimates of the cancer incidence and mortality in Europe in 2006. *Ann. Oncol.*, **18**, 581–592.
49. Morris, M., Platell, C., McCaul, K., Millward, M., van Hazel, G., Bayliss, E., Trotter, J., Ransom, D., and Iacopetta, B. (2007) Survival rates for stage II colon cancer patients treated with or without chemotherapy in a population-based setting. *Int. J. Colorectal Dis.*, **22**, 887–895.
50. Gill, S., Loprinzi, C.L., Sargent, D.J., Thome, S.D., Alberts, S.R., Haller, D.G., Benedetti, J., Francini, G., Shepherd, L.E., Francois, S.J., Labianca, R., Chen, W., Cha, S.S., Heldebrant, M.P., and Goldberg, R.M. (2004) Pooled analysis of fluorouracil-based adjuvant therapy for stage II and III colon cancer: who benefits and by how much? *J. Clin. Oncol.*, **22**, 1797–1806.
51. Mamounas, E., Wieand, S., Wolmark, N., Bear, H.D., Atkins, J.N., Song, K., Jones, J., and Rockette, H. (1999) Comparative efficacy of adjuvant chemotherapy in patients with Dukes' B versus Dukes' C colon cancer: results from four National Surgical Adjuvant Breast and Bowel Project adjuvant studies (C-01, C-02, C-03, and C-04). *J. Clin. Oncol.*, **17**, 1349–1355.
52. Figueredo, A., Charette, M.L., Maroun, J., Brouwers, M.C., and Zuraw, L. (2004) Adjuvant therapy for stage II colon cancer: a systematic review from the Cancer Care Ontario Program in evidence-based care's gastrointestinal cancer disease site group. *J. Clin. Oncol.*, **22**, 3395–3407.
53. Piironen, T., Laursen, B., Pass, J., List, K., Gårdsvoll, H., Ploug, M., Danø, K., and Høyer-Hansen, G. (2004) Specific immunoassays for detection of intact and cleaved forms of the urokinase receptor. *Clin. Chem.*, **50**, 2059–2068.
54. Stephens, R.W., Nielsen, H.J., Christensen, I.J., Thorlacius-Ussing, O., Sørensen, S., Danø, K., and Brünner, N. (1999) Plasma urokinase receptor levels in patients with colorectal cancer: relationship to prognosis. *J. Natl. Cancer Inst.*, **91**, 869–874.
55. Parkin, D.M., Bray, F., Ferlay, J., and Pisani, P. (2005) Global cancer statistics, 2002. *CA Cancer J. Clin.*, **55**, 74–108.
56. Pedersen, H., Grøndahl-Hansen, J., Francis, D., Østerlind, K., Hansen, H.H., Danø, K., and Brünner, N. (1994) Urokinase and plasminogen activator inhibitor type 1 in pulmonary adenocarcinoma. *Cancer Res.*, **54**, 120–123.
57. Pappot, H. (1999) The plasminogen activation system in lung cancer--with special reference to the prognostic role in "non-small cell lung cancer". *APMIS Suppl.*, **92**, 1–29.
58. Knoop, A., Andreasen, P.A., Andersen, J.A., Hansen, S., Laenkholm, A.V., Simonsen, A.C., Andersen, J., Overgaard, J., and Rose, C. (1998) Prognostic significance of urokinase-type plasminogen activator and plasminogen activator inhibitor-1 in primary breast cancer. *Br. J. Cancer*, **77**, 932–940.
59. Offersen, B.V., Pfeiffer, P., Hamilton-Dutoit, S., and Overgaard, J. (2001) Patterns of angiogenesis in nonsmall-cell lung carcinoma. *Cancer*, **91**, 1500–1509.
60. Almasi, C.E., Høyer-Hansen, G., Christensen, I.J., Danø, K., and Pappot, H. (2005) Prognostic impact of liberated domain I of the urokinase plasminogen activator receptor in squamous cell lung cancer tissue. *Lung Cancer*, **48**, 349–355.
61. Pedersen, H., Brünner, N., Francis, D., Østerlind, K., Rønne, E., Hansen, H.H., Danø, K., and Grøndahl-Hansen, J. (1994) Prognostic impact of urokinase, urokinase receptor, and type 1 plasminogen activator inhibitor in squamous and large cell lung cancer tissue. *Cancer Res.*, **54**, 4671–4675.
62. Bhoola, S. and Hoskins, W.J. (2006) Diagnosis and management of epithelial

ovarian cancer. *Obstet. Gynecol.*, **107**, 1399–1410.

63. Naora, H. and Montell, D.J. (2005) Ovarian cancer metastasis: integrating insights from disparate model organisms. *Nat. Rev. Cancer*, **5**, 355–366.

64. Konecny, G., Untch, M., Pihan, A., Kimmig, R., Gropp, M., Stieber, P., Hepp, H., Slamon, D., and Pegram, M. (2001) Association of urokinase-type plasminogen activator and its inhibitor with disease progression and prognosis in ovarian cancer. *Clin. Cancer Res.*, **7**, 1743–1749.

65. Rosen, D.G., Wang, L., Atkinson, J.N., Yu, Y., Lu, K.H., Diamandis, E.P., Hellstrom, I., Mok, S.C., Liu, J., and Bast, R.C. Jr. (2005) Potential markers that complement expression of CA125 in epithelial ovarian cancer. *Gynecol. Oncol.*, **99**, 267–277.

66. Nordengren, J., Fredstorp, L.M., Bendahl, P.O., Brünner, N., Ferno, M., Hogberg, T., Stephens, R.W., Willen, R., and Casslen, B. (2002) High tumor tissue concentration of plasminogen activator inhibitor 2 (PAI-2) is an independent marker for shorter progression-free survival in patients with early stage endometrial cancer. *Int. J. Cancer*, **97**, 379–385.

67. Cozzi, P.J., Wang, J., Delprado, W., Madigan, M.C., Fairy, S., Russell, P.J., and Li, Y. (2006) Evaluation of urokinase plasminogen activator and its receptor in different grades of human prostate cancer. *Hum. Pathol.*, **37**, 1442–1451.

68. Almasi, C.E., Christensen, I.J., Høyer-Hansen, G., Danø, K., Pappot, H., Dienemann, H., and Muley, T. (2011) Urokinase receptor forms in serum from non-small cell lung cancer patients: relation to prognosis. *Lung Cancer.*, **74**, 510–515.

14
Clinical Relevance of MMP and TIMP Measurements in Cancer Tissue

Omer Bashir, Jian Cao, and Stanley Zucker

14.1
Introduction

Tadpole (interstitial) collagenase, the first matrix metalloproteinase family member identified (renamed MMP-1), was discovered in 1962 in experiments designed to explain how the collagen-rich tail of the frog is resorbed during metamorphosis. The focus of MMPs for the next four decades was on the degradation of extracellular matrix (ECM) components (collagen, laminin, fibronectin, etc.). Following the identification of increased cancer cell levels of MMP-2 and MMP-9, which digest type IV collagen, the key component of basement membranes, these and other MMPs were implicated in cancer-related processes including invasion, metastasis, angiogenesis, and cell proliferation. Although several MMPs (MMP-1, -2, -7, -9, -11, -14) have been readily detected in most tumor types, the pattern of expression of other MMPs (MMP-3, -8, -13) varies considerably [1]. The saga continues to unfold with numerous other functions of MMPs being continually discovered.

In contrast to classical oncogenes, MMPs are not upregulated by gene amplification or activating mutations in cancer cells. The increase in MMP expression in tumors is probably due to transcriptional changes rather than genetic alteration. The concept that measurement of MMPs in tissues and biological fluids could be used as diagnostic tools for disease was a natural outgrowth of attribution of MMPs with disease-producing properties.

The goal of this chapter is to summarize the vast amount of literature that has accumulated on the subject of tumor MMPs and tissue inhibitors of metalloproteinases (TIMPs) as clinical diagnostic and prognostic tools in cancer. On the basis of numerous earlier observations describing MMPs and TIMPs in cancer tissue specimens, followed by the recent emphasis on analyzing MMP measurements as independent variables, this chapter focuses on more recent clinical studies employing univariate and multivariate statistical analyses. It is hoped that this compilation will help to clarify important variations in MMPs in different tumor types, leading to a more rational design of future clinical studies addressing therapeutic decisions based on MMP and TIMP expression in human tumors.

Matrix Proteases in Health and Disease, First Edition. Edited by Niels Behrendt.

14.2
MMP Structure

MMPs are a family of Zn^{2+}-dependent proteins and peptide hydrolases; 24 paralogs have been identified. These enzymes were initially named on the basis of their preferred substrate and subsequently based on an MMP numbering system in order of discovery. MMPs contain the following structural domains: (i) a signal peptide that directs MMPs to the secretory pathway; (ii) a prodomain that confers latency to the enzymes and is configured as a triple-helix globular domain; and (iii) a zinc-containing catalytic domain that is compact and spherical, divided by a shallow substrate-binding crevice into an upper and a lower subdomain. The catalytic domain has an extended zinc-binding motif, HEXXHXXGXXH, which contains three zinc-binding histidines and a glutamate that activates a zinc-bound H_2O molecule, providing the nucleophile that cleaves peptide bonds; and (iv) a hemopexin domain with a characteristic three-dimensional disc-like β-propeller structure, which mediates interactions with substrates and confers specificity to the enzymes [1].

Most MMPs are secreted as inactive zymogens with a cysteine embedded in a conserved PRCGXPD within the prodomain, folding into and inhibiting the catalytic zinc. Removal of the prodomain permits access of the substrate molecule to the active site cleft, which harbors a hydrophobic pocket as the main determinant of enzyme specificity [1].

A breakthrough in understanding the role of MMPs in cancer came with the discovery that cell-surface-bound MMPs display enhanced function in the pericellular environment [2] and later with the discovery of a group of intrinsic plasma-membrane-type matrix metalloproteinases (MT-MMPs). MT1-MMP (MMP-14), MT2-MMP (MMP-15), MT3-MMP (MMP-16), and MT5-MMP (MMP-24) contain short transmembrane and cytoplasmic domains [3]; MT4-MMP (MMP-17) and MT6-MMP (MMP-25) are tethered to the plasma membrane via a glycosylphosphatidyl inositol linkage.

14.3
MMP Biology and Pathology

Following secretion from cells, MMPs have been demonstrated to participate in many physiologic processes including tissue turnover and repair during blastocyst implantation, ovulation, postlactational involution, and bone resorption. A pathological role for MMPs in numerous organ systems has been proposed in arthritis, non-healing wounds, aortic aneurysms, congestive heart failure, brain injury, and degenerative diseases. In addition to ECM proteins (collagens, laminin, fibronectin, etc.), MMP substrates include other proproteases, protease inhibitors, antimicrobial peptides, clotting factors, chemotactic and adhesion molecules, hormones, growth factors, angiogenic factors, regulators of immunity, apoptosis, and cytokines as well as their receptors and binding proteins [4].

14.4
Natural Inhibitors of MMPs

Despite the presence of several natural inhibitors, the physiologic mechanisms that activate and inactivate MMPs in the pericellular environment are incompletely understood [5]. TIMP-1, -2, -3, and -4 comprise a four-member family of homologous MMP inhibitors that inhibit the activation and function of most MMPs [6]. The major exception is that TIMP-1 is relatively ineffective in inhibiting MT-MMPs. In addition to binding to the catalytic site of activated MMPs, TIMP-1 naturally binds to the hemopexin domain (PEX) domain of pro-MMP-9 and TIMP-2 binds to the PEX domain of pro-MMP-2 [7]. TIMPs have also been shown to have growth-promoting activities, independent of their MMP inhibitory function, which might help explain the correlation between elevated TIMP-1 levels and poor prognosis in cancer. The concentration of TIMPs in tissue and extracellular fluids generally far exceed the concentration of MMPs, thereby tending to limit proteolytic activity to focal pericellular sites [8].

14.5
Regulation of MMP Function

In vivo activity of MMPs is under tight control at several levels including gene expression, compartmentalization, proenzyme activation, inhibition by protease inhibitors, and endocytosis [9]. Most members of the MMP family share common cis-acting elements in their promoters [10]. As a result, they are often coexpressed in response to inductive stimuli or corepressed by inhibitors of gene expression.

14.5.1
MMPs in Cancer

Cancer progression is recognized to be a complex, multistage process in which the transformation from normal to malignant cells involves genetic changes that lead to numerous phenotypic alterations. Although serine, cysteine, aspartic acid, and metalloproteinases have been implicated in the invasive process, MMPs appear to exert the dominant effect and have been implicated in virtually all aspects of cancer progression and dissemination.

Hundreds of studies in experimental animals have demonstrated (i) that cancer progression (invasion and metastasis) correlates with enhanced production and secretion of MMPs by tumor cells and/or stromal cells; (ii) *in vivo* reduction of invasion and metastasis using natural (TIMPs) and synthetic MMP inhibitors, neutralizing antibodies, or antisense oligonucleotides; (iii) modulation of the invasive properties of cancer cells by transfection with the cDNA of MMPs and TIMPs; and (iv) alteration of tumor progression in mice endowed with genetically modified MMP production. In spite of this voluminous literature, the question remains: does the presence of high tumor concentrations of MMPs mean that they have a causative role in cancer progression/dissemination? As has been repeatedly stated, "correlation does not necessarily prove causation." It is still uncertain whether elevated levels of MMPs in tissues result in more functional activity of these proteinases as a critical component of cancer dissemination; molecular layers of complexity continue to baffle simplistic explanations of this process [11].

Another important problem in deciphering the mechanisms involved in cancer dissemination is that only a small minority of carcinoma cells in the primary tumors appear to undergo the alteration required for metastasis, for example, epithelial-to-mesenchymal transition (EMT), in the cancer stem cell population; thus, alteration of gene expression in such cells can be masked by the bulk of nonmetastatic cells. Hence, the transcriptomic and proteomic contribution of a minor population would be diluted by the whole, which may negate the utility of tissue measurements of MMPs for clinical assessment of prognosis or treatment-related decisions.

14.6 Cancer Stromal Cell Production of MMPs

The concept that cancer cells were solely responsible for producing MMPs in human tumors underwent revision when Basset *et al.* [12] employing *in situ* hybridization technology, reported that stromal fibroblasts within tumors, not the tumor cells themselves, were responsible for the production of MMP-11 in human breast cancer. Other studies demonstrated that the localization of MMP-1, -2, -3, -9, and MMP-14 mRNA was primarily in fibroblasts, especially in proximity to invading cancer cells in breast, colorectal, lung, prostate, ovarian, and head and neck cancers [13]. However, contrary reports described the localization of MMP mRNA in pancreatic, prostate, and brain carcinoma cells, rather than in fibroblasts within human tumor specimens [14].

Immunohistochemical (IHC) examinations of human cancer specimens have also reported the localization of MMP protein in both cancer cells and stromal cells. An explanation for the production of MMPs by reactive stromal cells in tumors came from the seminal discovery by Biswas *et al.* of extracellular matrix metalloproteinase inducer (EMMPRIN, CD147). EMMPRIN, a plasma membrane glycoprotein prominently displayed in all epithelial cancer cells, stimulates peritumoral fibroblasts to synthesize MMP-1, -2, -3, -9, and -14 [15].

14.7 Anticancer Effects of MMPs

Contrary to the dogma, some MMPs exert anticancer effects. Expression of these MMPs (MMP-8 and -12) provides a protective effect in cancer progression [16]. The capacity of MMP-12 and -9 to cleave plasminogen and generate angiostatin, which is a powerful inhibitor of tumor angiogenesis in mouse cancer models, provides an example of multipurpose proteases that display antitumor effects.

14.8
Tissue Levels of MMPs and TIMPs in Cancer Patients

A voluminous literature has reported increased tissue levels of latent and activated MMPs and TIMPs in different types of cancers and proposed potential usefulness of these measurements in the management of patients with cancer. It is critical to recognize, however, that few, if any, publications in the field have met the guidelines (*http://www.nature.com/nrclinonc/journal/v2/n8/pdf/ncponc0252.pdf*) required to establish MMPs as clinical tumor markers.

The remainder of this chapter focuses on recent developments in the use of MMPs and TIMPs as biomarkers in cancer.

14.8.1
Breast Cancer

Five groups of investigators [17–21], employing different techniques, have reported a correlation between high tumor tissue levels of TIMP-1 (tumor cells or stromal cells) and poor prognosis in breast cancer by multivariate analysis (Table 14.1). In a subgroup receiving adjuvant chemotherapy, high TIMP-1 expression correlated with shorter survival [19]. Nakopoulou *et al.* [17] demonstrated that TIMP-1 mRNA identified in the cytoplasm of stromal cells at the tumor invasive margin was associated with worse prognosis. Liss *et al.* [22] reported that high TIMP-4 levels were an independent prognostic marker for poor survival in patients with invasive ductal cancer. In contrast to TIMP-1 and -4 data, Kotzsch *et al.* [23] reported that TIMP-3 mRNA expression levels in breast cancer specimens were inversely related to prognosis; hence, patients with low tissue TIMP-3 levels had a sixfold higher mortality.

The relevance of tumor tissue measurements of MMPs in breast cancer survival has been controversial. Wu *et al.* [18] discussed the prognostic significance of MMP-9 in patients with breast cancer; before surgery. High tissue levels of MMP-9 was associated with lymph node metastasis and higher tumor grade, but not prognosis. Mylona *et al.* [26] evaluated MMP-9 in patients with invasive breast cancer. Stromal MMP-9 was more often observed in tumors expressing c-ERB-B2 and was identified as an independent variable with poor prognosis. Jobim *et al.* [24] examined the prevalence of MMP-2, MMP-9, TIMP-1, and TIMP-2 in tumor samples obtained from patients with clinically negative axillary lymph node and no distant metastasis. No correlation was found between MMP-2, MMP-9, TIMP-1, and TIMP-2 and the presence of tumor cells in sentinel lymph nodes. Shah *et al.* [25] employed gelatin zymography to assess MMP-2 and -9 in patients with stage 1–4 breast cancer. Multivariate analysis revealed a significant correlation between pro-MMP-2, total MMP-2, and activated MMP-2 in adjacent normal

Table 14.1 Breast cancer tissue MMPs and TIMPs before treatment: correlation with prognosis.

Author	Number of patients	MMP	TIMP	Assay method	Analysis	Poor prognosis
Wu *et al.* [18]	60	9 in tumor	1 in tumor	IHC, ISH, ELISA	Multivariate	High TIMP-1
Schrohl *et al.* [19]	2984	n.d.	1 in tumor cells	ELISA	Multivariate	High TIMP-1
Schrohl *et al.* [20]	525	n.d.	1 in tumor	ELISA	Multivariate	High TIMP-1 in chemo patients
Wurtz *et al.* [21]	341	n.d.	1 in invasive margin	ELISA	Multivariate	High TIMP-1
Nakopoulou *et al.* [17]	117	n.d.	1 in stromal cells	ISH/IHC	Multivariate	High TIMP-1
Kotzsch *et al.* [23]	205	n.d.	3 in tissue	PCR	Multivariate	Low TIMP-3
Liss *et al.* [22]	183	n.d.	4 in tissue	IHC	Multivariate	High TIMP-4
Jobim *et al.* [24]	95	2, 9	1, 2 in tissue	IHC	n.d.	n.d.
Shah *et al.* [25]	49	2, 9	n.d.	Zymo	n.d.	n.d.
Mylona *et al.* [26]	175	1, 9	n.d.	IHC	Multivariate	High stroma MMP-9
Tetu *et al.* [27]	539	2, 14 in stroma	2 in tumor and stroma	ISH, IHC	Multivariate	High MMP-14
Talvensaari- Mattila *et al.* [28]	453	2	n.d.	IHC	Multivariate	High MMP-2
Del Casar *et al.* [29]	124	1, 2, 7, 9, 13, 14 in stroma	1, 2, 3 in stroma	IHC tissue array	Multivariate	MMPs and TIMPs in fibroblasts
Rahko *et al.* [30]	178	9	1	IHC	n.d.	n.d.

n.d., not done; ISH, *in situ* hybridization; zymo, gelatin substrate zymography.

tissue and lymph node metastasis, but no correlation with survival was noted. Talvensaari-Mattila *et al.* [28] reported that MMP-2 protein in tumor cell cytoplasm was associated with a modestly shortened recurrence-free survival and overall survival.

Employing three different monoclonal antibodies directed at different epitopes, Chenard *et al.* [31] identified high levels of MMP-14 in peritumoral fibroblastic cells in invasive breast cancers, rather than in tumor cells themselves. Tetu *et al.* [27] evaluated MMP-14, MMP-2, and TIMP-2 levels and breast cancer prognosis in patient samples. MMP-2 and MMP-14 mRNA were detected primarily in stromal cells, whereas TIMP-2 mRNA was expressed in both stromal and cancer cells. High MMP-14 mRNA alone predicted for significantly shorter overall survival.

Other recent studies have emphasized the complexity of MMP and TIMP measurements in breast cancer. Del Casar *et al.* [29] reported an IHC study on tissue arrays from patients with invasive breast cancer. MMP-1, -2, -7, -9, -11, -13, -14, and TIMP-1, -2, and -3 were identified in fibroblasts at the center of the tumor and at the invasive tumor front. Multivariate analysis showed that a high profile of MMP and TIMP staining in both fibroblast populations was the most potent predictive factor of distant metastases.

Other studies have focused on MMP and TIMP measurements in early and preneoplastic breast cancer. Rahko *et al.* [30] noted higher levels of MMP-9 and TIMP-1 in patients with ductal carcinoma *in situ* than in invasive cancers, suggesting that overexpression of MMP-9 and TIMP-1 are early markers of breast carcinogenesis preceding tumor invasion. On the basis of the fact that women with atypical ductal hyperplasia of the breast had a five times higher incidence of developing invasive breast cancer, Poola *et al.* [32] employed gene expression analysis and IHC to demonstrate that MMP-1 is a candidate marker for the identification of breast lesions that can develop into cancer.

14.8.2
Gastrointestinal (GI) Cancer

14.8.2.1 Colorectal Cancer

Our previous review of colorectal cancer implicated MMP-9, -7, -1, -13, and MT1-MMP in colorectal cancer progression [33]. A correlation was demonstrated between IHC detection of MT1-MMP/MMP-2 and both the depth of tumor invasion and vascular invasion in human colon cancer, but not with distant metastases.

In our current update of the literature, five of six studies of patients with colorectal cancers evaluating tumor MMP-9 reported that patients with high tumor levels of MMP-9 had diminished survival (Table 14.2). Svagzdys *et al.* [34] evaluated MMP-2, MMP-9, and TIMP-2 and TIMP-3 in patients with rectal carcinoma before surgery; each of the antigens was identified in both tumor and stromal cells. Multivariate analysis showed that high expression of MMP-9 in tumor cells and low expression of MMP-9 in stromal cells were associated with increased mortality. Bendardaf *et al.* [35] evaluated tumor MMP-9 in patients with stage 1–4 colorectal

Table 14.2 Colorectal cancer tissue MMPs and TIMPs: correlation with prognosis.

Author	Number of patients	MMP	TIMP	Assay method	Analysis	Poor prognosis
Svagzdys *et al.* [34]	64	2, 9 in tumor and stroma	2, 3 in tumor and stroma	IHC	Multivariate	High MMP-9 in tumor
Bendardaf *et al.* [35]	359	9 in tumor	n.d.	IHC	Multivariate	High MMP-9
Cho *et al.* [36]	338	9 in tumor	n.d.	IHC	Multivariate	High MMP-9
Buhmeida *et al.* [37]	202	9 in tumor	n.d.	IHC	Multivariate	High MMP-9
Sutnar *et al.* [38]	40	7, 9 in liver metastasis	1, 2 in liver metastasis	PCR	Multivariate	High MMP-9 High TIMP-1
Roca *et al.* [39]	84	9, 7 in tumor cells	1, 2 in tumor	IHC	Multivariate	High TIMP-2
Sundov *et al.* [40]	152	2 in tumor	n.d.	IHC	Multivariate	High MMP-2
Asano *et al.* [41]	112	1, 10, 11, 15, 19 in tumor	1 in tumor	PCR	Multivariate	low MMP-15
Koskensalo *et al.* [42]	643	7 in tumor	n.d.	IHC	Univariate Multivariate	No correlation
Hilska *et al.* [43]	351	1, 2, 7, 13 in tumor and stroma	1, 2, 3, 4 in tumor and stroma	IHC	Multivariate	High TIMP-3 in stroma

n.d., not done.

cancer. Multivariate analysis showed MMP-9 to be an independent predictor of disease-free survival (defined as the length of time after treatment during which a patient survives with no sign of cancer), but not disease-specific survival (defined as the percentage of people in a study who have not died from cancer in a defined period of time). Bendardaf had previously reported that low MMP-1 and -14 expression levels were favorable survival markers in advanced colorectal cancer. Cho *et al.* [36] evaluated tumor MMP-9 in surgical specimens in patients with lymph node negative, large colorectal or locally invasive colorectal cancers before adjuvant chemotherapy. Multivariate analysis showed that high MMP-9 was predictive of shortened disease-free survival. Buhmeida *et al.* [37] evaluated tumor and normal tissue MMP-9 in untreated patients with stage 2 colorectal cancer. Multivariate analysis showed MMP-9 to be an independent predictor of shortened disease-free survival, but not disease-specific survival. Sutnar *et al.* [38] evaluated MMP-7, MMP-9, TIMP-1, and TIMP-2 by PCR in patients with colorectal cancer and liver metastases. High mRNA expression of MMP-9 and TIMP-1 correlated with shortened disease-free interval and overall survival. Islekel *et al.* [44] used gelatin zymography to demonstrate that the ratio of active pro-MMP-9 was also increased in colorectal cancer tissue. Roca *et al.* [39] evaluated MMP-7, MMP-9, and TIMP-1, TIMP-2 in previously untreated patients with colorectal carcinoma. All MMPs and TIMPs were overexpressed in tumor cells as compared to normal cells, but on multivariate analysis only TIMP-2 overexpression remained a significant prognostic factor for poor survival.

Sundov *et al.* [40] evaluated tumor MMP-2 in patients with stage 2 colon cancer. Multivariate analysis showed that high MMP-2 expression correlated with poor prognosis. Asano *et al.* [41] evaluated several MMPs and TIMP-1; all markers were increased in tumor tissue as compared to normal tissue. MMP-1, MMP-10, and MMP-11 mRNA were statistically higher in the primary lesions than in the metastatic liver lesions; the expression of TIMP-1 was higher in the metastases. MMP-15 expression was shown to be a significant independent prognostic predictor of survival by multivariate analysis. Koskensalo *et al.* [42] evaluated MMP-7 as a prognostic marker in patients with stage 1–4 colorectal cancer. High MMP-7 was associated with poor five-year survival, but during longer follow up the difference disappeared. Hilska *et al.* [43] evaluated MMP-1, -2, -7, and -13 and TIMP-1 to -4 in patients with stage 1–4 colorectal cancer. Multivariate analysis showed that high TIMP-3 in stroma was associated with shorter survival in the entire group of colorectal cancer patients.

14.8.2.2 **Gastric Cancer**

In contrast to increased levels of MMP-9 in lower gastrointestinal (GI) cancer, MMP-9 was not generally identified as a prognostic tumor marker in upper GI cancer (Table 14.3). Kubben *et al.* [45] evaluated MMP-2, -7, -8, and -9 and TIMP-1 and -2 by ELISA and substrate zymography in gastric tumor tissue as compared to normal tissue. Of these factors, only MMP-2 was identified as an independent prognostic factor by multivariate analysis. Mrena *et al.* [46] evaluated presurgical tissue MMP-2 and -9, both in tumor and stroma. Epithelial, but not stromal, MMP-2

Table 14.3 Gastric and pancreatic cancer MMPs and TIMPs: correlation with prognosis.

Author	Number of patients	MMP	TIMP	Assay method	Analysis	Poor prognosis
Kubben *et al.* [45] Gastric cancer	81	2, 9, 7, 8, in tumor	1, 2 in tumor	Zymo, ELISA, BIA	Multivariate	High MMP-2
Mrena *et al.* [46] Gastric cancer	342	2, 9 in tumor and stromal	n.d.	IHA	Univariate Multivariate	High MMP-2 n.a.
Alakus *et al.* [47] Gastric cancer	116	2 in tumor	2 in tumor	IHA	Univariate Multivariate	High MMP-2 n.a.
Fujimoto *et al.* [49] Gastric cancer	129	1 in tumor	n.d.	IHA	Univariate Multivariate	High MMP-1 n.a.
Chu *et al.* [50] Gastric cancer	286	9 in tumor	n.d.	IHA	Multivariate	High MMP-9
Koskensalo *et al.* [51] Gastric cancer	264	7 in tumor	n.d.	IHA	Univariate	High MMP-7
Mori *et al.* [48] Gastric cancer	68	14 in tumor cells	n.d.	RT-PCR IHC	Multivariate T/N ratio	High MMP-14
Giannopoulos *et al.* [52] Pancreatic cancer	49	2, 9 in tumor	2 in tumor and stroma	IHC	Univariate Multivariate	Low TIMP-2 n.a.
Jones *et al.* [53] Pancreatic cancer	103	1, 3, 7, 9, 2, 8, 11, 12, 14 in tumor	1, 2, 3 in tumor	IHC, PCR	Univariate Multivariate	High MMP-11 High MMP-7
Juuti *et al.* [54] Pancreatic cancer	127	2 in tumor	n.d.	IHC	Univariate Multivariate	High MMP-2 n.a.

n.d., not done; n.a., no association; zymo, gelatin substrate zymography; BIA, bioactivity assay; T/N., Ratio of tumor/normal tissue.

was associated with poor prognosis by univariate, but not multivariate analysis. Alakus *et al.* [47] reported that high MMP-2, but not TIMP-2, was associated with tumor stage and poor survival by univariate, but not multivariate analysis. Mori *et al.* [48] reported a significant correlation between high tumor versus normal tissue expression of MMP-14 and poor prognosis. Fujimoto *et al.* [49] reported that presurgical patients with high tumor MMP-1 and PAR-1 had poor survival by univariate, but not multivariate analysis. Chu *et al.* [50] reported that MMP-9 correlated with metastasis and was an independent prognostic factor for both overall and disease-free survival. Koskensalo *et al.* [51] reported a correlation in gastric cancer between tumor MMP-7 and prognosis by univariate, but not multivariate analysis.

14.8.2.3 Pancreatic Cancer

Giannopoulos *et al.* [52] evaluated MMP-2, MMP-9, and TIMP-2 (tumor and stroma) in pancreatic ductal and ampullary cancer (Table 14.3). In pancreatic ductal carcinoma, lower levels of glandular TIMP-2 were found in poorly differentiated cancer; an association with poor survival was noted by univariate, but not multivariate analysis. Jones *et al.* [53] evaluated MMP-1, -2, -3, -7, -8, -9, -11, -12, and MT1-MMP and TIMP-1, -2, and -3. Increased tumor expression of MMP-7, -8, -9, and -11 and TIMP-3 was noted as compared to normal tissue. On multivariate analysis, only MMP-7 predicted shorter survival. Juuti *et al.* [54] reported a correlation between high levels of epithelial MMP-2 and poor survival by univariate, but not multivariate analysis; no correlation was observed with stromal MMP-2.

14.8.2.4 Non-Small-Cell Lung Cancer (NSCLC)

Six reports described the use of IHC to assess tissue MMP and TIMP levels in patients with stage I–III non-small-cell lung cancer (NSCLC) before undergoing surgery [55–60] (Table 14.4). Most reports focused on MMP-2, MMP-9, and TIMP-1; distinction between MMP staining in tumor cells versus stromal cells was noted in some reports. Ishikawa *et al.* [55] reported that strong tumor cell MMP-2 expression was more frequent in adenocarcinomas and stromal MMP-2 expression was more frequent in squamous cell carcinoma (SCC). Multivariate analysis demonstrated stromal MMP-2 status to be an independent and significant prognostic factor only in early SCC, but not adenocarcinoma. There was no correlation between postoperative survival and MMP-2 or -9 status within tumor cells. Shou *et al.* [56] reported that tissue MMP-2 and -9 levels in NSCLC were significantly associated with poor survival. Multivariate analysis showed that MMP-2, but not MMP-9, was an independent characteristic related to tumor size and pathologic stage. Leinonen *et al.* [57] reported that high expression of MMP-2 in either tumor cells or stromal cells correlated with increased risk of tumor recurrence and worse prognosis (multivariate analysis). Passlick *et al.* [58] reported a relationship between high MMP-2 and poor prognosis that held up under multivariate regression analysis. Aljada *et al.* [59] evaluated MMP-2, MMP-9, and TIMP-1. Multivariate analysis revealed that TIMP-1 expression in tumor cells, but not MMP-2 or -9, correlated with poor survival.

Table 14.4 Lung (stages 1–3) cancer tissue MMPs and TIMPs before treatment: correlation with prognosis.

Author	Number of patients	MMP	TIMP	Assay method	Analysis	Poor prognosis
Ishikawa *et al.* [55]	218	2 in adeno cells Stromal 9 in SCC	n.d.	IHC	Multivariate	High stromal MMP-2 in early SCC
Shou *et al.* [56]	111	2, 9	n.d.	IHC	Univariate Multivariate	High MMP-9 High MMP-2
Leinonen *et al.* [57]	212	2 in stromal and tumor cells	n.d.	IHC	Multivariate	High MMP-2 in tumor cells and stromal cells
Passlick *et al.* [58]	193	2 in adeno and SCC	n.d.	IHC	Multivariate	High MMP-2 in node negative
Aljada *et al.* [59]	160	2, 9	1 in tumor cells	IHC	Multivariate	High TIMP-1
Mino *et al.* [60]	143	2, 9	TIMP-3 high in SCC	IHC	Multivariate	High MMP-2 low TIMP-3
Gouyer *et al.* [61]	116 pre-RT	1, 9	1, 2 in tumor	Northern blot	Multivariate	High TIMP-1
Simi *et al.* [62]	100	9 in adeno	1 in adeno and SCC	PCR	Multivariate	High MMP-9

n.d., not done; adeno, adenocarcinoma; SCC, squamous cell carcinoma.

Mino *et al.* [60] reported that high cancer tissue TIMP-3 expression in NSCLC patients who underwent complete resection was independently correlated by multivariate analysis with less nodal involvement/decreased pathological stage/better five-year survival. TIMP-3 was higher in patients with SCC as compared to adenocarcinoma. In a study limited to MMP-7, Liu *et al.* [63] investigated tumor samples from patients with stage IA–IIIB lung cancer. MMP-7 was significantly higher in SCC than in adenocarcinoma and was an independent factor predictive of survival by multivariate analysis.

Gouyer *et al.* [61] performed a study of patients before treatment with radiotherapy. MMP-1, MMP-9, and TIMP-1 levels, measured by Northern blot analysis, were significantly higher and TIMP-2 was lower in tumor as compared to normal lung tissues. Although MMP-1 and -9 showed a correlation with tumor size and stage, multivariate regression analysis failed to support a correlation with survival. In contrast, multivariate regression analysis demonstrated that TIMP-1 mRNA expression was an independent prognostic predictor for shortened survival. Shah *et al.* [64] employed multiplex array immunologic assays to compare MMP-1, -2, -3, -7, -8, -9, -12, and -13 levels in lung cancer tissue extracts. As compared to normal levels, MMP-1, -8, -9, and -12 were increased in SCC, but not in adenocarcinomas. Patients with high MMP-8 and -9 levels had a significantly higher recurrence rate (univariate analysis). Simi *et al.* [62] measured both tissue MMP-9 and TIMP-1 by RT-PCR in lung cancer patients. MMP-9 mRNA was significantly increased in adenocarcinoma, but not in SCC. Multivariate analysis showed MMP-9 to be independently associated with shortened survival in patients with operable lung cancer. TIMP-1 was increased both in adenocarcinoma and SCC tumors. Atkinson *et al.* [65] studied the relationship between MT-MMPs and by qRT-PCR in lung cancer tissues; clinical correlation was not examined. MT1-MMP was definitively expressed in 100% of lung cancers in comparison to low levels of detection in normal tissue.

14.8.3
Genitourinary Cancers

14.8.3.1 Bladder Cancer

Although several studies have been published [66–69], there is no consensus on the prognostic value of measuring MMPs and TIMPs in patients with bladder, renal, and prostate cancers (Table 14.5). Vasala *et al.* [66] explored the correlation between tissue MMP-9 and MMP-2 expression and prognosis. IHC overexpression of MMP-2 correlated with reduced disease-specific survival; multivariate analysis was not assessed in this study. Contrary to all prior studies in other types of cancers, multivariate analysis demonstrated that high MMP-9 expression predicted better prognosis [67]. Szarvas *et al.* [68] evaluated MMP-7 in patients with bladder cancer. Measuring qRT-PCR of tissue samples, revealed a 21-fold increase in mRNA expression in patients with metastatic bladder cancer compared to those without metastases. Multivariate analysis revealed that high tissue expression of MMP-7 in patients is a stage- and grade-independent predictor of decreased

Table 14.5 Bladder cancer tissue MMPs and TIMPs before treatment: correlation with prognosis.

Author	Number of patients	MMP	TIMP	Assay method	Analysis	Poor prognosis
Vasala *et al.* [66]	54	2 in tumor	n.d.	IHC	Univariate	High MMP-2
Gakiopoulou *et al.* [69]	106	n.d.	2 in tumor and stroma	IHC	Univariate	High TIMP-2
Vasala *et al.* [67]	87	9 in tumor	n.d.	IHC	Multivariate	Low MMP-9
Mohammad *et al.* [70]	50	1,2 in tumor	n.d.	Western blot, zymogram	Whitney–Wallis	High MMP-2 and LN metastasis
Szarvas *et al.* [68]	101	7 in tumor	n.d.	PCR, IHC	Multivariate	High MMP-7
Seargent *et al.* [71]	60	10 in tumor	n.d.	IHC, Western blot	Whitney–Wallis	Negative

n.d., not done; LN, lymph node.

metastasis-free and disease-specific survival. Gakiopoulou *et al.* [69] evaluated the relationship between TIMP-2 in tumor and stroma and prognosis. In univariate, but not multivariate analysis, patients with positive TIMP-2 expression in stromal cells had worse overall survival.

14.8.3.2 **Renal Cancer**

Three publications employing IHC evaluated MMP-2 and MMP-9 and two evaluated TIMP-1 and TIMP-2 in patients with nonmetastatic clear cell renal cancer (Table 14.6). Kawata *et al.* [72] reported that high MMP-9 correlated with nuclear grade, but not directly with survival. Kallakury *et al.* [73] reported that increased expression of MMP-9 and TIMP-1 correlated with poor survival on univariate analysis. On multivariate analysis, only high TIMP-1 was independently associated with poor prognosis. Takahashi *et al.* [74], employing gelatin substrate zymography, reported that the MMP-2 activation ratio and high MMP-9 were significant predictors of poor prognosis by univariate, but not multivariate, analysis. Kamijima *et al.* [75] evaluated the prognostic significance of MMP-2 and MMP-9 in transitional cell carcinoma of the renal pelvis and ureter. Neither MMP-2 nor MMP-9 was found to correlate with survival. Miyata *et al.* [76] evaluated MMP-7 and MMP-10 expression in tumor cells and endothelial cells. High MMP-7 expression in cancer cells was an independent predictor of poor survival. Univariate, but not multivariate, analysis showed that high MMP-10 predicted poor survival [77].

14.8.3.3 **Prostate Cancer**

Boxler *et al.* [78] identified increased MMP-2, -3, -7, -9, -13, and -19 levels in prostate cancer tissue as compared to normal tissue in patients undergoing radical prostatectomy (Table 14.7). Univariate, but not multivariate, analysis, revealed low MMP-9 levels in tumors to be predictive of diminished overall survival; this result is unexpected. Miyake *et al.* [79] evaluated MMP-2 and -9 in prostate cancer. In contrast to Boxler *et al.* [78] high tumor tissue levels of MMP-9 and -2 were identified as significant factors associated with tumor recurrence, but not survival (univariate, but not multivariate, analysis). Trudel *et al.* [80] evaluated MMP-2, MMP-14, and TIMP-2 in prostate tumor tissue. Multivariate analysis showed diminished survival in patients with low TIMP-2 expression in stromal cells. High MMP-2 expression by benign epithelial cells correlated with shorter disease-free survival in multivariate analysis. Ross *et al.* [81] evaluated MMP-2 and TIMP-2 in prostate cancer specimens; high coexpression of MMP-2 and TIMP-2 correlated with advanced tumor stage ($p < 0.05$) and disease recurrence ($p = 0.07$).

14.8.3.4 **Ovarian Cancer**

In ovarian cancer, all six reports evaluating MMP-9 revealed an association between high tissue levels of MMP-9 in either tumor or stromal cells and diminished survival [61–66] (Table 14.8). Survival correlations with MMP-2 and -14 were inconsistent. Lengyel *et al.* [82] studied both MMP-2 and MMP-9 (pro and active forms) in patients with advanced ovarian cancer preoperatively. Active MMP-2 was detected only in malignant tissue. No activated MMP-9 was found in either benign or tumor

Table 14.6 Renal cell cancer (RCC) MMPs and TIMPs before treatment: correlation with prognosis.

Author	Number of patients	MMP	TIMP	Assay method	Analysis	Poor prognosis
Kawata *et al.* [72]	120	2, 9 in tumor	1, 2 in tumor	IHC	Univariate, multivariate	High TIMP-2
Kallakury *et al.* [73]	153	2, 9 in tumor	1, 2 in tumor	IHC	Univariate Multivariate	High MMP-9 High TIMP-1
Takahashi *et al.* [74]	57	2, 9 in tumor	n.d.	Zymo	Univariate Multivariate	High MMP-2, -9 n.a.
Kamijima *et al.* [75]	69	2, 9	n.d.	IHC	Univariate, multivariate	Negative
Miyata *et al.* [76]	165	7 in tumor	n.d.	IHC	Multivariate	High MMP-7
Miyata *et al.* [77]	103	10 in tumor	n.d.	IHC	Multivariate, Univariate	Negative, High MMP-10

n.d., not done; n.a., no association; zymo, gelatin substrate zymography.

Table 14.7 Prostate cancer tissue MMPs and TIMPs: correlation with prognosis.

Author	Number of patients	MMP	TIMP	Assay method	Analysis	Poor prognosis
Boxler *et al.* [78]	278	2, 3, 7, 9, 13, 19 in tumor	n.d.	IHC	Univariate Multivariate	Low MMP-9 n.a.
Miyake *et al.* [79]	193	2, 9 in tumor	n.d.	IHC	Univariate Multivariate	High MMP-2, 9 n.a.
Trudel *et al.* [80]	189	2, 14 in tumor, stroma, and epithelial cells	2 in tumor, stroma, and epithelial cells	IHC	Multivariate	Low TIMP-2 Stromal cells High MMP-2 Epithelial cells
Ross *et al.* [81]	138	2 in tumor	2 in tumor	IHC	Univariate Multivariate	n.a. n.a.

n.d., not done; n.a., no association.

Table 14.8 Tissue MMPs and TIMPs in ovarian cancer: correlation with prognosis.

Author	Number of patients	MMP	TIMP	Assay method	Analysis	Poor prognosis
Lengyel *et al.* [82]	84	2 in tumor 9 in tumor	n.d.	Zymo, IHC	Univariate and multivariate	High pro-MMP-9
Kamat *et al.* [83]	90	2, 9, 14 in tumor, stroma	n.d.	IHC	Univariate and multivariate	High MMP-14 (t, s), high MMP-9 (s)
Alshenawy *et al.* [84]	120	9 in tumor, stroma	n.d.	IHC	Univariate	High MMP-9 (t, s)
Ozalp *et al.* [85]	45	9 in tumor, stroma	n.d.	IHC	Univariate	High MMP-9 (s)
Davidson *et al.* [86]	45	2 in tumor, stroma, 9 in tumor, stroma	2 in tumor, stroma	ISH	Univariate Multivariate	Primary tumor; high TIMP-2, MMP-9; metastasis high TIMP-2, MMP-2 and -14 High TIMP-2 (s) MMP-9 (t)
Sillanpaa *et al.* [87]	292	9 in tumor, stroma	n.d.	IHC	Univariate Multivariate	Low MMP-9 (t) High MMP-9 (s) n.a.
Perigny *et al.* [88]	100	2, 11 in tumor, stroma	n.d.	IHC	Univariate and multivariate	High MMP-2 (t)
Torng *et al.* [89]	84	2, 14 in tumor, stroma	2 tumor, stroma	IHC	Univariate Multivariate	Low MMP-2 (s) n.a.
Sillanpaa *et al.* [90]	284	7 in tumor, stroma	n.d.	IHC	Univariate and multivariate	Low MMP-7 (t)

n.d., not done; n.a., no association; t, tumor cells; s, stromal cells; zymo, gelatin substrate zymography.

tissue. In multivariate analysis, high pro-MMP-9, but not MMP-2, expression was associated with poor prognosis only in the subgroup with no residual mass following surgery. Kamat *et al.* [83] evaluated the relationship between MMP-2, -9, and -14 in both tumor and stroma in preoperative patients with epithelial ovarian cancer. In both univariate and multivariate analysis, high MMP-14 in tumor and stroma and MMP-9 in stroma were associated with poor prognosis. Ozalp *et al.* [85] evaluated MMP-9 (tumor and stromal) in patients with primary ovarian cancer. MMP-9 was expressed in both tumor and stromal tissue. Only univariate analysis was done and it revealed that only high stromal MMP-9 was associated with poor prognosis. Davidson *et al.* [86] evaluated MMP-2, -9, -14, and TIMP-2 in tumor and stromal tissues of patients with ovarian cancer. Univariate analysis revealed that high tumor MMP-9 and TIMP-2 in the primary tumor were associated with poor prognosis. In metastatic lesions, high tumor MMP-2 and MMP-14 and high stromal TIMP-2 were associated with poor prognosis. In multivariate analysis, high TIMP-2 in stroma and MMP-9 in primary tumor were associated with poor prognosis. Sillanpaa *et al.* [87] evaluated MMP-9 (tumor and stroma) in patients with epithelial ovarian cancer. Univariate, but not multivariate, analysis revealed that low MMP-9 in tumor and high MMP-9 in stromal tissue correlated with poor survival.

Perigny *et al.* [88] evaluated MMP-2 and MMP-11 in tumor and stromal tissue in patients with stage 3 ovarian cancer with peritoneal implants. MMP-2 and MMP-11 were expressed in tumor, stromal, and peritoneal implants. In multivariate analysis, high MMP-2 in tumor cells in peritoneal implants was an independent prognostic factor related to poor prognosis. Torng *et al.* [89] evaluated MMP-2, MMP-14, and TIMP-2 in tumor and stromal tissue in presurgical patients with ovarian adenocarcinoma. Univariate, but not multivariate, analysis revealed an association between low MMP-2 in stromal tissue and poor prognosis in ovarian endometeroid adenocarcinoma. MMP-14 was not associated with prognosis. Sillanpaa *et al.* [90] evaluated the prognostic significance of MMP-7 in tumor and stromal tissue. Univariate and multivariate analysis showed low tumor, but not stroma MMP-7 to be associated with poor survival.

14.8.4 Brain Cancer

Data on brain cancer MMP and TIMP measurements are limited. Kunishio *et al.* [91] employed IHC and RT-PCR to evaluate MMP-2 and MMP-9 expression in 40 patients with brain astrocytomas. Neither MMP-2 nor MMP-9 expression was related to histology. MMP-2 IHC staining was significantly higher in the invasive glioma group as compared to other astrocytic tumors. However, no correlation with prognosis was noted on univariate analysis. Jaalinoja *et al.* [92] employed IHC to evaluate MMP-2 in 101 patients with brain cancer. Most strong staining was observed in metastatic brain tumors and glioblastomas; oligodendrogliomas and pilocystic astrocytomas showed no staining. Univariate analysis revealed that high MMP-2 levels were associated with poor prognosis. Rorive *et al.* [93] employed IHC to evaluate tumor TIMP-4 and CD63 and their relationship to prognosis in 471 preoperative patients with brain astrocytomas of various types. In astrocytomas, both TIMP-4 and CD63 were overexpressed in pilocystic astrocytomas as compared to grade 2 astrocytomas. Using multivariate analysis in glioblastoma patients, high expression of TIMP-4, CD63, and high TIMP-4/CD63 were independently associated with poor prognosis.

14.9 Conclusions

MMPs and TIMPs are involved in numerous physiologic and pathologic processes. Although a great deal is known about the chemistry and biology of MMPs and TIMPs, their specific role in cancer remains incompletely understood. The unsolved mystery relates to uncertainty regarding the circumstances in which MMPs and TIMPs play an essential role in the invasive/metastatic process (target effect), versus acting as a host response in an attempt to prevent tumor dissemination (antitarget effect).

As compared to normal tissues, there is general agreement that most cancer tissues overexpress some MMPs and TIMPs, not only in the tumor cells but also in stromal/inflammatory cells. Each cancer type and subtype appears to express a different set of MMPs and TIMPs; tissue levels also vary at different stages of disease.

In spite of hundreds of publications on the subject, there is limited concordance between clinical reports regarding the levels of expression of MMPs and TIMPs in different cancer tissues (Tables14.1–14.8). On the basis of our review of the recent literature that employed critical statistical analysis of clinical outcome and adequate numbers of subjects, colorectal and ovarian cancers represent the two types of cancer in which there is general agreement that MMP-9 is increased in tumors, often with prominence in inflammatory and/or tumor cells, and this increase is independently correlated with poor prognosis. These data suggest that drugs designed to specifically inhibit MMP-9 should be tested in patients with colorectal and ovarian cancers. On the basis of the experience with other biologically active drugs in solid tumors, MMP activity within each tumor specimen should be assessed before initiating treatment in individual patients. High MMP-2 expression in NSCLC likewise is incriminated as an independent prognostic factor for poor prognosis; hence, these patients should be considered for treatment with a specific MMP-2 inhibitor. Increased tissue TIMP-1 levels in colorectal, breast, and lung cancers are also predictive of poor prognosis by multivariate analysis; the causal role of TIMP-1 in human cancer remains to be clarified. One possibility is that increased production of TIMP-1 in late-stage cancer may reflect a protective response by the host to limit the increased MMP activity and tissue degradation associated with cancer. Of interest, these predictive associations with colorectal cancers were not found in upper GI cancers, which emphasize the lack of uniformity between similar histologic types of tumors originating in different organs.

Regarding the lack of agreement on the tissue levels of MMPs and TIMPs with other cancer types, it is uncertain whether reported discrepancies are related to technical issues, of which there are many, or differences in patient sampling. In terms of tissue measurements of MT-MMPs, limitations of commercially available antibodies have probably limited investigators in the field, which is unfortunate since presently it appears that MMP-14 might be the best candidate for specific MMP inhibitory therapy in cancer [94]. MMP-7 is also strongly implicated in cancer progression, but has not received as much attention in the clinical literature as MMP-2 and -9.

We [9] and others have previously emphasized that future researchers will need to employ more critical approaches, rather than limited descriptive reports to identify practical medical applications for MMP and TIMP measurements. Recent reports have more diligently pursued clinical outcome evaluations in their MMP/TIMP studies, but progress in identifying the utility of quantifying MMP and TIMP levels in cancer tissue specimens regarding prognosis has been limited. On the basis of the cytotoxic chemotherapy currently being utilized in the treatment of cancer, tissue levels of MMPs and TIMPs have not proved useful in treatment decisions. It is hoped that MMP and TIMP tissue measurements will be more useful in this regard with the implementation of new biological-based chemotherapy.

Finally, we need to address the future role of using cancer tissue MMP and TIMP measurements in predicting response to treatment with specific MMP inhibitory drugs, which are in the cancer therapy pipeline, for example, a specific monoclonal antibody inhibitor of MMP-14 [94]. If MMPs do, in fact, play an important role in cancer dissemination in humans, it is logical to propose that MMP/TIMP measurements will be clinically useful in the future in selecting those patients for treatment with specific MMP inhibitors.

Acknowledgments

Grant support provided by a Merit Review Grant from the Department of Veterans Affairs, an NIH grant NIH-R01CA11355301A1, a Baldwin Breast Cancer Foundation grant, and a Walk-for-Beauty grant from the Research Foundation, Stony Brook University.

References

1. Zucker, S., Pei, D., Cao, J., and Lopez-Otin, C. (2003) in *Cell Surface Proteases* (eds S. Zucker and W.-T. Chen), Academic Press, pp. 1–74.
2. Zucker, S., Lysik, R.M., Wieman, J., Wilkie, D., and Lane, B. (1985) Diversity of human pancreatic cancer cell proteinases. Role of cell membrane metalloproteinases in collagenolysis and cytolysis. *Cancer Res.*, **45**, 6168–6178.
3. Seiki, M. (2002) The cell surface: the stage for matrix metalloproteinase regulation of migration. *Curr. Opin. Cell Biol.*, **14**, 624–632.
4. Dean, R.A. and Overall, C.M. (2007) Proteomics discovery of metalloproteinase substrates in the cellular context by iTRAC^{TM} labeling reveals a diverse MMP-2 substrate degradome. *Mol. Cell Proteomics*, **6**, 611–621.

5. Fu, X., Parks, W.C., and Heinecke, J.W. (2008) Activation and silencing of matrix metalloproteinases. *Cell Dev. Biol.*, **19**, 2–13.
6. Baker, A.H., Edwards, D.R., and Murphy, G. (2002) Metalloproteinase inhibitors: Biologic actions and therapeutic opportunities. *J. Cell Sci.*, **115**, 3719–3727.
7. Zucker, S., Schmidt, C.E., Dufour, A., Kaplan, R.C., Park, H.I., and Jiang, W. (2009) ProMMP-2: TIMP-1 complexes identified in plasma of healthy individuals. *Connect. Tissue Res.*, **50**, 223–231.
8. Sternlicht, M.D. and Werb, Z. (2001) How matrix metalloproteinases regulate cell behavior. *Annu. Rev. Cell Dev. Biol.*, **17**, 463–516.
9. Zucker, S., Doshi, K., and Cao, J. (2004) Measurement of matrix metalloproteinases (MMPs) and tissue inhibitors of metalloproteinases (TIMP) in blood and urine: Potential clinical applications. *Adv. Clin. Chem.*, **38**, 37–85.
10. Vincenti, M.P. and Brinkerhoff, C.E. (2007) Signal transduction and cell-type specific regulation of matrix metalloproteinase gene expression: Can MMPs be good for you? *J. Cell Physiol.*, **213**, 355–363.
11. Deryugina, E. and Quigley, J.P. (2006) Matrix metalloproteinases and tumor metastasis. *Cancer Metastasis Rev.*, **25**, 9–34.
12. Basset, P., Bellocq, J.P., Wolf, C., Stoll, I., Hutin, P., Limacher, J.M., Podhajcer, O.L., Chenard, M.P., Rio, M.C., and Chambon, P. (1990) A novel metalloproteinase gene specifically expressed in stromal cells of breast carcinomas. *Nature*, **348**, 699–704.
13. Nelson, A.R., Fingleton, B., Rothenberg, M.L., and Matrisian, L.M. (2000) Matrix metalloproteinases: Biologic activity and clinical implications. *J. Clin. Oncol.*, **18**, 1135–1139.
14. Still, K., Robson, C.N., Autzen, P., Robinson, M.C., and Hamby, F.C. (2000) Localization and quantification of mRNA for matrix metalloproteinase-2 (MMP-2) and tissue inhibitor of matrix metalloproteinase-2 (TIMP-2) in human benign and malignant prostatic tissue. *Prostate*, **42**, 18–25.
15. Zucker, S., Hymowitz, M., Rollo, E.E., Mann, R., Conner, C.E., Cao, J., Foda, H., Tompkins, D.C., and Toole, B. (2001) Tumorigenic potential of extracellular matrix metalloproteinase induce (EMMPRIN). *Am. J. Pathol.*, **158**, 1921–1928.
16. Martin, M.D. and Matrisian, L.M. (2007) The other side of MMPs: Protective roles in tumor progression. *Cancer Metastasis Rev.*, **26**, 717–724.
17. Nakopoulou, L., Giannopoulou, G., Stefanaki, K., Panayotopoulou, E., Tsirmpa, I., Alexandrou, P., Mavrommatis, J., Katsarou, S., and Davaris, P. (2002) Enhanced mRNA expression of tissue inhibitor of metalloproteinase-1 (TIMP-1) in breast carcinomas is correlated with adverse prognosis. *J. Pathol.*, **197**, 307–313.
18. Wu, Z.S., Wu, Q., Yang, J.H., Wang, H.Q., Ding, X.D., Yang, F., and Xu, X.C. (2008) Prognostic significance of MMP-9 and TIMP-1 serum and tissue expression in breast cancer. *Int. J. Cancer*, **122**, 2050–2056.
19. Schrohl, A.S., Look, M.P., Meijer-van Gelder, M.E., Foekens, J.A., and Brunner, N. (2009) Tumor tissue levels of Tissue Inhibitor of Metalloproteinases-1 (TIMP-1) and outcome following adjuvant chemotherapy in premenopausal lymph node-positive breast cancer patients: A retrospective study. *BMC Cancer*, **9**, 322.
20. Schrohl, A.S., Holten-Andersen, M.N., Peters, H.A., Look, M.P., Meijer-van Gelder, M.E., Klijn, J.G., Brunner, N., and Foekens, J.A. (2004) Tumor tissue levels of tissue inhibitor of metalloproteinase-1 as a prognostic marker in primary breast cancer. *Clin. Cancer Res.*, **10**, 2289–2298.
21. Wurtz, S.O., Christensen, I.J., Schrohl, A.S., Mouridsen, H., Lademann, U., Jensen, V., and Brunner, N. (2005) Measurement of the uncomplexed fraction of tissue inhibitor of metalloproteinases-1 in the prognostic evaluation of primary breast cancer patients. *Mol. Cell Proteomics*, **4**, 483–491.

22. Liss, M., Sreedhar, N., Keshgegian, A., Sauter, G., Chernick, M.R., Prendergast, G.C., and Wallon, U.M. (2009) Tissue inhibitor of metalloproteinase-4 is elevated in early-stage breast cancers with accelerated progression and poor clinical course. *Am. J. Pathol.*, **175**, 940–946.
23. Kotzsch, M., Farthmann, J., Meye, A., Fuessel, S., Baretton, G., Tijan, H., Schmitt, M., Luther, T., Sweep, F.C.G., Magdolen, V. *et al.* (2005) Prognostic relevance of uPAR-del14/5 and TIMP-3 in mRNA expression levels in breast cancer. *Eur. J. Cancer*, **41**, 2760–2768.
24. Jobim, F.C., Xavier, N.L., Uchoa, D.B., Cruz, D.B., Saciloto, M., Chemello, N., and Schwartsmann, G. (2009) Prevalence of vascular-endothelial growth factor, matrix metalloproteinases and tissue inhibitors of metalloproteinase in primary breast cancer. *Braz. J. Med. Biol. Res.*, **42**, 979–987.
25. Shah, F.D., Shukla, S.N., Shah, P.M., Shukla, H.K., and Patel, P.S. (2009) Clinical significance of matrix metalloproteinase 2 and 9 in breast cancer. *Indian J. Cancer*, **46**, 194–202.
26. Mylona, E., Nomikos, A., Magkou, C., Kamberou, M., Papassideri, I., Keramopoulos, A., and Nakopoulou, L. (2007) The clinicopathological and prognostic significance of membrane type 1 matrix metalloproteinase (MT1-MMP) and MMP-9 according to their localization in invasive breast carcinoma. *Histopathology*, **50**, 338–347.
27. Tetu, B., Brisson, J., Wang, C.S., Lapointe, H., Beaudry, G., Blanchette, C., and Trudel, D. (2006) The influence of MMP-14, TIMP-2 and MMP-2 expression on breast cancer prognosis. *Breast Cancer Res.*, **8**, R28.
28. Talvensaari-Mattila, A., Paakko, P., and Turpeenniemi-Hujanen, T. (2003) Matrix metalloproteinase-2 (MMP-2) is associated with survival in breast carcinoma. *Br. J. Cancer*, **89**, 1270–1275.
29. Del Casar, J.M., Gonzalez, L.O., Alvarez, E., Junquera, S., Marin, L., Gonzalez, L., Bongera, M., Vazquez, J., and Vizoso, F.J. (2009) Comparative analysis and clinical value of the expression of metalloproteases and their inhibitors by intratumor stromal fibroblasts and those at the invasive front of breast carcinomas. *Breast Cancer Res. Treat.*, **116**, 39–52.
30. Rahko, E., Kauppila, S., Paakko, P., Blanco, G., Apaja-Sarkkinen, M., Talvensaari-Mattila, A., Turpeenniemi-Hujanen, T., and Jukkola, A. (2009) Immunohistochemical study of matrix metalloproteinase 9 and tissue inhibitor of matrix metalloproteinase 1 in benign and malignant breast tissue--strong expression in intraductal carcinomas of the breast. *Tumour Biol.*, **30**, 257–264.
31. Chenard, M.-P., Lutz, Y., Mechine-Neuville, A., Stoll, I., Bellocq, J.-B., Rio, M.-C., and Basset, P. (1999) Presence of high levels of MT1-MMP protein in fibroblastic cells of human invasive carcinomas. *Int. J. Cancer*, **82**, 208–212.
32. Poola, I., DeWitty, R.L., Marshalleck, J.J., Bhatnagar, R., Abraham, J., and Leffall, L.D. (2005) Identification of MMP-1 as a putative breast cancer predictive marker by global gene expression analysis. *Nat. Med.*, **11**, 481–483.
33. Zucker, S. and Vacirca, J. (2004) Role of matrix metalloproteinases (MMPs) in colorectal cancer. *Cancer Metastasis Rev.*, **23**, 101–117.
34. Svagzdys, S., Lesauskaite, V., Pangonyte, D., Saladzinskas, Z., Tamelis, A., and Pavalkis, D. (2011) Matrix metalloproteinase-9 is a prognostic marker to predict survival of patients who underwent surgery due to rectal carcinoma. *Tohoku J. Exp. Med.*, **223**, 67–73.
35. Bendardaf, R., Buhmeida, A., Hilska, M., Laato, M., Syrjanen, S., Syrjanen, K., Collan, Y., and Pyrhonen, S. (2010) MMP-9 (gelatinase-B) expression is associated with disease-free survival and disease-specific survival in colorectal cancer patients. *Cancer Invest.*, **28**, 38–43.
36. Cho, Y.B., Lee, W.Y., Song, S.Y., Shin, H.J., Yun, S.H., and Chun, H.K. (2007) Matrix metalloproteinase-9 activity is associated with poor prognosis in T3-T4 node-negative colorectal cancer. *Hum. Pathol.*, **38**, 1603–1610.

37. Buhmeida, A., Bendardaf, R., Hilska, M., Collan, Y., Laato, M., Syrjanen, S., Syrjanen, K., and Pyrhonen, S. (2009) Prognostic significance of matrix metalloproteinase-9 (MMP-9) in stage II colorectal carcinoma. *J. Gastrointest. Cancer*, **40**, 91–97.
38. Sutnar, A., Pesta, M., Liska, V., Treska, V., Skalicky, T., Kormunda, S., Topolcan, O., Cerny, R., and Holubec, L. Jr. (2007) Clinical relevance of the expression of mRNA of MMP-7, MMP-9, TIMP-1, TIMP-2 and CEA tissue samples from colorectal liver metastases. *Tumour Biol.*, **28**, 247–252.
39. Roca, F. Mauro, Le.V., Morandi, A., Vaccaro, C., Quintana, G.O., Specterman, S., de Kier, Joffe, E.B., Pallotta, M., Puricelli, L.I., and Lastiri, J. (2006) Prognostic value of E-cadherin, MMP (7 and 9), and TIMPs (1 and 2) in patients with colorectal carcinoma. *J. Surg. Oncol.*, **93**, 151–160.
40. Sundov, Z., Tomic, S., Vilovic, K., Kunac, N., Kalebic, M., and Bezic, J. (2008) Immunohistochemically detected high expression of matrix metalloproteinase-2 as predictor of poor prognosis in Duke's B colon cancer. *Croat. Med. J.*, **49**, 636–642.
41. Asano, T., Tada, M., Cheng, S., Takemoto, N., Kuramae, T., Abe, M., Takahashi, O., Miyamoto, M., Hamada, J., Moriuchi, T. *et al.* (2008) Prognostic values of matrix metalloproteinase family expression in human colorectal carcinoma. *J. Surg. Res.*, **146**, 32–42.
42. Koskensalo, S., Louhimo, J., Nordling, S., Hagstrom, J., and Haglund, C. (2011) MMP-7 as a prognostic marker in colorectal cancer. *Tumour Biol.*, **32**, 259–264.
43. Hilska, M., Roberts, P.J., Collan, Y.U., Laine, V.J., Kossi, J., Hirsimaki, P., Rahkonen, O., and Laato, M. (2007) Prognostic significance of matrix metalloproteinases-1, -2, -7 and -13 and tissue inhibitors of metalloproteinases-1, -2, -3 and -4 in colorectal cancer. *Int. J. Cancer*, **121**, 714–723.
44. Islekel, H., Oktay, G., Terzi, C., Canda, A.E., Fuzun, M., and Kupelioglu, A. (2007) Matrix metalloproteinase-9, -3 and tissue inhibitor of matrix metalloproteinase-1 in colorectal cancer: relationship to clinicopathological variables. *Cell Biochem. Funct.*, **25**, 433–441.
45. Kubben, F.J., Sier, C.F., van Duijn, W., Griffioen, G., Hanemaaijer, R., van de Velde, C.J., van Krieken, J.H., Lamers, C.B., and Verspaget, H.W. (2006) Matrix metalloproteinase-2 is a consistent prognostic factor in gastric cancer. *Br. J. Cancer*, **94**, 1035–1040.
46. Mrena, J., Wiksten, J.P., Nordling, S., Kokkola, A., Ristimaki, A., and Haglund, C. (2006) MMP-2 but not MMP-9 associated with COX-2 and survival in gastric cancer. *J. Clin. Pathol.*, **59**, 618–623.
47. Alakus, H., Grass, G., Hennecken, J.K., Bollschweiler, E., Schulte, C., Drebber, U., Baldus, S.E., Metzger, R., Holscher, A.H., and Monig, S.P. (2008) Clinicopathological significance of MMP-2 and its specific inhibitor TIMP-2 in gastric cancer. *Histol. Histopathol.*, **23**, 917–923.
48. Mori, M., Mimori, K., Shiraishi, T., Fujie, T., Baba, K., Kusumoto, H., Haraguchi, M., Ueo, H., and Akiyoshi, T. (1997) Analysis of MT1-MMP and MMP-2 expression in human gastric cancers. *Int. J. Cancer*, **74**, 316–321.
49. Fujimoto, D., Hirono, Y., Goi, T., Katayama, K., and Yamaguchi, A. (2008) Prognostic value of protease-activated receptor-1 (PAR-1) and matrix metalloproteinase-1 (MMP-1) in gastric cancer. *Anticancer Res.*, **28**, 847–854.
50. Chu, D., Zhang, Z., Li, Y., Zheng, J., Dong, G., Wang, W., and Ji, G. (2011) Matrix metalloproteinase-9 is associated with disease-free survival and overall survival in patients with gastric cancer. *Int. J. Cancer.*, **129**, 887–895.
51. Koskensalo, S., Mrena, J., Wiksten, J.P., Nordling, S., Kokkola, A., Hagstrom, J., and Haglund, C. (2010) MMP-7 overexpression is an independent prognostic marker in gastric cancer. *Tumor Biol.*, **31**, 149–155.
52. Giannopoulos, G., Kaatzas, N., Tiniakos, D., Karakosta, T.D., Tzanakis, N., and Peros, G. (2008) The expression of matrix metalloproteinases-2 and -9 and their tissue inhibitor 2 in pancreatic ductal and ampullary carcinoma and

their relation to angiogenesis and clinicopathological parameters. *Anticancer Res.*, **28**, 1875–1881.

53. Jones, L.E., Humphreys, M.J., Campbell, F., Neoptolemos, J.P., and Boyd, M.T. (2004) Comprehensive analysis of matrix metalloproteinase and tissue inhibitor expression in pancreatic cancer: increased expression of matrix metalloproteinase-7 predicts poor survival. *Clin. Cancer Res.*, **10**, 2832–2845.
54. Juuti, A., Lundin, J., Nordling, S., Louhimo, J., and Haglund, C. (2006) Epithelial MMP-2 expression correlates with worse prognosis in pancreatic cancer. *Oncology*, **71**, 61–68.
55. Ishikawa, S., Takenaka, K., Yanagihara, K., Miyahara, R., Kawano, Y., Otake, Y., Hasegawa, S., Wada, H., and Tanaka, F. (2004) Matrix metalloproteinase-2 status in stromal fibroblasts, not in tumor cells, is a significant prognostic factor in non-small-cell lung cancer. *Clin. Cancer Res.*, **10**, 6579–6585.
56. Shou, Y., Hirano, T., Gong, Y., Kato, Y., Yoshida, K., Ohira, T., Ikeda, N., Konaka, C., Ebihara, Y., Zhao, F. *et al.* (2001) Influence of angiogenetic factors and matrix metalloproteinases upon tumour progression in non-small-cell lung cancer. *Br. J. Cancer*, **85**, 1706–1712.
57. Leinonen, T., Pirinen, R., Bohm, J., Johansson, R., and Kosma, V.M. (2008) Increased expression of matrix metalloproteinase-2 (MMP-2) predicts tumour recurrence and unfavourable outcome in non-small cell lung cancer. *Histol. Histopathol.*, **23**, 693–700.
58. Passlick, B., Sienel, W., Seen-Hibler, R., Wockel, W., Thetter, O., Mutschler, W., and Pantel, K. (2000) Overexpression of matrix metalloproteinase 2 predicts unfavorable outcome in early-stage non-small cell lung cancer. *Clin. Cancer Res.*, **6**, 3944–3948.
59. Aljada, I.S., Ramnath, N., Donohue, K., Harvey, S., Brooks, J.J., Wiseman, S.M., Khoury, T., Loewen, G., Slocum, H.K., Anderson, T.M. *et al.* (2004) Upregulation of the tissue inhibitor of metalloproteinase-1 protein is associated with progression of human non-small-cell lung cancer. *J. Clin. Oncol.*, **22**, 3218–3229.
60. Mino, N., Takenaka, K., Sonobe, M., Miyahara, R., Yanagihara, K., Otake, Y., Wada, H., and Tanaka, F. (2007) Expression of tissue inhibitor of metalloproteinase-3 (TIMP-3) and its prognostic significance in resected non-small cell lung cancer. *J. Surg. Oncol.*, **95**, 250–257.
61. Gouyer, V., Conti, M., Devos, P., Zerimech, F., Copin, M.C., Creme, E., Wurtz, A., Porte, H., and Huet, G. (2005) Tissue inhibitor of metalloproteinase 1 is an independent predictor of prognosis in patients with nonsmall cell lung carcinoma who undergo resection with curative intent. *Cancer*, **103**, 1676–1684.
62. Simi, L., Andreani, M., Davini, F., Janni, A., Pazzagli, M., Serio, M., and Orlando, C. (2004) Simultaneous measurement of MMP9 and TIMP1 mRNA in human non small cell lung cancers by multiplex real time RT-PCR. *Lung Cancer*, **45**, 171–179.
63. Liu, D., Nakano, J., Ishikawa, S., Yokomise, H., Ueno, M., Kadota, K., Urushihara, M., and Huang, C.L. (2007) Overexpression of matrix metalloproteinase-7 (MMP-7) correlates with tumor proliferation, and a poor prognosis in non-small cell lung cancer. *Lung Cancer*, **58**, 384–391.
64. Shah, S.A., Spinale, F.G., Ikonomidis, J.S., Stroud, R.E., Chang, E.I., and Reed, C.E. (2010) Differential matrix metalloproteinase levels in adenocarcinoma and squamous cell carcinoma of the lung. *J. Thorac. Cardiovasc. Surg.*, **139**, 984–990; discussion 990.
65. Atkinson, J.M., Pennington, C.J., Martin, S.W., Anikin, V.A., Mearns, A.J., Loadman, P.M., Edwards, D.R., and Gill, J.H. (2007) Membrane type matrix metalloproteinases (MMPs) show differential expression in non-small cell lung cancer (NSCLC) compared to normal lung: correlation of MMP-14 mRNA expression and proteolytic activity. *Eur. J. Cancer*, **43**, 1764–1771.
66. Vasala, K., Paakko, P., and Turpeenniemi-Hujanen, T. (2003) Matrix metalloproteinase-2 immunoreactive protein as a prognostic marker in bladder cancer. *Urology*, **62**, 952–957.

67. Vasala, K., Paakko, P., and Turpeenniemi-Hujanen, T. (2008) Matrix metalloproteinase-9 (MMP-9) immunoreactive protein in urinary bladder cancer: a marker of favorable prognosis. *Anticancer Res.*, **28**, 1757–1761.

68. Szarvas, T., Becker, M., vom Dorp, F., Gethmann, C., Totsch, M., Bankfalvi, A., Schmid, K.W., Romics, I., Rubben, H., and Ergun, S. (2010) Matrix metalloproteinase-7 as a marker of metastasis and predictor of poor survival in bladder cancer. *Cancer Sci.*, **101**, 1300–1308.

69. Gakiopoulou, H., Nakopoulou, L., Siatelis, A., Mavrommatis, I., Panayotopoulou, E.G., Tsirmpa, I., Stravodimos, C., and Giannopoulos, A. (2003) Tissue inhibitor of metalloproteinase-2 as a multifunctional molecule of which the expression is associated with adverse prognosis of patients with urothelial bladder carcinomas. *Clin. Cancer. Res.*, **9**, 5573–5581.

70. Mohammad, M.A., Ismael, N.R., Shaarawy, S.M., and El-Merzabani, M.M. (2010) Prognostic value of membrane type 1 and 2 matrix metalloproteinase expression and gelatinase A activity in bladder cancer. *Int. J. Biol. Markers*, **25**, 69–74.

71. Seargent, J.M., Loadman, P.M., Martin, S.W., Naylor, B., Bibby, M.C., and Gill, J.H. (2005) Expression of matrix metalloproteinase-10 in human bladder transitional cell carcinoma. *Urology*, **65**, 815–820.

72. Kawata, N., Nagane, Y., Hirakata, H., Ichinose, T., Okada, Y., Yamaguchi, K., and Takahashi, S. (2007) Significant relationship of matrix metalloproteinase 9 with nuclear grade and prognostic impact of tissue inhibitor of metalloproteinase 2 for incidental clear cell renal cell carcinoma. *Urology*, **69**, 1049–1053.

73. Kallakury, B.V., Karikehalli, S., Haholu, A., Sheehan, C.E., Azumi, N., and Ross, J.S. (2001) Increased expression of matrix metalloproteinases 2 and 9 and tissue inhibitors of metalloproteinases 1 and 2 correlate with poor prognostic variables in renal cell carcinoma. *Clin. Cancer Res.*, **7**, 3113–3119.

74. Takahashi, H., Natuo, O., Naroda, T., Nishitani, M.-A., Kanda, K., Kanayama, H.-O., and Kagawa, S. (2002) Prognostic significance of matrix metalloproteinase-2 activation ratio in renal cancer. *Int. J. Urol.*, **9**, 531–538.

75. Kamijima, S., Tobe, T., Suyama, T., Ueda, T., Igarashi, T., Ichikawa, T., and Ito, H. (2005) The prognostic value of p53, Ki-67 and matrix metalloproteinases MMP-2 and MMP-9 in transitional cell carcinoma of the renal pelvis and ureter. *Int. J. Urol.*, **12**, 941–947.

76. Miyata, Y., Iwata, T., Ohba, K., Kanda, S., Nishikido, M., and Kanetake, H. (2006) Expression of matrix metalloproteinase-7 on cancer cells and tissue endothelial cells in renal cell carcinoma: prognostic implications and clinical significance for invasion and metastasis. *Clin. Cancer Res.*, **12**, 6998–7003.

77. Miyata, Y., Iwata, T., Maruta, S., Kanda, S., Nishikido, M., Koga, S., and Kanetake, H. (2007) Expression of matrix metalloproteinase-10 in renal cell carcinoma and its prognostic role. *Eur. Urol.*, **52**, 791–797.

78. Boxler, S., Djonov, V., Kessler, T.M., Hlushchuk, R., Bachmann, L.M., Held, U., Markwalder, R., and Thalmann, G.N. (2010) Matrix metalloproteinases and angiogenic factors: predictors of survival after radical prostatectomy for clinically organ-confined prostate cancer? *Am. J. Pathol.*, **177**, 2216–2224.

79. Miyake, H., Muramaki, M., Kurahashi, T., Takenaka, A., and Fujisawa, M. (2010) Expression of potential molecular markers in prostate cancer: correlation with clinicopathological outcomes in patients undergoing radical prostatectomy. *Urol. Oncol.*, **28**, 145–151.

80. Trudel, D., Fradet, Y., Meyer, F., Harel, F., and Tetu, B. (2008) Membrane-type-1 matrix metalloproteinase, matrix metalloproteinase 2, and tissue inhibitor of matrix proteinase 2 in prostate cancer: identification of patients with poor prognosis by immunohistochemistry. *Hum. Pathol.*, **39**, 731–739.

81. Ross, J.S., Kaur, P., Sheehan, C.E., Fisher, H.A., Kaufman, R.A., and Kallakury, B.V. Jr. (2003) Prognostic significance of matrix metalloproteinase 2 and tissue inhibitor of metalloproteinase 2 expression in prostate cancer. *Mod. Pathol.*, **16**, 198–205.

82. Lengyel, E., Schmalfeldt, B., Konik, E., Spathe, K., Harting, K., Fenn, A., Berger, U., Fridman, R., Schmitt, M., Prechtel, D. *et al.* (2001) Expression of latent matrix metalloproteinase 9 (MMP-9) predicts survival in advanced ovarian cancer. *Gynecol. Oncol.*, **82**, 291–298.

83. Kamat, A.A., Fletcher, M., Gruman, L.M., Mueller, P., Lopez, A., Landen, C.N., Han, L., Gershenson, D.M., and Sood, A.K. Jr. (2006) The clinical relevance of stromal matrix metalloproteinase expression in ovarian cancer. *Clin. Cancer Res.*, **12**, 1707–1714.

84. Alshenawy, H.A. (2010) Immunohistochemical expression of epidermal growth factor receptor, E-cadherin, and matrix metalloproteinase-9 in ovarian epithelial cancer and relation to patient deaths. *Ann. Diagn. Pathol.*, **14**, 387–395.

85. Ozalp, S., Tanir, H.M., Yalcin, O.T., Kabukcuoglu, S., Oner, U., and Uray, M. (2003) Prognostic value of matrix metalloproteinase-9 (gelatinase-B) expression in epithelial ovarian tumors. *Eur. J. Gynaecol. Oncol.*, **24**, 417–420.

86. Davidson, B., Goldberg, I., Gotlieb, W.H., Kopolovic, J., Ben-Baruch, G., Nesland, J.M., Berner, A., Bryne, M., and Reich, R. (1999) High levels of MMP-2, MMP-9, MT1-MMP and TIMP-2 mRNA correlate with poor survival in ovarian carcinoma. *Clin. Exp. Metastasis*, **17**, 799–808.

87. Sillanpaa, S., Anttila, M., Voutilainen, K., Ropponen, K., Turpeenniemi-Hujanen, T., Puistola, U., Tammi, R., Tammi, M., Sironen, R., Saarikoski, S. *et al.* (2007) Prognostic significance of matrix metalloproteinase-9 (MMP-9) in epithelial ovarian cancer. *Gynecol. Oncol.*, **104**, 296–303.

88. Perigny, M., Bairati, I., Harvey, I., Beauchemin, M., Harel, F., Plante, M., and Tetu, B. (2008) Role of immunohistochemical overexpression of matrix metalloproteinases MMP-2 and MMP-11 in the prognosis of death by ovarian cancer. *Am. J. Clin. Pathol.*, **129**, 226–231.

89. Torng, P.L., Mao, T.L., Chan, W.Y., Huang, S.C., and Lin, C.T. (2004) Prognostic significance of stromal metalloproteinase-2 in ovarian adenocarcinoma and its relation to carcinoma progression. *Gynecol. Oncol.*, **92**, 559–567.

90. Sillanpaa, S.M., Anttila, M.A., Voutilainen, K.A., Ropponen, K.M., Sironen, R.K., Saarikoski, S.V., and Kosma, V.M. (2006) Prognostic significance of matrix metalloproteinase-7 in epithelial ovarian cancer and its relation to beta-catenin expression. *Int. J. Cancer*, **119**, 1792–1799.

91. Kunishio, K., Okada, M., Matsumoto, Y., and Nagao, S. (2003) Matrix metalloproteinase-2 and -9 expression in astrocytic tumors. *Brain Tumor Pathol.*, **20**, 39–45.

92. Jaalinoja, J., Herva, R., Korpela, M., Hoyhtya, M., and Turpeenniemi-Hujanen, T. (2000) Matrix metalloproteinase 2 (MMP-2) immunoreactive protein is associated with poor grade and survival in brain neoplasms. *J. Neurooncol.*, **46**, 81–90.

93. Rorive, S., Lopez, X.M., Maris, C., Trepant, A.L., Sauvage, S., Sadeghi, N., Roland, I., Decaestecker, C., and Salmon, I. (2010) TIMP-4 and CD63: new prognostic biomarkers in human astrocytomas. *Mod. Pathol.*, **23**, 1418–1428.

94. Devy, L., Huang, L., Noa, L., Yanamandra, N., Pieteters, H., Frans, N., Chang, E., Tao, Q., Vanhove, M., Lejeune, A. *et al.* (2009) Selective inhibitor of matrix metalloproteinase-14 blocks tumor growth, invasion, and angiogenesis. *Cancer Res.*, **69**, 1517–1524.

15
New Prospects for Matrix Metalloproteinase Targeting in Cancer Therapy

Emilie Buache and Marie-Christine Rio

15.1
Introduction

The importance of the tumor microenvironment has been increasingly recognized [1–6]. This is due to molecular changes in the tumor stroma during cancer development and progression. The matrix metalloproteinase (MMP) family currently consists of 24 enzymes, that, by definition, cleave at least one extracellular matrix (ECM) component. For several years, there has been considerable evidence implicating MMP activity in cancer-associated tissue remodeling [7–14]. Indeed, the expression of individual MMPs, as well as the number of different MMPs, increases during tumor progression. Since excess of MMP activity is associated with ECM destruction, and is presumably required for tumor extension and cancer cell colonization of adjacent and distant tissues, these findings gave the first rationale for targeting MMPs in cancer therapies. It has been proposed that inhibitors of matrix metalloproteinases (MMPIs) might be useful in the battery of tools to cure cancer. Significant efforts over the past 30 years were therefore dedicated to designing compounds to block MMP activity, leading to the discovery of numerous drugs inhibiting MMP activity *in vitro*. Moreover, numerous *in vivo* experiments have been conducted in rodents. Expression analyses provided evidence that high MMP levels are associated with malignancy. Further studies using transgenic mice overexpressing MMPs and MMP-deficient animals showed that MMPs play important functions in the tumor process, strongly supporting the idea that at least some of them can be interesting as therapeutic targets [15–17]. Finally, preclinical studies (animal tumor models) using several MMPIs have given promising evidence that MMPIs may be an effective treatment for cancer [18–20]. Despite these encouraging data, however, results from clinical trials have been disappointing as no real benefits were observed from MMPI treatments, and in a few cases, they have been detrimental [21–25]. The present chapter attempts to dissect out aspects of previous animal and human studies that may be helpful in making decisions about the future of MMPI drug development for the treatment of cancer, and will explore the latest research on the emerging concepts about using MMP targeting to design more potent cytostatic or cytotoxic anticancer drugs.

Matrix Proteases in Health and Disease, First Edition. Edited by Niels Behrendt.

15.2
Lessons Learned from Preclinical and Clinical Studies of MMPIs in Cancer and Possible Alternatives

We have to take into account the lessons from past failures to find ways in which future iterations of MMPIs could potentially be of significant therapeutic benefit. The main characteristics and results of the MMPI clinical trials conducted in cancer patients have recently been reviewed by Fingleton [26] and Dorman *et al.* [27]. These trials almost all failed for various reasons including lack of specificity/affinity/selectivity, ignorance of multifaceted MMP activities, incorrect schedule of drug delivery, and excess of negative side effects [23]. These defaults should be corrected to obtain new generations of efficient MMPIs.

15.2.1
Improve Specificity/Affinity/Selectivity

Tissue inhibitors of metalloproteinases (TIMPs) are endogenous MMPIs that control the physiological activities of MMPs in tissues. Four TIMPs have been identified (TIMP-1 to TIMP-4) [28]. The use of these endogenous MMPIs as drugs has been proposed. However, inappropriate MMP inhibition and/or activation, as well as off-target effects, show that TIMPs were not convenient MMPIs for therapeutic use [23, 29, 30]. The presence of a zinc atom in the MMP catalytic domain is a characteristic that has been extensively explored to design synthetic inhibitors. The use of a strong zinc-binding group such as the Zn^{2+}-chelating hydroxamate has led to the development of extremely potent MMPIs. Several other MMPIs exhibiting strong to weak Zn-binding groups and nonzinc-chelating MMPIs have also been developed. The characteristics and structures of these MMPIs have not been described in the present review since they have already been described in numerous reviews [31–34]. Unfortunately, all of these MMPIs displayed weak selectivity. The absence of selectivity was such for some that they also inhibited MMP-related proteinases such as ADAMs (a disintegrin and metalloproteinases) and ADAMTS (a disintegrin and metalloproteinase with thrombospondin motifs). This observation led to the design of new MMPIs that are specific to a particular MMP conformational stage. Thus, structural and computational information allow the optimization of inhibitors with a target MMP [35, 36]. Among the various types of MMPIs, phosphinic peptides were promising, but they have problems similar to those observed for other MMPIs [37]. Nevertheless, based on the metabolic stability of such compounds and their ability to target the tumor tissue, it is possible to envisage their radiolabeling for *in vivo* positron emission tomography (PET) imaging. Thus, 3D-structure-based designs have become essential in MMPI development [27]. One problem with computational drug design is the difficulty in treating the flexibility of the target. Different approaches can be used to overcome this problem, including new algorithms and molecular dynamics simulation. The interest in applying normal mode analysis (NMA) to drug design has been demonstrated with MMP-3 [38]. In most MMPs, the substrate-binding site consists of a primary hydrophobic S1′ subsite exhibiting a tunnel-shaped structure. Its depth is determined by the so-called S1′ specificity loop whose size and sequence vary between MMPs. In this context, MMPIs containing weak zinc-binding groups, such as phosphoryl or carboxylic groups, or no zinc-binding group are less constraining, making it easier to exploit the depth of the MMP S1′ cavities and to improve selectivity profiles [39]. Currently, a good selective inhibitor has been reported for MMP-12, which possesses a weak zinc-binding group, and a phosphoryl group or a carboxylate function [40]. High-throughput screening of chemical libraries has led to the discovery of a new class of MMPIs, which exhibit a lower molecular weight than MMPI containing a zinc-binding group, and does not possess a zinc-binding group. Without the zinc tether, these compounds can be tailored to target the depth of the S1′ cavity. Thus, an inhibitor designated compound-8 selectively inhibits MMP-13 but not MMP-1, -2, -3, -7, -8, -9, -10, -11, -12, -14, and -16. Compound-8 binds to the bottom of the S1′ cavity of MMP-13 and extends into an additional cavity termed the ≪ *S1′ side pocket* ≫ [41]. A similar inhibitor was recently described for MMP-8 [42]. Finally, several natural MMPIs have been described, the most common corresponding to doxycyclin, a tetracyclin analog, that inhibits MMP activity. It is the only marketed MMPI (Periostat) and is used to treat adult periodontitis [43]. Nonantimicrobial chemically modified tetracyclins have been found to be efficient in preclinical tumor models and tested in preliminary clinical trials in patients with advanced breast, prostate, and lung cancers [44]. Several other molecules mainly issued from plants (e.g., genistein, nobiletin, curcumin, and catechins) that might be of potential interest are under consideration for MMPI development (reviewed in Mannello [45]). Thus, although progress has been made to design highly specific MMPIs in order to avoid the kind of problems encountered with broad-spectrum MMPIs, only very few highly selective MMPIs are available to date.

15.2.2
Increase Knowledge of Multifaceted Activities for a given MMP

15.2.2.1 Target an Active MMP

Proteolysis is a very effective mechanism for adapting the proteins to maintain tissue homeostasis. However, this is an irreversible change. This implies that MMPs should be active at the right time and place to efficiently respond to a defined physiopathological process. Activity levels should therefore be spatially and temporally tightly controlled. In this context, MMPs might be subdivided into two groups depending on their activity. Most of them are synthesized as pre-enzymes and are secreted as proenzymes that can be activated on proteolytic removal of the propeptide [46]. A key step in regulating the activity of these MMPs is therefore the conversion of zymogen into an active proteolytic enzyme. The other MMPs (MMP-11, -21, -28, and membrane-type (MT)–MMPs) are exceptions to this rule and are activated intracellularly through a furin-dependent process [47, 48]. Thus, they are directly secreted or inserted into the plasma membrane as an active form. MMPs from both types have been shown to be overexpressed in cancers. In order to develop efficient MMPIs, the goal is therefore to determine when and where each MMP exerts its proteolytic function in the tumor microenvironment [14, 49, 50]. To address this question, imaging probes based on MMP-specific activities have been developed using fluorescent resonance energy transfer (FRET), PET, single photon emission computed tomography (SPECT), and magnetic resonance imaging (MRI) (reviewed in Scherer *et al.* [51]). Results obtained with these various approaches have confirmed that MMP activities are increased in tumors compared with normal tissues [52, 53]. Thus, cell-penetrating fluorescent peptides activated by proteolysis have allowed the visualization of MMP-2 and MMP-9 activity at the interface between tumor and stroma in mouse xenograft tumor models [54]. Thus, imaging techniques are fundamental tools for *in vivo* monitoring of the potential effectiveness of drugs designed to target active MMPs in cancer.

15.2.2.2 **Fully Characterize the Spatio-Temporal Function of Each MMP: the MMP-11 Example**

MMPs have long been associated with cancer progression because of their ability to degrade ECMs, which facilitates invasion and metastasis. However, recent studies revealed that this view of MMP function, although not wrong, is somewhat reductionist, at least for some, as exemplified below with MMP-11. On one hand, in human carcinomas, it is well established that MMP-11 expression by nonmalignant peritumoral fibroblast-like cells is a stromal factor of bad prognosis [55–62]. Moreover, MMP-11 is one of the factors often found in association with cancer and/or tumor invasion using high-throughput approaches [63]. It was found to be associated with tumor invasion in a stromal gene expression profiling [64]. MMP-11 is part of a gene expression signature that distinguishes invasive ductal breast carcinomas (IDCs) from ductal carcinoma *in situ* tumors (DCIS) [65]. Finally, significantly increased MMP-11 serum levels are positively correlated with metastases in gastric cancer patients [66]. Thus, all these studies provide evidence that MMP-11 is a pejorative factor in human cancers. On the other hand, MMP-11 function during natural cancer history, from primary tumor development to metastasis dissemination, dissected in several mouse models of tumorigenesis using wild type, nude, or MMP-11-knockout (KO) mice, highlighted unexpected cues, pointing to an extremely complex function. Indeed, host MMP-11 is a key player during local invasion, favoring cancer cell survival in connective tissues through an antiapoptotic function [15, 67–70]. However, although MMP-11 favors primary tumor development, MMP-11 deficiency leads to unexpected higher numbers of metastases, indicating that the cancer cells developing in MMP-11-deficient stroma have an increased potential for hematogenous dissemination [70]. The impact of MMP-11 on the dynamics of metastatic development was further studied by real-time imaging of developing metastases in living mice using X-ray computed tomography system (microCT) imaging. Lung metastases were found to occur earlier and to grow faster in wild-type mice, but in lower numbers compared with MMP-11-KO mice. Moreover, MMP-11-KO mice, but not wild-type mice, developed miscellaneous metastases in the liver, adrenal gland, mammary gland, and ovary. Collectively, these results revealed significant spatiotemporal variability of host MMP-11 impact, pointing out its paradoxical role in favoring the local invasion and the onset and growth of metastases, but in limiting cancer cell dissemination and the metastasis foci number [71]. Thus, the same MMP can play activator or repressor functions depending on the invasive step. It cannot be excluded that this observed variability results from the "dominant" function(s) of another protein/proteinase that counterbalances the MMP function. Similar unexpected and favorable MMP effects in tumors have also been reported for other MMPs [72–74]. How does one reconcile such paradoxical effects to design an efficient MMPI strategy to treat patients? It should be noted that, although animal models have revealed a two-sided function for some MMPs, similar data can never be detected in human patients, since the observed incidence of a factor is established at diagnosis. Globally, in most patients, the increased expression of MMPs is associated with a worse prognosis, meaning that the sum of the yin and the yang MMP effects are globally detrimental [75]. Thus, tumor microenvironmental factors, such as MMPs, are crucial for the entire tumor progression. Nevertheless, an identical MMP may display paradoxal positive or negative functions. Therefore, before reconsidering MMPIs as anticancer drugs, it is essential to define carefully the spatiotemporal function of each MMP at the various stages of cancer progression.

15.2.3 Minimize Negative Side Effects

MMPs participate in various normal biological processes such as angiogenesis, embryonic development, wound healing, ovulation, and nerve growth. They play important roles in normal cell functions including proliferation, apoptosis, and migration. We must therefore keep in mind that MMPIs should target those MMPs not involved in basic physiological functions to avoid toxic side effects. It has been shown that, in addition to ECM proteins, many MMPs can cleave numerous and miscellaneous non-ECM substrates, including notably other proteinases and proteinase inhibitors [76, 77]. In addition, MMPs have been shown to act on proteins playing a key role in tissue homeostasis via cleavage of cell–cell adhesion molecules and cell–ECM interactors. MMP-3 and MMP-9 overexpression lead to E-cadherin cleavage, which favors epithelial–mesenchymal transition (EMT), and laminin receptor is a physiological substrate of MMP-11 [78]. MMPs also play important functions at the cancer cell–connective cell interface. For example, MMP-11 induces the delipidation of cancer-associated adipocytes (CAAs) [79–84]. Further, MMPs interfere with immunological factors. MMP-7 cleaves Fas ligand and therefore interferes with the induction of apoptosis in malignant cells, which lowers chemotherapy effects [85]. Finally, MMPs are involved in growth factor (GF) pathways via cleavage of latent GFs, GF-binding proteins, and/or growth factor receptors (GFRs). MMP-9 regulates the bioavailability of vascular endothelial growth factor (VEGF), a potent inducer of tumor angiogenesis [17]. Similarly, MMP-11 favors the release of insulin-like growth factor-1 (IGF-1) from insulin-like growth-factor-binding protein-1 (IGFBP-1) [86]. Recently, MMP-15 was shown to cleave the extracellular domain (ECD) of erbB2 (HER2) [87], a member of the EGFR family known to play a pejorative role in several cancers [88]. The impact of MMPIs on these various MMP functions explains, at least partially, the dramatic side effects in cancer patients treated with such drugs. The most frequent and severe side effect is the development of the musculoskeletal syndrome characterized by arthralgia and myalgia [26, 89]. Thus, we need to have a better understanding of the myriad effects of MMPs before developing drugs that alter MMP activity in patients with cancer. Moreover, the failure of clinical trials also highlights the insufficient detection of side effects such as musculoskeletal syndrome in preclinical models. Therefore, the predictability of preclinical models for the efficiency and safety of MMPIs must be improved.

15.2.4
Optimize MMPI Administration Schedule

Until now the majority of the randomized therapeutic trials designed to test MMPIs in patients have been performed in advanced-stage cancers. By contrast, numerous animal experiments have focused on primary tumor development, but few on metastatic development. These preclinical studies showed that MMPIs mainly have an effect in the early steps of the disease. Thus, it might be expected that early rather than late inhibition of MMPs will yield a more beneficial response. In this context, the use of sensitive MMP probes will help determine the time of MMP activities and the optimal time of MMPI administration. Another reason for the failure of many MMPIs is the inaccurate estimation of drug dosage to prevent toxicity (often musculoskeletal side effects). Interestingly, animal models have shown different antitumor efficacies for the same drug depending on the sequence of administration. For example, a phosphinic MMPI (RXPO3) that potently inhibits *in vitro* MMP-11, MMP-8, and MMP-13, but not MMP-1 and MMP-7, gave paradoxical results on the growth of mouse primary tumors, depending on the dose used and on the treatment schedule [90]. Achieving an optimal therapeutic index, the ratio of dose required for efficacy versus that for toxicology, therefore remains a challenge [91]. The mode of drug delivery, in local areas or by systemic routes should also be more carefully studied. Finally, MMPIs represent an entirely different therapeutic modality from proven anticancer agents in that they target normal peritumoral cells rather than cancer cells since most of the MMPs are secreted by normal stromal cells and not by cancer cells. Potential cellular targets include cancer-associated fibroblasts (CAFs), CAAs, infiltrating macrophages/histiocytes, and tumor endothelial cells. It is well established that cancer-associated stromal cells not only participate in the tissue structure but also contribute actively to tumor nutrition and progression. The therapeutic potential of targeting the tumor stroma has been proved in several preclinical and clinical studies, most notably using drugs directed against endothelial cells [92]. Thus, MMPI treatments should be cytostatic rather than cytotoxic. Collectively, these findings suggest that MMPIs should be used as adjuvant therapy with minimal residual disease after initial cytotoxic treatment rather than in advanced disease. The most useful MMPIs should be those that can be applied as frequently and repeatedly as possible to maintain the tumor residual in dormancy [93].

15.3 Novel Generation of MMPIs

Increased understanding of MMP biology has revealed novel ways to inhibit MMPs and several drugs are on the horizon [26, 94].

15.3.1 Target the Hemopexin Domain

The difficulty of distinguishing between MMPs lies in the marked sequence similarity between their catalytic domains. Some authors raise the question of whether other MMP domains might be appropriate targets for innovative therapies. A genetic dissection of the functions of the catalytic and hemopexin domains was performed using a canonical MMP in *Drosophila melanogaster*, an organism with only two MMPs that function nonredundantly. Although the catalytic domain appears to be required for all MMP functions including ECM remodeling, the hemopexin domain is required specifically for tissue invasion, a finding with potential implications for inhibitor therapies [95]. Indeed, it has been shown that the MMP-9 hemopexin domain mediates epithelial cell migration in a transwell chamber assay [96]. Other functions have been also reported for the hemopexin domain. Importantly, it has been reported to play a role in signaling pathways. Thus, several MMPs interact with surface receptors such as integrins. For example, the MMP-9 hemopexin domain induces signals for B-CLL (B-chronic lymphocytic leukemia) cell survival via its binding to the docking receptors a4b1 integrin and CD44v [97]. MMP-2 binds to integrin avb3 via its hemopexin domain. MT1-MMP regulates VEGFR-2 cell surface localization and forms a complex with VEGFR-2 and Src that is dependent on the MT1–MMP hemopexin domain and independent of its catalytic activity. The signaling cascade dependent on the MT1–MMP–VEGFR-2-Src complex activates Akt and mTOR, ultimately leading to increased VEGF-A transcription [98]. Finally, the MMP-12 hemopexin domain has been shown to play an essential role in antimicrobial function [99]. Thus, the development of selective inhibitors to the MMP hemopexin domain may provide an attractive option since they exhibit excellent MMP specificity. In this context, a cheminformatics-based drug design approach for the identification and characterization of inhibitors targeting the hemopexin domain of MMP-13 has identified several potential selective inhibitors of MMP-13. Thus, the hemopexin domain represents an interesting target to develop new MMPIs. The systematic cheminformatics-based drug design approach can be applied for the search of other potent molecules, capable of selectively inhibiting MMPs [100].

15.3.2 Antibodies as MMPIs

Among the other proposed alternatives to inhibit MMPs, antibody-based biotherapeutic agents may offer the desired selectivity and potency. One of the major tasks for the future is the development of antibodies specific to single MMPs with no cross-reactivity with other MMPs. In this context, *in vivo* inhibition of MMP-2 using a blocking antibody significantly reduced the number of metastases and tumor weight in a xenograft mouse model [101]. A monoclonal antibody directed against MMP-14 successfully inhibits the migration and invasion of endothelial cells in collagen and fibrin gels [102]. Similar approaches targeting MMP-14 have also been shown to inhibit angiogenesis, slow tumor progression, and block proMMP-2 processing [103]. MMPIs selectively targeting MT–MMPs with therapeutic antibodies are currently under evaluation (reviewed in [104]. These types of MMPIs have several disadvantages, which include intravenous administration rather than oral, production of neutralizing antibodies by the patients themselves, and widespread negative side effects. Nevertheless, combination therapy with several drugs including antibody-based MMPIs represents a promising approach that may produce a synergistic antitumor effect and a survival benefit for patients.

15.3.3
Immunotherapy

Vaccination against stromal antigen is a feasible approach for anticancer therapy [105]. Several MMPs have been identified as tumor-associated antigens that can serve as target candidates for cancer immunotherapy. Such possibilities have been tested in animal models. The antitumor effects of a vaccine against MMP-2 has been reported. Vaccination prolonged the survival of cancer-bearing mice, and the antitumor activity was dependent on CD4+ and CD8+ T cells [106]. MMP-11, identified as a broadly expressed tumor-associated antigen, represents a good target candidate for cancer immunotherapy. MMP-11 vaccine has been shown to induce a cell-mediated and antibody-dependent immune response that give significant antitumor protection to mice with colon cancer in prophylactic and therapeutic settings [63]. Finally, a T-cell epitope specific for MMP-7 has also been proposed as a candidate for vaccine development [107]. Thus, vaccines targeting a few MMPs have been successful in mice. These data indicate that MMPs are valuable candidates for antigen-specific immunotherapy. The potency of this therapeutic approach remains to be tested in humans.

15.4
Exploit MMP Function to Improve Drug Bioavailability

In a general manner, a cause of anticancer therapy failure is poor drug bioavailability. The pharmacodelivery of therapeutic agents to the tumor site is a crucial field of modern anticancer research, which promises to concentrate bioactive molecules onto neoplastic lesions while sparing normal tissues. One possible version of these new drugs is antibody based. In this context, MMPs can be regarded as antigens to direct drugs into tumors via specific antibodies. Moreover, MMPs have an additional interest compared with other putative tumor-associated antigens.

Indeed, their catalytic function can be exploited to locally activate cytotoxic agents [108]. Thus, MMPs might serve to improve anticancer therapies by (i) improving tumor-targeting of various types of drugs and/or (ii) providing intratumor MMP-activation of anticancer drugs.

15.5 Conclusion

Although MMPs are clearly stroma-derived bad prognosis predictors in human tumors, the development of MMPIs for cancer treatment is more challenging than originally thought. In this context, the present review highlights the extreme difficulties of defining the right MMP to be targeted, as well as the correct time and place to use MMPIs in order to treat cancers successfully. Several avenues are under consideration to develop MMPIs targeting the MMP enzymatic domain or the hemopexin domain. There are also attempts to design drugs that take advantage of cancer-associated MMP antigens. Finally, new approaches based on the use of MMPs for improving cytotoxic or cytostatic drug bioavailability, via improved delivery and/or local drug activation, are also promising. Thus, despite the initial setbacks, recent progress in our understanding of MMP biology and in emerging ways to inhibit MMPs, gives new hope to the future for MMPIs in cancer treatment.

Acknowledgments

We thank Susan Chan for the helpful discussion. This work was supported by funds from the Institut National de la Santé et de la Recherche Médicale, the Centre National de la Recherche Scientifique, the Hopital Universitaire de Strasbourg, the Association pour la Recherche sur le Cancer and the Ligue Nationale Française contre le Cancer (LNCC; Equipe labellisée 2010), and the Comités du Haut-Rhin et du Bas-Rhin. Buache E. was recipient of LNCC fellowship.

References

1. Bissell, M.J., Radisky, D.C., Rizki, A., Weaver, V.M., and Petersen, O.W. (2002) The organizing principle: microenvironmental influences in the normal and malignant breast. *Differentiation*, **70**, 537–546.
2. Radisky, D., Muschler, J., and Bissell, M.J. (2002) Order and disorder: the role of extracellular matrix in epithelial cancer. *Cancer Invest.*, **20**, 139–153.
3. De Wever, O. and Mareel, M. (2003) Role of tissue stroma in cancer cell invasion. *J. Pathol.*, **200**, 429–447.
4. Freije, J.M., Balbin, M., Pendas, A.M., Sanchez, L.M., Puente, X.S., and Lopez-Otin, C. (2003) Matrix metalloproteinases and tumor progression. *Adv. Exp. Med. Biol.*, **532**, 91–107.
5. Mueller, M.M. and Fusenig, N.E. (2004) Friends or foes – bipolar effects

of the tumour stroma in cancer. *Nat. Rev. Cancer*, **4**, 839–849.
6. Maquart, F.X., Bellon, G., Pasco, S., and Monboisse, J.C. (2005) Matrikines in the regulation of extracellular matrix degradation. *Biochimie*, **87**, 353–360.
7. Basset, P., Bellocq, J.P., Wolf, C., Stoll, I., Hutin, P., Limacher, J.M., Podhajcer, O.L., Chenard, M.P., Rio, M.C., and Chambon, P. (1990) A novel metalloproteinase gene specifically expressed in stromal cells of breast carcinomas. *Nature*, **348**, 699–704.
8. Basset, P., Okada, A., Chenard, M.P., Kannan, R., Stoll, I., Anglard, P., Bellocq, J.P., and Rio, M.C. (1997) Matrix metalloproteinases as stromal effectors of human carcinoma progression: therapeutic implications. *Matrix Biol.*, **15**, 535–541.
9. Birkedal-Hansen, H. (1995) Proteolytic remodeling of extracellular matrix. *Curr. Opin. Cell Biol.*, **7**, 728–735.
10. Itoh, Y. and Seiki, M. (2006) MT1-MMP: a potent modifier of pericellular microenvironment. *J. Cell Physiol.*, **206**, 1–8.
11. Lochter, A., Sternlicht, M.D., Werb, Z., and Bissell, M.J. (1998) The significance of matrix metalloproteinases during early stages of tumor progression. *Ann. N. Y. Acad. Sci.*, **857**, 180–193.
12. Jodele, S., Blavier, L., Yoon, J.M., and DeClerck, Y.A. (2006) Modifying the soil to affect the seed: role of stromal-derived matrix metalloproteinases in cancer progression. *Cancer Metastasis Rev.*, **25**, 35–43.
13. Nagase, H., Visse, R., and Murphy, G. (2006) Structure and function of matrix metalloproteinases and TIMPs. *Cardiovasc. Res.*, **69**, 562–573.
14. Kessenbrock, K., Plaks, V., and Werb, Z. (2010) Matrix metalloproteinases: regulators of the tumor microenvironment. *Cell*, **141**, 52–67.
15. Masson, R., Lefebvre, O., Noel, A., Fahime, M.E., Chenard, M.P., Wendling, C., Kebers, F., LeMeur, M., Dierich, A., Foidart, J.M. *et al.* (1998) In vivo evidence that the stromelysin-3 metalloproteinase contributes in a paracrine manner to epithelial cell malignancy. *J. Cell Biol.*, **140**, 1535–1541.
16. Sternlicht, M.D., Bissell, M.J., and Werb, Z. (2000) The matrix metalloproteinase stromelysin-1 acts as a natural mammary tumor promoter. *Oncogene*, **19**, 1102–1113.
17. Bergers, G., Brekken, R., McMahon, G., Vu, T.H., Itoh, T., Tamaki, K., Tanzawa, K., Thorpe, P., Itohara, S., Werb, Z., and Hanahan, D. (2000) Matrix metalloproteinase-9 triggers the angiogenic switch during carcinogenesis. *Nat. Cell Biol.*, **2**, 737–744.
18. Goss, K.J., Brown, P.D., and Matrisian, L.M. (1998) Differing effects of endogenous and synthetic inhibitors of metalloproteinases on intestinal tumorigenesis. *Int. J. Cancer*, **78**, 629–635.
19. Noel, A., Hajitou, A., L'Hoir, C., Maquoi, E., Baramova, E., Lewalle, J.M., Remacle, A., Kebers, F., Brown, P., Calberg-Bacq, C.M., and Foidart, J.M. (1998) Inhibition of stromal matrix metalloproteases: effects on breast-tumor promotion by fibroblasts. *Int. J. Cancer*, **76**, 267–273.
20. Rudolph-Owen, L.A., Chan, R., Muller, W.J., and Matrisian, L.M. (1998) The matrix metalloproteinase matrilysin influences early-stage mammary tumorigenesis. *Cancer Res.*, **58**, 5500–5506.
21. Zucker, S., Cao, J., and Chen, W.T. (2000) Critical appraisal of the use of matrix metalloproteinase inhibitors in cancer treatment. *Oncogene*, **19**, 6642–6650.
22. Overall, C.M. and Lopez-Otin, C. (2002) Strategies for MMP inhibition in cancer: innovations for the post-trial era. *Nat. Rev. Cancer*, **2**, 657–672.
23. Coussens, L.M., Fingleton, B., and Matrisian, L.M. (2002) Matrix metalloproteinase inhibitors and cancer: trials and tribulations. *Science*, **295**, 2387–2392.
24. Pavlaki, M. and Zucker, S. (2003) Matrix metalloproteinase inhibitors (MMPIs): the beginning of phase I or the termination of phase III clinical trials. *Cancer Metastasis Rev.*, **22**, 177–203.

25. Noel, A., Jost, M., and Maquoi, E. (2008) Matrix metalloproteinases at cancer tumor-host interface. *Semin. Cell Dev. Biol.*, **19**, 52–60.
26. Fingleton, B. (2008) in *The Cancer Degradome* (ed. D.R. Edwards), Springer Science, pp. 757–783.
27. Dorman, G., Cseh, S., Hajdu, I., Barna, L., Konya, D., Kupai, K., Kovacs, L., and Ferdinandy, P. (2010) Matrix metalloproteinase inhibitors: a critical appraisal of design principles and proposed therapeutic utility. *Drugs*, **70**, 949–964.
28. Visse, R. and Nagase, H. (2003) Matrix metalloproteinases and tissue inhibitors of metalloproteinases: structure, function, and biochemistry. *Circ. Res.*, **92**, 827–839.
29. Baker, A.H., Zaltsman, A.B., George, S.J., and Newby, A.C. (1998) Divergent effects of tissue inhibitor of metalloproteinase-1, -2, or -3 overexpression on rat vascular smooth muscle cell invasion, proliferation, and death in vitro. TIMP-3 promotes apoptosis. *J. Clin. Invest.*, **101**, 1478–1487.
30. Clutterbuck, A.L., Asplin, K.E., Harris, P., Allaway, D., and Mobasheri, A. (2009) Targeting matrix metalloproteinases in inflammatory conditions. *Curr. Drug Targets*, **10**, 1245–1254.
31. Whittaker, M., Floyd, C.D., Brown, P., and Gearing, A.J. (1999) Design and therapeutic application of matrix metalloproteinase inhibitors. *Chem. Rev.*, **99**, 2735–2776.
32. Cuniasse, P., Devel, L., Makaritis, A., Beau, F., Georgiadis, D., Matziari, M., Yiotakis, A., and Dive, V. (2005) Future challenges facing the development of specific active-site-directed synthetic inhibitors of MMPs. *Biochimie*, **87**, 393–402.
33. Nuti, E., Tuccinardi, T., and Rossello, A. (2007) Matrix metalloproteinase inhibitors: new challenges in the era of post broad-spectrum inhibitors. *Curr. Pharm. Des.*, **13**, 2087–2100.
34. Tu, G., Xu, W., Huang, H., and Li, S. (2008) Progress in the development of matrix metalloproteinase inhibitors. *Curr. Med. Chem.*, **15**, 1388–1395.
35. Maskos, K. and Bode, W. (2003) Structural basis of matrix metalloproteinases and tissue inhibitors of metalloproteinases. *Mol. Biotechnol.*, **25**, 241–266.
36. Rao, B.G. (2005) Recent developments in the design of specific Matrix Metalloproteinase inhibitors aided by structural and computational studies. *Curr. Pharm. Des.*, **11**, 295–322.
37. Dive, V., Georgiadis, D., Matziari, M., Makaritis, A., Beau, F., Cuniasse, P., and Yiotakis, A. (2004) Phosphinic peptides as zinc metalloproteinase inhibitors. *Cell Mol. Life Sci.*, **61**, 2010–2019.
38. Floquet, N., Marechal, J.D., Badet-Denisot, M.A., Robert, C.H., Dauchez, M., and Perahia, D. (2006) Normal mode analysis as a prerequisite for drug design: application to matrix metalloproteinases inhibitors. *FEBS Lett.*, **580**, 5130–5136.
39. Devel, L., Czarny, B., Beau, F., Georgiadis, D., Stura, E., and Dive, V. (2010) Third generation of matrix metalloprotease inhibitors: Gain in selectivity by targeting the depth of the S1′ cavity. *Biochimie*, **92**, 1501–1508.
40. Dublanchet, A.C., Ducrot, P., Andrianjara, C., O'Gara, M., Morales, R., Compere, D., Denis, A., Blais, S., Cluzeau, P., Courte, K. *et al.* (2005) Structure-based design and synthesis of novel non-zinc chelating MMP-12 inhibitors. *Bioorg. Med. Chem. Lett.*, **15**, 3787–3790.
41. Engel, C.K., Pirard, B., Schimanski, S., Kirsch, R., Habermann, J., Klingler, O., Schlotte, V., Weithmann, K.U., and Wendt, K.U. (2005) Structural basis for the highly selective inhibition of MMP-13. *Chem. Biol.*, **12**, 181–189.
42. Pochetti, G., Montanari, R., Gege, C., Chevrier, C., Taveras, A.G., and Mazza, F. (2009) Extra binding region induced by non-zinc chelating inhibitors into the S1′ subsite of matrix metalloproteinase 8 (MMP-8). *J. Med. Chem.*, **52**, 1040–1049.
43. Golub, L.M., Sorsa, T., Lee, H.M., Ciancio, S., Sorbi, D., Ramamurthy, N.S., Gruber, B., Salo, T., and Konttinen, Y.T. (1995) Doxycycline inhibits neutrophil (PMN)-type matrix

metalloproteinases in human adult periodontitis gingiva. *J. Clin. Periodontol.*, **22**, 100–109.

44. Lokeshwar, B.L. (2010) Chemically modified non-antimicrobial tetracyclines are multifunctional drugs against advanced cancers. *Pharmacol. Res.*, **63**, 145–150.
45. Mannello, F. (2006) Natural bio-drugs as matrix metalloproteinase inhibitors: new perspectives on the horizon? *Recent Pat. Anticancer Drug Discov.*, **1**, 91–103.
46. Sternlicht, M.D. and Werb, Z. (2001) How matrix metalloproteinases regulate cell behavior. *Annu. Rev. Cell Dev. Biol.*, **17**, 463–516.
47. Pei, D. and Weiss, S.J. (1995) Furin-dependent intracellular activation of the human stromelysin-3 zymogen. *Nature*, **375**, 244–247.
48. Bassi, D.E., Mahloogi, H., and Klein-Szanto, A.J. (2000) The proprotein convertases furin and PACE4 play a significant role in tumor progression. *Mol. Carcinog.*, **28**, 63–69.
49. Bregant, S., Huillet, C., Devel, L., Dabert-Gay, A.S., Beau, F., Thai, R., Czarny, B., Yiotakis, A., and Dive, V. (2009) Detection of matrix metalloproteinase active forms in complex proteomes: evaluation of affinity versus photoaffinity capture. *J. Proteome Res.*, **8**, 2484–2494.
50. Dabert-Gay, A.S., Czarny, B., Devel, L., Beau, F., Lajeunesse, E., Bregant, S., Thai, R., Yiotakis, A., and Dive, V. (2008) Molecular determinants of matrix metalloproteinase-12 covalent modification by a photoaffinity probe: insights into activity-based probe development and conformational variability of matrix metalloproteinases. *J. Biol. Chem.*, **283**, 31058–31067.
51. Scherer, R.L., McIntyre, J.O., and Matrisian, L.M. (2008) Imaging matrix metalloproteinases in cancer. *Cancer Metastasis Rev.*, **27**, 679–690.
52. Bremer, C., Tung, C.H., and Weissleder, R. (2001) In vivo molecular target assessment of matrix metalloproteinase inhibition. *Nat. Med.*, **7**, 743–748.
53. Littlepage, L.E., Sternlicht, M.D., Rougier, N., Phillips, J., Gallo, E., Yu, Y., Williams, K., Brenot, A., Gordon, J.I., and Werb, Z. (2010) Matrix metalloproteinases contribute distinct roles in neuroendocrine prostate carcinogenesis, metastasis, and angiogenesis progression. *Cancer Res.*, **70**, 2224–2234.
54. Olson, E.S., Aguilera, T.A., Jiang, T., Ellies, L.G., Nguyen, Q.T., Wong, E.H., Gross, L.A., and Tsien, R.Y. (2009) In vivo characterization of activatable cell penetrating peptides for targeting protease activity in cancer. *Integr. Biol. (Camb)*, **1**, 382–393.
55. Chenard, M.P., O'Siorain, L., Shering, S., Rouyer, N., Lutz, Y., Wolf, C., Basset, P., Bellocq, J.P., and Duffy, M.J. (1996) High levels of stromelysin-3 correlate with poor prognosis in patients with breast carcinoma. *Int. J. Cancer*, **69**, 448–451.
56. Delebecq, T.J., Porte, H., Zerimech, F., Copin, M.C., Gouyer, V., Dacquembronne, E., Balduyck, M., Wurtz, A., and Huet, G. (2000) Overexpression level of stromelysin 3 is related to the lymph node involvement in non-small cell lung cancer. *Clin. Cancer Res.*, **6**, 1086–1092.
57. Hahnel, E., Dawkins, H., Robbins, P., and Hahnel, R. (1994) Expression of stromelysin-3 and nm23 in breast carcinoma and related tissues. *Int. J. Cancer*, **58**, 157–160.
58. Kawami, H., Yoshida, K., Ohsaki, A., Kuroi, K., Nishiyama, M., and Toge, T. (1993) Stromelysin-3 mRNA expression and malignancy: comparison with clinicopathological features and type IV collagenase mRNA expression in breast tumors. *Anticancer Res.*, **13**, 2319–2323.
59. Kossakowska, A.E., Huchcroft, S.A., Urbanski, S.J., and Edwards, D.R. (1996) Comparative analysis of the expression patterns of metalloproteinases and their inhibitors in breast neoplasia, sporadic colorectal neoplasia, pulmonary carcinomas and malignant non-Hodgkin's lymphomas in humans. *Br. J. Cancer*, **73**, 1401–1408.

60. Muller, D., Wolf, C., Abecassis, J., Millon, R., Engelmann, A., Bronner, G., Rouyer, N., Rio, M.C., Eber, M., Methlin, G. *et al.* (1993) Increased stromelysin 3 gene expression is associated with increased local invasiveness in head and neck squamous cell carcinomas. *Cancer Res.*, **53**, 165–169.
61. Polette, M., Clavel, C., Birembaut, P., and De Clerck, Y.A. (1993) Localization by in situ hybridization of mRNAs encoding stromelysin 3 and tissue inhibitors of metallo-proteinases TIMP-1 and TIMP-2 in human head and neck carcinomas. *Pathol. Res. Pract.*, **189**, 1052–1057.
62. Soni, S., Mathur, M., Shukla, N.K., Deo, S.V., and Ralhan, R. (2003) Stromelysin-3 expression is an early event in human oral tumorigenesis. *Int. J. Cancer*, **107**, 309–316.
63. Peruzzi, D., Mori, F., Conforti, A., Lazzaro, D., De Rinaldis, E., Ciliberto, G., La Monica, N., and Aurisicchio, L. (2009) MMP11: a novel target antigen for cancer immunotherapy. *Clin. Cancer Res.*, **15**, 4104–4113.
64. Ma, X.J., Dahiya, S., Richardson, E., Erlander, M., and Sgroi, D.C. (2009) Gene expression profiling of the tumor microenvironment during breast cancer progression. *Breast Cancer Res.*, **11**, R7.
65. Hannemann, J., Velds, A., Halfwerk, J.B., Kreike, B., Peterse, J.L., and van de Vijver, M.J. (2006) Classification of ductal carcinoma in situ by gene expression profiling. *Breast Cancer Res.*, **8**, R61.
66. Yang, Y.H., Deng, H., Li, W.M., Zhang, Q.Y., Hu, X.T., Xiao, B., Zhu, H.H., Geng, P.L., and Lu, Y.Y. (2008) Identification of matrix metalloproteinase 11 as a predictive tumor marker in serum based on gene expression profiling. *Clin. Cancer Res.*, **14**, 74–81.
67. Noel, A.C., Lefebvre, O., Maquoi, E., VanHoorde, L., Chenard, M.P., Mareel, M., Foidart, J.M., Basset, P., and Rio, M.C. (1996) Stromelysin-3 expression promotes tumor take in nude mice. *J. Clin. Invest.*, **97**, 1924–1930.
68. Boulay, A., Masson, R., Chenard, M.P., El Fahime, M., Cassard, L., Bellocq, J.P., Sautes-Fridman, C., Basset, P., and Rio, M.C. (2001) High cancer cell death in syngeneic tumors developed in host mice deficient for the stromelysin-3 matrix metalloproteinase. *Cancer Res.*, **61**, 2189–2193.
69. Wu, E., Mari, B.P., Wang, F., Anderson, I.C., Sunday, M.E., and Shipp, M.A. (2001) Stromelysin-3 suppresses tumor cell apoptosis in a murine model. *J. Cell Biochem.*, **82**, 549–555.
70. Andarawewa, K.L., Boulay, A., Masson, R., Mathelin, C., Stoll, I., Tomasetto, C., Chenard, M.P., Gintz, M., Bellocq, J.P., and Rio, M.C. (2003) Dual stromelysin-3 function during natural mouse mammary tumor virus-ras tumor progression. *Cancer Res.*, **63**, 5844–5849.
71. Brasse, D., Mathelin, C., Leroux, K., Chenard, M.P., Blaise, S., Stoll, I., Tomasetto, C., and Rio, M.C. (2010) Matrix metalloproteinase 11/stromelysin-3 exerts both activator and repressor functions during the hematogenous metastatic process in mice. *Int. J. Cancer*, **127**, 1347–1355.
72. Overall, C.M. and Kleifeld, O. (2006) Tumour microenvironment – opinion: validating matrix metalloproteinases as drug targets and anti-targets for cancer therapy. *Nat. Rev. Cancer*, **6**, 227–239.
73. Lopez-Otin, C. and Matrisian, L.M. (2007) Emerging roles of proteases in tumour suppression. *Nat. Rev. Cancer*, **7**, 800–808.
74. Kruger, A., Kates, R.E., and Edwards, D.R. (2010) Avoiding spam in the proteolytic internet: future strategies for anti-metastatic MMP inhibition. *Biochim. Biophys. Acta*, **1803**, 95–102.
75. Rio, M.C. (2005) From a unique cell to metastasis is a long way to go: clues to stromelysin-3 participation. *Biochimie*, **87**, 299–306.
76. McCawley, L.J. and Matrisian, L.M. (2001) Matrix metalloproteinases: they're not just for matrix anymore!. *Curr. Opin. Cell Biol.*, **13**, 534–540.
77. VanSaun, M.N. and Matrisian, L.M. (2006) Matrix metalloproteinases and cellular motility in development and disease. *Birth Defects Res. C Embryo Today*, **78**, 69–79.

78. Fiorentino, M., Fu, L., and Shi, Y.B. (2009) Mutational analysis of the cleavage of the cancer-associated laminin receptor by stromelysin-3 reveals the contribution of flanking sequences to site recognition and cleavage efficiency. *Int. J. Mol. Med.*, **23**, 389–397.
79. Andarawewa, K.L., Motrescu, E.R., Chenard, M.P., Gansmuller, A., Stoll, I., Tomasetto, C., and Rio, M.C. (2005) Stromelysin-3 is a potent negative regulator of adipogenesis participating to cancer cell-adipocyte interaction/crosstalk at the tumor invasive front. *Cancer Res.*, **65**, 10862–10871.
80. Motrescu, E.R. and Rio, M.C. (2008) Cancer cells, adipocytes and matrix metalloproteinase 11: a vicious tumor progression cycle. *Biol. Chem.*, **389**, 1037–1041.
81. Motrescu, E.R., Blaise, S., Etique, N., Messaddeq, N., Chenard, M.P., Stoll, I., Tomasetto, C., and Rio, M.C. (2008) Matrix metalloproteinase-11/stromelysin-3 exhibits collagenolytic function against collagen VI under normal and malignant conditions. *Oncogene*, **27**, 6347–6355.
82. Dirat, B., Bochet, L., Escourrou, G., Valet, P., and Muller, C. (2010) Unraveling the obesity and breast cancer links: a role for cancer-associated adipocytes? *Endocr. Dev.*, **19**, 45–52.
83. Andarawewa, K.L. and Rio, M.C. (2008) in *The Cancer Degradome* (ed. D.R. Edwards), Springer Science, pp. 353–364.
84. Mueller, M., and Fusenig, N. (2011) The role of Cancer-Associated Adipocytes (CAA) in the dynamic interaction between the tumor and the host, Springer SBM, Springer Science, pp. 111–123.
85. Mitsiades, N., Yu, W.H., Poulaki, V., Tsokos, M., and Stamenkovic, I. (2001) Matrix metalloproteinase-7-mediated cleavage of Fas ligand protects tumor cells from chemotherapeutic drug cytotoxicity. *Cancer Res.*, **61**, 577–581.
86. Manes, S., Mira, E., Barbacid, M.M., Cipres, A., Fernandez-Resa, P., Buesa, J.M., Merida, I., Aracil, M., Marquez, G., and Martinez, A.C. (1997) Identification of insulin-like growth factor-binding protein-1 as a potential physiological substrate for human stromelysin-3. *J. Biol. Chem.*, **272**, 25706–25712.
87. Carey, K.D., Schwall, R.H., and Sliwkowski, M.X. (2006) Inhibiting HER2 shedding with MMP antagonists. Patent WO/2006/086730.
88. Molina, M.A., Codony-Servat, J., Albanell, J., Rojo, F., Arribas, J., and Baselga, J. (2001) Trastuzumab (herceptin), a humanized anti-Her2 receptor monoclonal antibody, inhibits basal and activated Her2 ectodomain cleavage in breast cancer cells. *Cancer Res.*, **61**, 4744–4749.
89. Abbenante, G. and Fairlie, D.P. (2005) Protease inhibitors in the clinic. *Med. Chem.*, **1**, 71–104.
90. Dive, V., Andarawewa, K.L., Boulay, A., Matziari, M., Beau, F., Guerin, E., Rousseau, B., Yiotakis, A., and Rio, M.C. (2005) Dosing and scheduling influence the antitumor efficacy of a phosphinic peptide inhibitor of matrix metalloproteinases. *Int. J. Cancer*, **113**, 775–781.
91. Peterson, J.T. (2006) The importance of estimating the therapeutic index in the development of matrix metalloproteinase inhibitors. *Cardiovasc. Res.*, **69**, 677–687.
92. Folkman, J. (2002) Role of angiogenesis in tumor growth and metastasis. *Semin. Oncol.*, **29**, 15–18.
93. Demicheli, R. (2001) Tumour dormancy: findings and hypotheses from clinical research on breast cancer. *Semin. Cancer Biol.*, **11**, 297–306.
94. Corbitt, C.A., Lin, J., and Lindsey, M.L. (2007) Mechanisms to inhibit matrix metalloproteinase activity: where are we in the development of clinically relevant inhibitors? *Recent Pat. Anticancer Drug Discov.*, **2**, 135–142.
95. Glasheen, B.M., Kabra, A.T., and Page-McCaw, A. (2009) Distinct functions for the catalytic and hemopexin domains of a Drosophila matrix metalloproteinase. *Proc. Natl. Acad. Sci. U. S. A.*, **106**, 2659–2664.

96. Dufour, A., Zucker, S., Sampson, N.S., Kuscu, C., and Cao, J. (2010) Role of matrix metalloproteinase-9 dimers in cell migration: design of inhibitory peptides. *J. Biol. Chem.*, **285**, 35944–35956.

97. Redondo-Munoz, J., Ugarte-Berzal, E., Terol, M.J., Van den Steen, P.E., Hernandez del Cerro, M., Roderfeld, M., Roeb, E., Opdenakker, G., Garcia-Marco, J.A., and Garcia-Pardo, A. (2010) Matrix metalloproteinase-9 promotes chronic lymphocytic leukemia B cell survival through its hemopexin domain. *Cancer Cell*, **17**, 160–172.

98. Eisenach, P.A., Roghi, C., Fogarasi, M., Murphy, G., and English, W.R. (2010) MT1-MMP regulates VEGF-A expression through a complex with VEGFR-2 and Src. *J. Cell Sci.*, **123**, 4182–4193.

99. Houghton, A.M., Hartzell, W.O., Robbins, C.S., Gomis-Ruth, F.X., and Shapiro, S.D. (2009) Macrophage elastase kills bacteria within murine macrophages. *Nature*, **460**, 637–641.

100. Kothapalli, R., Khan, A.M., Basappa , Gopalsamy, A., Chong, Y.S., and Annamalai, L. (2010) Cheminformatics-based drug design approach for identification of inhibitors targeting the characteristic residues of MMP-13 hemopexin domain. *PLoS ONE*, **5**, e12494.

101. Kenny, H.A. and Lengyel, E. (2009) MMP-2 functions as an early response protein in ovarian cancer metastasis. *Cell Cycle*, **8**, 683–688.

102. Galvez, B.G., Matias-Roman, S., Albar, J.P., Sanchez-Madrid, F., and Arroyo, A.G. (2001) Membrane type 1-matrix metalloproteinase is activated during migration of human endothelial cells and modulates endothelial motility and matrix remodeling. *J. Biol. Chem.*, **276**, 37491–37500.

103. Devy, L., Huang, L., Naa, L., Yanamandra, N., Pieters, H., Frans, N., Chang, E., Tao, Q., Vanhove, M., Lejeune, A. *et al.* (2009) Selective inhibition of matrix metalloproteinase-14 blocks tumor growth, invasion, and angiogenesis. *Cancer Res.*, **69**, 1517–1526.

104. Devy, L. and Dransfield, D.T. (2011) New strategies for the next generation of matrix-metalloproteinase inhibitors: selectively targeting membrane-anchored MMPs with therapeutic antibodies. *Biochem. Res. Int.*, **2011**, 191670.

105. Hofmeister, V., Schrama, D., and Becker, J.C. (2008) Anti-cancer therapies targeting the tumor stroma. *Cancer Immunol. Immunother.*, **57**, 1–17.

106. Yi, T., Tian, L., Su, J.M., Liu, J., and Wei, Y.Q. (2004) Constructing the tumor cell vaccine based on homologous matrix metalloproteinase-2 and exploring its effects. *Sichuan Da Xue Xue Bao Yi Xue Ban*, **35**, 301–304.

107. Yokoyama, Y., Grunebach, F., Schmidt, S.M., Heine, A., Hantschel, M., Stevanovic, S., Rammensee, H.G., and Brossart, P. (2008) Matrilysin (MMP-7) is a novel broadly expressed tumor antigen recognized by antigen-specific T cells. *Clin. Cancer Res.*, **14**, 5503–5511.

108. Liu, S., Wang, H., Currie, B.M., Molinolo, A., Leung, H.J., Moayeri, M., Basile, J.R., Alfano, R.W., Gutkind, J.S., Frankel, A.E. *et al.* (2008) Matrix metalloproteinase-activated anthrax lethal toxin demonstrates high potency in targeting tumor vasculature. *J. Biol. Chem.*, **283**, 529–540.

Index

Matrix Proteases in Health and Disease, First Edition. Edited by Niels Behrendt.

r

s

Edited by Niels Behrendt

Matrix Proteases in Health and Disease

Related Titles

De Clercq, Erik (ed.)

Antiviral Drug Strategies

2011

ISBN-13: 978-3-527-32696-9

Ghosh, A. K. (ed.)

Aspartic Acid Proteases as Therapeutic Targets

2010

Hardcover

ISBN: 978-3-527-31811-7

Wang, B.

Drug Design of Zinc-Enzyme Inhibitors

Functional, Structural, and Disease Applications

2009

Hardcover

ISBN: 978-0-470-27500-9

Permyakov, E.

Metalloproteomics

2009

Hardcover

ISBN: 978-0-470-39248-5

Mayer, R. J., Ciechanover, A. J., Rechsteiner, M. (eds.)

Protein Degradation Series

4 Volume Set

2007

Hardcover

ISBN: 978-3-527-31878-0